Juvenile Primates

Juvenile Primates

Life History, Development, and Behavior

Edited by

MICHAEL E. PEREIRA

LYNN A. FAIRBANKS

New York   Oxford

OXFORD UNIVERSITY PRESS

1993

Oxford University Press

Oxford   New York   Toronto
Delhi   Bombay   Calcutta   Madras   Karachi
Kuala Lumpur   Singapore   Hong Kong   Tokyo
Nairobi   Dar es Salaam   Cape Town
Melbourne   Auckland

and associated companies in
Berlin   Ibadan

Published by Oxford University Press, Inc.
200 Madison Avenue, New York, New York 10016

Oxford is a registered trademark of Oxford University Press

Library of Congress Cataloging-in-Publication Data
Juvenile primates : life history, development, and behavior / edited
by Michael E. Pereira and Lynn A. Fairbanks.
p. cm.   Includes bibliographical references and index.
ISBN 0-19-507206-5
1. Primates—Infancy.
2. Primates—Behavior.
3. Primates—Development.
4. Social behavior in animals.
I. Pereira, Michael Eric. 1956–
II. Fairbanks, Lynn A.
QL737.P9J88   1993   599.8'0439—dc20
91-46182

9 8 7 6 5 4 3 2 1

Printed in the United States of America
on acid-free paper

To the juveniles of today's and tomorrow's worlds.

Foreword

This is not a balanced volume; indeed, it is designed as a counterbalance to bias in previous primate studies. Given that perhaps one-third to one-half of primates are juveniles, they have not received the attention they logically deserve. Here, then, as redress, is a bookful of chapters about juveniles of all the major divisions of the Order.

There are practical reasons for the imbalance. In the field, juveniles are always the last animals you learn to recognize. Adult males are frequently battle scarred, and their style is conspicuous. Older females have also accumulated distinctive scars, and adult females can be recognized early because they are obviously pregnant or lactating, which allows you to learn more subtle characteristics in association. Infants are recognizable by association with their mothers. But the juveniles? In forests they may be impossible to see for much of the time—a small juvenile independently running on top of a thick branch in the canopy is invisible simply because light does not go round corners. Then, they tend to flock together in fast-moving play groups in which it is hard to attach the right tail to the right head for more than a few seconds at a time.

The immature genitalia of furry juveniles can be impossible to see—I have spent whole days following a juvenile waiting for it to urinate in view, so that I could see what sex it was. And then, once you are confident you can recognize the gestalt of a juvenile, it grows and changes. If your study is not continuous, you have to begin all over again next year. Small wonder that people have started with questions about adults, or about mothers and infants. Then, there is the question of time. Whatever the significance of the juvenile period, it is clearly a period of transition. That means that most questions about

juveniles require knowing about the point of departure and the destination of the individual juvenile. That in turn means that any study that produces more than the vaguest generalizations about juveniles must at least be long enough to follow an individual through the juvenile period and into adulthood. To accumulate a sufficient sample size from a small population, the time required will probably be much in excess of that, as several cohorts must be followed through.

Much of the research reported in this book is drawn from many years of study, often requiring cooperation over extended periods by successive teams of observers. We must salute the overcoming of formidable logistical, financial, and political problems all over the world that have made this collection possible. Beyond that achievement is the question of continuity in such circumstances. When you collect data over a decade or more, interests and priorities change, new methods become possible or fashionable, and it is truly an achievement if data from the beginning and the end are comparable for even the simplest measures. Coordination of studies in different places by different teams is even more difficult. Contributors are to be congratulated for heroic, and successful, efforts in both these areas.

The book begins with the theoretical questions surrounding juvenility in mammals in general and primates in particular. Why should an individual waste an extended period being a juvenile when it might be killed before it even tries to breed? Are the characteristics of juveniles determined by the ultimate selection of adults, or is the juvenile period under its own adaptive constraints? Is juvenility in all its aspects the primary response to selection pressure, or are juvenile characteristics secondarily adaptive, juvenility itself being required by the constraints of growth trajectories determined by other life stages? We are presented with a mass of correlations—which is causal? The authors of the first section lead us skillfully through a maze of correlations fraught with possibilities for circular argument. It becomes clear that studies of the juvenile period are in fact the key to life history studies of individuals, needed to confirm or reject so many adaptationist theories.

The juvenile period is taken to begin with weaning, and hence with weaning conflict, which seems to be a theoretical necessity in sociobiology but has proved remarkably difficult to observe and pin down among real animals. We should perhaps be suspicious of theoretical necessities that fit too closely with our cultural ideology: the idea of weaning conflict fits neatly into a western puritanical view of the child as dangerously demonic, requiring suppression and control, and at the same time as being helpless and aimless, requiring moulding and guidance, if it is to become an acceptable adult citizen. We have long known that nonhuman primate parents are regrettably prone to spare the rod and so run the risk, in this view, of spoiling the child. The juvenile primates we meet in this book are by no means passively waiting to grow up, under adult guidance. They are actively and independently interacting with their environment, which is not always the same as that perceived by adults. They are, for example, vulnerable to smaller predators, and sometimes they may be able to exploit a rather different range of foods from those accessible to adults. Protection from predators usually requires being in a group, and being in the center if the group is large enough to have a center. The agenda of the group is set by its adults, and this may preclude its juveniles from obtaining all the food that might otherwise be available to them in their habitat. Here is the source of conflicting pressures on juveniles, which may be resolved, eventually, by emigration, but very rarely by expulsion.

The second section of this book teaches that generalizations such as the one I just attempted should be made with extreme caution. The environment determined by their adults presents juveniles of different species with diverse problems, and they respond with equally diverse solutions. In particular, we see differences in behavior between the sexes, which can be seen as appropriate precursors for the adult behavior of each species, but which develop before physiological differences in requirements become apparent. At the same time, adults behave differently toward juvenile males and females long before their sex is relevant to the adults' own reproductive activities.

Here we enter semantically dangerous waters. Several authors describe the behavior of juveniles in terms of their future reproductive success, writing of "decisions" made by juveniles, which are adaptive in those terms. At some point we are going to have to look very carefully at these "decisions." Are they decisions in the cognitive sense used in normal parlance, or are they "decisions" in some evolutionary shorthand for a supposed process of selection and adaptation over many generations? In the cognitive sense, the decisions would require a degree of foresight most certainly not apparent in juveniles of our own species, although we generally flatter ourselves as being especially good at cognition. Yet if these are "shorthand decisions," someone will have to take the time to spell out at least once in longhand exactly how the presumed natural selection would work; it is not at all obvious to me.

The third section continues the theme of the juvenile period as one of preparation for adulthood, when skills, and especially social skills are learned, and relationships are established that will be important later. Milton provides a refreshing corrective to facile assumptions. She reviews a serendipitous experiment, a population of spider monkeys founded by juveniles and showing all the characteristics described of normal spider monkey groups without any benefit of an extended period of learning from adults. Catastrophes are normal, and animals must be robust in their behavioral development: too fine an adjustment to learning from a "normal" environment could lead to extinction.

The fourth section concerns aggression and dominance, and not surprisingly is mostly about macaques, which seem to be obsessed with the subject. Japanese macaques have mastered rank at a basic level before they are weaned, using a variety of learning methods; juvenile rhesus macaques develop and refine their appreciation of rank relationships as they outgrow the tolerance of adults. Juvenile long-tailed macaques use reconciliation following aggression among themselves in the same way as adults. These studies of the development of dominance relationships are moving toward an appreciation of dominance as something more than an expression of or substitute for aggression among gregarious primates. Dominance can also be the obverse of affection in these species, as in our own, a point I first saw developed by Yi-Fu Tuan (1984). It seems that gregarious lemurs, on the other hand, may not be able to appreciate transitive dominance, a finding I found particularly interesting because it is rare to see evidence of less developed abilities in the so-called lower primates.

The final section applies some of the comparative methods and theories derived from nonhuman primates to human populations. It is discouraging to those who would still ask, what is natural or normal or primitive in modern human behavior. Even the hunter-gathering lifestyle, in its modern manifestations, is not necessarily a relic of some original state of the species, so much as an appropriate and rather successful response to contemporary economics and politics. In Africa the devastation produced by the Rinderpest epidemic at the end of the last century may well have led to the local adoption of hunter-gathering by mixed farmers (Kjekshus 1977). There is no one social system associated with hunter-gathering, including the role of chil-

dren in society. Some of the differences can be correlated neatly with differences in the local ecosystems; others might equally well be derived from the influences of neighbors, or even be residual from earlier life-styles (as urban school holidays are residual from the farming calendar and the seasonal need for child labor in the United States today).

Looking at a wider range of cultures, it seems that people are similar to other primates in that little girls generally spend more time with infants than little boys, but it is by no means clear why this is so; the whole lengthy process of transition from infant to adult is a flexible interaction between physiology of the individual, constraints and opportunities offered by the environment, and social expectations of adults and peers.

We are brought around again to admitting that it is not, in fact, possible to study juveniles separately from other life stages. This book is an important corrective to the idea that the converse might be true—that we may be able to study adults and infants and leave aside juveniles.

T. E. Rowell

Preface

In a sense, this project began in late August 1980, when I traveled to Amboseli, Kenya, to conduct research on mother and infant baboons at the long-term study site maintained by Jeanne and Stuart Altmann. On arriving, I discovered that several females expected to be pregnant in projected study groups had not yet conceived. Several other females had lost new infants and a few others had disappeared themselves. The planned research was out of the question. Fortuitously, each of two baboon groups offered ample cohorts of young and old juveniles for research. Earlier, I had spent 2 years investigating play behavior and juvenile vocal communication in a captive group of Japanese macaques. That experience had caused me to approach biological questions from developmental perspectives. Circumstances in Amboseli returned me from plans to study infants to three additional years of research on the social behavior of juvenile monkeys, this time 20 young savanna baboons.

In 1985, I accepted a position at the Duke University Primate Center that afforded me the opportunity to study forest-living groups of lemurs—including, of course, their juvenile members (Chapter 20). Within a couple of years, funding was secured for longitudinal research on behavioral development in relation to social organization, and long-standing plans to organize an edited volume on juvenile primates were dusted off. Seeking a collaborator with whom to develop ideas, as well as share the editorial burden, I contacted Lynn Fairbanks, one of only a handful of primatologists who maintained a long-term focus on behavioral development and was actively engaged in research on juvenile social behavior.

From the beginning, we worked toward two goals we felt were critically important. First, the volume was to offer as broad a comparative vision as possible. Primatology had sustained negligible interest in juveniles (Chapter 1), and that which had been done comprised research on a single subfamily of Old World monkeys. Also, little had been done to relate research on primate development to the burgeoning literature on life history evolution in animals. Second, the volume was to be founded on a base of previously unpublished empirical work with group-living animals. To address the existing lack of knowledge, our prescription was for the book to contribute by describing the naturally occurring behavior of juveniles, if it accomplished nothing else. Further details on the conceptual and empirical objectives we had set for the project are discussed in our introduction (Chapter 1) and epilogue (Chapter 24).

Our first hurdle was to discover who had new quantitative data on juveniles. After contacting many primatologists known for research on diverse taxa and for interest in behavioral development, we gathered the names of nearly 20 scientists who were interested in contributing to one of two organizational symposia. The symposia, each entitled "The Socioecology of Juvenile Primates," were convened at the 1990 meeting of the American Society of Primatologists and the 1990 Congress of the International Primatological Society. Most of those who planned contributions were also able to accept invitations to author chapters for this volume. By late summer 1990, the authors of Sections I and V and those of a few additional chapters on primates also had accepted invitations to contribute.

This brings us to our third organizational goal for the project and our first set of acknowledgments. To enhance our chances of achieving a high-quality presentation of contributed works, we required of all articles critical review by peers. One of the truly memorable aspects of editing this volume was the enthusiasm, efficiency, and professionalism applied by those asked to criticize chapters, often under time limitations bordering on the unreasonable. For so undertaking these chores, we thank S. Altmann, J. Bernardo, N. Blurton Jones, M. Boyce, D. Candland, T. Caro, B. Chapais, D. Cheney, A. Clark, M. Clarke, M. Cords, C. Crockett, R. Fagen, B. Galef, K. Glander, A. Harcourt, H. Harpending, P. Harvey, M. Hauser, J. Horrocks, S. Hrdy, F. Johnston, H. Kaplan, P. Kappeler, S. Lindstedt, J. Lockhard, R. Martin, J. McKenna, K. Milton, J. Moore, N. Peacock, M. Raleigh, D. T. Rasmussen, A. Richard, J. Robinson, C. van Schaik, R. Seyfarth, J. Silk, E. Smith, E. O. Smith, B. Smuts, V. Sommer, C. Stanford, S. Stearns, K. Strier, M. Symington, S. Tardif, P. Turke, J. Walters, F. White, and R. H. Wiley.

Each of us, of course, also owes individual gratitude to particular persons and agencies.

(MEP) I thank Terri Pyer for unwavering support and understanding, not only throughout the recent period, but also since 1980. And special thanks go to Tiana and Torin Pyer-Pereira for being themselves—wonderfully flexible and inspirational young juveniles. P. Kappeler, P. Klopfer, S. Nowicki, C. van Schaik, L. Vick, and D. Watts contributed time, criticism, enthusiasm, and friendship. The National Institute of Child Health and Human Development provided the fiscal support (R29-HD23243) for my research, including that for my assistant, L. Martin, whose parallel interest in juveniles was an invaluable resource. The administrative and technical staffs of the Duke University Primate Center must be acknowledged for their caring and able maintenance of an irreplicable colony of extraordinary primates. This includes my gratitude to Elwyn Simons, whose vision and energy have promoted the persistence of prosimian primates.

(LAF) I thank Steve Cole for understanding and for helping take care of our twin sons, Ben and Jeff, who at the age of 13 have managed to survive the juvenile period in spite of having an often distracted mother. I also thank all my colleagues at the Nonhuman Primate Research Facility for sharing my enthusiasm about juvenile primates, especially M. McGuire, M. Raleigh, K. Blau, J. Kusnitz, M. Wortz, D. Crumley, M. Heeb, G. Morton, and D. Diekmann. D. Torigoe and T. Cronin helped with the logistical problems of space, computers, mailing, and electronic networks. And special thanks to J. Silk, who served as a valuable critic and sounding board for ideas throughout this project.

Durham, N.C.   M. E. P.
Los Angeles, Calif.   L. A. F.
December 1991

Contents

Contributors

Filippo Aureli
Ethology and Socioecology
University of Utrecht
Utrecht, the Netherlands

Bernard Chapais
Department of Anthropology
University of Montreal
Montreal, Quebec
Canada

Carolyn M. Crockett
Regional Primate Research Center
University of Washington
Seattle, Washington

National Zoological Park
Smithsonian Institution
Washington, D.C.

Frans B. M. de Waal
Yerkes Regional Primate Research Center
  and Psychology Department
Emory University
Atlanta, Georgia

Carolyn Pope Edwards
College of Human Environmental Sciences
University of Kentucky
Lexington, Kentucky

Robert Fagen
School of Fisheries and Ocean Sciences
Juneau Center for Ocean Sciences
University of Alaska, Fairbanks
Juneau, Alaska

Lynn A. Fairbanks
Department of Psychiatry and Biobehavioral
  Sciences
University of California
Los Angeles, California

Carole Gauthier
Department of Anthropology
University of Montreal
Montreal, Quebec
Canada

Paul H. Harvey
Department of Zoology
University of Oxford
Oxford, England

Charlotte K. Hemelrijk
Laboratory of Comparative Physiology
University of Utrecht
Utrecht, the Netherlands

Louis A. M. Herremans
Laboratory of Comparative Physiology
University of Utrecht
Utrecht, the Netherlands

Julia A. Horrocks
Department of Biology
University of the West Indies
Cave Hill, St. Michael, Barbados

Wayne Hunte
Bellairs Research Institute of McGill
  University
St. James, Barbados

Charles H. Janson
Department of Ecology and Evolution
State University of New York
Stony Brook, New York

Nicholas Blurton Jones
Departments of Education, Anthropology,
  and Psychiatry
University of California
Los Angeles, California

Katharine Milton
Department of Anthropology
University of California
Berkeley, California

Leanne T. Nash
Department of Anthropology
Arizona State University
Tempe, Arizona

Timothy G. O'Brien
Department of Ecology and Evolutionary
  Biology
Princeton University
Princeton, New Jersey

Wildlife Conservation International
New York Zoological Society
Bronx, New York

Mark D. Pagel
Department of Mathematics
Queen Mary College
London, England

Michael E. Pereira
Duke University Primate Center
Durham, North Carolina

Theresa R. Pope
Department of Biological Anthropology
  and Anatomy
Duke University
Durham, North Carolina

Anne E. Pusey
Department of Ecology, Evolution,
  and Behavior
University of Minnesota
Minneapolis, Minnesota

Lal Singh Rajpurohit
Department of Zoology
University of Jodhpur
Jodhpur, India

John G. Robinson
Wildlife Conservation International
New York Zoological Society
Bronx, New York

Thelma Rowell
Department of Zoology
University of California
Berkeley, California

Daniel I. Rubenstein
Department of Ecology and Evolutionary
  Biology
Princeton University
Princeton, New Jersey

Volker Sommer
Institut für Anthropologie
University of Göttingen
Göttingen, Germany

Elisabeth H. M. Sterck
Laboratory of Comparative Physiology
University of Utrecht
Utrecht, the Netherlands

Karen B. Strier
Department of Anthropology
University of Wisconsin
Madison, Wisconsin

Carel P. van Schaik
Department of Biological Anthropology
  and Anatomy
Duke University
Durham, North Carolina

Maria A. van Noordwijk
Laboratory of Comparative Physiology
University of Utrecht
Utrecht, the Netherlands

David P. Watts
Department of Biological Anthropology
  and Anatomy
Duke University
Durham, North Carolina

Carol M. Worthman
Department of Anthropology
Emory University
Atlanta, Georgia

Juvenile Primates

1

What Are Juvenile Primates All About?

MICHAEL E. PEREIRA and LYNN A. FAIRBANKS

Watch! The 5-year-old arriving at his playgroup just now—the one carrying the unusual toy—will receive special treatment from his peers. First, he is surrounded. Even after avoiding all attempts by others to touch the toy, he remains the secure center of attention. The oldest girl, also 5 years old, prompts him for a plot for fantasy play and helps set the group in motion. Over the next 10 minutes, that girl holds the toy longer than any other child. Her 3-year-old brother touches it at a rate that rivals that of its owner, while few others achieve more than fleeting contact. Some children's attention begins to stray before a neighbor's ball flies into the yard. Peripheral group members lead a merry explosion to investigate and the toy is abandoned along with the youngest boy and girl. The boy instantly reaches for it, but his effort is displaced by his partner's shoulder and elbow. Sitting back, the boy shows only partial resignation as the girl grants him a limited view. When the older children return from their escapade, the girl, without prompting, offers the toy back—to its owner— and says, "I care it."

Kids. This book is about kids. In particular, it focuses on juvenile primates—their physical growth and maturation, social development, and species and sex differences in behavior prior to reproduction. The project is meant to redress the historical circumstance that, after 35 years of modern research on primate behavior and ecology, the juvenile period remains the most neglected phase of life histories (Pereira & Altmann 1985). This situation is ironic because protracted development, including extended juvenility or delayed sexual maturation, is the life history feature that best distinguishes the order Primates among mammals (Schultz 1956; Clark 1971; Harvey & Clutton-Brock 1985).

Juvenile primates are of particular concern to physical anthropologists because comparisons of development among primates and between primates and other mammals offer to illuminate primate natural histories considerably (cf. Shea 1986; Watts 1990; Rasmussen & Tan 1992). Juvenile primates also engage psychologists concerned with patterns of behavioral development in humans and other animals; sociobiologists, who study the evolution and development of social behavior; and other evolutionary biologists, whose theory maintains that natural selection favors the earliest possible reproduction in animal life histories.

What, then, are juvenile primates—primate

kids—all about? From the adult vantage in human society, we view them as innocent and incomplete, uninformed and intensely curious, energetic and virtually carefree. The play scenario reconstructed above serves to introduce not only starting points and next steps in our subject matter, but also one major approach to their scientific investigation: noninterventive behavioral observation, often in the natural environment. Whereas for most readers the scene's familiarity stems from its origin with our own species, students of the nonhuman primates will recognize many of its central elements from their daily manifestation among the juveniles in groups of lemurs, monkeys, and apes.

Juveniles are innocent, are they not? Before answering, one really must watch a weanling monkey groom its mother's chest for several minutes, put her to sleep, and access "unapproved" time on the nipple (Fig. 1.1)! Immatures readily assimilate rules for behavior, and at times they appear equally adept at operating and *appearing* to operate within them.

Incomplete? "Different" is a better description: juveniles are forerunners specialized for the tasks of surviving the wait until reproduction and of using that time wisely. Uninformed? Curious? Without question, juvenile primates are among the most curious creatures on Earth— their very business is about becoming informed.

Energetic? Another unqualified yes; this is true for all juvenile mammals (Fagen 1981). But, carefree? Not really. Virtually any selection of the following chapters will testify that life is no party for immature primates. Rather, diversity of risk is a basic element of juvenile lives, such that major aspects of growth, behavioral development, and even social organization appear to have evolved partly to modulate risk for juveniles (cf. Dittus 1977; Silk et al. 1981; Pereira 1988a; van Schaik & de Visser 1990). Most interesting, perhaps, is that some of the danger inherent in youngsters' lives derives from the fact that immature primates must try to succeed in a fundamentally adult world (see Janson & van Schaik, Chapter 5, this volume).

Previous information on juvenile social behavior in primates came largely from quantitative research on three topics: social play (e.g., Smith 1978a; Symons 1978), interaction with infants (e.g., Hrdy 1976; Fairbanks 1990), and acquisition of dominance (e.g., Chapais 1992; Pereira 1992). These data, together with

results on other aspects of juveniles' relationships with peers and adults (Cheney 1978a, 1978b; Silk et al. 1981; Hayaki 1983, 1985a; Baker-Dittus 1985; Colvin 1985; Colvin & Tissier 1985; Fairbanks & McGuire 1985; Pereira & Altmann 1985; Johnson 1987, 1989; Pereira 1988a, 1988b, 1989; Hiraiwa-Hasegawa 1989), allowed some provisional explanatory schemes to be pondered (e.g., Pereira & Altmann 1985; Walters 1987a; Pereira 1988a, 1992). They were limited, however, by the fact that most of the initial studies were on macaques, baboons, or vervet monkeys—primates that share pervasive commonalities in their phylogeny and natural histories. We lack information on how developmental strategies differ across taxa. And the small amount of work done on juveniles overall limits our understanding of how prereproductive behavioral tactics shift with socioecological circumstance within species.

We therefore sought in this volume to cover as broad a range of the order Primates as possible, with most chapters reporting new data from natural populations or socially housed groups. These works are complemented by an opening set of chapters that contrasts schedules of growth and maturation in primates with those observed in nonprimates, and by concluding chapters on comparable aspects of development in humans. By providing a comparative database on juvenile development, we hope to initiate a period during which information from juveniles can contribute significantly toward our understanding of both behavioral development in general and the natural histories of primates in particular.

Both classical ethology (e.g., Lorenz 1950; Tinbergen 1951) and modern evolutionary theory have led behavioral biologists to conceive of males and females as subtypes whose conflicting interests must be dovetailed before members of either group can reproduce effectively (e.g., Smuts 1985). Investigators of primates must recognize juveniles as yet a third subtype. Juvenile modes of behavior differ substantially from those of mature males and females, and in the majority of primate species the lives of juveniles and adults are intertwined through cohabitation in social groups. Patterns of complementation and conflict in juvenile–adult social relationship contribute fundamentally to the organization of primate groups and populations

a.

b.

Fig. 1.1. Weanling monkeys use diverse tactics to access their last hours of suckling time. (a) The Amboseli baboon mother Slinky guarding her nipples against her 1-year-old son, Sluggo. Sluggo waited and occasionally groomed his mother until she finally took a nap, allowing Sluggo limited time on the nipple (b). (Photos, M. E. Pereira)

(Pereira 1992). Individually and collectively, the following chapters provide initial glimpses into the ways that juvenile and adult behavior reciprocally influence one another.

Ironically, the development and natural histo-ry of juvenile primates remain poorly understood, although development has been a major focus for research in primatology and animal behavior over the years (e.g., Altmann 1980; Immelman et al. 1981; Bateson & Klopfer

1982). We turn now to a selective overview of the history of behavioral primatology that explains how juveniles and juvenility came to be neglected in this field. Several more thoroughgoing accounts of the history of behavioral primatology are available, written from diverse backgrounds to meet a variety of objectives (e.g., Gilmore 1981; Haraway 1989; Parker 1991; Loy & Peters 1991; see also Hinde 1982; Dewsbury 1984). Interested readers should refer to them for more comprehensive summaries.

HOW DID JUVENILES GET OVERLOOKED?

Origins and Field Work

Whereas humans have studied animals throughout history (Mountjoy 1980), most scientific historians regard the publication of Darwin's (1872) treatise on human and animal emotion as a pivotal juncture for all kinds of research on behavior (e.g., Dewsbury 1984). Darwin's writings certainly influenced the limited research on primates that preceded the Second World War, which focused largely on the great apes (Kohler 1925; Yerkes & Yerkes 1929; Nissen 1931; Bingham 1932; Schulz 1936). Monkeys of the Old and New Worlds and the lesser apes also received some attention before the war (Pocock 1928; Zuckerman 1932; Carpenter 1934, 1935, 1940; Hill 1936).

Like the founders of classical ethology (Beer 1973), the pioneers of primatological field work were naturalists, preferring discovery in the field to the confines of laboratories. Their methods did not derive from those of the ethologists, but shared with them a heritage (Carpenter 1964; also J. Altmann 1974). Most important, ethologists, with their well-known emphasis on species-typical behavior (Lorenz 1950; Tinbergen 1951), and the first primatologists shared interests in the potential phylogenetic significance of behavioral research and in the contributions the work might make to understanding or even ameliorating the human condition (Yerkes 1932; Carpenter 1964; but see Zuckerman 1932). Yerkes (1916, 1932) insisted from the outset that the road to discovery lay in weaving together simultaneous commitments to field work and to research that required captive settings. His postdoctoral students, including Nissen (1931), Bingham (1932), and Carpenter (1934, 1935, 1940), helped to bring that philosophy into reality.

The New Anthropology. After the Second World War, a handful of academics worldwide trained the majority of scientists that ultimately gave rise to modern behavioral primatology. In the United States, two were most important: Sherwood Washburn, an anthropologist, and Harry Harlow, a psychologist. Washburn championed the integration of primate studies into American physical anthropology and ultimately sponsored 15 or more students who in turn are thought to have trained about half of today's American behavioral primatologists (Gilmore 1981). Like Yerkes and Carpenter before him, Washburn was devoted to the idea that the nonhuman primates, through both analogy and homology, stand to teach us as much about humans and their origins as do fossils and cross-cultural studies (Washburn 1951; DeVore & Washburn 1963). He maintained that hypotheses in the "new anthropology" must derive from field work. Accordingly, his early graduate students, Irven DeVore and Phyllis Jay, conducted two of the first modern field studies, on savanna baboons (*Papio cynocephalus:* Washburn & DeVore 1961; DeVore 1963) and Hanuman langurs (*Presbytis entellus:* Jay 1963b).

The order of the day for anthropologists was to understand the adaptive functioning of social systems (e.g., Radcliffe-Brown 1956). This topic simultaneously commanded the attention of several field biologists, such as John Crook, whose work on birds and primates (Crook 1965, 1970, 1972; Crook & Gartlan 1966) was influential in both behavioral ecology and primatology (e.g., Eisenberg et al. 1972; Wilson 1975). For his part, Washburn devoted much of his considerable energy to the idea that adult male aggression and dominance hierarchies maintained order in primate societies (Washburn & Avis 1958; Washburn & DeVore 1961; Washburn & Hamburg 1965, 1968). Subsequently, throughout the 1960s and 1970s (e.g., Sade 1967; Hausfater 1975; Packer 1979a, 1979b; Bernstein 1981), and even to this day (Harcourt & de Waal 1992; Silverberg & Gray 1992), an enormous proportion of primatological research has focused on aggression and dominance relations.

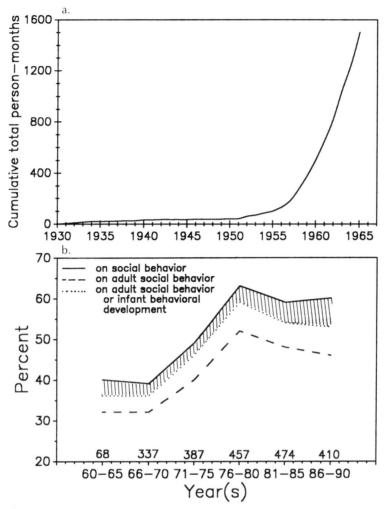

Fig. 1.2. (a) The rapid expansion of research on primate behavior began in the late 1950s, as reflected here by the cumulative total person-months devoted to field studies of primate behavior and ecology. The amount of research conducted between 1962 and 1965 alone was greater than all the research that preceded it. (Adapted with permission from Altmann 1967) (b) Proportions of all primatological research reported in four journals that concerned social behavior, adult social behavior, and infant or adult social behavior. The small remaining proportion (hatched) was research concerned in any way with juveniles (research on adolescents pooled with that on adults). Note also sharp increase in proportion of primatological research devoted to social behavior in the 1970s. Data are from scans of all titles published from 1960 through 1990 in the journals *Primates, Folia Primatologica, Animal Behaviour,* and *Ethology* (formerly *Zeitschrift für Tierpsychologie*). Total numbers of titles displayed along abscissa; note, not all issues of *Primates* could be reviewed for 1960 through 1965.

Research on Behavioral Development

If Washburn trained half the mentors of today's American behavioral primatologists, Harry Harlow accounted for most of the other half. In the late 1950s, Harlow turned his attention to infant attachment and the interplay between maternal attributes and behavioral development in rhesus monkeys (Harlow 1959). Over the next decade, his laboratory produced or promoted many of the American primatologists working with captive primates who would contribute immensely to the international explosion of primatology that began late in the 1950s (Fig. 1.2). R. Goy, E. Hansen, W. Mason, G. Mitchell, M. Novak, L. Rosenblum, G. Sackett, A. Schrier, B. Seay, and S. Suomi are only some of the students and visiting colleagues who emerged in this period (Suomi & Leroy 1982; Leroy, personal communication). Harlow's new interest fostered more than two decades of intensive research on primate mother–infant relations and behavioral development.

Even as Harlow turned to his new work, an ethologist named Robert Hinde was diverting much of his research attention from birds to primates (Hinde 1985). He met the London psychoanalyst John Bowlby in the mid-1950s, when Bowlby was developing new views on the roles of infant attachment and separation anxiety in human development (Bowlby 1969, 1973). Bowlby sought the input of an ethologist for his weekly seminar. Soon thereafter, he helped Hinde obtain the funding needed to establish a colony of rhesus monkeys at the Madingley (Ornithological!) Field Station with the objective of studying the long-term behavioral effects of mother–infant separation. Consequently, Hinde met Harlow at an American conference (1957), and on his return to Madingley, Bowlby initiated his written correspondence with Harlow (Harlow Primate Laboratory files; Leroy, personal communication).

After initial descriptive studies (Hinde & Rowell 1962; Rowell & Hinde 1962), Hinde and his colleagues launched a decade of research on infant development and the dynamics of mother–infant relations in captive rhesus macaques. In so doing, they set new standards in the use of quantitative methods. Early on, they developed an index to monitor which of two individuals is primarily responsible for the time they spend together (the percentage of all approaches between A and B effected by A minus the percentage of all departures between A and B effected by A; Hinde & Spencer-Booth 1967) and demonstrated its superiority over the other measures commonly used at the time (Hinde 1969; Hinde & Atkinson 1970). Colleagues appreciated the analysis and many adopted use of the index in their own studies of primate mother–infant relations. Since then, the index has been used to analyze other classes of social relationship in

primates (e.g., Harcourt 1979a; Deputte 1983; Stein 1984a; Colvin & Tissier 1985; Smuts 1985; Pereira, Chapter 20, this volume) and mother–infant relations in nonprimate mammals (e.g., Taber & Thomas 1982; Lickliter 1984).

Selective historical overviews such as this risk offense to those whose favorite early researchers go unmentioned or underemphasized, as well as those who would have had their favorites recognized for other aspects of their work. Hinde, for example, has been celebrated recently for his myriad contributions to the field of animal behavior, including his promotion of primate field work by Goodall, Clutton-Brock, Cheney, Seyfarth, and others (Bateson 1991). It is, in fact, important to note at this juncture that many other scientists whose writings flourished in the 1960s contributed enormously to the development of behavioral primatology. It is difficult to resist pausing to highlight contributions by S. Altmann, Bernstein, Goodall, Goy, Hall, van Hooff, Itani, Kummer, Marler, Mason, Rosenblum, Rowell, Sackett, Sade, Struhsaker, and others.

In the present context, however, it would be difficult to overemphasize the work of one other scientist. Jeanne Altmann's early work on infant development and mother–infant relations in free-ranging baboons (Altmann 1980) is exemplary of the primatological tradition of using methods and information from research on captive animals to maximize the value of field studies. Her current research on primate development and in other areas continues to set standards for the field (e.g., Altmann & Samuels 1992; Moses et al. 1992).

The Two Big Topics

Broadly speaking, then, two topics dominated behavioral primatology throughout its first two decades: aggression and dominance relations, particularly among males, and infant development and mother–infant relations. Substantial work on primate foraging ecology complemented these foci (e.g., S. Altmann 1974; Clutton-Brock 1977a). In this initial phase, researchers were attempting to characterize primate social behavior and, later in the period, reproductive strategies. This required focus on adults. The primates most often studied were macaques, baboons, and great apes, all of which show sexual size dimorphism. As in other dimorphic animals, the reproductive strategies of at least some males seemed to depend on their relative success in behaviorally dominating their peers (e.g., Goodall 1968; Hausfater 1975; Packer 1979a). And, as in all other mammals and most birds, female reproductive success clearly depended on effective maternal behavior (e.g., Goodall 1968; Altmann 1978, 1980). Also, one major premise underlying much research in this early period was that the earliest experiences of primates played an overriding role in shaping lifelong patterns of behavior (see Fairbanks & Pereira, Chapter 24, this volume).

Thus, in the first 25 years of behavioral primatology, large amounts of detailed information accumulated on the social behavior of adults and infants, while far fewer data were gathered on juveniles (Pereira & Altmann 1985) or adolescents (Caine 1986). Research on mother–infant relations, spurred by attachment theory and the enthusiasm surrounding research by Harlow, Hinde, Mason, Rosenblum, and others, so dominated ontogenetic thinking that the study of behavioral development in primates became virtually synonymous with the study of infants, a bias still detectable in recent reviews (Chism 1991). The volume of research on later phases of primate development began to rival that on infancy only after 1980 (Fig. 1.2).

PRIMATOLOGY AND BIOLOGY

In many ways, the period from the late 1960s through the mid-1970s constituted the heyday of primatology, especially in the United States. It saw two milestone developments for the field: (1) a giant step taken toward uniformity in noninterventive observational methods, and (2) integration of the theories of evolutionary biology as a conceptual framework for research. Before then, several field primatologists had implemented robust quantitative methods and biological thinking in their work (e.g., Carpenter 1940; Altmann 1962, 1965; Struhsaker 1967; Kummer 1968), but this was a period of rapid transition during which the research of large proportions of primatologists began to reflect an appreciation of the principles of natural selection at the level of the individual (Darwin 1859; Fisher 1930; Haldane 1932; Williams 1966), sexual selection (Darwin 1871; Fisher 1930;

Trivers 1972), population biology (Lotka 1925; Wright 1931; Haldane 1932; Dobzhansky 1937; MacArthur & Wilson 1967), kin selection and inclusive fitness (Fisher 1930; Haldane 1932; Hamilton 1964; Maynard-Smith 1964), and life-history theory (Cole 1954; MacArthur & Wilson 1967; Gadgil & Bossert 1970).

Observational Methods

Soon after joining one of the earliest field studies of baboon behavior and ecology (Altmann & Altmann 1970), Jeanne Altmann recognized needs for fuller appreciation of the sampling methods available to those who would conduct behavioral research and for use of common terminology. Some methods had limitations that seemed to have gone unrecognized; others had been underutilized. Differences among related studies, either in methods chosen or the names given them, had severely limited the extent to which their results could be compared effectively. To help with this, Altmann (1974) published an article that became a handbook for behavioral biologists, creating a standard methodological toolbox and vocabulary for research on behavior.

Behavioral Ecology and Sociobiology

After ethology flourished in the 1950s, the 1960s witnessed the merging of research on animal behavior with the more traditional animal ecology of prior decades, giving rise to the new subdiscipline of behavioral ecology. This was required by deepening appreciation for "the adaptive complex" among organismal biologists (e.g., Tinbergen 1951; Lack 1954, 1968; Crook 1964). No longer could animals be viewed as anonymous automatons best described demographically and studied only as participants in intergroup and interspecific competition and predator–prey relations. With consistent emphasis on proximate causation, roles of experience, individual variation, and mother–offspring relations, along with the traditional concerns of evolutionary function and phylogeny, the field of animal behavior was a principal catalyst for the development of "life-history thinking" in biology (see also Cole 1954).

Primatology was not only affected by this transition, it also influenced it. Primatologists' descriptions of behavior involved levels of detail exceeding those from many other animal studies (e.g., Altmann 1962; Andrew 1963; Struhsaker 1967; Goodall 1968; van Hooff 1970; see also Chadwick-Jones 1991), which facilitated initial understanding of interrelated patterns of behavior. Well-conceived field experiments also brought attention to the field (e.g., Hall 1965; Kummer 1971; Waser 1977b).

Most important, longitudinal observation of known individuals was a hallmark of the best-established field projects (e.g., Kawai 1965; Goodall 1968; Kummer 1968; Sade 1972a; Hausfater 1975). Attention to individual differences was required by both primates' general flexibility of behavior and their long generation times, which resulted in slow accumulation of independent data. But also, major topics in social behavior, such as nepotism and dominance relations, would have been beyond study without individual recognition of subjects. Long-term studies highlighted the need to consider the implications of cross-sectional results from the life-span perspective (e.g., Saunders & Hausfater 1978). These aspects of primate research enabled the field to contribute to the expansion of thought on optimal and alternative reproductive strategies (e.g., Semler 1971; van Rhijn 1973; Wiley 1974a; Hrdy 1977; Packer 1977; Clutton-Brock & Harvey 1978; Altmann 1980).

The impact of Wilson's *Sociobiology* (1975) on animal behavior, behavioral ecology, and primatology is well known. In this treatise, Wilson synthetically reviewed far-ranging aspects of population biology and behavioral ecology along with innumerable case studies in ethology, all in relation to the theories of sexual selection, parental investment, kin selection, and reciprocal altruism. For primatologists, and anthropologists studying preindustrial cultures, the result was more than a decade of renewed vigor in research on the distribution of aggression and altruism within and between groups (e.g., Cheney 1977; Hrdy 1977; Konner 1977a; Kurland 1977; Seyfarth 1977; Altmann 1979; Chagnon & Irons 1979; Silk et al. 1981; Hinde 1983; Smuts et al. 1987; Dunbar 1988).

The integration of the principles of behavioral ecology and sociobiology was enormously productive in guiding research on primates. But, as for animal behavior in general (Bateson 1991), it also stymied the expansion of research on development. Just as the practitioners of the "Modern Synthesis" neglected research on physical development in their excitement to merge genetics with systematics in the early 1900s (Buss 1987), those of the "New Synthesis" overlooked the significance of behavioral development in their rush to substantiate that systems of behavior evolve. The earlier focus on adult behavior and parent–infant relations became still more pronounced as these were the most accessible correlates of short-term reproductive success. The long lives of primates relative to those of research grants reinforced the inclination to infer the adaptiveness of modal behavior patterns from cross-sectional data. Students of kinship, altruism, and competition tended to disregard age as an independent variable, partly because few had enough information to consider fairly how it modulates behavioral expression.

JUVENILE PRIMATES

Now that the adult behavior and reproductive strategies of the best-known primates are beginning to be understood, research can be focused productively on juveniles, adolescents, and developmental strategies. Our present stage of knowledge should also facilitate the simultaneous study of adults and immatures in those primates for which behavior at all ages remains yet poorly known. This book is an effort to help launch this phase of primatology. The remainder of this chapter introduces major issues in the study of juvenile primates and the organization of the volume.

Primates exhibit, by far, the longest juvenile periods among mammals of a given size class (Schultz 1956; Clark 1971; Harvey & Clutton-Brock 1985). With this project, we begin to illuminate life-historical correlates and likely functions of primate juvenility by examining how size, behavior, social relationships, and other conditions of existence change for immature primates between weaning and maturation (e.g., Fig. 1.3). Bringing together researchers concerned with the life histories, development, and prereproductive behavior of primates, we investigate the selective pressures and constraints that determine the length and developmental objectives of the juvenile period. Through comparative treatment of diverse species, we reveal how juvenile behavioral tactics

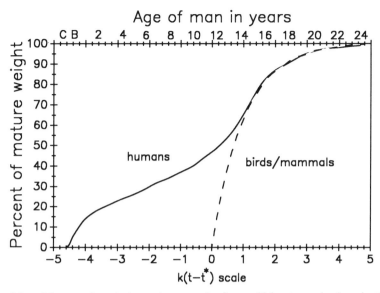

Fig. 1.3. Brody's weight–growth equivalence plot comparing farm and laboratory animals against humans. For the animals, percentage of mature weight attained is plotted against standardized values representing the product of instantaneous growth rate (k) and age ($t - t^*$) between conception and adulthood. The human curve is similarly derived and then shifted to the left until its inflection point, representing the end of puberty, coincides along the abscissa with that of the animals' curve. This reveals the dramatic difference in juvenile growth rates between humans and nonprimate homeotherms. Several large anthropoid primates are known now to show growth trajectories similar to that of humans (see Janson & van Schaik, Chapter 5, this volume). Brody's animal data came from cows, swine, sheep, rabbits, Cornish fowl, albino rats, albino mice, guinea pigs, common pigeons, and ring doves. (Adapted from Brody 1945)

function in primates to promote the immediate goals of growth and survival and the long-term goal of development into competent adults.

The theory of life-history evolution asserts that natural selection favors the earliest possible reproduction in animal life histories (see Charlesworth 1980, Chapter 2). A central problem in research, then, is to understand how delayed maturation can be adaptive (e.g., Mertz 1971; Wiley 1974a; Roff 1981, 1984, 1986; Charnov 1990, 1991). This topic is addressed in Part I, "Why Be Juvenile?" which provides a theoretical framework for the volume. Four chapters focus specifically on evolutionary determinants of the duration of the juvenile period in a variety of species, including invertebrate and vertebrate ectotherms, birds, nonprimate mammals, and primates.

The variation in adult behavior and social organization that exists among primates provides a rich array of circumstances with which to test hypotheses on the adaptedness of juvenile behavior and development. As yet, we have little ability to further our understanding of primates through consideration of this diversity because the large majority of existing data comes from a single subfamily of Old World monkeys (Cercopithecinae). The chapters in Part II, "Growing into Different Worlds," offer descriptions of juveniles in several species that were heretofore unavailable. They provide data on the emergence of species- and sex-typical behavioral repertoires and explore how adult male and female reproductive strategies can complement and conflict with juvenile strategies for survival and development.

The chapters in Part III, "Developing Skills and Relationships for Later Use," investigate how juveniles exploit opportunities to promote their own physical and social welfare and how the activities and relationships developed during the juvenile years can influence adult competence. They consider roles of intrinsic preferences, social learning, and practice in the development of skill in areas such as foraging, fighting, and infant handling. They also examine the hypotheses that juveniles seek to develop advantageous relationships with particular group members and that relationships formed during the juvenile years have effects that extend into adulthood.

Group living exacerbates aspects of resource competition and thereby generates some social conflict (Alexander 1974). Among the mechanisms used by group-living animals to mediate conflict are aggression, submission, and reconciliation. As they develop, all primates learn about and engage in social conflict and most enter into dominance relations. Juveniles' success in this area is of particular importance because of the basic difficulties they face in competition with older, more experienced group members. The chapters of Part IV, "Managing Social Conflict and the Development of Dominance Relationships," strive toward fuller understanding of the cercopithecine system of matrilineally organized dominance relations and compare the system with that of a prosimian primate sharing basic aspects of natural history with the Cercopithecinae.

In Part V, "Comparative Socioecology of Childhood," the themes of earlier chapters are revisited in three studies of children developing in different cultures. Parental investment strategies, the development of sex-typical behavior, and developmental factors influencing age at maturation illuminate cross-cultural similarities and differences in child development. These chapters also illustrate how commonalities pervade the interests and approaches of anthropologists, biologists, and psychologists due to the researchers' common concern with development.

In the epilogue, we attempt to extract general themes from the volume and suggest directions for future research on primate development. Experts maintain that 90% of the unsolved problems of biology have to do with development and behavior (Maynard-Smith 1986; Buss 1987; Bateson 1991). It is our hope that rather than simply summarizing a field or phase of research, this volume will help to increase interest in the investigation of juveniles, development, and behavioral mechanisms in life history among anthropologists, biologists, and psychologists alike. Clearly, we are all keenly interested to understand fully the evolution and ontogeny, functions and causes of behavior.

ACKNOWLEDGMENTS

J. Altmann, I. Bernstein, D. Candland, M. Cords, E. Hansen, and especially H. Leroy generously assisted in reconstructing the history of relationships, among people and ideas, in anthropology, psychology and zoology in the 1950s and '60s. We also thank D. Candland, B. Chapais, P. Harvey, P. Klopfer, T.-J. Pyer, C. van Schaik, J. Walters, R. H. Wiley, and D. T. Rasmussen for criticism of an early draft of this chapter. MEP was supported by the National Institute of Child Health and Human Development (R29-HD23243) and the Duke University Primate Center (DUPC Publication No. 556).

Part I

WHY BE JUVENILE?

It has long been recognized that the life histories of primates are unusually protracted in comparison with those of nonprimate mammals (Brody 1945; Schulz 1956, 1969). That is, primate rates of development are generally retarded, leading to long periods of gestation, infancy, juvenility, and maturity. In conjunction with this ontogenetic profile, all anthropoid and many prosimian species show large brains relative to their body size (Jerison 1973; Radinsky 1982). These distinctive features—developmental retardation and large brains—are usually interpreted conjointly as indicative of an evolutionary trend away from relative invariability of behavior toward increased dependence on learning and extreme behavioral plasticity (e.g., Schulz 1969; Sacher 1975). The extended juvenile period in primates has been presumed to function to provide developing individuals the time needed to learn effective behavior, including the complexities of primate social behavior (e.g., Schulz 1969; Poirier & Smith 1974; Gavan 1982).

Are there so many difficult things for young primates to learn that they must spend roughly 25% of their postweaning life expectancies learning them before maturing and initiating their reproductive careers? And why are juvenile primates so small? That is, why do primates delay growth? This life-history characteristic seems most peculiar in relation to the fact that one major function of juvenility in animals is to maximize opportunity for growth. If the need for learning were the central issue for primates, would it not better be accomplished at larger sizes, to guard against predation and the potential for conspecific aggression? Collectively, the chapters of this volume begin to suggest some answers to these and other gnatty questions relating to the evolution of development in primates.

Part I presents four chapters that examine the evolution of animal schedules of development, with particular focus on mammals. Evolutionary theory asserts that organisms should evolve that allocate resources among growth, maintenance, and reproduction during development in a manner that maximizes reproductive potential across the life span. Mathematical models of evolution in stationary and especially expanding populations suggest that natural selection favors early onset of reproduction (Charlesworth 1980). Accelerated maturation allows for early reproductive effort, which in turn safeguards lifetime reproductive success against the possibility of accidental death and allows lineages to expand rapidly within populations. On the other hand, early reproductive effort should be selected against if concomitant factors, such as small size, rapid growth, or inadequate experience, lead individuals to experience disproportionately heavy or long-lasting costs of living. Delay of maturation provides immatures the time and/or energy needed to grow large well and become experienced, traits especially helpful in K-selected populations, like those of most primates, where adult mortality is relatively low.

In Chapter 2, Pereira introduces life-history theory in greater detail and reviews roles for the juvenile period in the lives of invertebrates, ectothermic vertebrates, and birds. He points out that many organisms besides primates exhibit extended juvenility and presents a body of comparative evidence, featuring natural and artificial experiments, strongly supporting the theory of life-history evolution. The review illuminates the general function of juvenility in animals—modulation of growth and reproduction. The chapter concludes with a discussion of basic roles for experience in development: the cueing of conversions (e.g., sexual maturation) and, more generally, feedback regulation. A major frontier indicated for future research is the full characterization of experiential regulation of development in animal life histories.

Chapters 3, 4, and 5 investigate the nature of the juvenile period in mammals. Each considers the traditional components of natality, ontogeny, and mortality—for example, neonatal and adult body size, age and size at maturation, age-specific fecundity and mortality—but interpretations vary as to which of the intercorrelated variables of life histories drive the evolution of mammalian and primate life cycles.

Pagel and Harvey first explain the traditional alternative approaches to understanding species differences in life histories, before reexamining their own and other recent comparative analyses to consider specific implications for the Primates. The allometric or design-constraint view maintains that the length of the juvenile period is set by the time it takes to grow to adult size. One limitation of this view is that it does not explain the diversity of animal sizes. Also, it fails to acknowledge that many important correlations remain among life-history traits after the effects of body size or metabolic rate are removed analytically. From the perspective of demographers, death is certain for all organisms and those experiencing high rates of mortality should reproduce early in life. Some combination of increased survivability and/or fecundity must be the advantage gained when maturation is delayed. Pagel and Harvey conclude that the relatively large body sizes of most primates, via constraints on rates of growth, inevitably extend the juvenile period. But also, they note that selective advantages accrue to individuals that use juvenile time "wisely." Several modes of learning are available to primates across development (cf. Galef 1976, 1988), including opportunities for practice. Selection favoring relatively large brains may have helped establish an evolutionary cycle that stabilizes or extends the juvenile period in primates (cf. Sacher & Staffeldt 1974; Sacher 1975, 1982).

In Chapter 4, Rubenstein presents a classical demographic model that explores relationships among age of maturation, juvenile mortality, and adult fecundity and mortality. He then tests predictions from the model using data from a variety of mammals. A main conclusion of the work is that the timing of maturation should depend on the sensitivities of adult fecundity and mortality to rates of development (cf. Wiley 1974a; Roff 1986). In stable populations, selection should favor early maturation whenever fecundity and mortality are little affected by accelerated development. When fecundity or mortality is compromised, development should be retarded, maturation delayed. Next, Rubenstein reviews intraspecific case histories to illustrate how different types of early experience can produce differential sensitivities in life histories. For example, daughters of dominant female horses appear able to breed at earlier ages without increased risk of mortality or reduced fecundity. The author concludes that some individuals within populations can accelerate maturation because developmental circumstances buffer them against the usual negative effects of rapid development.

Finally, Janson and van Schaik take a novel approach and consider evolutionary implications of the anomolous scheduling of growth in primates. A mammalian life-history feature yet known only in anthropoid primates is that not only maturation but also somatic growth is deferred during development. Whereas nonprimate mammals show maximal growth rates soon after birth, anthropoid growth velocities descend to modest values several days to weeks after birth and a damped growth trajectory is maintained across the juvenile period (Brody 1945; Watts 1986). Even provisioned primates, among the larger species, attain adult size and mature only after several years. At puberty, the suppression of growth rates is lifted, whereupon all males yet studied and the females of many species show a distinctive growth spurt (Watts 1986, 1990). The adolescent growth spurt coin-

cides with reactivation of the hypothalamic–pituitary–gonadal hormonal axis. Because rates of neonatal growth in primates are also associated with gonadal steroid secretion (Hobson et al. 1980, 1981), it appears that at least some anthropoids have exploited the normative mammalian schedule for gonadal endocrine activity as mechanism to schedule suppressed growth across prereproductive development (see also van Wagenen 1947, 1949).

The most unusual aspect of growth in anthropoid primates, then, is not so much the adolescent spurt, as often thought, but rather the suppression of growth between early infancy and puberty (Brody 1945). Janson and van Schaik argue that this growth schedule is best viewed as an adaptive response to the unique socioecology of juvenile primates, functioning to maximize individuals' chances of surviving to adulthood. Juvenile primates are group-living, independent foragers whose daily travel and other patterns of maintenance activity are determined largely by adult behavior. Along the way, their small size and inexperience render them poor foraging competitors and thus particularly vulnerable to malnourishment during periods of nutrient shortage. Group-living and related behavior, however, also effectively shelter immature primates from predation, making retarded growth a feasible way for them to minimize the instantaneous risk of starvation. Janson and van Schaik present a mathematical model of growth effects on juvenile survivorship and show that whenever mortality increases more than linearly with increases in growth rate, a single nonmaximal growth rate can be identified that would provide the highest juvenile fitness—that is, the greatest chance of surviving to maturity.

2

Juvenility in Animals

MICHAEL E. PEREIRA

This chapter introduces the basic hypotheses and methods of life-history theory, a biological subdiscipline important to any effort to understand a particular phase of an organismic life cycle, such as the juvenile period in animals. By delaying maturation, many animal species exhibit longer juvenile periods than do others of similar size, whereas theory expects natural selection to favor the earliest possible reproduction in life histories. A central problem in life-history theory is to identify the various conditions under which delayed maturation, or extended juvenility, can be adaptive. Toward this end, a selective review of juvenility in the Animal Kingdom is presented. Because the remaining chapters of this section purposely focus on mammals, other taxa are highlighted here. In particular, many examples are taken from the invertebrates and ectothermic vertebrates (amphibians, reptiles, fishes), which represent the vast majority of animal taxa and a great diversity of life-styles. Finally, two basic roles for individual experience during development are discussed, and the general functions of juvenility in animal life histories are summarized.

LIFE-HISTORY THEORY

The theory of life-history evolution is that natural selection favors organismic life cycles in which resources are allocated among growth, maintenance, and reproduction in relation to age or size in a manner that maximizes the reproductive potential across individual life spans; that is, Fisher's (1930) reproductive value is max-imized at each age (Williams 1966; Schaffer 1974). Whereas the roots of this idea are moderately old (Fisher 1930; Cole 1954; also Medawar 1946, 1952; Lack 1954; Williams 1957), the recent surge of interest in life-history evolution did not begin until the late 1960s (e.g., MacArthur & Wilson 1967; Gadgil & Bossert 1970; Charnov & Schaffer 1973; Wilbur & Collins 1973; review by Charlesworth 1980).

The life-history variables of primary interest to evolutionary biologists are those that characterize natality, ontogeny, and mortality in populations. These include duration of gestation or incubation, offspring size and number, age and size at independence, at sexual maturation, and at first reproduction, rate of reproduction, and age and size effects on fecundity and mortality. Comparative correlational analyses called allometric analyses depict the manner in which changes in rate or magnitude for one life-history variable relate to changes in another across taxa (cf. Hartnoll 1982; Fleagle 1985; Shea 1983, 1986 on developmental allometry). The technique has been used repeatedly in recent research to investigate the degree and consistency with which various life-history variables scale in relation to brain size, body size, metabolic rate, and even ecological factors such as seasonality of resource abundance or proportion of leaves in the diet (e.g., Sacher 1959, 1975; Sacher & Staffeldt 1974; Western 1979; Clutton-Brock & Harvey 1980; Martin 1981; Calder 1984; Martin & Harvey 1985; Lindstedt & Swain 1988; Pagel & Harvey 1988a, Chapter 3, this volume; Rasmussen & Izard 1988; Ross 1988; Zeveloff & Boyce 1988; McNab & Eisenberg 1989; Martin & MacLarnon 1990; Promislow & Harvey 1990).

Particular interest in size and metabolic rate stems from the idea that these variables determine an animal's speed of living (Lindstedt & Calder 1981). All organisms take their place on a fast–slow continuum of physiological time and developmental rate (Calder 1984; Schmidt-Nielsen 1984; Sibly & Calow 1986). Species whose members attain independence at young ages typically also show relatively early sexual maturation, heavy investment in reproduction, and high rates of mortality—thus short lifespans. Most are small animals showing relatively high rates of metabolism. Species near the other extreme of the continuum grow slowly, mature late, invest less in any one reproductive effort, and live long lives. These tend to be large animals and ones with relatively low metabolic rates.

The systematic correlations between body size and metabolic rate, on the one hand, and most other life-history variables, on the other, have led several researchers to propose that size and metabolic rate represent basic constraints on life-history evolution (e.g., Western 1979; Western & Ssemakula 1982; Calder 1984; Schmidt-Nielsen 1984). Others object, pointing out that many important correlations remain among life-history characters after the effects of size or metabolic rate are held constant statistically (e.g., Pagel & Harvey 1988a, Chapter 3, this volume; Partridge & Harvey 1988). These investigators suggest, instead, that the major variables respond to natural selection on life histories much as do the other traits. Reasonable versions of the two points of view, however, are not mutually exclusive (cf. Rose 1983, 1991; Charnov 1991), and proponents of each agree that the pervasive correlations observed among large sets of life-history variables within and across major taxa constitute basic support for the theory of evolved life-history strategies (e.g., plants: Lacey 1986; fish: Pauly 1980; birds: Ricklefs 1984; Bennett & Harvey 1988; Saether 1988; mammals: Harvey & Zammuto 1985; Sutherland et al. 1986).

The other major approach to illuminating general patterns among diverse life histories is mathematical modeling. Most models derive from the Euler–Lotka or characteristic equation of mathematical demography, which relates age-specific probabilities of survival and re-production to rates of increase in a population or lineage (e.g., Rubenstein, Chapter 4, this volume). Some for declining populations suggest that deferred reproduction should evolve in life histories (e.g., Mertz 1971; cf. Wiley 1974a), whereas most for expanding populations indicate that natural selection should strongly favor early reproduction (e.g., Cole 1954; Lewontin 1965; Charnov & Schaffer 1973; Caswell & Hastings 1980). In stable populations, age at maturity should be adjusted in relation to the effects of other variables (e.g., juvenile mortality, later fecundity) so as to maximize lifetime reproductive success (Wiley 1974a; Lande 1982; Roff 1984, 1986; review by Charlesworth 1980). Among otherwise equivalent strategies of life history, however, those that invest early in reproduction are expected to experience a selective advantage by more often avoiding chance factors of mortality, such as harsh weather and disease (Williams 1957; Rose 1983). The pivotal role of age at maturation in life histories has led much work in allometric analysis and modeling to focus directly on this variable (Roff 1981, 1983, 1984; Harvey & Zammuto 1985; Stearns & Koella 1986; Sutherland et al. 1986; Promislow & Harvey 1990; Charnov 1990, 1991).

Age at maturation is the life-history variable that, along with age at independence, delimits the juvenile period (see below). Many animals mature sexually and initiate reproduction only after a lengthy juvenile period. Many of these, though, are large animals, for which the feature is not unexpected, because (1) there are absolute limits to rates of growth, (2) relative rates of growth tend to decline with increases in body size, and (3) growth and reproduction are antagonistic processes (Kleiber 1961; Case 1978; Ricklefs 1983, 1984; Stearns 1984; Sibly & Calow 1986; Bronson 1989). But when animals exhibit longer juvenile periods than do closely related taxa or other animals of comparable size—as do primates—they are considered to exhibit delayed maturation, and consequently deferred reproduction. One of the central problems in life-history theory is to illuminate the various conditions under which deferred reproduction can be adaptive (e.g., Wilbur & Collins 1973; Wiley 1974a, 1974b, 1981; Roff 1981, 1984; Wiley & Rabenold 1984; Stearns & Koella 1986; Charnov 1990; Reznick et al. 1990).

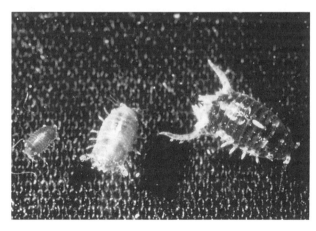

Fig. 2.1. Juveniles come in all sizes. As in all primates, this 2-year-old juvenile male savanna baboon (*Papio c. cynocephalus;* to right, top panel) is less than one-half the size of his adolescent male associate (6.5 years old). In contrast, the juvenile wattled jacana (*Jacana jacana;* to left, center panel), as in most other birds, attained adult proportions long before maturation. Finally, in species like the marine crustacean *Paracerceis sculpta* (bottom panel), which show more than one type of male, a juvenile of the largest morph, prior to complete maturation, is much larger than fully mature individuals of the smaller morphs. (Baboons, M. Pereira; jacanas, J. Walters; crustaceans, S. Shuster, with permission for reproduction from *Nature*)

WHAT IS A JUVENILE?

For biologists, the definitive characteristics of juvenility concern survivability and reproductive capacity: *juveniles are animals that would be likely to survive the death of their caretaker or loss of parental provisions (e.g., yolk sac) but have not yet matured sexually.* Contrary to popular thought, neither body size nor growth rate is a broadly definitive character (Fig. 2.1). Juveniles can be small relative to adults, as in mammals; indistinguishable from adults in size, as in many birds (Lack 1968;' Ricklefs 1983); or even much larger than many adults, as can occur in fish and other ectotherms (e.g., Warner & Hoffman 1980; Hartnoll 1982; Jonsson & Hindar 1982; Conlan 1989). Whereas growth is remarkably slow in most juvenile primates, growth rates in most other animals are maximal prior to maturation (Brody 1945; Case 1978). Although insects, birds, mammals, and other animals terminate growth before or just after maturing, many other taxa extend growth across the life span (Charlesworth 1980; Roff 1984; Kozlowski & Uchmanski 1987).

Primates are not the only animals distinguished by extended juvenility. The most dramatic examples, perhaps, occur in semelparous species, which delay maturation and reproduce only once in their lives (Cole 1954). The well-known Pacific salmon (*Oncorhynchus* spp.) mature after 2 to 3 years of feeding and growing at sea and then conduct long, often arduous migrations to freshwater spawning grounds, where they expire after exhaustive reproductive effort (Schaffer & Elson 1975). The homopteran cicadas, or harvest flies, also delay maturation remarkably. Dog-day cicadas (*Tibicen* spp.) take 2 to 5 years to mature, while the periodic cicadas (*Magicicada*) emerge to reproduce only after 13 or 17 years of feeding underground, depending on the species (Lloyd & Dybas 1966a, 1966b). Many iteroparous or repeatedly reproducing species are also noted for delayed maturation, such as the pelagic seabirds (e.g., Ashmole 1963; Goodman 1974). In the Atlantic fulmar (*Fulmarus glacialis*), for example, modal age of first reproduction in females is about 12 years (Ollason & Dunnet 1988).

Ectotherms, or "cold-blooded" animals, offer many more examples. Ectothermic growth, of course, depends strongly on environmental factors such as ambient temperature (e.g., Vollrath 1987). Sometimes, however, poor conditions truncate rather than extend prereproductive phases of development, or enhance rather than diminish the extent of growth (Alm 1959; Wilbur & Collins 1973; Brett 1979). The mountain dusky salamander (*Desmognathus ochrophaeus*), for example, normally matures after 1 larval year and 3 to 5 more as a juvenile. The members of one population studied at high elevation add a year to juvenility but also mature at larger sizes than do counterparts in a nearby population at lower elevation (Tilley 1977,

1980). Other ectotherms delay maturation under optimal conditions (Stearns & Koella 1986; Reznick 1990). Turtles (Testudinata) are paragons of the phenomenon, with females often maturing only after 10 to 30 years of life (Wilbur & Morin 1988). Likewise, primitive spiders (Aranaea: Orthognatha) are important case studies. In particular, some "tarantulas" (Theraphosidae), raised in captivity and amply fed, require up to 13 years to reach sexual maturity (Baerg 1938, 1963; Stradling 1978).

Extended juvenility per se is enigmatic to evolutionary theory, which asserts that individuals whose developmental strategies maximize reproductive potential across the life span will be those that contribute the most representatives to subsequent generations. For several reasons, natural selection favors early maturation whenever shifts in the timing of maturation and first reproduction would leave other aspects of the life history unaffected. First, earlier maturation reduces the duration of juvenility, thereby acting to reduce the probability of prereproductive death. Second, earlier maturation could expand reproductive careers, increasing lifetime reproductive success through increased opportunity. Third, earlier reproduction would lead a lineage to expand in a population by geometric progression when early breeders' offspring themselves begin breeding early; that is, generation time would be reduced (Cole 1954; Lewontin 1965).

WHY DO JUVENILES EXIST: TRADE-OFFS IN ANIMAL LIFE HISTORIES

Shifts in the timing of maturation rarely leave other life-history traits unaffected. In the simplest cases, early reproduction would be reproduction out of season: most seasonal breeding functions to minimize adult or offspring mortality due to the timing of reproduction in the annual cycle (Boyce 1979). More generally, organisms have finite amounts of energy and nutrients to allocate among growth, maintenance, and reproduction (Gadgil & Bossert 1970), and investments toward growth and reproduction, in particular, are demonstrably antagonistic (Sibly & Calow 1986; Bronson 1989). In species showing indeterminate growth, like many invertebrates and the ectothermic vertebrates, reproduction usually is associated with a retardation or hiatus in growth (Kozlowski & Uchmanski 1987). Some fish even break down somatic tissue to support gonad development (Roff 1982, 1983). In the determinate growers, such as insects, birds, and mammals, growth invariably ceases before or just after maturation (Brody 1945; Sadlier 1969; Roff 1981; Ricklefs 1983).

Thus, early maturation often leads to relatively small adult size. And small size can be a liability, as it is often associated with increased risk of mortality, reduced fecundity, and reduced offspring size and viability (e.g., Boyce 1981; Roff 1981, 1986; Stearns & Koella 1986; Partridge 1988). Moreover, early maturation frequently forecasts a short life, as both rapid growth and reproduction itself effectively wear animals out (Dickerson 1954; Eklund & Bradford 1977; Rose 1983, 1984; Reznick 1985; Nur 1988a, 1988b; Ollason & Dunnet 1988; Partridge 1988; Crowl & Covich 1990; see also Williams 1957 and Kirkwood & Rose 1991). On the other hand, animals that delay maturation risk loss of reproductive opportunity while their size passes the point after which further growth confers no significant advantage. And though maturation leads to the costs of reproduction, successful individuals must start sometime. Thus, the timing of maturation plays a basic role in many of the cost–benefit trade-offs that are an essential and inescapable aspect of organismic existence.

The variation observed in the timing of maturation among animal species, among populations, and between the sexes conforms well to the expectations of life-history theory (MacArthur & Wilson 1967; Gadgil & Bossert 1970; Stearns 1976, 1977; Horn 1978). When juvenile mortality is high relative to that of adults, for example, or when reproductive populations are essentially saturated, full maturation tends to be delayed, often in favor of additional growth or experience (e.g., Reznick et al. 1990). Conversely, when juvenile mortality is low relative to adult mortality, or when reproductive opportunities are abundant, maturation occurs early and at small sizes (e.g., Etter 1989). Also, commitment to growth and postponement of reproduction are observed in relation to the occurrence of harsh seasons in annual cycles (e.g., Wise 1984). The following review of life-history research informs us in more detail as to the diversity of contexts for shifts in the timing of maturation, and thus to the general functions of juvenility in animals.

Hardness and Size: Early Investment Options

The ectotherms provide myriad examples where the timing of maturation and related aspects of development clearly are life-history tactics specific to the members of a taxon, population, sex, or size class (e.g., Lawlor 1976; Roff 1981, 1984; Sastry 1983; Potts & Wooton 1984; Stearns & Koella 1986). The construction of shells by turtles, molluscs, and other invertebrates, for instance, demands resources, and its utility usually increases, but sometimes decreases with investment. In a later section, I discuss a barnacle that forgoes a mode of hardness and a snail that forgoes large size whenever possible. But, when greater hardness and/or size promote survival, early reproduction is sacrificed for potentially long reproductive lives (e.g., Gibbons & Semlitsch 1982). Turtles of many species, for example, can survive for more than a century (Wilbur & Morin 1988). In fact, 30 years of demographic records for one marked population of box turtles (*Terrapene carolina*) revealed no effects of age on adult rates of mortality (Stickel 1978; see also Wilbur 1975; Turner et al. 1984).

When mortality risk consistently increases with size or other age correlates, maturation and reproduction at earlier ages and smaller sizes are favored. Etter (1989), for example, demonstrated that rates of mortality increased with degree of exposure (wave energy) across closely situated populations of intertidal whelks (*Nucella lapillus*) and that the additional mortality was incurred primarily by large adults. Compared with counterparts living on protected shores, the whelks on wave-swept shores matured at smaller sizes and deposited twice as many egg capsules, each containing twice as many offspring. Individual hatchlings were only about 25% smaller than those from protected shores, suggesting that in these populations resources are diverted from growth to increased reproductive effort to offset the higher rates of adult mortality.

Etter (1989) exploited a "natural experiment"—one conducted by environmental circumstances—to illuminate life-history evolution. Other important tests of theory become available when experimental interventions can be performed with free-living organisms whose basic conditions of existence are well understood. Reznick et al. (1990), for example, manipulated a population of guppies (*Poecilia reticulata*) in a manner that actually induced the evolution of predicted adjustments in life histories (see also Reznick & Bryga 1987). Across local populations, the study animals show naturally occurring life-history differences, associated with differences in predation, that accord with theoretical expectations. Guppies living where their predominant predator is the pike cichlid (*Crenicichla alta*), which takes primarily large individuals, mature at earlier ages and show greater reproductive effort than counterparts living where their predominant predator is the killifish (*Rivulus hartii*), which takes small guppies.

In 1976, the investigators translocated guppies from a population living with *C. alta* to a guppy-free site containing *R. hartii* but no *C. alta*. Repeated sampling of the introduced population over the following 12 years documented the predicted changes in life-history features, including delay of maturation, increases in adult and offspring size, and decrease in female investment per reproductive effort. Rearing of fish from the original and introduced populations through two generations in common captive environments revealed the differences to be of genetic origin. This field experiment and its complementary laboratory work have provided the strongest evidence to date in support of the hypothesis that age-specific patterns of mortality play a fundamental role in guiding life-history evolution (see also Rose & Charlesworth 1981; Luckinbill et al. 1984; Rose 1984).

The evolving guppies also support the specific hypothesis that, whenever possible, animals "hurry through" early phases of vulnerability via rapid growth (Sibly & Calow 1986; cf. Janson & van Schaik, Chapter 5, this volume). For most readers, a more familiar example of this may be growth in altricial birds. Nest-bound hatchlings in these species grow roughly twice as fast as do most mammals (Case 1978). Indeed, birds may be the only determinate growers in which young juveniles (fledglings) are adult in size. Their rapid growth is likely to have evolved in part because of nestlings' and fledglings' vulnerabilities to predators and harsh seasons. Birds whose nesting habits expose

young to relatively great environmental risk show high growth rates and early fledging in comparison with species with well-protected or inaccessible nests, where nestlings show slow growth and reluctance to fledge (Lack 1968; Case 1978).

Ydenberg (1989) presented a finer-grained analysis for the Alcidae, a family of pelagic seabirds (e.g., murrelets, puffins) that exhibits diversity in the timing of migration by young from land nests to the open ocean. Intra- and interspecific patterns indicated that fledging for these birds is a developmental conversion sensitive to trade-offs among rates of growth and risks of mortality in the nest and at sea, respectively, all in relation to the ultimate need to complete development before the onset of winter.

The adaptedness of avian growth rates is, indeed, a complicated matter (see also below). Some have suggested that birds simply must attain adult proportions to be able to fly. But strong flight by the superprecocial megapodes within 1 day of hatching (Fritch 1962, cited in Skutch 1976, p. 234) shows that this can be accomplished before adult size is attained. Without rejecting the importance of nestling vulnerability, Ricklefs (1984) presented evolutionary models and a compelling review of data that suggest that rapid growth may function in many bird species largely to maximize the rate of reproduction (cf. Blurton Jones, Chapter 21, this volume).

Sexual Selection: Differences Between the Sexes and Among Morphs

Evolutionary interactions between schedules for growth and the timing of maturation also relate to differences in behavior and reproduction between the sexes and among individuals. Males often mature later than females in polygynous systems, for example, where older, often larger individuals monopolize male reproductive opportunity (Charnov 1982); that is, sexual selection is intense (Darwin 1871; Trivers 1972). In this circumstance, males trying to reproduce early might increase their risk of death due to effects of intrasexual aggression. Also, by retarding rates of growth, early efforts could actually delay first reproduction or permanently diminish reproductive potential for young males. Evolutionary theory predicts male deferral of reproduction in these cases. Its costs are thereby avoided while growth is promoted and experience is garnered (Caughley 1966; Williams 1966; Gadgil & Bossert 1970; Schaffer 1974; Pianka & Parker 1975) or a position in a queue for attainment of full reproductive status is secured (Wiley 1974b, 1981; see below). Female life histories in polygynous species, by contrast, are typically less variable, with reproductive success relating primarily to absolute rather than relative size and often environmental rather than competitive experience.

Warner (1984) provided the first direct demonstration that deferred reproduction can function to allow rapid attainment of effective reproductive size. He studied a population of bluehead wrasse (*Thalassoma bifasciatum*) in which the intensity of sexual selection differed markedly between small and large reefs. At small reefs, one or a few territorial males occupied all spawning sites; at large reefs, small, group-spawning males occupied sites preferred by females, while large males occupied separate territories. Young males at small reefs spent less time and energy on reproduction and grew twice as fast as young males at large reefs. In contrast, no difference in female growth rates was evident between the two site classes. Moreover, young males transferred between large reefs continued to enjoy high mating success, whereas such males translocated from large to small reef sites were never successful in their reproductive activities.

In many ectotherms, there are not only two sexes, but also two or three types of males (morphs) among which differences in the timing of maturation are associated with differences in size, shape, coloration, and/or behavior (fish: e.g., Warner & Hoffman 1980; Jonsson & Hindar 1982; Gross 1984; crustaceans: e.g., Hartnoll 1982; Sastry 1983; Conlan 1989; Shuster 1989). The phenotypic differences support competition using different reproductive strategies.

In the intertidal isopod *Paracerceis sculpta*, for example, three genetically determined male morphs are large (alpha), moderate (beta), and small (gamma) in size (Shuster 1989; Fig. 2.1). Beta males mature earlier than do alpha males, terminating growth at smaller sizes; gamma males shift puberty to yet an earlier molt, becoming essentially active testes (Shuster, unpublished data). During breeding, alpha males occupy the oscula of separate sponges, admit females, and resist entry by other males. Beta males mimic female morphology and sexual behavior to access guarded spongocoels, whereas the tiny gamma males occupy spongocoels surreptitiously. As the density of females increases, the reproductive success of alpha males declines relative to that of beta and gamma males, suggesting that density-dependent sexual selection maintains the three morphs. Shuster (1989) noted that reduced generation time may also help to maintain the beta and gamma genotypes (but see Shuster & Wade 1991).

Predictable Unpredictability: Roles for Experience

The impact of experience in shaping animal life histories is an important area for further research, particularly with mammals, for which relatively little in this area is yet known (cf. Pagel & Harvey, Chapter 3; Rubenstein, Chapter 4; Fagen, Chapter 13, this volume). If the timing of maturation related strictly to effects of size, then juvenile adaptations would relate exclusively to predator avoidance, foraging, and competition and effects of time apart from growth would be insignificant. For many animals, however, effects of age or season on the timing of maturation remain after size effects are removed analytically or ruled out by circumstance (e.g., Alm 1959; Policansky 1983; Stearns & Koella 1986; Pagel & Harvey, Chapter 3, this volume). Roles for experience in development always merit investigation (see below), but they especially demand attention whenever persistent diachronic effects are observed.

Developmental Conversion. Wise (1976, 1984) discovered bimodal rates of development in filmy dome spiders (Linyphiidae: *Neriene radiata*) that reflect remarkable effects of seasonal constraints. In several north temperate populations, early-spring-maturing females produce hatchlings that themselves mature by early summer (~3 months of development). The latter produce autumn hatchlings that overwinter as young juveniles (early instars) and mature late in the following spring (~7 months). Late maturers' progeny postpone maturation, grow large, overwinter, and mature early the following spring (~7 months). This developmental strategy appears entrained to the changes in photoperiod that occur around the summer solstice

(Wise 1984), and it allows lineages to complete three generations every 2 years. Life-history trade-offs include spring hatchlings' eventual sacrifice of high fecundity for early first reproduction and large offspring size, both of which likely promote successful overwintering by offspring.

This and other research shows that life-history trade-offs involving discrete adjustments in the timing or form of development can be conditional. In these cases, developmental conversions (Smith-Gill 1983) depend on individual experience for their expression (see also Wilbur & Collins 1973). An elegant series of studies on the acorn barnacle (*Chthamalus anisopoma*) provides a different example. Lively (1986a, 1986b) showed that this normally conical crustacean responds to the presence of a carnivorous snail (*Acanthina angelica*) by developing an attack-resistant, bent-over shell shape. Bent morphs, however, also grow more slowly and are less fecund than conical morphs. Thus, the developmental strategy includes a conditional sacrifice of reproductive potential for survivability, which is clearly adaptive in an environment that is heterogeneous with regard to risk of predation. Lively (1986b) discusses other conditional trade-offs that result in adult dimorphisms in insects and other invertebrates. More recently, Crowl and Covich (1990) demonstrated that chemical evidence of predation by crayfish (*Orconectes virilis*) causes the members of a population of freshwater snail (*Physella virgata*) to double their adult size by doubling the length of their juvenile period. In so doing, the snails grow out of the size range taken by the predator.

A third class of conditional conversion is sex reversal. Protandric (male to female) and protogynous (female to male) sex reversals have evolved independently in several orders of fish (Smith 1975) and in many other ectotherms. In these cases, the timings of two conversions— maturation and sex reversal—have evolved to effect trade-offs between growth and reproductive opportunity. One conversion is responsive to information about the social environment. In species where intense male reproductive competition is mediated by size, developing individuals reproduce first as females. Whenever male abundance relative to that of females diminishes, the largest females switch sex to begin reproducing as males (Shapiro 1984; cf. Warner

et al. 1975; Warner & Hoffman 1980). In species where reproductive competition is intense for neither sex and size enhances primarily female reproductive capacity, individuals begin their reproductive careers as males and defer female reproduction until after having attained much larger size (Charnov 1982). In some such species, males actually begin their reproductive careers living as virtual organs inside females (e.g., Foighil 1985).

One important role for experience in animal life histories, then, is the cuing of developmental conversions, including transitions to and from torpor and metamorphosis from one body shape or plan to another (e.g., maturation or sex change) (Shapiro 1984; Warner 1984; Wise 1984; Lively 1986b; Alford & Harris 1988; Crowl & Covich 1990; see theory by Wilbur & Collins 1973; Gould 1977; Oster & Alberch 1982; Smith-Gill 1983). In some cases, the facultative response to social or ecological stimuli enables individuals to make trade-offs only if challenges must be engaged or opportunities can be exploited that might not have been encountered in a given lifetime. In other cases, the environmental input influences only the timing of the conversion from one ontogenetic trajectory to another.

Regulatory Feedback and Learning. The more general role for experience is that environmental feedback is an essential regulatory aspect of all development—morphological, physiological, and behavioral (Waddington 1940; Schneirla 1957; Lehrman 1970; Maynard-Smith et al. 1985; Johnston 1987). Genotype–environment interactions can induce continuous variation in physical characters via quantitative changes in rates or durations of development (e.g., Crowl & Covich 1990; Reznick 1990; reviews and theory by Wilbur & Collins 1973; Stearns & Koella 1986). When the focus shifts to cognitive aspects of behavior, the relevant developmental processes and changes are called learning. Social learning, trial-and-error learning, and practice are all thought to play important roles in life histories, particularly those of birds and mammals (e.g., Galef 1976, 1988; Mason 1979a; Johnston 1982; Boyd & Richerson 1985). In addition, we should consider that processes analogous to learning phenomena probably also guide aspects of anatomical, physiological, and noncognitive behavioral development (Holt 1931; Schneirla 1957; Galef 1981). The familiar "adolescent sterility" in primates (Short 1976), for example, may be explained partly by the need for physiological "practice."

It is important to note that one regulatory role for experiential feedback in development can be to induce either early or late maturation, depending on the precise character of the developmental process at issue and the potential magnitude of reproductive costs for a given individual. Many have argued for species with parental provisioning, for example, that individuals should not mature until adequate foraging skills have been acquired (e.g., Wooller & Coulson 1977; Johnston 1982). But, many developmental milestones must be attempted to be achieved, including fully competent reproduction in birds (Saether 1990). Because organisms serve their own best interests over those of offspring (Trivers 1974), we can expect to see early attempts at reproduction whenever the environmental costs of maturation and reproduction are low (see Rubenstein, Chapter 4, this volume). Conversely, when reproductive effort is likely to jeopardize the survivability or future fecundity of a young animal—or even an experienced individual (e.g., Thomas & Coulson 1988)—we should expect maturation or reproduction to be suspended.

The large amount of theoretical, experimental, and comparative work already accomplished on development in birds (e.g., Ricklefs 1983, 1984; Bennett & Harvey 1988; Saether 1988, 1990) make this group of animals particularly important for future research on experiential regulation of life histories. While many birds develop rapidly (Case 1978), protracted development, including deferred reproduction, occurs not only in the large, long-lived species, such as seabirds and raptors, but also in many tropical species and in the males of many polygynous and communally breeding species (Wiley 1974a, 1981; Ricklefs 1984; Wiley & Rabenold 1984; Lyon & Montgomerie 1986).

For the males of polygynous and communally breeding birds, in particular, juvenility appears poorly explained by the lines of reasoning advanced in this chapter for other taxa. Unlike the ectotherms, most birds appear to *begin* juvenility having already achieved adult size. On the surface, then, these instances of deferred reproduction seem unlikely to be tactics to pro-

mote physical development. Evolutionary inferences based on single, gross measures of development (e.g., body weight), however, risk error due to neglect of variation in things such as energy allocation, degrees of allometric growth, patterns of fat deposition, and metabolic efficiency. Attainment of adult physical status, in most cases perhaps, means considerably more than can be interpreted from a body-weight or wing-length measurement. Investigation of subsidiary indices of physical development will be an important area for future research on birds and mammals.

Deficiencies in behavioral skill, particularly in the areas of foraging (e.g., Dunn 1972; Johnston 1982) and agonistic competition (e.g., Lyon & Montgomerie 1986), are often proposed as causes for deferred reproduction in birds. As suggested above, ineptitude during the reproductive effort would have to threaten the survivability or future fecundity of first- or second-year birds disproportionately for this to be so, most likely in conjunction with other challenges such as disease, predators, or harsh seasons. In particular, deferred reproduction might often be favored when lack of breeding skill or experience jeopardizes younger birds' chances of overwintering. Further complicating matters for investigators is the converse possibility—that behavioral success on the wintering grounds, through physiological feedback, determines the level of reproductive effort put forth the following breeding season.

Social Constraints. To argue that the inferior competitiveness of young birds selects for deferred reproduction, however, may be to place the evolutionary cart before its horse. Wiley (1981), having studied several avian systems in which young males appear capable of competing while they actually do not try (Wiley 1974b; Wiley & Harnett 1976; Wiley & Rabenold 1984), rejects the argument, which indeed seems better suited to the indeterminate growers and at least some mammals, where delay may be needed to attain the requisite size or experience (cf. Warner 1984; Janson & van Schaik, Chapter 5, this volume). Unless age-related physical differences or some effect of experience can be demonstrated, as discussed above, some deferral of reproduction in birds may be due to solely social constraints.

With an extensive analysis of social behavior,

mortality rates, and reproductive success for dispersing and nondispersing individuals, Wiley and Rabenold (1984) showed how participation in the communal breeding system of the stripe-backed wren (*Campylorhynchus nuchalis*) might represent evolutionarily stable strategies. In the study population, males and females appear to maximize their lifetime reproductive potential by deferring reproduction and joining queues for the acquisition of breeding status. Positions in queues are strictly correlated with age; older group members occupy positions toward the heads of queues. All nonbreeding adults assist their breeding pair raise its broods. Breeding pairs with two or more helpers enjoy annual reproductive success six times greater than that experienced by solitary pairs and trios. Ultimately, surviving males "inherit" breeding status in their natal groups, while females compete with senior females from other groups for breeding vacancies in adjacent territories.

Dominance hierarchies are not readily quantified within groups of stripe-backed wrens (Wiley & Rabenold 1984; Wiley, personal communication). Thus an intriguing possibility is that this system of communal breeding would simply break up—that is, group members would disband and all birds' reproductive success would decline—if junior adults began trying to jump queues when breeding opportunities became available (Wiley & Rabenold 1984). Proximate mechanisms for the stabilization of transitive hierarchies of priority are known in animals, however, such as correlations of dominance rank with size or degree of experience (e.g., Geist 1971; Byers 1986) and hierarchy-reinforcing third-party aggression (see Chapais & Gauthier, Chapter 17, this volume). In relation to the latter phenomenon, in particular, future research could reveal species in which purely social proximate constraints force young animals to defer reproduction and enter queues for the attainment of reproductive status (Pereira & Weiss 1991 discuss possible primate examples).

CONCLUSION

Even a selective overview of literature such as this provides overwhelming evidence in support of the theory of adaptive evolution of life histories. Most important, the growing body of pre-dicted results from well-designed experimental manipulations and well-documented natural experiments (e.g., Rose & Charlesworth 1981; Warner 1984; Lively 1986a, 1986b; Etter 1989; Reznick et al. 1990; Rose 1991) critically enhances our ability to interpret the natural variation in life histories observed across undisturbed populations. Animals of all kinds shift the timing of maturation, or duration of juvenility, in response to the major socioecological variables that influence individuals' abilities to survive and reproduce in their population.

The general function of animal juvenility is modulation of growth and the onset of reproduction. In many cases, it functions to maximize the rate and/or extend the duration of growth, thereby enabling an animal to escape the period during which small size renders it particularly vulnerable to predation or virtually ineligible to compete for reproductive opportunity. In others, extension of juvenility maximizes individuals' chances to survive harsh seasons, again by avoiding the costs of reproduction and accessing the buffering effects of large size. Conversely, when small size entails little cost or when large size is penalized by the environment, juvenility often is abbreviated or does not occur in a life history. Juvenility is also diminished when adult size can be attained by or soon after the exhaustion of parental provisions, as in many birds and mammals, or when early reproductive effort does not compromise further growth, future fecundity, or longevity.

In comparing life histories, the duration of juvenility can be judged only in relation to each group's typical rate of growth, frequency and magnitude of reproductive effort, and life expectancy. That growth rate, reproductive effort, and longevity are actually linked to the timing of maturation is underscored by research confirming costs of growth and reproduction (e.g., Eklund & Bradford 1977; Reznick 1985; Gustaffson & Sutherland 1988; Nur 1988a, 1988b; Crowl & Covich 1990), including genetic correlations between developmental rates and senescence (Rose 1983, 1984, 1991; Luckinbill et al. 1984; cf. Luckinbill & Clare 1986).

A major frontier of research that remains is to characterize roles of experience in regulating the development of animal life histories. Ecological and social stimuli are known to cue the occurrence of developmental conversions such as maturation, sex change, torpar, and other animal metamorphoses. But little is yet known about the metabolic and neurophysiological processes underlying these phenomena. Likewise, most research attempting to evaluate the reproductive decision making of maturing and fully adult animals has relied on single measures of physical status, thereby neglecting developmental variation in energy allocation, degrees of allometric growth, patterns of fat deposition and depletion, and metabolic efficiency. Also, it has not always been appreciated that the need for experiential feedback can favor either early or late maturation, depending on the character of the crucial developmental process and potential reproductive costs for a given individual. For birds and mammals in particular, the mechanisms and processes of physical and behavioral development will have to be illuminated further before hypotheses on the adaptedness of the timing of maturation and reproduction can be evaluated satisfactorily. If this can be accomplished, we will be in position to determine whether in some cases purely social constraints, ultimate or proximate, cause deferred reproduction and queuing for attainment of reproductive opportunity.

ACKNOWLEDGMENTS

I thank J. Bernardo, S. Marshall, S. Shuster, and G. Uetz for references to published articles and access to unpublished data on amphibians, crustaceans, reptiles, and spiders. For leads to literature on bioenergetics, birds, development, or learning, I thank P. Klopfer, S. Nowicki, S. Peters, C. van Schaik, J. Walters, and R. Wiley. I thank D. Roff for writing clearly on development and evolution and R. Fagen for discussions and correspondence on the same topic. T.-J. Pyer created much of the time I needed to conceive and compose this chapter. B. Chapais, P. Harvey, P. Klopfer, D. T. Rasmussen, J. Walters, and R. H. Wiley criticized its first draft and L. Fairbanks criticized several. While writing, I was supported by the National Institute of Child Health and Human Development (R29-HD23243) and the Duke University Primate Center (DUPC Publication No. 557).

3
Evolution of the Juvenile Period in Mammals

MARK D. PAGEL and PAUL H. HARVEY

Reproductive quiescence before adulthood is found in every major taxonomic group from plants to animals. Mammals often have lengthy juvenile periods, sometimes extending over a decade. Among mammals, primate juvenile periods are particularly long. And yet, the juvenile period, as a time when an individual is not reproducing, poses interesting problems for life history theory.

This chapter is about the evolution of the length of the juvenile period in primates and other mammals. Our approach is comparative in that we are interested in the evolutionary forces responsible for patterns observed across species. The length of the juvenile period varies positively with the age at maturity and with life span in mammals, and so the selective or other forces that may delay the age at which sexual maturity is reached or that set the length of life may also apply in large measure to variation among species in the length of the juvenile period. A positive correlation between the length of the juvenile period and the age of maturity in mammals is not necessarily automatic, however. It is true that only those species with late ages at maturity can have long juvenile periods. But under at least some views of the evolution of life histories, a normally long-lived species—that is, a species that would typically also have a long juvenile period—could mature right after weaning. Thus, we need explanations for the empirical trends.

Our hope in presenting a review of theory and the comparative evidence is to arrive at a better understanding of which of three explanations for

juvenile periods is most plausible: that is, are juvenile periods best understood as a consequence of seasonal constraints and the length of time required to reach adult size, as a consequence of natural selection favoring individuals that spend an extended period of time before adulthood learning social and behavioral skills that affect age-specific mortality and later reproductive success, or as some combination of the two.

THE JUVENILE PERIOD AS PHENOTYPIC LIMBO

The juvenile period in mammals is a time during which offspring are deemed too old to receive direct nourishment by their mothers and too young to get on with the business of reproduction. As with many aspects of social behavior and development, definitional disputes abound about just exactly how to demarcate the juvenile period. Our definition (see also Pereira & Altmann 1985) takes the juvenile period to be that part of the young mammal's life between weaning and sexual maturity. Whether the time of weaning is a time of complete independence from the mother has little effect on the arguments to follow. Reaching sexual maturity, then, represents an escape from a sort of phenotypic limbo, a time when an individual is able to play, socialize, even fight, but not reproduce.

Some species spend a longer time in limbo than others. Weaning, the beginning of the juvenile period, occurs at that point during juvenile growth and development when it pays the mother to reduce or terminate investment in the current offspring and increase investment in future offspring (Trivers 1974). The essential question seems to be: Why should offspring delay reproduction until sometime after the age at which they are weaned? There are many possible reasons for remaining a juvenile over an extended period. Although the juvenile period is a time of transition to adult size, there may be more to being an adult than just being big. The juvenile period may be a time to learn about such things as social systems, parental care, and aggression (Pereira & Altmann 1985). The time required to learn these social and behavioral skills may be the reason that juvenile periods exist. Without juvenile periods, individuals as adults might not be as good at staying alive, and might not be as successful at raising young. The emphasis in this sort of explanation for the length of the juvenile period is on the time required to learn skills that affect later birth and death rates, and not on growth per se.

However, not all mammals, or even all primates, have prolonged juvenile periods. For example, 5-week-old preweaned *Mustela erminea* can have fertile matings (Muller 1970). So we should not assume that social and behavioral events of the juvenile period are important or even necessary evolutionarily for successful reproduction. To argue that they, and not the time required to grow to adult size, are responsible for the length of the juvenile period, the skills acquired during the juvenile period must be shown to improve reproductive efficiency sufficiently to compensate for the increased risk of death before sexual maturity is reached.

An alternative view of the events of the juvenile period may have the age of maturity set rigidly by growth laws. According to this view, species that are larger in size mature later because it takes them longer to grow to adult size. We suppose that during this period of growth the juvenile is acquiring the physiological machinery for reproduction (why mammals generally wait until adult size before reproducing is a question of importance in its own right). If we ignore why some species are large, but accept that there is some upper limit to the rate at which an animal can grow, then juvenile periods just happen; their lengths are incidental consequences of growth rates and have no necessary evolutionary significance. We can also incorpo-

rate here the constraint of seasonality: however fast an organism grows, it may have to delay the onset of sexual maturity until an appropriate season for adequate resources to raise the offspring (Rutberg 1987; Kiltie 1988). Behaviors such as play might be important or might merely be ways of passing time in limbo, but, given rigid growth laws and constraints of seasonality, they are not determinants of the length of the juvenile period.

Yet another perspective on juvenile periods is that their length, although set by growth laws and seasonal constraints, might enable or even encourage the evolution of social and behavioral skills that affect reproductive success. Individuals that make use of the juvenile period may be better off than those that just pass the time doing nothing (Pereira & Altmann 1985). For example, play-aggression during the enforced juvenile period may yield adults better able to protect their young. Janson and van Schaik (Chapter 5, this volume) suggest that juvenile primates have evolved slow growth rates and possibly some of the behavior of the juvenile period as a way of reducing their own (i.e., juvenile) mortality. These views give more credence to the idea that the social and behavioral events of the juvenile period are selectively important in their own right, but nevertheless still places them in a secondary role to body size as a determinant of the length of the juvenile period.

Having outlined these alternative views on the evolution of the juvenile period, we describe in the second section two contrasting theoretical views of the evolution of the age at maturity, and present data from primates and other mammals to illustrate the ideas. The third section concentrates on a model that can predict the quantitative form of the relationship between the age at maturity and body size in mammals. The rate of adult mortality is the key selective force in this model. Much of the emphasis in attempts to understand delayed maturity and hence long juvenile periods in mammals has been on patterns of age-specific fecundity. We suggest, however, that the principal determinant of delayed maturity is not so much age-specific fecundity, but the length of the reproductive life span, which is largely determined by body size. Accordingly, the fourth section describes ecological and other causes of variation in the rates of extrinsic mortality among adults, and shows their links to body size. In the last section we suggest how to choose among the various theoretical explanations for variation in the length of the juvenile period.

EVOLUTION OF THE AGE AT MATURITY

Two types of theory dominate explanations for variation in the age at which maturity is reached. Allometric explanations view variation in life-history habits as inevitable consequences of growth laws. Demographic explanations rely on age-specific patterns of fecundity and mortality to set optimal life-history tactics.

The Allometric View

The relationship between the size of an organism and the age at which it first reproduces is so regular in nature as to be given the status of an empirical law. A reasonable guess about an organism's generation time can be made from knowledge of body size alone for haploid organisms, diploid organisms, asexuals, sexuals, bacteria, plants, birds, insects, mammals, and reptiles (Bonner 1965). Among mammals, a suite of other life-history traits—including neonatal size, length of gestation, age at weaning, length of the juvenile period, and life span—also vary positively with body size across species. The size of the litter also varies with body size, but negatively. Together these relationships describe a fast–slow continuum of reproductive habits that covaries with body size. Thus species that are small in size typically have large litters, and their offspring reach maturity at an early age. Large mammals typically have the opposite characteristics.

Regularity in the relationships of various traits with body size has encouraged some researchers to attribute variation in life-history habits among organisms to Laws of Size and Growth. Lindstedt and Swain (1988, p. 93) provide this muscular statement of the allometric point of view:

Body size affects virtually all aspects of an animal's physiology and morphology as well as its life history. Hence when these variables are expressed as body-mass-dependent functions, the noise of interspecific variation is replaced with predictable qualitative and quantitative patterns. We feel that these patterns proclaim the presence of constraints on animal design.

Like morphology and physiology, life history traits may be best understood when examined through the perspective of size-dictated patterns.

The suspicion that variation in life-history tactics might be interpretable as stemming from allometric design constraints related to body size arose not just from observations such as those reported by Bonner (1965), but also from the empirical observation that many different traits having to do with rates or timing also scale allometrically with body size. Lindstedt and Calder (1981) and Calder (1984), for example, report that variables as diverse as maximum life span, gestation length, age to maturity, cardiac cycle, blood volume circulation, glomerular filtration rate, respiratory cycle, inulin clearance rate, and even the plasma half-life of methotrexate all scale allometrically with body size.

Perhaps the most bewitching aspect of these timing relationships is that each of the variables scales roughly to the one-fourth power of body size: not only do timing variables tend to change with changes in body size, there seems to be a single scaling relationship that can explain all of them. Lindstedt and Swain (1988, p. 96) conclude that there must be some "common body-size-dependent clock to which all of these [timing] events are entrained," a timing device that Lindstedt (1985) christened the Periodengeber. Interspecific variation in life-history timings is attributed to the fact that all animals are run by a common physiological clock that uses different units of time as a function of body size: one tick of the clock of a large animal will encompass many ticks of the clock for a small one when measured in common units of time. Elephants and shrews reach maturity at very different absolute ages but after approximately the same number of units of time according to their respective physiological time scales. Thus, according to proponents of this view, elephants and shrews have the same or very similar life histories when expressed in units of physiological time.

Despite its utility for organizing the diversity of life histories, the limitation of the design-constraint view is that it does not provide an explanation for that diversity—it does not tell us why elephants bother to be so large. Additional theory is required to understand what sets the age and size at maturity. Nevertheless, once we know why elephants are so large, growth laws, if they exist, can tell us why it takes elephants as long as it does to achieve their size. However, the design-constraint view must confront awkward empirical facts. There is much variation in life histories that is uncorrelated with body size (e.g., Harvey & Clutton-Brock, 1985; Harvey & Zammuto, 1985; Lindstedt & Swain 1988; see also Watts & Pusey, Chapter 11, this volume), and life-history variables often correlate even after statistically removing the effects of size (Harvey & Clutton-Brock, 1985; Harvey & Zammuto, 1985; Harvey et al. 1989).

The Demographic View

The demographic view of the timing of the age at maturity can be summarized with one axiom and one assumption. The axiom is very simple: individuals cannot reproduce after they die. This immediately places an upper bound on the age of sexual maturity, and explains, in a very coarse way, some of the interspecific variation: species that have high rates of mortality will have early ages of reproduction (Horn 1978). Short-lived species simply cannot afford the risk of death that would accompany a long juvenile period. Although the axiom helps to explain the habits of shrews and other small short-lived animals, it does not explain why elephants wait so long to mature. This requires the assumption that there must be some advantage to delaying reproduction until sometime after the time of weaning. Neither the axiom nor the assumption is concerned with body size, and yet together they can potentially explain the tendency for age at maturity to covary with body size. If larger animals have low rates of prematurational mortality, they may be able safely to extend their juvenile periods. Such an extension will be rewarded if during the juvenile period they acquire skills or capabilities that improve their reproductive success over what it would have been otherwise.

A more precise formulation of the demographic approach is probably not necessary for understanding the qualitative comparative patterns, but is nevertheless useful for illustrating just how the juvenile period may influence life-history variation. Consider the following very simplified model in which animals grow until adulthood, after which fecundity and rates of adult mortality are fixed (Sutherland et al. 1986). Let F be the birth rate of daughters per unit time (fecundity) for an average mature

female, S_m is the fraction of daughters that survive to maturity, L_m is the expected length of the adult life span (assumed to be given by $1|M$ where M is the fixed adult rate of extrinsic mortality), and R is the expected number of daughters produced over a female's lifetime:

$$R = S_m F L_m \qquad (1)$$

If the population is stable and not growing then $R = 1$. Rearranging Equation 1, the rate of adult mortality is seen to relate to fecundity and the probability of reaching maturity:

$$M = S_m F \qquad (2)$$

This very simple demographic model of ecological compensation can potentially explain some of the comparative patterns of life-history variation. For example, a decrease in life expectancy (i.e., increased adult mortality, M) is predicted to be compensated by an increase in fecundity, an increase in survivorship to maturity, or both. One way of increasing survivorship to maturity, other things held constant, is to reduce the age of maturity. Conversely, using Equation (1), a female that extends her juvenile period, and thereby exposes herself to increased risk of death (i.e., lower S_m), will be rewarded for that risky behavior if she can increase her fecundity or lower her age-specific mortality sufficiently. The model does not always specify which of the possible changes will occur in response to changes in one of the other variables but points out that demographic considerations alone lead us to expect that there will be some relationships among life history variables.

Demographic models make no necessary assumptions about body size or rates of development. Body size may turn out to be an important feature of life histories, but only insofar as it influences age-specific patterns of fecundity and mortality. For example, Harvey and Zammuto (1985), using an important data set reported by Millar and Zammuto (1983), studied life expectancy and age at maturity in mammals. The two variables are positively related such that the species array themselves along a fast–slow continuum of large animals with long life spans and late ages at maturity to small animals with short life spans and early reproduction ($r = 0.98, p < 0.001$). However, the fast–slow continuum remained even after statistically removing the effects of body size from both variables ($r = 0.89$, $p < 0.001$)). This result did not change when life expectancy was measured from the age at maturity rather than from birth (Sutherland et al. 1986). Adult mortality rate, insofar as it is reflected in life span, would appear to influence (or be influenced by) the age at maturity, independently of body size.

In primates, two studies have found significant relationships between life span and age at maturity, controlling for body size (Harvey & Clutton-Brock 1985; Ross 1988). Read and Harvey (1989) also report significant correlations among a suite of life-history traits in mammals, even when the effects of body size are removed. One result of interest is that the length of the juvenile period is strongly positively correlated with age at maturity, independently of body size ($r = 0.89$, $p < 0.001$). Some correlation between these two measures is expected merely because age at maturity is partly a function of the length of the juvenile period. Nevertheless, it is possible for an animal to be weaned late, have a short juvenile period, and then mature.

More direct evidence relating life histories to age-specific rates of mortality comes from data reported by Promislow and Harvey (1990). The authors calculated rates of mortality from life tables of natural populations for a range of mammal species. Adult mortality was negatively correlated with adult body weight, and with length of gestation, age at weaning, and age at maturity. Adult body weight may be a surrogate for adult mortality rate, which could account for its correlations with life history variables (Western 1979; Western & Ssemakula 1982). However, mortality rates also predict variation in life histories that is independent of body size. Most relevant to theoretical explanations of life histories is that adult mortality rate be negatively correlated with age at maturity even after removing the effects of adult body size (Wiley 1974a; Horn 1978).

Rates of mortality, then, are consistently correlated with the fast–slow continuum, and provide a straightforward explanation for why age at maturity and the length of the juvenile period decrease with body weight and with life expectancy. Lacking from these correlations, however, is any explanation of why animals at the slow end of the life history continuum delay their reproduction.

Life-History Trade-Offs and the Juvenile Period

Harvey et al. (1989, p. 24) suggested that "if there are mortality costs associated with reproduction, and if reproductive efficiency increases with age, then animals with lower nonreproductive mortality rates should be selected to delay reproduction." Promislow and Harvey (1990) report information from 15 mammal species in which litter size, prenatal and neonatal survival, or prenatal and neonatal weight increase with increasing maternal age within the species studied. These data are consistent with the suggestion of Harvey et al. but do not bear directly on the issue of whether females who breed earlier die earlier. Another way that delayed maturity will be selected for is if by delaying the age at which she first breeds a female's subsequent fecundity is higher than that of females who do not delay reproduction (Wiley 1974a). Gustaffson and Part (1990) have shown, in a study of a bird (*Ficedula albicollis*), higher fecundity among individuals that first breed a year after others in the same population, but the effect was not large enough to make delaying a better strategy. Indeed, we probably should not expect to find such effects, except over very short time periods, since natural selection would be expected to move the population to the optimal strategy. Where we do find such effects they may be entirely phenotypic or result from fluctuating environmental influences.

In theory, however, if we consider an evolving population, whether maturity is delayed will depend on the functions relating lifetime fecundity and the probability of juvenile death before reaching maturity to age at maturity (Fig. 3.1). Factors that promote delayed maturation create juvenile periods. Some of the ways that fecundity or survivorship are improved with age probably depend on only physiological or developmental changes to the female. Such changes could be independent of precisely what females learn as juveniles. On the other hand, having an extended time of, for example, social learning during the juvenile period may give females of social species skills that improve either their own or other offsprings' chances of survival.

Although demographic theory and trade-offs between age and fecundity can potentially explain the timings of life-history events, there

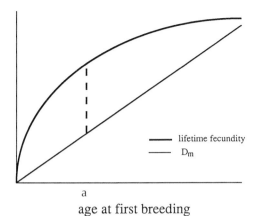

age at first breeding

Fig. 3.1. Lifetime fecundity for animals that survive to maturity and the probability of death before reaching maturity (D_m) versus age at maturity. Natural selection will set the age at maturity (a) at that point where the difference between the two curves is maximal. (After Charnov 1990)

remain two limitations to the demographic explanations discussed earlier. One is that the demographic approach does not explain the regularity of the allometric exponents linking life-history traits to body size—the 0.25 phenomenon. The second is that in explaining delayed maturity, the emphasis has been on trading-off early and late fecundity. Thus extrapolating to evolutionary trends across species we expect later ages at maturity to be associated with higher fecundity. But they are not: species with later ages at maturity generally have lower fecundity. So this trade-off view fails to capture the across-species trend. The next section describes a model that addresses the first issue. A later section takes up the across-species question.

PREDICTING THE FORM OF LIFE-HISTORY RELATIONSHIPS

Charnov (1991) has taken an important step toward incorporating growth relationships into life-history models, in an attempt to explain the quantitative and qualitative patterns of life-history covariation (see related work by Stearns & Koella 1986; Stearns & Crandall 1981, 1984). Charnov begins with the demographic and population assumptions recorded in the model of

Equations (1) and (2), plus the assumption that rate of growth per unit time in mammals scales with body size to the 0.75 power:

$$dW/dt = AW^{0.75} \tag{3}$$

where W stands for weight, and A is a taxon specific growth coefficient. Integrating Equation (3) and setting time zero equal to the age at weaning, Charnov gets the following expression in a, the age at maturity:

$$a = \frac{1 - (W_0/W_a)^{0.25}}{0.25A} W_a^{0.25} \tag{4}$$

where W_a and W_0 refer, respectively, to body weight at the ages of maturity and weaning. On the assumption that the ratio of weaning weight to weight at maturity is a constant, Equation (4) shows that the age at maturity can be expected to scale to the 0.25 power of body weight at maturity—that is, adult body weight. In the previous section we reported that suites of life-history traits including age at weaning and age at maturity scale to the 0.25 power of adult body weight. This supports but does not explain Charnov's assumption about the constant ratio of weights at these two times. Charnov's assumption also automatically generates an increasingly long juvenile period among animals with later ages at maturity.

Next Charnov assumes that at maturity, an animal redirects all of its potential growth into reproduction. Given the same assumption about the ratio of weight at weaning to weight at maturity being a constant, the following result for F, fecundity per unit time is obtained:

$$F = \frac{A}{C(W_0/W_a)} W_a^{-0.25} \tag{5a}$$

where C is a constant relating to the efficiency of transfer of energy from growth to reproduction. Thus fecundity per unit time is expected to decline with increasing weight at maturity. Intuitively, this result follows from the assumptions about growth rate, and the ratio of the two weights. Females that mature later will be bigger, and they will wean larger offspring. The absolute amount of energy available to put into offspring increases to the 0.75 power of body size, but W_0 and W_a covary in direct proportion to each other. Thus the amount of energy available to put into increasingly larger offspring is a declining function of adult size, a result that fits with the across-species data showing fewer number of offspring per unit time are produced by larger species.

A further result bears directly on the difficult issue of delaying maturity. Charnov wants to solve for the optimal age at maturity as a function of the adult mortality rate. The result of Equation (5a) shows that fecundity declines with increasing weight, and hence with increasing age at maturity. Other things equal, under these conditions it pays not to delay maturity, but to breed earlier. And yet, the observed empirical patterns show that species with lower fecundity do breed later. To find a solution that has larger animals breeding later, Charnov drops the assumption that the weights at weaning and maturity are in constant proportion, and allows W_a to vary independently of W_0. This yields an equation in which fecundity increases with increasing size at maturity:

$$F = \frac{A}{CW_0} W_a^{0.75} \qquad (5b)$$

This is properly thought of as a within-species result, and accords with the view cited earlier on fecundity and age at first breeding within species. If fecundity is now optimized in what is essentially the model of Equation (1), a relationship between weight at maturity and mortality arises such that

$$M = 0.75 \, AW_a^{-0.25} \qquad (6)$$

where M, as before, is the adult mortality rate. Because weight and age at maturity vary positively, Charnov is able to obtain delayed maturity as adult mortality rate decreases.

This is an important result, since a species's age at maturity in Charnov's model is determined by the rate of adult mortality. Charnov divides mortality prior to maturity into two components: one is preweaning rates of mortality, and the other is postweaning, which is assumed to descend to the adult rate sometime before maturity is reached. The importance of this division is that varying the age at maturity has no effect on preweaning survivorship, and thus the age at maturity becomes a function of adult mortality. It is possible that females that delay maturity become better at raising their offspring and so reduce the preweaning mortality of their offspring, but the model intentionally ignores such effects.

Charnov's model yields an intriguing set of results because, from specified assumptions about rates of growth and the relationship between weight at weaning and weight at maturity, the model makes predictions about life histories that are not only qualitatively but also quantitatively in accord with the empirical data. These results do not derive from body size per se, which is a function of time in the model, but from selection acting on the age at maturity. Adult body size is determined secondarily from the age at maturity on the assumption that all energy up until the time is put into growth. The model is also able to generate qualitative predictions about life-history covariation in the absence of variation in body size. For example, suppose size is held constant in Equations (4), (5), and (6). Variation in A, the taxon-specific growth parameter, will cause a positive correlation between rates of mortality and fecundity (i.e., higher adult mortality is predicted to be compensated for by increased reproductive output), and negative relationship between adult mortality and age at maturity, and a negative relationship between fecundity and age at maturity. Stearns and Crandall (1984) also emphasize how age at maturity can vary independently of body size given variation in the availability of resources required for growth.

Lifetime Fecundity and Body Size

The components of lifetime fecundity can provide an alternative to the explanations for delayed maturity that emphasise trade-offs between early and late fecundity. There are two components to lifetime fecundity as portrayed in Figure 3.1: age-specific fecundity and the number of breeding attempts throughout the adult life span. Age-specific fecundity can stay the same or even decline, and yet lifetime fecundity increase, if the number of breeding attempts is sufficiently increased. This may explain why across species, later ages at maturity can be associated with lower fecundity: species with late ages at maturity can offset the loss of fecundity by having longer reproductive life spans. This section focuses on body size as a determinant of reproductive life span.

Imagine that the rate of adult mortality decreases with increasing body size at maturity, across species, and that selection acts on adult body size—assume, for example, that larger size is selected for as a predator defence mechanism. We suppose that within some range growth laws dictate the age at which an animal can reach a given size, and thus the age at maturity is set by the time required to grow to adult size. This differs from Charnov's view that selection sets age at maturity directly and that body size adjusts according to Equation (4). Following Charnov, however, we assume that the amount of energy a mother can transfer to her developing offspring is set by her body size, and that her offspring must eventually grow to some fixed proportion of her size before they can be weaned. It is conceivable that juveniles enter the adult niche, in the sense that they can forage and fend for themselves, when they reach a given proportion of adult weight, and that proportion does not differ markedly among species. Indeed, that proportion might be expected to be determined by the feeding and antipredator niches occupied by different species. Maturity will be reached at a later age among animals that are larger as adults.

Fecundity per unit time is expected to decrease with increasing body size according to this view, for the same reasons as in Equation (5a). The trick is to see how delayed maturation can, nevertheless, be favored. The increased costs that are paid by delaying maturation, lower survivorship to maturity and possibly lower fecundity, will have to be offset by the increase in the total number of reproductive attempts arising from $1/M$ or life span for those organisms that do survive. What must be true is that expected lifetime reproductive success increases either at the same rate or faster than expected survivorship to maturity and fecundity fall, creating a line of demographic equilibrium. Such a line is implicit in the model of Equation (1) and seems plausible on ecological grounds: above a certain size the rate of adult mortality must be very low and so life span potentially becomes very long.

This verbal model of the evolution of the age at maturity allows for the social and behavioral events of the juvenile period to play an important role, despite the fact that the age at maturity is largely set by growth requirements. One way for a female's daughters to postpone maturation and yet not pay the price of lower survivorship (due simply to the longer time spent as a juvenile) is for them to learn things during their juvenile periods that promote their survival and make them better parents. Longer juvenile periods act as enabling factors for the evolution of behavioral and social skills. These skills may in turn promote even further delay of maturation. The skills might be as simple as fighting ability or as sublime as the ability to build social coalitions. Reducing adult mortality may be the primary reason to become larger and thereby to delay maturation; making use of the juvenile period may be the way to offset the inevitable demographic costs of doing so.

ECOLOGICAL INFLUENCES ON LIFE-HISTORY VARIATION

The mortality in the models that we have been discussing is properly thought of as externally imposed mortality acting on adults and possibly on the immediate prematurity times of life. Extrinsic mortality is external to the organism itself and includes death due to predation, accidents, disease, and starvation due to fluctuating environmental conditions. To the extent that environments can be expected to vary systematically in the rates of extrinsic mortality they impose, we expect different life histories to be associated with them. Much recent comparative work on life history relationships has emphasized that life history traits covary independently of body size. This is true, but the size related variation should not be overlooked. The verbal model just described gives body size a central role: it is assumed to be the principal influence on the rate of extrinsic mortality. There is good reason to believe that this will often be true, although it need not be. In instances such as accidents, or cases in which we compare two different environments inhabited by the same species, extrinsic mortality may be uncorrelated with body size.

Size-Related Mortality

Suppose that death due to predation and death due to starvation both decline as body size increases. Small animals can be preyed on by lots of larger predators, but larger animals will have comparatively fewer predators. Similarly, smaller animals require food more frequently than larger animals, and so may be at greater risk of starving to death due to random environmental fluctuations. Risk of death due to disease may increase with population density, which probably declines with increasing body size. If these three sources of mortality constitute a large proportion of total adult mortality, then we expect smaller animals to live faster lives than larger ones, to avoid dying before they reproduce. This ecological perspective on life history covariation can be used potentially to explain the allometric patterns linking life span, age at maturity, and mortality to body size.

We can use data from Calder (1984, Table 11-4) to illustrate these ideas. Calder reports data on life span, average adult mortality, and body size in 15 bird species ranging in size from 3.5-g hummingbirds to the 8.5-kg albatross. Maximum recorded life span scales allometrically to the 0.18 power of body size in this sample:

$$\text{life span} = 0.82(\text{adult body weight})^{0.18} \qquad (7)$$

Life span, of course, also varies with average adult mortality, and average adult mortality varies allometrically with body size in this group:

$$\text{life span} = 0.95(\text{adult mortality rate})^{-0.44} \qquad (8)$$

$$\text{adult mortality rate} = 0.86(\text{adult body weight})^{-0.34} \qquad (9)$$

These two equations can be combined to predict the scaling exponent in Equation (7). Substituting Equation (9) for adult mortality rate in Equation (8), the predicted allometric relationship linking life span to adult body size is

$$\text{life span} = 0.085(\text{adult body weight})^{0.15} \qquad (10)$$

The predicted exponent of 0.15 is very close to the observed value of 0.18.

A similar analysis can be performed on the data reported by Millar and Zammuto (1983) on life expectancy at maturity, age at maturity, and body weight in mammals. Taking the reciprocal of life expectancy at maturity as the average adult mortality rate, the relevant equations are

$$\text{age at maturity} = -0.09(\text{adult body weight})^{0.27} \qquad (11)$$

$$\text{age at maturity} = -0.69(\text{adult mortality rate})^{-1.127} \qquad (12)$$

$$\text{adult mortality rate} = -0.56(\text{adult body weight})^{-0.23} \qquad (13)$$

Combining equations yields

$$\text{age at maturity} = 0.38(\text{adult body weight})^{0.25} \qquad (14)$$

The predicted exponent is very close to the observed exponent of 0.27.

These analyses, one from birds and the other from mammals, emphasize the idea that the allometric relations observed between life-history variables and body size may have their origin in the scaling of mortality rates on body size. This scaling may arise from selection setting optimal life histories, with body size adjusting according to Charnov's ideas, or by selection acting on body size, with life histories being set by growth laws, as outlined earlier.

Ecological Variation Independent of Variation in Size

Ecological influences on life-history traits need not be correlated with body size across species. However, few influences that are independent of size have been found (Millar 1981; Harvey & Clutton-Brock 1985; Gittleman 1986). Litter sizes of small mammals increase with latitude (Lord 1960; Spencer & Steinhof 1968). Also in small mammals, diet is associated with differences in life histories. Mace (1979) reports that folivorous small mammals have short gestation lengths, wean their young early, and have short life spans for their body size.

A study by Ross (1988) provides a rare example of an ecological influence that can act independently of variation in body size. Ross found generally slower life histories (relative to body size) for tropical-rain-forest primates compared with savannah-dwelling primates. Ross's interpretation of this result is that tropical environments may be more stable than savannah environments. If this is correct, species living in tropical environments may experience lower rates of mortality for their size than species inhabiting an environment in which environmental fluctuations may add to the risk of death.

CHOOSING AMONG ALTERNATIVE EXPLANATIONS FOR THE EVOLUTION OF JUVENILE PERIODS

We have discussed three different views of the evolution of juvenile periods in mammals: juvenile periods are unavoidable consequences of growth laws and seasonal breeding constraints;

juvenile periods are set solely by demography and life-history trade-offs; and selection for body size at maturity sets the age at maturity via growth laws and the juvenile periods that arise enable the evolution of social and behavioral traits that may further affect the age at maturation.

Sole adherence to a strict growth law interpretation is incomplete in that it gives no explanation for why animals are big or small, or somewhere in the middle.

Demographic considerations along with various assumptions about trade-offs can explain the fast–slow continuum of life-history habits. Delayed maturation can evolve if it increases lifetime fecundity (Wiley 1974a). Delayed maturation automatically gives rise to extended juvenile periods, and so they need not necessarily be given any evolutionary significance beyond their length of time per se; that is, an extended period of time during which females do not reproduce may somehow (perhaps through some unspecified physiological effect) allow females to increase their lifetime reproductive success. On the other hand, the social and behavioral events of the juvenile period may in fact be responsible for a female's longer reproductive life span. Imagine that through juvenile play, individuals are more skillful at avoiding predators as adults (see Fagen, Chapter 13, this volume). Alternatively, by delaying maturity, females may learn social skills that enable them to garner social support needed for effective reproduction.

Demographic models are an attractive alternative to the growth-law view because they do not necessarily rely on body size to derive their predictions. And yet, life-history traits do vary strikingly with body size, and so we are forced to ask what it is about body size that is so important for the evolution of life histories. If delayed maturation and thus long juvenile periods reflect solely the importance of such things as play, grooming, social learning, and hunting practice, but not body size, there need not be any covariation of life histories with body size, and no reason for juveniles to growth throughout the juvenile period (it might be argued that the length of the juvenile period is set by the time required for learning, and that individuals may as well grow during this time. But this implies some advantage to larger size, and thus learning cannot be the whole story). Interestingly, juvenile primates do not grow very fast during the early juvenile period (see Janson & van Schaik, Chapter 5, and Worthman, Chapter 23, this volume). Small in size, superefficient warrior species that could fight off predators of all sizes, and thus live long lives, can exist in our imaginations. But the fact they do not exist in reality suggests that maybe there is an advantage to size per se. Body size would seem to be the most obvious and the most important determinant of adult mortality rate: large size translates into lower rates of predation, reduced risks of starvation from short-term fluctuations in the food supply, and possibly lower rates of disease transmission from living at lower population densities.

Our view, then, is that adult body size, through its effects on reducing the rate of adult mortality, is the primary reason why delayed maturation is associated with large body size across species. If we are willing to accept upper limits to the rates of growth, then delayed maturity and long juvenile periods arise automatically as a result of selection for larger size and we need not give them any special evolutionary significance. However, and this seems the most plausible view of the evolutionary significance of the juvenile period, individuals that make the most of their time in phenoyptic limbo should be better prepared for the rigors of adult life. Behaviors such as play and mock-hunting, or the learning of social skills may evolve because of the juvenile period and even reinforce it. Their influence may be on adult mortality rate, fecundity, or juvenile survivorship. Where such influences are important, we expect age at maturity to be delayed even further than the point set by selection acting, via adult mortality rates, on adult body size.

This perspective on life history-variation suggests that the variation related to size and that which is independent of size may largely derive from different sources. Most size-related variation in life histories probably is due to the effects of body size per se on adult mortality [the variable M in the model of Equation (1)]. Much of the variation that remains after controlling for size is probably attributable to the skills and knowledge that juveniles acquire during their apprenticeship to become adults.

ACKNOWLEDGMENTS

We thank Ruth Mace, Sean Nee, and Anne Pusey for their comments on an earlier draft of this manuscript. M.P. was supported by a grant from the Commission of European Communities.

4

On the Evolution of Juvenile Life-Styles in Mammals

DANIEL I. RUBENSTEIN

Growing up is not easy. Small size and inexperience guarantee that youngsters will be more vulnerable to predation and less able to compete for critical resources than adults. Yet the costs of not overcoming these inherent problems are high. Since youngsters at some point must strike out on their own and attempt to reproduce for themselves, what transpires while growing up should have a dramatic effect on subsequent reproductive success. Whereas enriched ontogenies might enhance adult opportunities, impoverished circumstances, by limiting growth and experience, might hinder them. Small size will reduce competitive ability and limit fecundity. Moreover, small size could force individuals to mature either very early because of their deficiencies or very late as they delay attempting to compensate for them. If breeding begins too early before necessary skills are mastered or when fecundity is directly limited by size, reproductive success will be reduced. Similarly, if breeding is delayed too long, even if initial deficiencies are completely eliminated, opportunities will have been lost. Consequently, it is not surprising that natural selection has tailored the age of maturity, or the age of first reproduction, so that it maximizes an organism's Darwinian fitness (Charlesworth 1980; Stearns & Koella 1986).

Age of first reproduction is one of many life-history traits that appear to be highly correlated. Many studies on a variety of mammalian groups (e.g., Millar, 1977, 1981; Western 1979;

Harvey & Clutton-Brock 1985; May & Rubenstein 1985; Boyce 1988; Ross 1988; Promislow & Harvey 1990) show that fast breeding species tend to be small, to have short gestation and nursing periods, and to produce large litters of small young, whereas slow breeding species tend to be large, to show prolonged developments, and to produce a few large young. In all instances these traits scale allometrically, not geometrically, with body size (Calder 1984; Reiss 1989). The theory of r- and K-selection argues that selection has acted on different species to fashion life histories that allow them to succeed in different demographic or ecological conditions. Extensions of the theory (Charnov & Schaffer 1973; Caswell 1983; Horn & Rubenstein 1984) and analysis of empirical data (Harvey & Zammuto 1985; Promislow & Harvey 1990) suggest that once body size is held constant, species with high unavoidable juvenile mortality relative to adult mortality should tend to delay the onset of sexual maturity.

Although the onset of breeding almost always occurs at the end of the juvenile period, the age of maturity and the length of the juvenile period are not synonymous. Ontogeny entails a gradual lessening of dependence on parents. In mammals, weaning begins a phase of the juvenile period when direct nutritional dependence on mother's milk ceases. Yet until the onset of breeding, parental assistance in terms of protection from predators and competitors and aid in developing important motor and social skills continues. Moreover, since so much of parental investment and care is intended to prevent mortality among young, features of the juvenile period, as well as the forces that shape its duration, must be examined.

It is the aim of this chapter to investigate the features that constitute and determine the length and function of the juvenile period. We will begin by developing a simple model that shows what factors influence the optimal timing of sexual maturity. Then we will examine how mortality patterns in mammals—especially those of rodents, carnivores, primates, and ungulates—help shape juvenile life-styles in order to evaluate the predictions of the model. Finally, we shift from inter- to intraspecific case studies to determine how different types of parenting produce different juvenile life-styles and whether they affect adult reproductive success and longevity.

THE THEORY

In growing populations the intrinsic rate of growth, r, also known as the Malthusian parameter, is usually considered the appropriate measure of fitness (Charlesworth 1980). It is related to other life-history features as

$$r = \ln R_0 / T_c \tag{1}$$

where R_0 is the average number of female offspring produced over a female's lifetime, and T_c is the "cohort generation time," which corresponds closely with our intuitive notion of generation length and is roughly proportional to the time it takes to reach sexual maturity (May & Rubenstein 1985). Since the lifetime reproductive success term contributes to fitness logarithmically, hence weakly, fitness in an *expanding* population ($R_0 > 1$) depends mostly on the age of first reproduction. Individuals that mature early have an advantage.

Most extant mammal populations, however, are rarely found increasing in size. Many simply persist at the more or less constant levels they reached after their initial increase following colonization. Slight changes in population density tend to be self-correcting, since increasing numbers slow reproduction and increase mortality and decreasing numbers do just the opposite. In such populations fitness is best characterized by

R_0, the average number of daughters a female produces in her lifetime (Lande 1982; Charnov 1986). If $l(x)$ is the probability that a newborn female is alive at time x and $b(x)$ is the number of daughters she produces at time x, then

$$R_0 = \Sigma \; l(x) \cdot b(x) \tag{2}$$

If we assume that survival can be divided into juvenile and adult components, and that birth and death rates remain fairly constant throughout adulthood, then

$$R_0 = b(a) \cdot S(a)/M(a) \tag{3}$$

Here, a depicts the age of first reproduction, b represents a female's birth rate in terms of daughters per year, S represents the fraction of these daughters that survive to maturity, and M is the instantaneous adult mortality rate. All are functions of a, since delaying reproduction could potentially result in more robust, mature, and skilled adults that then could have higher than average birth rates while experiencing lower than average mortality for both themselves throughout their adult lives and for their offspring as they mature. This relationship, although simple, captures the important life-history trade-offs and reveals that the lifetime reproductive success of a female is essentially the product of the number of daughters a female produces during each of $1/M$ years and the proportion of these that survive to reproduce.*

Following Charnov (1990) it can be shown that there is an optimal age of first reproduction, $a*$, and that it depends primarily on the magnitude of juvenile mortality and on the way in which changing the timing of first reproduction affects adult mortality and fecundity. If juvenile survival, $S(a)$, is composed of an early compo-

---

*For R_0 to be an appropriate fitness measure the population must not be growing ($R_0 = 1$). Since the birthrate $b(\cdot)$ and mortality functions $S(\cdot)$ and $M(\cdot)$ are as much functions of population density as they are of first reproduction (a), it is important to determine how changes in density affect these functions and, in turn, if they affect the optimal age of maturity. Charnov (1990) has shown that as long as the negative affects of increasing density increase the mortality mostly on younger juveniles, or increase the mortality or decrease the fecundity uniformly for all aged adults, then density dependence is not likely to affect how natural selection affects the optimal age of first reproduction.

nent, S_0, that is fixed and depends on characteristics of the organism and ecological circumstances that are largely outside of its or the parents' control, and a later component, $e^{-Q(a)}$, where $Q(a)$ is the amount of juvenile mortality between the early phase and maturation, then

$$R_0 = F(a) \cdot S_0 e^{-Q(a)} \qquad (4a)$$

where lifetime fecundity is defined as $F(a) = b(a)/M(a)$. In this case, lifetime reproductive success is simply the product of the number of offspring born in a lifetime and the likelihood that each survives to independence. Or lifetime reproductive success can be defined as

$$R_0 = b(a) \cdot S_0 e^{-Q(a)}/M(a) \qquad (4b)$$

if instantaneous adult mortality and fecundity remain explicitly defined. Maximizing R_0 with respect to a is the same as maximizing $\ln R_0$, which yields

$$d \ln R_0/da = \\ d \ln F(a)/da - Q'(a) \qquad (5a)$$

or

$$d \ln R_0/da = \\ d \ln b(a)/da - Q'(a) - d \ln M(a)/da \\ (5b)$$

Setting Equations (5a) or (5b) equal zero and solving reveals that lifetime reproductive success is maximized in the former case when, for a given change in the age of maturity, the rate of change in lifetime fecundity equals the rate of change in juvenile mortality (see Charnov 1990; Pagel & Harvey, Chapter 3, Fig. 3.1, this volume) and in the latter case when the rate of change in the instantaneous birth rate equals the sum of the rates of change in the juvenile and the logarithm of adult mortality rates. In the first case, this occurs when the difference between $\ln F(a)$ and $Q(a)$ is maximized and means that for any given lifetime fecundity curve, as the slope of the juvenile mortality curve increases, the optimal age of maturity decreases. In the second case, it means that the effect of adult mortality will be weaker, because it is logarithmically adjusted, than that of juvenile mortality. In both cases, however, determining the optimal age of maturity will depend on how $b(\cdot)$, $Q(\cdot)$, and $M(\cdot)$ change with a, the age of maturity. To do this, we need to specify some particular forms of these functions.

If we assume for simplicity that the instantaneous juvenile mortality rate either remains fairly constant during the juvenile period or represents the average of a rate that declines with age, then $Q(a)$ can be represented by $q \cdot a$ where q represents the constant instantaneous rate (after the high initial rate S_0), or the average of the declining rate, for the juvenile period, and a represents the length of the juvenile period. Clearly, for any given level of juvenile mortality, the longer the juvenile period the lower the likelihood that a youngster will survive to maturity. What sets the level of q, however, is less clear. Perhaps as Janson and van Schaik (Chapter 5, this volume) propose, levels of predation or competition with adults play important roles: Juvenile mortality should be low for large-bodied species, or for those that live in relatively predator-free environments, or in habitats where resources can be acquired with little competition because they are abundant and evenly distributed.

If we also assume that the adult mortality rate is a function of the juvenile rate and that delaying the onset of reproduction can reduce the instantaneous rate of adult mortality, then it is reasonable to define the adult mortality rate as $M(a) = qe^{-ma}$, where m is the mortality exponent that describes how much a particular change in a will change M. At $m = 0$, the mortality rate of adults is unaffected by age at which maturity occurs and is that of juveniles. When $m > 0$, delaying puberty is accompanied by decreasing rates of adult mortality, but when $m < 0$ the opposite occurs. As m increases, a given delay in the age of maturity will lead to an increasing difference between the juvenile and adult rates. If, in part, levels of adult mortality incorporate "costs of reproduction," then larger values of m reflect the fact that delaying maturity can lower these costs.

Finally, if we assume that mature parents are better breeders than younger, more naive ones, then birth rates should increase with increases in the age of maturity as $b(a) = b_0 a^k$ where b_0 is the characteristic birth rate of an average individual, and k represents the fecundity exponent that describes how much a given change in the age of maturity alters the characteristic, or average, fecundity. If $k = 0$, fecundity is unaffected by changes in the age of maturity, and since k is likely to be proportional to body size or experience, a $k = 0$ implies that factors other than the timing of maturity or certain features of the phenotype are shaping birth rate. When $k > 0$, however, the larger it gets, the more dramatic will be the effect of a given a on birth rate.

It should be noted that although these variable and exponents depict species-specific characteristics, they represent averages based on responses of individuals to their environment. And these responses are themselves likely to vary depending on features of personal or parental phenotype as well as social status. Typically large values of the exponents k and m indicate that, on average, fecundity and adult survival are extremely *sensitive* to changes in the age of maturity for a particular class of individuals. If the variation among individuals within a population is not too great and the values of these exponents are large, then negative values of the exponents are unlikely. However, when the values of the exponents lie close to zero, it would not be surprising to observe some individuals exhibiting those that are negative.

Substituting these functions into Equation (5b), differentiating them with respect to a, setting the equation equal to zero, and then solving for the optimal age of maturity yields

$$a^* = k/(q - m) \qquad (6)$$

From this relationship it is clear that the optimal age of maturity increases as the impact that age

of maturity has on fecundity, as measured by the sensitivity exponent, k, increases. Even if one of the consequences of this sensitivity is disproportionately diminished reproduction for animals that begin breeding when very young, selection will favor delay as long as the delay ultimately has a marked effect on boosting fecundity. It is also clear from this relationship how the optimal age of maturity is affected by the sensitivity of adult mortality rate to the timing of maturity. Increases in the impact that age of maturity has on adult mortality, as measured by the sensitivity exponent, m, increases the age of maturity as long as the rate of juvenile mortality, q, is held constant and $m < q$. Since increases in the sensitivity of adult mortality, m, result in lower rates of adult mortality, age of maturity ultimately increases as the rate of adult mortality decreases.

It is less clear, however, how the age of maturity is affected by changes solely in levels of juvenile mortality. At first glance, it appears that for a given m, any increase in the rate of juvenile mortality will lower the age of first reproduction. Holding m constant, however, does not hold the adult mortality rate constant since $M(\cdot)$ is also a function of q (Fig. 4.1). In fact, for a constant m, increases in juvenile mortality are accompanied by increases in adult mortality. Thus increases in q represent increases in mortality over the entire lifetime, and, not surpris-

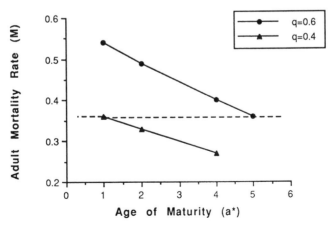

Fig. 4.1. Relationship between instantaneous rates of adult (M) and juvenile mortality (q). In these hypothetical relationships $m = 0.1$ and $k = 0.6$. Note that although m is held constant for either level of juvenile mortality, the rate of adult mortality changes. The dashed line depicts a constant rate of adult mortality and shows that as the rate of juvenile mortality increases, so does the optimal age of first reproduction (a^*). Conversely, for any level of juvenile mortality, increases in the rate of adult mortality lowers the optimal age of first reproduction.

ingly when this occurs, selection appears to favor accelerating the onset of reproduction. We can examine the sole affect that changing levels of juvenile mortality can have on the optimal age of maturity by dividing juvenile by adult mortality. As Figure 4.1 shows, increases solely in the rate of juvenile mortality tend to delay the onset of reproduction. Thus, overall, longer juvenile periods are favored when juvenile rates of mortality *relative* to those of adults increase, or adult fecundity is markedly enhanced by postponing the onset of sexual maturity.

The structure of the model is built on the premise that delays in maturity lower adult mortality rates, enhance adult fecundity, and lower the chances of juveniles surviving to adulthood, given a fixed juvenile survival rate. Thus it is not too surprising to find that when adult mortality is no different from juvenile mortality ($m = 0$) increases in the levels of juvenile mortality select for shortening the period during which juveniles are at risk by lowering the age of first reproduction. Such an acceleration of puberty will be offset only when shortening the juvenile period markedly diminishes future fecundity. In fact, if maturity is reached at 2 years of age, for example, and is to remain unchanged in the face of environmentally induced age-dependent changes in mortality and birth rate, then an increase in the rate of juvenile mortality by 0.10 would have to be accompanied by a 0.20 increase in k. This would mean that the species-specific birth rate would have to be increased by 0.15 to offset a 0.10 increase in juvenile mortality. And for the age of maturity to remain unchanged at higher ages of maturity, the fecundity benefit would require an even larger value of k to counteract a given increase in a species' rate of juvenile mortality.

If the model is expanded to make the instantaneous juvenile mortality rate, q, a function of the age of maturity rather than a constant, then age of maturity need not automatically decline when q is increased in the absence of offsetting increases in fecundity. Such a function implies that the length of the juvenile period actually changes the prospects for juvenile survival either by the actions of the juveniles themselves or by those of their parents. The rate of juvenile mortality could be represented as $q(a) = q_0/a^z$ where q_0 represents a baseline level of juvenile mortality before the effects of delay and z is the experience exponent ($z < 1$) that deter-

mines how strong the delay effects are. In turn, $Q(a) = q_0 a^{1-z}$. For the special case, $m = 0$,

$$a^* = [k/q_0(1 - z)]^{1/(1-z)} \tag{7}$$

Clearly, when changing the age of maturity has no effect on the rate of juvenile mortality ($z = 0$), this expanded expression for the optimal age of maturity reduces to the original form. But as z increases, the optimal age of first reproduction increases markedly. Thus even if environmental pressures increase the baseline level of juvenile mortality, as long as the experiential benefits of delay are sufficiently strong, no change in the optimal age of maturity need occur.

THE PATTERNS

Interspecific Comparisons

Using data compiled by Promislow and Harvey (1990) for a wide array of mammals, Harvey and Clutton-Brock (1985) and Ross (1988) for primates, and Bekoff et al. (1987) for carnivores, it is possible to test some of the predictions of the models. Figure 4.2a shows, as predicted, that as absolute rates of juvenile mortality increase, age of maturity decreases. But as the model also predicts, and Figure 4.2b shows, when juvenile mortality rates relative to those of adults increase, the opposite occurs [$r^2 = 0.22$; $F(1, 21) = 5.76$; $p < 0.03$]. As with most life-history variables, age of maturity and rates of both juvenile and adult mortality scale allometrically with body size. Whereas age of maturity increases as body size increases (Fig. 4.3; coeff. $= +0.22$), similar increases in size are associated with lower rates of juvenile and adult mortality [coeff. $= -0.24$; $F(1, 38) = 29.3$; $p < .0001$]. Even after removing the effect of body size from each of these critical mortality variables, the overall patterns remain the same.* Thus, regardless of body size, species

*Since many aspects of life history scale with body size, body size must be removed from the analysis before the unconfounded relationship between the two life history variables can be measured. In this chapter two methods have been employed to do this depending on the form of the published data. In one, residuals from the regressions of each life history variable versus adult body weight were regressed against each other. If the slope of the regression equation was significantly different from zero the relationship between the relative, stan-

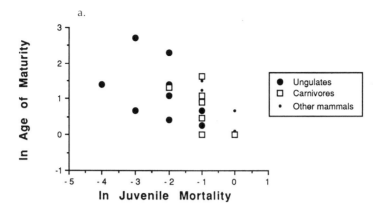

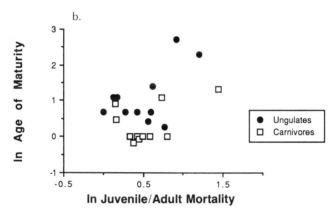

Fig. 4.2. Relationship between adult and juvenile mortality rates and age of first reproduction. (a) As the rate of juvenile mortality increases, the age of maturity decreases [$\ln(a) = \ln(q) + 0.03$]. Although ungulates tend to have lower rates of juvenile mortality than carnivores, the relationship is the same. (b) When adult mortality rate is held constant, then as the rate of juvenile mortality increases so does the age of maturity [$\ln(a/wt) = 0.80$ $\ln(q/wt/M/wt) - 9.4$]. Ungulates show a slight lengthening of puberty in relation to standardized levels of juvenile mortality relative to carnivores.

that have high levels of juvenile mortality tend to breed relatively early in life [$F(1, 37) = 9.28$; $p < 0.005$]. Only after the effect of adult mortality is removed is there a tendency for heightened

dardized, or weight-specific variables (as they are often referred to) were considered significant. In the other method, the ratio of each life history variable when divided by adult body weight was regressed against each other. Again, if the slope of the regression was significantly different from zero, the relationship between the two relative life history variables was considered significant.

levels of juvenile mortality to delay the onset of puberty [$F(1, 37) = 3.1$; $p < 0.04$ (one-tailed)].

Unfortunately, it is difficult to use interspecific data to examine directly the *sensitivity* of fecundity to changes in the age of first reproduction (values of k). However, an indirect test can be made if we assume that for sexually dimorphic species where sexual competition among males is intense, small increases in size or experience will greatly increase a male's chance of not only breeding, but of acquiring a disproportionate share of mates. Then if in-

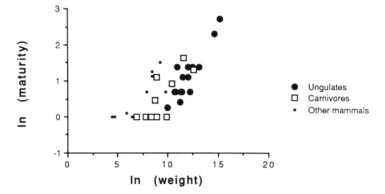

Fig. 4.3. Relationship between body size and age of first reproduction. The relationship is allometric, with large-bodied species having longer juvenile periods than smaller species [$\ln(a) = 0.22 \ln(wt) - 1.4$]. On a weight-specific basis, however, the length of the juvenile period is relatively shorter for larger species.

creases in size can be attributed to the lengthening of the juvenile period, more sexually dimorphic species should show stronger sensitivities of age of maturity on fecundity, and hence have higher k values, than less sexually dimorphic species. And since the optimal age of maturity increases as values of k increase, the optimal age of maturity for males should also increase as the degree of sexual dimorphism increases across species. As Figure 4.4 shows, this is the case for both primates [$F(1, 14) = 3.45$; $p = 0.04$ (one-tailed)] and carnivores

[$F(1, 7) = 3.94$; $p = 0.05$ (one-tailed)]. When size is removed from the analysis, species in which males are relatively larger than females have relatively delayed puberties.

Some other life-history features that are related to reproduction and help shape species-specific patterns of fecundity appear also to covary with age of maturity. As Figure 4.5 shows, species with relatively long nursing periods and relatively small litters have relatively delayed ages at maturity, whereas those with relatively heavy neonates or relatively heavy litters do not. But if

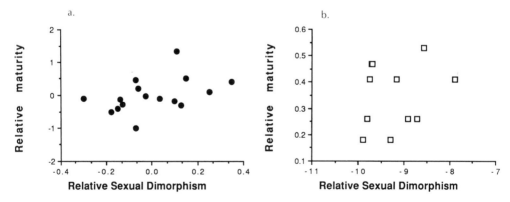

Fig. 4.4. Relationship between relative sexual size dimorphism and relative age of maturity. (a) Species of primates in which males are larger than females have relatively longer juvenile periods than more monomorphic species [$\ln(y) = 1.4 \ln(x)$]. Relative maturity is measured as the deviations, or residuals, of the relationship between ln Age of Maturity and ln Female Body Weight. (b) The same relationship holds for carnivores. As the degree of sexual dimorphism increases, so does the age of first reproduction. Relative age of maturity is measured as the ln of the ratio of age at maturity divided by female body weight and relative size dimorphism is measured as the ln of the ratio of male to female body weight [$\ln(y) = 3.05 \ln(x) - 9.98$].

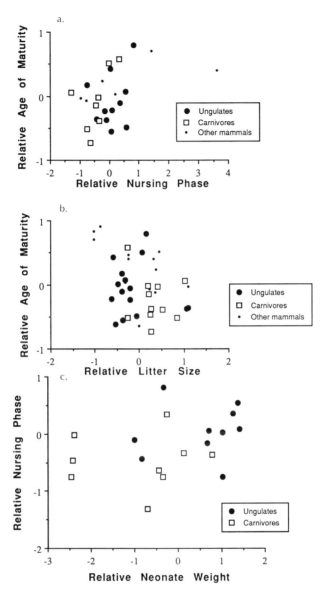

Fig. 4.5. Relationship between the relative length of the juvenile period and the relative value of various life-history traits. (a) The relationship between the relative age of maturity and the relative length of the dependent phase, or nursing period, is positive for both carnivores and ungulates [$\ln(y) = 0.20 \ln(x)$; $F(1, 26) = 6.03$; $p < 0.05$]. Thus species with relatively long nursing periods for their body size also have relatively long delays before beginning to reproduce. (b) For species with relatively larger litters, maturity is reached at relatively early age [$\ln(y) = -0.29 \ln(x) + 0.005$; $F(1, 37) = 5.5$; $p < 0.03$]. (c) The relative size of the neonate explains little of the variation in the weight-specific, or relative length of the juvenile period [$F(1, 23) = 0.22$; $p < 1.0$]. Not depicted is the nonsignificant relationship between relative litter weight and relative age of first reproduction [$F(1, 33) = 0.32$; $p < 0.6$]. Relative values are derived from the residuals of the regression of each life-history variable and the ln of female body size.

we subtract the length of the nursing period from the time it takes to reach maturity and call this remaining time the "independent juvenile phase," then a somewhat surprising trend emerges. Although it is true that larger species, which have longer nursing periods [$F(1, 26) = 11.2; p < 0.005;$ coeff. $= +0.24$] and heavier [$F(1, 37) = 109.7; p < 0.0001;$ coeff. $= +1.01$] but fewer young [$F(1, 33) = 196.1; p < 0.0001;$ coeff. $= -0.18$], have longer independent juvenile phases [$F(1, 20) = 30.0; p < 0.0001;$ coeff. $= +0.20$] than smaller species, neither of these *relative,* or weight-specific, life-history variables covaries significantly with the *relative* length of this independent period. In fact, not even rates of juvenile mortality relative to those of adult mortality account for any significant variation in the relative length of the independent juvenile period [$F(1, 15) = 1.91; p < 0.20$].

When we examine exactly which species exhibit relatively long independent juvenile periods after controlling for body size, no obvious pattern emerges. The five species with the longest relative juvenile periods are elephant (*Loxodonta africana*), brown bear (*Ursus americanus*), river otter (*Lutra canadensis*), Japanese macaque (*Macaca mulatta*), and mountain goat (*Ovis canadensis*). They represent five different families, range widely in body size, and have markedly different foraging styles and ecological requirements. And although these five species are polygynous, over the entire sample there is no statistically significant relationship between the magnitude of sexual selection, as measured by degree of sexual size dimorphism, and the length of the independent juvenile period [$F(1, 15) = 1.91; p < 0.20$]. Among primate species, for example, those in which males are relatively larger than females display relative independent juvenile periods that are somewhat longer than more monomorphic species. The pattern, however, is not statistically significant, as only 0.12 of the variation in the relative length of the juvenile period is explained by the degree of sexual size dimorphism.

These patterns suggest that life-history traits that are physiologically tied to features of reproduction are closely tied to the age at which breeding commences, but not to the length of time that youngsters spend developing after ceasing to rely solely on mother's milk for nour-

ishment. Moreover, even juvenile mortality, the factor most important in determining when adulthood begins has apparently little influence on determining how long the independent phase of the juvenile period should last. It is interesting to note that the dependent (nursing) phase tends to comprise about 0.16 ± 0.02 of the total juvenile period (0.11 for ungulates, 0.21 for carnivores, 0.27 for female primates, and 0.19 for male primates). And although each of these juvenile phases scales allometrically with body size [independent phase: $F(1, 25) = 30.0; p < 0.0001$] and the length of each is positively correlated with the other ($r^2 = 0.71; p < 0.001; n = 24$), the proportion of either with respect to the whole does not scale with body size [proportion of dependent phase: $F(1, 18) = 0.97; p < 0.40$)]. In fact, there is only a weak and statistically insignificant positive relationship between the *relative,* or weight-specific, lengths of the dependent and independent phases of the juvenile period [$F(1, 22) = 3.1; p < 0.10$]. This lack of a relationship might account for the finding that *relative* length of the independent juvenile period does not covary with relative age of maturity or any other life-history variable, whereas the length of the *relative* dependent period does.

That deviations from the shorter of the two juvenile phases should be most important in accounting for deviations in the onset of sexual maturity is intriguing and suggests that physical aspects of infantile maturation are of primary importance in timing the decision to begin reproducing. This also implies that variation in the independent juvenile phase is much less important in initiating this critical life-history transition to adulthood and raises questions about its ultimate functional significance. Is it really the critical period during which a superior adult is built? Or is it just a period during which a surviving juvenile is made more competent as a juvenile?

Intraspecific Comparisons

To answer these questions and fully appreciate how juvenility is related to subsequent reproductive success, variation within single species must be examined. It has long been appreciated in mammals that reproductive success increases with increasing parity and age, at least until senescence sets in. In moose (*Alces alces*)

(Saether & Haagenrud 1983), gazelles (*Gazella cuvieri, G. dorcas,* and *G. dama*) (Alados & Escos 1991), white-footed mice (*Peromyscus leucopus*) (Fleming & Rausher 1978), and red fox (*Vulpes vulpes*) (Allen 1984), older females tend to produce larger litters. Larger young are produced by heavier mothers in white-footed mice (Myers & Masters 1983), and moose (Saether & Haagerud 1983), by older mothers in the gazelles (Alados & Escos 1991) and by older, heavier and more dominant females in red deer (*Cervus elaphus*) (Clutton-Brock et al. 1986). And larger young, or those born to older females, tend to have higher survival prospects in each of these species. Prenatal mortality tends to decline in older red fox females (Allen 1984), and adult prospects for survival or breeding in a second consecutive year increase as female condition improves in red deer (Clutton-Brock et al. 1983). Exactly why these age effects occur is unclear, but presumably they are associated with the fact that older individuals are more experienced, are larger, or are in better physiological condition.

With maturity apparently comes enhanced reproductive efficiency. But is the improvement primarily due to benefits gained during the juvenile period, and, if so, is it primarily the result of the juvenile period being lengthened? Unfortunately, answering these questions is difficult because few studies have separated the generalized effects of aging on reproduction from the specific effects of changing the age at which reproduction first begins. A few studies, however, have drawn this distinction and are reviewed below.

Rodents. In the Mongolian gerbil (*Meriones unguiculatus*), age of maturity is bimodally distributed. Whereas some females become receptive just prior to weaning, others delay the onset of reproduction until after they have gained independence from their mothers. Clark et al. (1986) have shown that the early-maturing subset of the population matures on average about 20 days earlier and about 20 g lighter than the late-maturing subset. Although first litters born to early-maturing females are larger than those born to late-maturing females, but not significantly so, early-born first litters contain significantly more females and early-maturing daughters than those born to late-maturing females. Over their lifetime, early-maturing females produce significantly more litters and

slightly more than twice as many surviving young than later-maturing females. In part, this is because they begin breeding early. But it is also because early-maturing females continue breeding to later ages, even though late-maturing females live slightly longer. Early-maturing females also invest much less time suckling their pups and are less solicitous in caring for their pups than late-maturing females. That early-maturing females become receptive at lower weights shows that they do not grow faster than late-maturing females and suggests that sexual maturation is independent of morphological development.

Studies on the mechanisms of puberty control in other rodents reveal that variations in timing the onset of sexual maturity as exhibited by the Mongolian gerbils is common in rodent populations and could produce bimodal distributions of age at maturity in strongly seasonal environments. Many chemical compounds secreted by rodents and primates are known to serve as signals. Some attract males to females (Keverne 1976; O'Connell et al. 1981), induce males to mount estrus females (Singer et al. 1980), or advertise reproductive and social status (Preti et al. 1976). Others play a role in priming sexual behavior. Perhaps the best known are those that block pregnancy when a recently mated female detects the scent of a strange reproductively capable male (Bruce 1959) or those that cause dense populations of females to remain anestrus until exposed to a sexually active male (Whitten 1959). More recently, however, Vandenbergh and Coppola (1986) have shown that chemical compounds of mice can enhance or retard the onset of female puberty. Juvenile females housed with mature males, or to a lesser extent those exposed to only male scents, reach puberty about 20 days before females reared in isolation or in all female groups. This effect occurs in both laboratory colonies and natural populations and seems to be mediated via androgen levels, since castrated males fail to accelerate the onset of puberty and low-ranking males induce a weaker effect than do dominants. Since exposure to lactating and pregnant females also accelerates puberty, but only by 4–5 days (Drickamer & Hoover 1979), factors other than androgens must be involved. Normally, however, female mice reared in the presence of other nonpregnant or nonlactating females exhibit delays in the onset of sexual maturity (Vanderbergh et al. 1972). Clearly, male and female pheromones can interact, sometimes in opposite ways. When they act simultaneously, Drickamer (1982) has shown that the effect of female pheromones on *prepubertal* females takes precedence over those of males. That these effects also occur in other rodents and primates suggests that the phenomenon of puberty modulation is widespread (Vandenbergh & Coppola 1986).

These studies on both house mice and Mongolian gerbils show that age of maturity can change as social conditions change. But why might flexibility in timing the onset of maturity be adaptive? In other words, what about the environment could be maintaining maturational polymorphisms in the population. Delaying the onset of reproduction has traditionally been viewed as a way in which females can increase their survival prospects when the chances of successfully rearing young are already low (Vandenbergh & Cappola 1986). It is thought that by postponing breeding when environmental conditions are harsh, females can wait in an energetically favorable state for conditions to improve or for dispersal opportunities to arise. However, given that the life expectancy of an adult feral house mouse, for example, is about 1 month, the opportunity costs of delaying would appear to be excessively high. Moreover, given that the likelihood of the environment changing during a delay of only 5–10 days is so small, it seems unlikely that a delay in the onset of breeding would be the result of selection to reduce *adult* mortality. It is much more likely that delayed puberty would be favored by selection because it would increase prospects for *juvenile* survival. That late-maturing Mongolian gerbil females, those that would experience the most crowded and resource depauperate conditions in the wild, do in fact care for their young more vigorously suggests that selection may be operating in this way. Thus it appears that flexibility in timing the onset of reproduction is favored when environments fluctuate and that delay will be favored when longer juvenile periods actually lower juvenile mortality rates.

For rodents in general and Mongolian gerbils in particular, if delaying puberty occurs under harsh conditions and in accord with the predictions of the models, then either late-maturing females should show greater sensitivity to the affects of a on fecundity or adult mortality (large values of k and m), or (and perhaps more important given the above adaptationist explanation) the offspring of late-maturing females should experience lower levels of juvenile mortality than those of early-maturing females. For the latter to be the case, $z > 0$ so that $q(\cdot)$ can be lower for late as opposed to early-maturing females.

From the study by Clark and her co-workers (1986) there is enough intraspecific variation in the life-history data to estimate values of the sensitivity exponents. First, the litter sizes of late-maturing females do not increase as a result of the delay; hence, k will be very small. Second, given that early-maturing females live only slightly shorter lives than delaying females, m will be small, if not zero. Since early- and late-maturing classes of females do not differ significantly with respect to these two parameters, according to the models neither factor will have much influence in determining the optimal age of maturity. With respect to juvenile mortality, however, small differences exist, and because they are correlated with significant differences in parenting behavior, they appear to be able to account for the puberty delay that late-breeding females exhibit. Given that $q(\cdot)$ is quite large, varying from about 0.02/day for the early-maturing females to about 0.01/day for the late-maturing ones, it is not surprising that Mongolian gerbils exhibit short juvenile periods. But more importantly, the slightly smaller $q(\cdot)$ of late-maturing females is sufficient to account for the delay in their maturity that accompanies breeding during the winter. Under harsh winter conditions, increases in interbirth interval, shifts to the production of more sons than daughters, and devotion of more time to suckling and caring for the young appear to reduce the rate of juvenile mortality, all of which would make $z > 0$. According to the models, in a population where size is invariant and adult mortality is similar for both classes of adult females, a reduction solely in juvenile mortality rate such as this should lead to increases in the optimal age of maturity. This is what is seen under these seminatural experimental conditions. If the life histories of Mongolian gerbils and the physiological mechanisms that underlie them are representative of most rodents, then we see that the setting of the age of first reproduction for these kinds of small mammals follows the simple biological trade-offs contained in the models.

Ungulates. Some of the best data on mammalian life-history patterns come from long-term studies on ungulates and show that timing the onset of reproduction does have profound affects on fecundity and rates of mortality.

Clutton-Brock and his co-workers have followed the fates of many cohorts of red deer (*Cervus elaphus*) inhabiting Rhum, an island off the west coast of Scotland. Dominance rank is the most important factor affecting lifetime reproductive success of females. High-ranking females breed sooner (3.3 vs. 3.6 years), produce calves at a faster rate (0.8/year vs. 0.7/year), have a greater proportion of them survive to adulthood (0.8 vs. 0.7), and live longer (13 vs. 12 years) than their more subordinate counterparts (Clutton-Brock et al. 1986). Overall, high-ranking females rear more offspring to 1 year of age (6.2 vs. 4.4) over a lifetime than do low-ranking females (Clutton-Brock et al. 1986). Dominance is affected by female body weight, and this is a function of a female's own birth weight, which in turn is influenced by spring temperatures. High temperatures seem to increase fetal growth rates by increasing the amount of available vegetation (Albon et al. 1987).

Although Clutton-Brock and his co-workers have not explicitly determined the effect that altering the age of first reproduction has on either initial or lifetime reproductive success, they have shown that differences in early development have major effects on female status as adults. Even after controlling for differences in birth weight, daughters that are weaned relatively early for their weight attain significantly lower adult ranks than those that are nursed longer (Clutton-Brock et al. 1986) Thus, delaying the age of independence significantly enhances a female's dominance rank and ultimately her reproductive opportunities.

In terms of the models, the red deer is a species in which values of k and z will be high, since small delays in puberty enhance a female's dominance and her ability to increase her fecundity as well as the survival rates of her offspring. As a result, the red deer should be a species that exhibit relatively long juvenile periods, and they do (red deer lie above the regression for ungulates shown in Fig. 4.3). Also, given that dominance differences among females affect the survival prospects of their newborn and yearling sons more than their daughters (Clutton-Brock

et al. 1986), then in terms of the models $q(\cdot)$ should be lower and z should be greater for males than for females. As a result, maturity for males should be reached later than for females. This is indeed the case. Females mature on average at 4 years of age, whereas males often begin breeding as late as 6 years of age.

The age of first reproduction can have dramatic consequences on lifetime reproductive success in feral horses (*Equus caballus*) as well. On a barrier island off the east coast of North America, a population of horses has lived free from human demographic interference for over 100 years. From 1973 to 1986, the population has remained fairly constant, consisting of about 100 animals (Rubenstein 1986). During that time, the reproductive fates of over 70 females and their daughters have been followed. Only a subset of 18, however, has survived to breeding age, then another 3 years, and has lived during a period when population size was stable. As Table 4.1 shows, for this subset of females the onset of reproduction varies, but most females ($n = 9$) reach puberty at 4 years of age, 2 years later than do the youngest females. The timing of maturity has a strong effect on the prospects of survival to 1 year of age of foals conceived at the first breeding attempt [Table 4.1; $F(2, 14) = 6.8; p < 0.01$]. For females, that delay until at least 4 years of age, survival of their young to 1 year of age is 0.89, whereas for those that breed for the first time at 2 years of age, none of their young survives.

When subsequent reproductive episodes—those involving young born during the first, second, and third years after initiating reproduction—are also considered, the significant survival differences of foals born to mothers initiating reproduction at different ages diminish, but remain statistically significant [Table 4.1; $F(2, 14) = 5.6; p < 0.01$]. The survival rate of foals born during a 3-year period subsequent to females initiating breeding at age 2 increases from zero to 0.20, whereas the survival of foals born during the same 3-year period to females waiting until at least age 4 to breed for the first time decreases from 0.89 to 0.80. This survival rate is only slightly greater than the 0.70 for those born during the 3-year period to females beginning to reproduce at age 3. When yearly per capita reproductive success is compared, these same small but statistically significant differences are maintained [Table 4.1;

Table 4.1. Different Ages of First Reproduction and Their Effects on Various Measures of Reproductive Success and Survival for Feral Horses Inhabiting Shackleford Banks, N.C.

	Age to maturity (years)					
	2	3	4	5	*F* value	*p*
Number of females bearing young during 1 year	3	5	9	1		
Number of young surviving to 1 year of age	0	2	8	1		
Proportion surviving to 1 year of age	0	0.40	0.89	1.0	6.8	< 0.01
		(0.24)	(0.11)			
Number of young born during 3 years	5	10	19	2		
Per capita yearly birth rate	0.56	0.67	0.70	0.67		
Number of young surviving to 1 year of age	1	7	15	2		
Proportion surviving	0.20	0.70	0.80	1.0	5.6	< 0.01
	(0.17)	(0.12)	(0.08)			
Per capita reproductive success per year	0.11	0.47	0.56	0.67	7.3	< 0.01
	(0.09)	(0.08)	(0.08)			
Number of expected breeding opportunities	11	10	9	8		
Expected lifetime reproductive success	1.2	4.7	5.0	5.4		
Number of original females dying within 6 years after beginning to breed	2	2	1	0		
Mean age at death	6.0	7.5	10.0			

Note: Analysis of variance performed only on females beginning to breed at ages 2, 3, and 4. Sample sizes are based on the number of breeding females in the cohort. Standard deviations are in parentheses below the means.

$F(2, 14) = 7.3; p < 0.01$]. Given that females live on average for 12 years, we can estimate the expected lifetime per capita reproductive success of females that initiate reproduction at each of these different ages, live to age 12, but lose a breeding episode for every year they delay (Table 4.1). Although the expected lifetime reproductive success of females breeding at very young ages is markedly different from those that delay, the difference between those that delay until 3 years of age (4.7) and those that delay until 4 or greater (5.0) is much smaller.

Delayed breeding enhances expected lifetime reproductive success because postponing puberty beyond the point where females are physiologically able to reproduce lowers juvenile mortality. In terms of the models, whenever $z > 0$, $q(\cdot)$ declines and so does the optimal age of maturity. But does delay also have an affect on fecundity and adult mortality? With respect to fecundity, Table 4.1 shows that yearly per capita birth rate varies little, ranging from 0.56 for females beginning to breed at age 2 to 0.70 for those delaying until age 4. In terms of the model, the fecundity sensitivity exponent, k, will be positive but small. If we examine the survivorship of females that actually begin breeding at different ages, we see that 6 years

after the onset of reproduction those that start early have a lower probability of surviving to the end of the 6-year interval than those that delay (Table 4.1). Whereas only 0.1 of those delaying breeding for the first time until at least age 4 perish, of those initiating reproduction at age 2, 0.67 die. As a consequence, female age of death decreases as the age of first reproduction decreases (Table 4.1). In terms of the model adult mortality $M(\cdot)$ declines with increases in the age of maturity, suggesting that $m > 0$. Thus for horses, each of the critical variables is affected by the timing of maturity and in ways that suggest that as a species, the optimal age of reproduction should be delayed well past the physiological age when reproduction becomes possible. Most females do in fact appear to delay 2 years past the first possible date, yet approximately 0.45 of the females breed before this modal age. Why should this variation be maintained if the benefits of delay can be seen in each of the three critical life-history dimensions?

Acceleration of puberty could be favored by natural selection as long as the sensitivity of some females to the effects of age of maturity on fecundity or either juvenile or adult mortality is small. For this to be the case, the exponents k,

Table 4.2. Number of Females Beginning to Breed Early or Late Depending on Their Dominance Rank

Female dominance	Age of first reproduction (years)	
	2 or 3	4 or 5
High	5	5
Low	3	5

Note: $\chi^2 = 0.28$; $p < 0.60$.

m, and z in the models would have to be close to zero for this class of females, and possibly even negative in species where the average values were not large to begin with. Such small values would tend to accelerate their optimal age of maturity. At least for Shackleford horses, dominant females or females that reside in harems with dominant males tend to exhibit reduced sensitivities of adult mortality to changes in the timing of age of maturity. Although dominant and subordinate females show no significant differences in their propensity to breed early or late in life (Table 4.2; $\chi^2 = 0.28$; $p < 0.6$), four of the five of the females dying within 6 years after beginning to breed are early breeders (≤ 3 years of age) and, of these, three are subordinates. This suggests that the mortality of females of high rank is less affected by the timing of the initial bout of reproduction than that of females of low rank.

In addition, the rank of the male with which a female associates also appears to be an important determinant in timing the initiating reproduction. Table 4.3 shows that significantly more females that begin breeding at ages 2 or 3 live in groups tended by dominant males versus ones tended by subordinates [$\chi^2 = 2.81$; $p < 0.05$ (one-tailed)]. And here, too, rank of the male influences the *sensitivity* of adult female mortality to changes in the age of first reproduction. Whereas five of seven (0.71) females of dominant males survive if they breed

Table 4.3. Number of Females Beginning to Breed Early or Late Depending on the Rank of the Male with Whom They Associate

Male dominance	Age of first reproduction (years)	
	2 or 3	4 or 5
High	7	5
Low	1	5

Note: $\chi^2 = 2.81$; $p < 0.05$ (one-tailed).

at 2 or 3 years of age, the one female associating with a subordinate male does not. Yet for the 10 females that delay breeding, all those with dominant males survive the interval ($n = 5$), and 0.6 of the five females with subordinate males also do so.

Clearly, there is a cost to breeding early, but it is much less for those females that are dominant or associate with dominant males. Why should this be so? Since dominant females have more access to limited water supplies than subordinate females (Rubenstein 1993) and those associating with dominant males are harassed less and have more time to graze (Rubenstein 1986), it appears that favorable ecological circumstances can lower the costs of reproduction so that adult mortality rates become unaffected by changes in timing the onset of sexual maturity. For some females, the removal of critical life-history traits from an ordinarily strong maturational sensitivity means that selection no longer opposes their taking advantage of opportunities to begin reproducing early in life because reproduction has become virtually cost-free. Offspring of dominant mothers or of those associating with dominant males derive material benefits, while their mothers avoid costs that are typically harmful when ecological resources are more limiting. At least in this horse population, subordinate females and those associating with low-ranking males show greater sensitivities to changes in age of first reproduction than do dominants, or than those bonded to dominant males. For these subordinate females, selection will favor delaying the onset of reproduction. For those experiencing more favorable ecological circumstances, selection should favor an earlier than average age of first reproduction. In this feral horse population, both tendencies occur and in the ways predicted by the models.

Pinnipeds. Lifetime reproductive success has been collected on northern elephant seals (*Mirounga angustirostrisa*), and although there are strong age effects on breading and rearing success, sex differences are large and the results of timing the onset of reproduction on these life-history attributes are equivocal. LeBoeuf and Reiter (1988) show that breeding begins in males at age 6, on average, and yearly reproductive output increases until age 11, dropping to zero by age 13. For females, they show the pattern to be different. Although breeding can begin at 2 years of age, most females wait until ages 3 or 4, with only a few stragglers delaying until 5 or even 6 years of age. Four- and 5-year-old females are the most prolific breeders, but those between the ages of 4 and 7 wean the most pups. Unlike males, however, female fecundity varies little with age after a female has given birth for the first time.

Only for females does there appear to be a strong relationship between the age at which reproduction is initiated and a variety of factors that affect lifetime reproductive success. Both pup survival to 1 year of age and adult survivorship increases as females delay the onset of breeding (Reiter & LeBoeuf 1991). These relationships result from the fact that although the weight of pups at birth is greater for females breeding for the first time at age 3 as opposed to age 4, the ability of mothers to invest in pups without reducing their own prospects of survival increases as mothers grow older and increase their reserves of blubber. By projecting the growth rate of hypothetical populations composed either all of females commencing breeding at age 3 or all of females beginning at age 4, Reiter and LeBoeuf show that the strategy of delay to age 4 is the superior, since it spreads faster under high-density conditions, irrespective of the level of juvenile survival (40 vs. 80%). It loses its advantage, becoming equivalent to the early-breeding strategy, only when density is low and juvenile survival is high. The authors suggest that the early-breeding strategy is maintained in the population because neighboring populations typically exhibit population cycles that are out of phase, and in recently colonized and expanding populations selection allows early breeders to do well, although in crowded, or declining, ones late breeders have the advantage. As with the red deer and the Shackleford horses, there are ecological conditions where the sensitivity of the life history components, in this case m and z, to changes in age of maturity is reduced and with the lowering of the costs of breeding early, female puberty can be accelerated as predicted by the models.

For males, however, there are no significant relationships between the age of first reproduction and the number of young sired in a lifetime [$F(1, 6) = 1.8; p < 0.7$] or the number of these that are weaned [$F(1, 6) = 1.36; p < 0.7$]. Although about 0.12 of the variation in yearly rate of reproductive success is explained by the age

of a male's first breeding attempt, the tendency for males to sire more young per year if they delay the onset of breeding is not statistically significant [$F(1, 6) = 1.34; p < 0.3$]. In this small sample, the male initiating breeding at the youngest age breeds for the longest period, has the highest lifetime reproductive success, and, with respect to yearly offspring production, comes in second to a male delaying breeding until age 8. In terms of the model, the sensitivity exponents should be small and hence breeding should occur early. But empirically this is not the case; on average, males delay breeding 2 more years past the time when females typically begin breeding (LeBoeuf & Reiter 1988). Since dominance and experience play the major role in determining male breeding success (LeBoeuf & Reiter 1988), it appears that knowledge and skills gained once reproduction commences is more important than acquiring these attributes while growing. Perhaps the apparent limited importance of the juvenile period derives from the fact that breeding takes place on land, whereas most of the juvenile period, and its affects on the sensitivity parameters, occurs in the sea.

Primates. Perhaps more attention has been focussed on the juvenile lives of primates than on any other group of mammals. Detailed studies range from the physiological to the social and to the reproductive consequences of ontogeny, and they have been performed on laboratory as well as on wild populations.

Baboons (*Papio cynocephalus*) reveal a common pattern in which dominance rank has a profound affect on female reproduction (Altmann et al. 1988). Daughters born to mothers of high rank tend to have accelerated menarche and conceive about 200 days earlier than those born to low-ranking mothers. Over a female's lifetime, this acceleration amounts to an increase of 0.5 infants. Although age affects fertility up to about 6 years of age, the survival prospects of offspring after the first breeding attempt remain constant throughout life. Moreover, the age of first conception has no affect on the survival chances of the first born young. Since 0.79 of the variance in lifetime reproductive success is accounted for by life span and no ecological factors seem to strongly affect it (Altmann et al. 1988), it appears that breeding early is the key to enhancing a female's reproductive success in yellow baboons.

For other primate species, dominance seems to operate in ways similar to the baboons, accelerating puberty and hence enhancing lifetime reproductive success. Drickamer (1974) has shown for rhesus macaques that daughters of high-ranking females breed earlier than those of lower rank, and Gouzoules et al. (1982) have shown for Japanese macaques that age of maturity increases as rank decreases. Yet for other species, age of maturity seems little affected by social or ecological factors, such as for bonnet macaques where all females give birth at the same age (Silk et al. 1981), even though timing of puberty often still influences female lifetime reproductive success. In vervet monkeys (*Cercopithecus aethiops*), for example, age and dominance rank have little effect on female lifetime reproductive success (Cheney et al. 1988). Birth rate does not decline with age, and primiparous females have the same number of surviving offspring as do multiparous ones. Similarly, dominants do not produce more young than do subordinates, nor do their young have higher probabilities of survival. Dominants also do not reproduce at an earlier age than subordinates, nor do they have shorter interbirth intervals. The major determinants of differences among females in lifetime reproductive success appear to be juvenile survival prospects and adult longevity. And juvenile mortality rate, in particular, is correlated with age of first reproduction. For those females delaying reproduction the most, survival to adulthood is least. In troops where reproduction is delayed past five (5.7 and 5.1 years), the chances of infants dying before reaching maturity is on average 0.63 and 0.71, respectively. In the troop where reproduction commences on average at 4.4 years of age, juvenile survivorship is 0.53. Coupled with these changes in rates of juvenile mortality are changes in adult mortality. For troops delaying puberty past 5 years of age, adult yearly mortality is higher (0.17 and 0.22) than it is for females in the troop breeding at the earlier age (0.10).

Although these results are derived from a small number of troops and are not statistically significant, the overall pattern indicates that there is a connection between ecology, as measured in terms of resource availability, and timing the onset of reproductive maturity. Since those troops with the earliest ages of first reproduction have the lowest levels of mortality, selection apparently favors early reproduction when possible. This suggests that "those that can, do; while those that can't don't," and the "dos" and "don'ts" are determined by the severity of ecological circumstances. Apparently, daughters being raised under harsh conditions have extended juvenile periods and suffer high levels of juvenile mortality. The delay seems to be nonadaptive, since it has little affect in reducing subsequent levels of adult mortality. Since daughters with access to resources are able to accelerate puberty without incurring significant survival costs, selection at least in vervet monkeys is reinforcing traits that enhance competition, ultimately maintaining variation in the timing of puberty.

The tendency for ecological circumstance to influence the onset of reproduction is also seen in at least one group of humans (*Homo sapiens*). In her long-term study on the Kipsigis, an agropastorilist people of southern Kenya, Borgerhoff-Mulder (1988) shows that men marry at around 18 years of age and, if wealthy enough, take additional wives later in life. She also demonstrates that women marry at about 16 years of age after a period of seclusion during which they are fattened. About 0.76 of babies survive to reproductive age, and whereas men in this polygynous society father about 12 offspring in a lifetime, each woman bears about 6. For men, wealth, as measured in terms of the size of herds and plots of land, is the major determinant of reproductive output. Wealthy men marry younger and have more offspring, and more of their sons marry polygynously than do the sons of poorer individuals. For women, wealth is also important but for a different reason: reproductive success is influenced more by the likelihood of children surviving to adulthood than by the length of the reproductive period, even though younger brides have on average about 2.7 more surviving offspring than do their older counterparts. It appears that young brides are better nourished during their early development and simply become better mothers.

Thus, at least for primates, there is one clear generalization about how selection alters the length of the juvenile period. When females are capable of breeding at a young age, their lifetime reproductive success is enhanced because their fertility remains high, while either their own or an offspring's chances of survival also remain high. In yellow baboons and several species of macaques, a mother's rank seems to separate those that have access to resources and can begin breeding early without incurring survival costs from those that lack access and cannot. In vervet monkeys, neither age nor dominance affects birth rate or survival, but ecological circumstances, as determined by group and home-range residence, seems to separate the "haves" from the "have-nots." And for at least one group of humans, early breeding also seems to enhance the survival prospects of offspring, despite the fact that adolescents typically show depressed fertility (Jain 1969) and a physiological susceptibility for fetal or infant loss (Leridon 1977; Miller & Stokes 1985). For the Kipsigis, those females who can breed early seem to be superior and, as a result, seem to be able to overcome these inherent physiological problems, thus creating a long reproductive lifetime during which they produce many robust offspring. Although the mechanistic details are less well known for the other primates, the pattern exhibited by human females seems likely to be representative. In terms of the models, all these examples have one feature in common: when sensitivities of fertility and survival to age of maturity are reduced (exponents close to zero) for some classes of females, those females should and do show early ages of maturity.

DISCUSSION

The theory developed to understand the organization of the juvenile period was built on the premise that fecundity and mortality rates, those of both juveniles and adults, are affected by changes in the age of first reproduction (for a different approach, see Stearns & Koella 1986). Each life-history variable was assumed to be a nonlinear function of age of maturity, and, at least for those depicting mortality, delay could either enhance or diminish the rates. Traditionally, it has been assumed that with delay comes increased size, resources, and experience and that all of these can augment fertility and juvenile survival, while reducing the rate of adult mortality. Only because delayed breeding incurs costs associated with missed opportunities and the fact that death is inevitable, is delay thought to have limits. Thus the existence of this trade-off ensures that for any species an optimal age of maturity will exist.

What emerges from the intraspecific case studies is the fact that on a species-specific basis, those showing life-history variables with strong sensitivities to the timing of sexual maturity tend to delay puberty, whereas those showing weaker sensitivities accelerate it. Rodents, with their normally short life spans, should typically have exponents that are close to zero because delaying puberty would incur large opportunity costs. That they breed early, in accord with the predictions of the model, is only reassuring because the empirical results show that fecundity and mortality in fact show little sensitivity to the timing of age of maturity. The converse applies to the longer lived species. For two species, red deer and horses, where the data are reported in ways so that species-specific tendencies could be evaluated, juvenile mortality shows a strong sensitivity to parental age of maturity. Again, it is reassuring to observe that both species (if we assume that zebras behave like horses, their close kin) show average or greater than average tendencies to delay reproduction for species of their size (Fig. 4.3). It is interesting to note, however, that in addition to these commonalities, adult mortality for horses also shows a strong age sensitivity, whereas for red deer it is fecundity that shows an additional strong maturational sensitivity. This suggests that although the same general rules apply to determining the length of the juvenile period, different ecological or phylogenetic features determine exactly how selection operates and on what life-history components.

Additional insights into how the length of the juvenile period is shaped emerge from examining how life-history consequences of changing the onset of puberty affect the different classes of individuals that compose populations. In every case study analyzed in this chapter, some individuals in a population bred significantly earlier than a modal age of maturity. In each case, these individuals belonged to groups, cohorts, or matrilines in which sensitivities of fecundity, but mostly of juvenile mortality, to changes in the age of maturity were reduced. For rodents z was lower for late breeders than for early breeders; for red deer z was lower for females than for males; for horses z was lower for dominant females and females associating with dominant males; for elephant seals z was lower for cohorts free from competition; and for primates z was low for females that either had dominant mothers or were able to achieve access to otherwise limited resources. It is interesting to note that in each case the sensitivity that is most important is the one that measures how much better offspring do as a result of their mother's delay and is the one that is left unconsidered in Charnov's (1990) original model.

Thus although there might be a modal, or typical, age of maturity, one that is initially influenced by body size (Pagel & Harvey, Chapter 3, this volume) and that lies close to the optimal age predicted by the theory, there seems to be alternative reproductive investment strategies that enable some to change the "rules of the game," thus allowing selection to maintain variation, and at times discrete polymorphisms, within populations. Even in the red deer, where rank produces overall sensitivities of fertility and survival to timing of maturation, the sexes show different sensitivities to the effects of pubertal age on juvenile survival. What emerges from these studies is the sense that phenotype, in relation to features of the physical and social environment, does affect investment "decisions," which ultimately alter the levels of juvenile mortality. What is striking is the fact that those females that can reduce the mortality of their offspring seem to be able to do so without increasing their own chances of dying. As Charnov (1990) and Pagel and Harvey (Chapter 3, this volume) suggest, the optimal age of maturity is more a function of mortality than fecundity, and at least for certain species, some "supermoms" seem to be able to develop.

But this conclusion raises two interesting problems. First, what do these within-species comparisons mean for the interspecific ones where the shape parameters cannot be estimated but are assumed to be positive? On a macroevolutionary time scale, the appropriate level of variation is the level that reflects large-scale rearrangements of one life-history variable with another. Thus it is appropriate to assume that levels of mortality are *unavoidable* and that they represent rates that account for all the parents have already attempted to do. Given this assumption, the absolute rates of mortality then assume more prominence in the models and shape the optimal pattern. That this occurs is clearly seen from the results, which indicate that after controlling for body size, species with high levels of juvenile mortality tend to breed relatively early in life, unless the effect of adult mortality rate is also controlled. Only then do relatively high levels of juvenile mortality lead to delaying the onset of maturity. This striking result was first illustrated by Promislow and Harvey (1990) and shows that the conventional wisdom applies only when many other important effects are controlled. That the age of first reproduction scales more closely with the length of the dependent rather than the independent phase of the juvenile period suggests that physiology places limits on how rapidly any mammal can develop. But at least to a limited extent within a species, if certain classes of individuals can remove themselves from the constraints, then they can speed up the process and breed relatively early without incurring excessive costs.

Second, if selection does indeed favor at least some members of a population advancing the age of first reproduction, then what functions do juvenile periods serve when they are longer than the time it takes to become physiologically competent to reproduce? If individuals that ultimately obtain the highest lifetime reproductive success can do so with juvenile periods that are shorter than average, what does this mean for the notions that the juvenile period performs a valuable role in developing a better adult and that "more is better" because it gives youngsters more time to grow or gain experiences? The importance of this problem is underscored by the interspecific finding that the relative length of the independent juvenile phase, where this enhancement process is supposed to take place, is not correlated with other relative life-history features, including the relative age of maturity.

Answering these questions is not easy. The debate on the function of early development has often been heated and inconclusive (Bateson 1978, 1981; Klopfer 1981; Fagen 1984; Martin 1984). Due to its conspicuousness, play has often been the vehicle that has been used for gaining insights into the adaptive value of juvenile behavior. Although the above demographic and life-history analysis has been able to show that the juvenile period in the abstract has effects on subsequent reproductive success, ethologists have had a hard time demonstrating that particular features of play, or early development in general, are responsible for adult reproductive success (Bekoff & Beyers 1985). In fact, Martin and Caro (1985) argue that it might even be pointless trying to find such a relationship. They suggest that play may indeed serve to enhance social and cognitive skills and to develop motor proficiency, as has often been assumed. But not because play prepares a juvenile to become a better adult, but because the benefits of play help make a better juvenile, one that is competent presently and will increase its chances of surviving, not necessarily one that will be better able to cope with the unique problems that beset adults. And if play is not very costly as they contend, then it need not provide large benefits to be maintained by selection (cf. Fagen, Chapter 13, this volume).

Evidence supporting this view comes from the detailed studies of play. Cuvier's gazelles (*Gazella cuvieri*), a species we have already identified as one where maturity and experience influence reproductive success, provides one such example. Gomendio (1988) shows that fawns participate in a variety of types of play, but that each type has a typical time course. Whereas locomotor play is frequently engaged in when fawns are very young, it drops off quickly and disappears from the repertoire long before weaning. Play fighting and sexual play, on the other hand, do not appear until about 1 month of age, and although they increase with age, they, too, vanish before weaning. If each of these different types of play evolves to enhance some aspect of later adult existence, then why does each exhibit the particular time course that it does? Play that involves partners, for example, is often thought to develop social and cognitive skills that will certainly be of use during adulthood. Why, then, should these most socially interactive forms of play vanish just prior to weaning when the conditions and peers are likely to be most similar to those that will be encountered during adulthood? While play at any time during the juvenile period can have preparatory effects, it also seems likely, and even more forcibly so, that juveniles engage most vigorously in these types of social play during periods of social uncertainty or whenever new social millieus appear. For gazelles, this is likely to be midway through the juvenile period when reliance on hiding gives way to becoming active and integration into the social group creates many novel social situations. This is precisely when social and sexual play peaks during the life of a gazelle.

The early life-style of the gazelle is just one example that can be counterbalanced to some extent by others, such as horses (Rubenstein 1982) and elephants (Lee 1986), where play does seem to take different directions depending on the juvenile's future role in adult society. But it does highlight the point that the juvenile period seems mostly about producing a competent juvenile that will live to see another day. Thus our finding that the relative length of the independent phase of the juvenile period seems to bear little relation to any other life-history variable when also measured on a relative basis should not be too surprising. And it compliments the arguments of Janson and van Schaik (Chapter 5, this volume), who suggest that because juvenile primates are extraordinarily ecologically incompetent, they lengthen the juvenile period more than most mammals to reduce risk of mortality, not primarily to acquire experience since mastery of adult skills is often completed long before maturity is reached. Obviously exceptions exist, and, by highlighting them, the authors conclude that rapid development will be favored when ecological circumstances reduce competition, or niche overlap, between juveniles and adults. Overall, selection seems to be favoring the shortest possible period in which all the necessary developmental hurdles can be crossed without increasing immediate or future costs. Usually, however, only a subset of any population will find itself in such favorable circumstances.

Even the most basic allometric analysis shows that although large mammals have longer juvenile periods than smaller ones, the fact that the exponent describing the power function is less than one indicates that on a weight-specific basis, the larger species are exhibiting relatively shorter times to first breeding. Getting it right is apparently what counts, but deciding how long this should take is variable and, although constrained by some major physiological processes, seems to depend mostly on the sensitivity of an individual's fecundity and mortality to changes in the age of maturity. What, then, shapes these sensitivities? Much appears to depend on what one begins life with, where one's place in society is, and what parents, especially mothers, have to offer.

5
Ecological Risk Aversion in Juvenile Primates: Slow and Steady Wins the Race

CHARLES H. JANSON and CAREL P. VAN SCHAIK

Most studies on primates consider the juvenile period as a kind of life-history limbo, a time between the safety of infantile dependence and the complex world of independent adult life. Because juveniles cannot breed (by definition), one main focus in the study of juvenile primates is how they develop adult skills that will eventually help them increase their fitness. This approach suggests that the length of the juvenile period is constrained by the complexity of adult behavior and that the unusually long juvenile period in primates is determined by the time needed to learn these skills (e.g., Poirier & Smith 1974). Although we do not question the fact that primates rely greatly on behavior learned as juveniles, we wish to offer an alternative to the "adult constraints" model for the causes of their extended juvenile period.

We view the juvenile period in primates as one of great ecological risk. As the juvenile becomes less dependent on its mother, it is forced to fit into an ecological niche determined by conspecific adults when it has neither the size nor the experience to do so easily. The outcome is a high potential risk of juvenile death, either by starvation or by predation. Although a juvenile cannot contribute to its own reproductive success, it can affect its fitness by maximizing the chance that it reaches the size or age necessary for breeding. It could do this either by growing rapidly, and so minimize the time spent in the vulnerable juvenile stage, or by growing

slowly, and so reduce the risk of death per unit time even at the expense of a longer juvenile period. Much evidence from primates suggests that they follow the latter strategy, and we provide a simple model to predict under what conditions such a risk-averse policy can be adaptive. If our "juvenile risks" model is correct and slow growth in juveniles is an adaptive response to ecological risk, then we suggest that some of the social and even morphological attributes of primates may be related to the extended length of the juvenile period instead of the other way around. In any case, we hope to show that much of both juvenile and adult primate behavior is related directly to maximizing the survivorship of juveniles.

A note on terminology is in order. Juvenility ends when sexual maturity is reached (and growth tends to slow down; however, the link between sexual maturity and growth in primates is variable and complex: Watts 1990). It is often thought to begin when a young individual can survive the death of its mother (Pereira & Altmann 1985; Walters 1987a). In practice, especially the beginning is quite gradual, in terms of independent locomotion (e.g., Altmann 1980; Chalmers 1980a), locomotory abilities (Doran 1989), and feeding. Weaning, often depicted as a brief period of intense conflicting demands between mother and offspring (e.g., Trivers 1974), is often gradual as well (Hauser 1986); occasional nursing sometimes continues until the mother's next offspring is born (e.g., Goodall 1986).

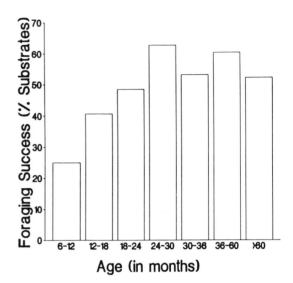

Fig. 5.1. Invertebrate foraging success (percentage of substrates searched at which food was ingested) of different age classes of brown capuchins (Janson, unpublished data). Data for age classes between 6 and 36 months are longitudinal samples on four juveniles. Data for older classes are cross-sectional data on two and seven individuals for 36–60 and >60 months, respectively.

JUVENILE ECOLOGY: MISFITS IN AN ADULT NICHE

Juvenile Foraging Success

Much descriptive evidence suggests that juvenile foraging success is less than that of adults (e.g., Altmann 1980; Boinski & Fragaszy 1989). Quantitative data on foraging reinforce this impression. For instance, in brown capuchins, the percentage of substrates searched (visually inspected and manipulated) that yield invertebrate food items increases from 25% in infants 6–12 months old to over 50% by 3 years of age (Fig. 5.1) (Janson, unpublished data). Similarly, in white-fronted capuchins (*Cebus albifrons*), the percentage of substrates searched that yield insect prey increases from 23.1% in infants less than 1 year of age to 40.0% in juveniles 1–3 years of age up to 44–58% in adult females and males, respectively (Janson, unpublished data). In weeper capuchins (*Cebus olivaceus*), juvenile foraging success is 43% compared with 48% for adult females and 79% for adult males (de Ruiter 1986). In long-tailed macaques (*Macaca fascicularis*), juvenile foraging success per substrate is 51% compared with 58% for adult females, although adult males obtain prey in only 44% of substrates (van Schaik 1985). Overall, then, juveniles have 80–90% of the foraging success on invertebrates of adult females.

The evidence on juvenile success in foraging for invertebrates is sometimes biased by juvenile mixing of "play" activities with serious foraging. However, juvenile foraging efficiency is demonstrably lower than that of an adult even when the juvenile is clearly eating a food item or a sequence of similar food items such as fruits. Across eight species of New and Old World primates, juvenile ingestion rates on vegetal foods vary from 36 to 109% that of adult females, with a median of 70% (Table 5.1). Although some of these values may be influenced by social competition for food (see below), the fact remains that juveniles usually ingest food at lower rates than adults.

Juveniles appear to have a harder time ingesting leaves than fruits: the median intake rate (relative to adults) for leaves was only 44%, but for fruits was 81% (Table 4.1). This difference probably arises because the rate-limiting step in ingesting leaves is mastication (because of its effect on digestibility: Kay & Scheine 1979), while at least for small fruits, ingestion is limited by the picking rate (van Schaik & van Noordwijk 1986; Janson, unpublished data). The strength of juvenile jaws probably limits mastication more than arm length limits picking rate.

The low foraging success of juveniles may force them to invest considerably more time foraging to meet their metabolic demands than

Table 5.1. Ingestion Rates of Juvenile Primates Relative to Those of Adult Females of the Same Species

Species	Food type	Juvenile relative feeding rate (%)
Long-tailed macaques[a]	5 fruit spp.	81
Toque macaques[b]	Fig fruit	59
Yellow baboons[c]	Sedge corms	75
Yellow baboons[c]	Tamarind seeds	95
Brown capuchins[d]	4 fruit spp.	84
Mantled howlers[e]	2 leaf spp.	36.5
Mantled howlers[e]	1 fruit sp.	52
Siamang[f]	Leaves	44
White-handed gibbon[f]	Leaves	64
Gorilla[g]	5 leaf spp.	109

[a]van Schaik and van Noordwijk (1986).
[b]Dittus (1977).
[c]Rhine and Westlund (1978).
[d]Janson (1985b).
[e]Milton (1984a).
[f]Raemaekers (1979).
[g]Watts (1988).

would adults of comparable size. Assuming that juveniles average 50–70% of adult female size, they should consume 60–80% as much as an adult female (if we are correct in assuming that within-species metabolic rates scale to body size approximately as do between-species rates; cf. Kleiber 1961). We ignore the possible complicating effects of juvenile growth and female reproduction, since both activities act to increase nutrient demands for the relevant age–sex class and tend to cancel each other out. Despite their smaller metabolic demands, juveniles generally spend considerably more than the expected 60–80% of adult female time foraging, the median value being 105% (Table 5.2). This increased foraging effort is consistent with lower juvenile foraging competence.

Juvenile Food Choice

If juveniles have relatively more food available to them than do adults, a juvenile might be able to feed itself as easily as an adult despite the latter's demonstrably higher average foraging success. However, several lines of evidence suggest that juveniles have no advantage over adults in relative food availability. First, although juvenile primates generally follow the lead of adults, tasting most of what the adults eat, they avoid certain large or tough fruit that are routinely eaten by conspecific adults (Terborgh 1983; van Schaik & van Noordwijk 1986). These large fruits contribute disproportionately to total energy intake because they provide higher rates of pulp intake than do small fruits (Fig. 5.2; see also van Schaik & van Noordwijk 1986; Malenky 1990). Juvenile primates will also often avoid the larger or tougher of the foraging substrates used by adults to find insects (e.g., Janson & Boinski 1992). Furthermore, juveniles may require practice in extracting or preparing certain animal foods (e.g., caterpillars with stinging hairs: Boinski & Fragaszy 1989). Although on occasion juveniles explore a variety of foods not eaten by adults, the majority of these foods are rarely eaten again and contribute very little to juvenile food intake (e.g., Watts 1985a; Whitehead 1986).

Juveniles on occasion can use foraging sites not available to adults because of the latter's large size. For instance, in brown capuchins (*Cebus apella*), only old juvenile and subadult animals commonly use the technique of hanging on to the tips of palm leaflets with hands and feet and thus moving along *beneath* the palm frond, searching visually for any pupae or other prey hidden in the folds of the leaflets (Janson, personal observation). In Barbary macaques (*Macaca sylvanus*), juveniles spend more time foraging in trees than do the larger adults, and the more time juveniles spend in the trees, the more their diet differs from that of adults (Menard 1985; Menard & Vallet 1986). Similar niche separation might occur in other terrestrial primates, although it has not been reported to date (Watts 1985a). In addition, juvenile primates may be the first to discover and master a novel foraging skill (such as potato washing in provisioned Japanese macaques [*Macaca fuscata*]: Kawamura 1959). It is fair to say, however, that juvenile diets are usually a subset of the adult diet in the same group, with the possible exception of terrestrial forest-living primates.

If juveniles have both lower foraging success and a more restricted diet than adults, they may have more difficulty meeting their daily nutrient demands. A test of this hypothesis is to compare the total foraging time allocation of adults and juveniles. We have shown that juveniles spend more time foraging than expected based on their presumed energetic demands, but the question here is whether juveniles have to forage more

Table 5.2. Juvenile Time Budgets Relative to Those of Adults

Species	Total time obtaining food		Foraging on invertebrates		Feeding on fruits and flowers	
	Adf (%)	Adm (%)	Adf (%)	Adm (%)	Adf (%)	Adm (%)
Cebus albifrons[a]	95	131	86, =	229, >	94, =	76, <
Cebus apella[a]	78	110	78, <	168, >	78, <	72, <
Cebus olivaceus[b]	121	116	124, >	121, >	112, <	121, <
Cebus olivaceus[c]	122	154				
Cebus olivaceus[d]	109	195				
Cercopithecus mitis[e]	74	129	102, >	632, >	64, <	87, <
Cercopithecus sabaeus[f]	103	125	200, >	160, >	124, <	96, <
Cercopithecus cephus[g]			>	>		
Hylobates klossi[h]	113	106				
Hylobates agilis[i]	110	110				
Gorilla gorilla[j]	86	72				
Macaca fascicularis[k]	98	102	131, >	169, >	82, <	78, <
Macaca mulatta[l]	90	217				
Macaca sinica[m]	117	177				
Macaca sylvanus[n]			>	>	<	<
Papio cynocephalus[o]	132	134				
Presbytis rubicunda[p]	104	179			161, >	160, <
Symphalangus syndactylus[q]	107	110				
Median for all species	105	127	113	169	94	87
Comparisons			6/8 >	8/8 >	6/8 <	8/8 <

Note: Comparison of absolute percentage of time devoted to all activities related to acquiring food, foraging on invertebrates, and feeding on fruits and flowers. The time allocated by juveniles is expressed as a percentage of the equivalent values for adult females (Adf) and males (Adm). In addition, for invertebrate foraging and fruit-plus-flower feeding, the relative amount of feeding time devoted to them as a proportion of total time spent acquiring food is compared qualitatively (< means juveniles are lower, > means juveniles are higher, = means that results were contradictory in different groups).
[a]van Schaik and van Noordwijk (unpublished data).
[b]Robinson (1986).
[c]de Ruiter (1986).
[d]Fragaszy (1990).
[e]Rudran (1978).
[f]Harrison (1983).
[g]Quris et al. (1981).
[h]Whitten (1980).
[i]Gittins (1982).
[j]Fossey and Harcourt (1977).
[k]van Schaik and van Noordwijk (1986, unpublished data).
[l]Camperio-Ciani and Chiarelli (1988).
[m]Dittus (1977).
[n]Menard (1985) and Menard and Vallet (1986).
[o]Post et al. (1980).
[p]Salafsky (1988) (seeds not included as fruits).
[q]Chivers et al. (1975).

than adults. In fact, juveniles spend a greater percent of their time foraging than either adult females or males, although the difference is statistically significant only for adult males (Table 5.2, Wilcoxon signed-ranks test, two-tailed, $p = 0.001$). The absolutely higher time allocation to foraging for juveniles compared with adults should make juveniles more vulnerable to starvation when food availability is low.

Consequences of Juvenile Foraging Incompetence

Juvenile death rates can sometimes be linked with relative food scarcity. For instance, during a severe drought, juvenile baboon mortality rates became more than four times greater than those for nonreproducing adult and subadult males (18 vs. 4% in a 6-month period: Hamilton

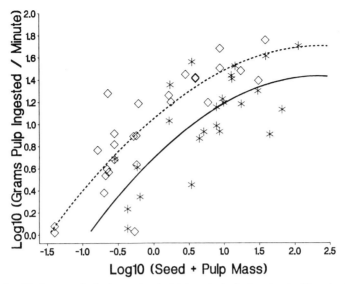

Fig. 5.2. Relationship between fruit size and pulp intake rate in fruits eaten by Peruvian brown capuchins (Janson, unpublished). Diamonds refer to fruit species without husks protecting the ripe pulp, stars to fruits with a closed husk.

1985). When provisioning was stopped for one group of Japanese macaques, juvenile mortality increased 30.1% more than did adult female mortality (Mori 1979, Table 1).

The potential vulnerability of juveniles to food shortages is also reflected in adult social strategies to reduce the impact of within-group food competition on their juvenile offspring. Because of their small size and consequent subordinate status, juveniles are intrinsically least able to gain access to aggressively contested food. Adults may aid or protect juvenile kin at contested food sources (e.g., Janson 1985a; Whitten 1987). When a female cannot by herself protect her offspring, she may enlist the aid of related females (e.g., Silk et al. 1981) or of an adult male, through either a social alliance or preferential mating (Janson 1984; Smuts 1985; see also Pereira 1989), or she may leave the offspring's potential father in charge of protecting it (van Noordwijk & van Schaik 1988).

The impact of aggressive competition for food on juvenile survival can be substantial. For instance, in papionines, juvenile females receive more aggression than do juvenile males (especially from unrelated females: e.g., Cords & Aureli, Chapter 19, this volume) and have

lower food intake in fruit trees and substantially higher death rates than do juvenile males (Dittus 1977, 1980; van Schaik & de Visser 1990). Similarly, the offspring of high-ranking long-tailed macaque females have higher juvenile growth rates than do the offspring of low-ranking females, suggesting that at least some juveniles have difficulty obtaining adequate nutrition (van Schaik et al., in preparation).

Adult primates may use other means to reduce the vulnerability of juveniles to food shortages. Adults may directly subsidize the food intake of juveniles by sharing (chimpanzees [*Pan troglodytes*]: Silk 1978; marmosets [*Saguinus* spp]: Terborgh 1983), although more commonly the adult simply allows the juvenile to feed in close proximity to it (Nicolson & Demment 1982; Janson 1985a; Pereira 1988a). Females may time reproduction so that weaning occurs when "weaning foods" are most available to their juvenile offspring (Altmann 1980). Finally, adults may reduce a juvenile's energy requirements by carrying it occasionally, especially at times of rapid group movement (Nicolson 1987). Although infrequent, such carrying may help a juvenile survive periods of energetic stress.

Fig. 5.3. Time allocated to vigilance by various age–sex classes in three species of capuchin monkeys. The open bars are for one group of brown capuchins, *Cebus apella* and the single-hatched bars are the average for two groups of white-fronted capuchins, *Cebus albifrons* (in Manu National Park, Peru; van Schaik and van Noordwijk, unpublished). These figures are based on scan samples, and no meaningful error term can be attached. The cross-hatched bars are the average values of three studies of weeper capuchins, *Cebus olivaceus*, in Hato Masaguaral, Venezuela (Robinson 1981; de Ruiter 1986; Fragaszy 1990), collected using a different scan sampling technique. The exact definition of vigilance varied between the two study sites precluding direct comparison of values.

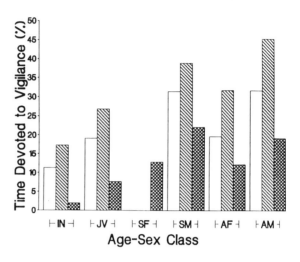

Juvenile Predation Risk

By several criteria, juveniles are not well equipped to avoid predators. First, rates of scanning or vigilance always increase in the order infants, juveniles, adult females, adult males, with the rates of subadults tending to resemble those of the same-sex adults (Fig. 5.3; see also Janson 1990b). Although this pattern could be caused by a preference by juveniles to stay where predation risk is low, juveniles also do not vary their rate of vigilance in obvious response to predation risk (e.g., at periphery vs. center of group; Janson 1990b). Second, juveniles do not detect or recognize predators as well as adults. In a test of predator detection using a model snake, juvenile capuchins had far lower detec-

tion rates than did adults (van Schaik & van Noordwijk 1989). Juvenile vervets (*Cercopithecus aethiops*) and ringtailed lemurs (*Lemur catta*) often alarm to harmless animals (Seyfarth & Cheney 1986), but fail to react appropriately to representational alarm calls given by adults (Seyfarth & Cheney 1980, 1986; cf. Macedonia 1991).

Finally, demographic data suggest that immature animals are significantly more vulnerable to predators than are adults (see also Cheney & Wrangham 1987). In nearly every demographic study of wild primates, infant death rates are higher than those of any other age class except extremely old individuals (e.g., Struhsaker 1973; Dittus 1977; Altmann 1980; Robinson 1986; but see Dunbar 1980). Infant mortality

Table 5.3. Estimated Rates of Predation on Immature (Infant and Juvenile) and Adult Age Classes Relative to Class Representation

Species, study site	Predation (%)		Group (%)		Relative predation rate immature/adult
	Immature	Adults	Immature	Adult	
Macaca fascicularis, Ketambe	75	25	50	50	3.00
M. sinica, Polonnaruwa	76	23	64	36	1.86
Cebus apella, Manu Park	70	30	40	60	3.50
C. olivaceus, Masaguaral	90	10	59	41	6.25
Papio cynocephalus, Amboseli	69	36	46	54	2.25
Cercopithecus aethiops, Samburu	91	9	37	63	17.22
C. aethiops, Amboseli	63	37	56	44	1.34

Note: The ratio of inferred predation on immatures versus adults, divided by the ratio of percentage immatures to adults in groups sampled. Data from Cheney and Wrangham (1987) (*Papio* cited as given even though they sum to > 100%) and associated review chapters, except for *Macaca fascicularis* (van Schaik et al. 1983a). Species were included if we had data for both percentage of predation events by age class and average group composition in same population.

has various causes independent of predation. However, estimated predation patterns show that infants and juveniles succumb to predation frequently relative to their abundance (Table 5.3). In one case, direct observation shows without doubt that young infants were the nearly exclusive target of predators (Boinski 1988). Also, patterns of juvenile mortality in relation to group size vary, depending on the presence of natural predators (van Schaik 1983). In populations without predators, where stronger food competition in larger groups should take its toll, juveniles in large groups survive less well than those in small groups. In contrast, when predators are present, juveniles in larger groups survive better than those in smaller groups, presumably because larger groups are safer.

CAUSES OF JUVENILE INCOMPETENCE: SIZE OR EXPERIENCE?

This review of the literature amply demonstrates that juveniles have a lower foraging efficiency and thus spend more time acquiring food than expected, and are also more vulnerable to predators. It is important to distinguish the relative importance and time course of the two major causes of juvenile incompetence: their smaller body size or strength (either through physiological constraints or directly) and their lack of experience.

Physiological Constraints

Physiological constraints (digestion, growth) may restrict the diet of juveniles, and thus make it more difficult to survive in an adult-determined niche. Juveniles of low foraging competence might prefer invertebrate foraging over fruit feeding if the rate of energy gain was greater for invertebrates than for fruit. In fact, however, invertebrate foraging usually yields substantially less energy per unit time than does feeding on fruit (Janson 1985a). Alternatively, juveniles may prefer more easily digested or protein-rich foods than do adults (Clutton-Brock 1977b). For instance, among frugivorous/insectivorous primates, juveniles devote a higher proportion of foraging time to insects and less to fruit than do adult females (Table 5.2), whereas their foraging success on invertebrates relative to adult females is only marginally better than that on fruit (Table 5.1). In one southeast Asian forest, larger fruits contain less protein (van Schaik & van Noordwijk 1986); larger fruits may also be more fibrous. The juvenile emphasis on invertebrate foraging is consistent with higher protein and digestibility requirements per unit body weight for juveniles than for adult females. These constraints, however, are directly or indirectly related to body size.

Body Size

A juvenile's ability to feed itself changes as it grows. Bigger animals require more food, but are also stronger, which may help in obtaining more food. We already noted that larger animals in a given species could take larger and harder fruits, and ate those fruits and leaves faster. Brown capuchins show a gradual increase in the diameter of branches searched for ant nests as they age (Janson & Boinski 1992). Also, in white-fronted capuchins, the age–sex classes vary in their techniques of opening *Astrocaryum* palm nuts and in their success at it. Juveniles open the nuts by banging them against a palm frond (success rate, 33%), adult females use one nut as an anvil against which to hit a second one (success rate, 43%), and adult males bite them open (success rate, 60%) (Terborgh 1983; Janson & Terborgh, unpublished data). This increasing success may be important to fitness, as *Astrocaryum* palm nuts are a crucial resource for white-fronted capuchins during times of food scarcity (Terborgh 1986).

Larger juveniles are probably less vulnerable to predators than are smaller individuals. Smaller species appear vulnerable to a broader range of predators than do large species; for example, small tamarins alarm at any of 10 hawk species larger than about 200 g, whereas spider monkeys alarm at only the two largest eagles (Terborgh 1983; Janson, unpublished data).

Effects of Experience

With age, juvenile foraging skills improve markedly. Although part of this change is due to changes in size, much is likely due to changes in experience. Experience can improve juvenile foraging skills in several ways: by direct observation, by trial and error, and by practice.

Direct Observation. Young primates are intensely curious about what their mothers eat: infants commonly sniff at their mothers' mouths while they feed and touch some of their foods (e.g., Nicolson & Demment 1982). This curiosity extends into the juvenile period, but becomes less obvious as juveniles spend more time at greater distances from their mothers. Juveniles may learn food habits from individuals other than the mother; in monogamous marmosets, juveniles are often provisioned directly by the male (Terborgh 1983), who will relinquish even the largest insect prey to a begging juvenile. In wild *Saguinus imperator,* one juvenile apparently learned to recognize a particular species of mushroom as edible by following the male (Janson, unpublished data). This species of mushroom (probably order Tremellales) grows only on bamboo stems and was eaten by only the group's male, but ignored by the female and older siblings. The juvenile frequently sniffed the mushroom while the male was eating it, and would go to the place on the bamboo stalk where the mushroom had been attached. In the following year, the former juvenile found and ate the bamboo mushrooms by itself, even though the adult male had moved to a new social group and other group members still ignored this protein food source.

Juveniles may learn from adults not only what foods are palatable, but also what foods are distasteful or even toxic. For instance, immature squirrel monkeys probably learn to avoid stinging caterpillars by specific adult warning calls (Boinski & Fragaszy 1989). Similarly, Glander (1981) found that only adult howler monkeys ever sampled (potentially toxic) leaves from previously unused trees. Although many examples of observational learning by juvenile primates exist (Cambefort 1981; Whitehead 1986; Hauser 1988; Pereira 1988a), Galef (1988) has emphasized the role of social facilitation rather than imitation (but see Hauser 1988).

Trial and Error. Field observers have noted frequent "play"-feeding by juvenile primates of items not consumed by adults (e.g., Whitehead 1986). Although juveniles almost never consume such fruits, this behavior may lead occasionally to the addition of new foods to the diet. With age, juveniles try new foods less often, although even adults occasionally sample novel foods (Watts 1985a; Whitehead 1986). Even

after juveniles learn what is edible, they must still learn what items are most profitable or nutritious (assuming that adult diets are dictated by some form of constrained optimization) (e.g., Altmann & Wagner 1978; Post 1984). Hauser (1987) has shown that infants start with little preference among the major foods used by adults, but juveniles come to prefer these foods in the same order as adults. Although the effects of age, size, experience, and social facilitation cannot be evaluated separately in these studies, the evidence is consistent with a gradual learning of adult foraging behaviors.

Practice. Juveniles may require extensive practice to become efficient at handling particular foods once they are an established part of the diet. For instance, young squirrel monkeys take several months to learn the various techniques used by adults when eating caterpillars: removal of stinging or irritating hairs by rubbing, picking up stinging caterpillars with a tuft of tail hair, or eviscerating the caterpillar and discarding the gut contents (Boinski & Fragaszy 1989). Juvenile brown capuchin monkeys will practice breaking open any dead twig they find, but as adults they are highly selective for tree species, size, and possibly age of branch (Janson, unpublished data). Juvenile chimpanzees require several years practice to perfect termite-fishing (Goodall 1986).

Predation avoidance is a skill that is usually best learned socially rather than by practice or trial and error. Mineka et al. (1984) showed experimentally that juvenile rhesus monkeys would not react fearfully to snakes unless they observed adults react fearfully to them.

Size or Experience?

The importance of experience in mastering foraging techniques is clear, but how much experience is really needed? In squirrel monkeys, juveniles apparently mastered adult foraging techniques well before they reached adult size (Boinski & Fragaszy 1989). Similarly, in brown capuchins, juveniles 2–3 years old had foraging success per substrate that was as high as that of adults (Fig. 5.1), yet they would not reach adult size for another 2 years. In vervet monkeys, juvenile foraging preferences were nearly adult-like and much different from those of infants (Hauser 1987). While yearling baboons have low probabilities of catching grasshoppers, 2 year olds are equally proficient as 3 year olds (S. Altmann, personal communication). All these examples suggest that improvements in foraging efficiency level off well before the end of the juvenile period.

Sex differences in foraging styles and choices among adults can already be observed among juveniles (Robinson 1986; van Noordwijk et al., Chapter 6, this volume). These sex differences can arise through observational learning, a possibility consistent with the tendency for juveniles to associate with like-sex adults (e.g., Dittus 1977; Robinson 1981; Pereira 1988a). However, sex differences in foraging technique can develop without close association between juveniles and adults (Boinski 1988; Boinski & Fragaszy 1989, Milton, Chapter 12, this volume). Thus delayed maturity is not likely a direct result of the need to acquire (sex-specific) competence in foraging.

Even if juveniles acquire sex-specific adult-like foraging skills before reaching adult size, their smaller body size may put them at a foraging disadvantage. Such a size effect is especially noticeable in species that use very large, tough food items or foraging substrates, in which juvenile substrate use is substantially restricted compared to adults (e.g., capuchin monkeys: Terborgh 1983). The size effect on foraging is reduced in species that use small or easily manipulated substrates. For instance, in Costa Rican squirrel monkeys, juveniles apparently mastered techniques for insect foraging only a few months after weaning, when they were still much smaller than adults (Boinski & Fragaszy 1989).

Body size may also be more important than experience in reducing juvenile predation. Even if a juvenile learns to recognize a predator, it is still vulnerable to its attack. For instance, juvenile vervets show essentially adult-like responses to alarm calls by 7 months of age (Seyfarth & Cheney 1986), but they remain vulnerable for several more years to a broader range of predators than adults (Cheney & Seyfarth 1981), presumably because of small body size. Although increased experience in recognizing and avoiding predators can reduce predation rates at a given body size, only large body size can reduce the minimum predation rate.

In sum, the need to improve ecological performance by experience probably does not explain rates of juvenile growth in primates, since both foraging and antipredator skills seem adult-like several years before adult size is attained. Increasing body size may lead to a greater ability of juveniles to feed themselves in a few primate species, but the ability to deter or survive predator attacks is likely to increase with body size in all but the very largest primates.

COPING WITH ADVERSITY: ECOLOGICAL RISK AVERSION

How do primates cope with the ecological difficulties of the juvenile period? We suggest that juvenile primates adopt a risk-averse policy: try to minimize the risk of death from either starvation or predation. To reduce the risk of starvation, juvenile primates appear to reduce their metabolic needs by suppressing growth rates. First, anthropoid primates have low juvenile growth rates compared with mammals of similar body size (Case 1978). Second, even within species, juvenile growth rates are suppressed. Growth rates generally decrease with increasing body size (e.g., Case 1978), yet juvenile growth rates in primates are often less than those of larger subadults (i.e., those species with a pronounced adolescent growth spurt) (e.g., van Wagenen & Catchpole 1956; Coelho 1985; van Schaik et al., in preparation). Third, juvenile primates of many species do not dramatically increase growth rate in response to increased food. For instance, van Schaik et al. (in preparation) compared long-tailed macaque growth in the wild and in captivity. Captive juveniles had growth rates (g d^{-1}) only 20–30% greater than wild juveniles, whereas captive adolescents doubled the growth rates of their wild counterparts (Fig. 5.4). Similarly, juvenile (1–6-year-old) female Japanese macaques grew only 30% faster when freely provisioned than when provisioning was severely curtailed (Mori 1979). By contrast, among baboons, both captive (Altmann & Alberts 1987) and provisioned wild (S. Altmann, personal communication) juvenile baboons grow much faster than wild ones (by 50–100%). Although this indicates substantial variability within the anthropoids, even the highest observed growth rates of these papionines are still lower than those of nonprimates (see Case 1978).

Juvenile primates appear to minimize mor-

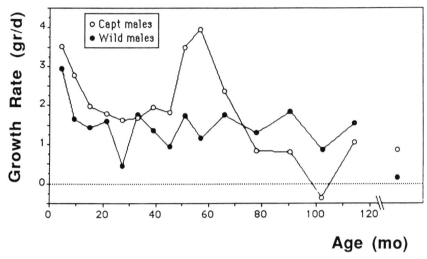

Fig. 5.4. Growth rates of captive and wild male long-tailed macaques, based on mixed longitudinal (captive) and predominantly cross-sectional data. Food supply is a major difference between the two conditions. (From van Schaik et al., in preparation)

tality due to predation by avoiding high predation risk. First, juvenile capuchin monkeys use exposed branches (where a raptor could strike more easily) less than do adults (Fig. 5.5). Second, the increased group cohesion in groups of long-tailed macaques after experimental exposure to a python was caused mainly by infants and juveniles (van Schaik & Mitrasetia 1990). Similarly, juvenile baboons associate with adult males most often when predation risk is high (Rhine 1975; Rasmussen 1983), and young juveniles form more cohesive cohorts than do old juveniles (Pereira 1988a). Third, juvenile long-tailed macaques prefer large foraging parties over small ones only when feline predators are present and regardless of average party size in their population (Fig. 5.6). It is especially interesting that the juvenile preference for large foraging parties was essentially identical in two Sumatran populations with predators (Ketambe, Mangrove), despite the fact that the mean party size at Ketambe was more than 12 individuals, while in the Mangrove site, it was fewer than 6, similar to the mean party size on Simeulue,

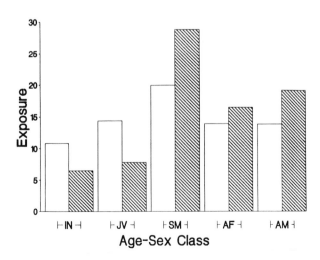

Fig. 5.5. Mean percentage of time spent by different age–sex classes on substrates open to approach by aerial predators (large diameter branches or at outside of tree crown). Based on scan samples of one group of brown capuchins (open bars) and one of white-fronted capuchins (single-hatched bars) in Manu National Park, Peru (van Schaik and van Noordwijk, unpublished data).

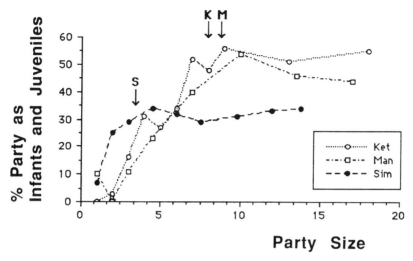

Fig. 5.6. Percentage of individuals in foraging parties of long-tailed macaques that were infants or juveniles (van Schaik & van Noordwijk 1985a, unpublished data). Arrows indicate party sizes at which stable representation is reached. Both Ketambe and Mangrove (near Madan) are on Sumatra, and have feline predators. The island of Simeulue lacks these. Number of parties sampled: Ketambe, >100; Mangrove, 51; Simeulue, 71.

which does not have predators. Thus the preference of juveniles for large parties on Sumatra is not just a reflection of larger overall party sizes there.

Juvenile primates often cannot minimize starvation and predation risk simultaneously: individuals are safest from predators in the center of a large group, where crowding increases food competition and reduces intake (Robinson 1981; van Schaik 1989; Janson 1990b). Although this trade-off potentially affects all group members, it is likely more of a dilemma for juveniles, which are more vulnerable to predation and less efficient as foragers than adults. Several lines of evidence suggest that juveniles sacrifice access to food in favor of greater protection from predators. First, they prefer larger foraging parties even when these ensure higher food competition (Fig. 5.6). Second, in large groups, they often form distinct subgroups that are often closer to the protected spatial center of the group than are mothers with infants (Rhine 1975; Robinson 1981; Janson 1990a). Third, the only study where potential food intake and predation risk (estimated by vigilance rates) were measured for juveniles in different spatial locations in a group concluded that juveniles prefer locations of minimum predation risk, not of maximum food intake (Janson 1990b). Within the constraint of avoiding predation, juveniles,

especially young juveniles, may choose adult neighbors that allow them relatively high access to food (Robinson 1981; Janson 1985a; Pereira 1988a; van Noordwijk & van Schaik 1988).

THE EVOLUTION OF LOW GROWTH RATES: AN ECOLOGICAL MODEL

Why do primates grow more slowly than other mammals of similar body size, even in the best conditions? The traditional explanation was that low growth rates are a nonadaptive result of a long developmental period, needed to learn how to deal with the complex social and ecological challenges of adulthood. We have emphasized the juvenile period as one of great ecological risk. Lack of foraging experience is not a major determinant of juvenile ecological competence, but even if protracted development were advantageous for social learning, a juvenile could as easily gain some experience while nearly adult in size with far less ecological risk. Many birds with polygynous mating systems reach adult size a few weeks after hatching, whereas sexual (social) maturity is delayed for more than a year (Pereira, Chapter 2, this volume).

Given this ecological vulnerability, one might expect that the juvenile period would be as short as possible (e.g., Williams 1966). Yet most juvenile primates grow far more slowly than they are physiologically capable of; witness the adolescent growth spurt of males (and sometimes females). If increased growth rates necessarily increase ecological risk, then low growth rates can be an optimal strategy, even if maturation must be delayed (Case 1978; Ricklefs 1984). In a nongrowing population, the timing of sexual maturation and consequent reproduction is adaptively neutral except as it affects juvenile survival (Cole 1954). Because primates are large, long-lived animals, their populations (in nature) should rarely deviate much from a long-term average zero population growth rate (even if they are not usually at demographic equilibrium) (Dunbar 1987). Thus natural selection may favor low juvenile growth rates and a delay in reaching adult size *if such delays increase a juvenile's survival to maturity*.

The Basic Model

To explore this possibility we present a mathematical model. We shall assume that mortality rates are an increasing function of the growth constant—the fraction or multiple of an individual's metabolism devoted to growth (as opposed to maintenance). In particular, we will allow mortality to be an exponential function of the growth constant. Using juvenile survival as the measure of fitness, we will then show that natural selection will favor a maximal growth constant if mortality increases linearly or less than linearly with the growth constant. If mortality increases more rapidly than linearly with the growth constant, then natural selection favors a single, nonmaximal value of the growth constant (see also Case 1978). This basic result is retained even if larger juveniles are assumed to be less vulnerable to predation.

There is an obvious advantage to fast growth: if some "target" adult size defines the end of juvenility, then the faster an animal grows, the less time it spends as a vulnerable juvenile. Specifically, the *growth rate* at time t, $g(t)$, equals $d[s(t)]/dt$, the change in size at time t. If size change occurs by directing a fixed fraction of an individual's total metabolism toward growth (see Charnov 1990), then growth rate can be written as an allometric function of size: $d[s(t)]/dt = ks(t)^a$, where k is the *growth constant* and a is the *growth exponent*, which we

shall assume is fixed by factors external to the model. When integrated, this equation yields

$$s(T)^{1-a} - s(0)^{1-a} = (1 - a)kT$$

or the time to maturity

$$T = [s(T)^{1-a} - s(0)^{1-a}]/k(1 - a)$$

Clearly, the greater the growth constant k, the smaller the time to maturity.

There are potential costs to a high growth rate as well. Individuals with a high fixed growth rate should be more vulnerable to starvation if food abundance falls, and feeding more or just faster may require being more exposed to predators (van Schaik 1989; Janson 1990b). In either case, we may expect the *death rate* at time t, $m(t)$ to be a function of the growth rate $g(t)$. Specifically, let $m(t)$ depend on the growth rate in the following way: $m(t) = c + bk^n$ (see Case 1978). In this form, c measures the *unavoidable juvenile mortality* (which is not affected by variation in the growth rate), and b, k, and n index avoidable mortality (adaptively adjusted through the growth rate). The reason we let $m(t)$ depend on k rather than on the full expression for $g(t)$ is that k represents the relative proportion of total metabolism devoted to growth *regardless of how metabolism scales to body size*. Thus if faster growth requires more foraging effort, this is likely to be reflected in k, rather than in the allometric scaling itself. As b, k, or n increases, the faster the death rate increases with the growth rate.

We can combine the time to maturity and the time-specific death rate into a single measure of fitness. Because we have fixed the size at maturity, we need not worry about the effects of an individual's size at maturation on its future fitness. Thus in a nongrowing population at equilibrium, the only important measure of juvenile fitness is survival to maturity. Survival to any age x can be expressed as $l_x = e^{-\int m(t)dt}$. For $x = T$ (age of maturity) and the functional forms given above, we have $l_T = \exp - \{\int (c + bk^n) dt\}$. We wish to know for what value(s) of the growth constant k is l_T maximal? Maximizing l_T is equivalent to maximizing

$$\log(l_T) = -\{\int (c + bk^n) dt\} = -T(c + bk^n) = -R(c + bk^n)/k$$

where $R = [s(T)^{1-a} - s(0)^{1-a}]/(1 - a)$. Thus $d[\log(l_T)]/dt = -R[bk^n(n - 1) - c]/k^2$. The latter expression equals zero only if $R = 0$ (impossible unless infants are born the size of adults) or $bk^n(n - 1) - c = 0$, which reduces to $k^* = [c/b(n - 1)]^{1/n}$. A quick check on the signs for larger or smaller k is sufficient to show that k^* is a maximum as long as $n > 1$. If $n \leqslant 1$, then fitness is always increased by increasing k.

Thus as long as the death rate increases allometrically with the growth rate with an exponent greater than one, some nonmaximal growth rate will provide the highest juvenile fitness (survival to maturity). The value of this growth rate, k^*, increases as sources of mortality independent of the growth rate (measured by c) increase relative to those that depend on the growth rate (indexed by b).

Before accepting the conclusion that natural selection may favor nonmaximal growth rates, it is important to consider whether the condition that mortality increases faster than linearly with growth rate is reasonable. Assume that a juvenile's chance of death equals the likelihood that the energy supply rate in the environment falls below its total metabolic rate (i.e., the juvenile starves). The close relationship between growth rate and mortality (Case 1978) suggests that organisms have only a limited flexibility in adjusting growth rate to food supply. Even if they could, the basal metabolic rate of an individual with a high potential growth rate is likely to be greater than in an individual with a low maximal growth rate, because basal metabolism is known to be linked to energetic capacity (Bennet 1978). If the distribution of energy supply rates in the environment is stochastic with a normal distribution, then the probability that it will fall below the total metabolic rate is just the integral of that normal distribution from 0 (or minus infinity) up to the total metabolic rate. This integral increases faster than linearly with increases in the upper limit of integration (total metabolic rate) (Fig. 5.7). Thus these simple assumptions are sufficient to make mortality increase more rapidly than linearly with the growth constant. Note that this conclusion still holds even if death is not certain to occur when the environmental supply rate falls below the juvenile's metabolic rate, but occurs only with some fixed probability under that condition. Even if mortality from starvation were to increase only linearly or less with growth rate, the increased predation risk that seems to be an inevitable correlate of increasing food acquisition could make mortality increase faster than linearly with growth rate.

Effects of Body Size and Experience on Mortality

This model may understate the benefits of growing fast because we assume that, in the absence of growth, the mortality rate, $m(t)$, is a constant c, independent of body size. This assumption is clearly unrealistic for predation rate and possibly for foraging competence. The greater the

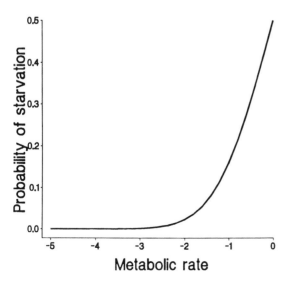

Fig. 5.7. Expected probability that a juvenile will starve as a function of its metabolic rate. Metabolic rate is expressed as the number of standard deviations below the mean energy supply rate in the juvenile's home range, assuming a normal distribution over time of energy supply rates.

survival costs of small body size, the more rapid growth rates ought to be.

To incorporate an ecological benefit of larger body size in the model, we let $m(t) = (c + bk^n) + d(e^{-qs(t)})$, where the first term in parentheses represents the previous equation for mortality, and the additional mortality term decreases exponentially with increasing body size. Although we can write down the fitness equation for juveniles with this new mortality function, we cannot solve explicitly for the maximum in the general case of the growth rate's being an allometric function of body size (as above). However, we can solve for the maximum explicitly in the special case of $a = 0$ (growth rate is a constant independent of body size). In this case,

$$\log(l_T) = -\int\{(c + bk^n) + d(e^{-q[s(0)+kt]})\}\, dt$$
$$= -(c + bk^n)R/k - dQ(1 - e^{-qR})/qk$$

where $Q = e^{-qs(0)}$ and $R = s(T) - s(0)$. Differentiating with respect to k yields

$$\{dQ(1 - e^{-qR})/q - R[bk^n(n - 1) - c]\}/k^2$$

Setting the numerator to zero and solving gives (for $n > 1$):

$$k^* = \{[c + dQ(1 - e^{-qR})/qR]/[b(n - 1)]\}^{1/n}$$

This solution differs only by a constant from the previous model, which lacked the ecological benefit to being large. Thus, as before, if $n \leq 1$, juvenile survival always increases with an increasing growth constant; but if the death rate increases with growth rate faster than linearly ($n > 1$), there will be a nonmaximal growth rate that maximizes fitness. In this case, the optimal growth rate will always be greater than in the previous model without the ecological benefit to larger size. However, this increased growth rate is due only to the fact that the unavoidable juvenile mortality is increased from an average of c to $c + [dQ(1 - e^{-qR})/qR]$.

Although we have concluded that experience does not play a major role in explaining long juvenility, we can examine whether incorporating an effect of experience on mortality affects the relationship between k and n. If we assume that mortality will decrease with time (because of increased competence due to learning) but is independent of body size, this change in mortality can be represented as fe^{-rt}. Thus, by anal-

ogy with preceding arguments, the log of survival can be written as $\log(l_T) = -\int\{c + bk^n + fe^{-rt}\}\, dt$. Integrating the right-hand side and differentiating with respect to k gives

$$R\{fe^{-rR/k} - [bk^n(n - 1) - c]\}/k^2$$

where $R = [s(T)^{1-a} - s(0)^{1-a}]/(1 - a)$. Setting this expression equal to 0, we obtain the formula

$$c - b(n - 1)Rk^n + fe^{-rR/k} = 0$$

Although solving for the optimal k^* is difficult in practice, inspection of this equation shows several facts. First, when mortality is exactly proportional to growth rate (i.e., $n = 1$), no positive value of k solves the equation, in fact, the optimal k is infinite. Thus the effect of decreased mortality through increased experience (as expressed in this model) is not sufficient to outweigh the benefits of a shorter juvenile period: growth rates are expected to be maximal. Second, if $n < 1$, the optimal k remains infinite. Third, if $n > 1$, there is a finite value of k that solves the equation, and this value maximizes survival. Thus, as before, nonmaximal growth rates are expected only when mortality increases faster than linearly with growth rate. As was the case for size-dependent mortality, the difference in the optimal k^* between this model and first one is due entirely to the difference in average unavoidable juvenile mortalities between the models.

Another possible effect of experience that is not included in the model, is on the individual's future reproductive success (see Charnov 1990). Including this factor would favor delayed maturity, but would not necessarily favor slow growth rates. Another limitation of this model is that we assume that the growth constant k is a constant rather than a parameter that could vary with time or size. Relaxing this assumption complicates the model, but the essential relation between juvenile growth rates and unavoidable mortality rates is unchanged (Janson, unpublished).

In conclusion, incorporating the effects of body size and experience on mortality does not change the basic result of the model: natural selection favors a single, nonmaximal value of the growth constant if mortality increases more than linearly with the growth constant.

DISCUSSION

The preceding model supports the notion that juvenile primates favor a slow-growth strategy that minimizes the risk of death by either starvation or predation prior to adulthood even though the juvenile period is longer than it needs to be physiologically. If this assessment is correct, several questions arise. First, what makes (anthropoid) primates different from most other mammals? Second, what are the consequences of a long developmental period? Third, how does this explanation for long juvenility compare with current life-history models?

Differences Between Primates and Other Animals

Primates have lower growth rates than other homeothermic animals, if body size is taken into account. The optimal growth rate increases with increases in the ratio of c/b, where c is the level of unavoidable mortality (independent of growth rate), and b is the parameter relating increases in growth rate to increased mortality. By examining causes of variation in c and b, we may gain a better understanding of the place of primates.

Unavoidable mortality, c, is determined by the temporal variability in the food supply and by the level of unavoidable predation. The social life of diurnal primates creates the opportunity for social protection of vulnerable immatures (e.g., Terborgh & Janson 1986; van Schaik 1989), which reduces actual predation rates even in the presence of predators, and so makes low growth rates possible. Thus within a given dietary "guild" (e.g., frugivore/insectivore), high relative growth rates should be found not only in small-bodied species, but also in those that are solitary, live in small social groups, or live in habitats in which even social vigilance has limited effectiveness (e.g., savannahs). The importance of baseline mortality levels is strongly suggested by the many aspects of mammalian life histories that are related to juvenile or adult mortality rates (e.g., Promislow & Harvey 1990).

Lower growth rates are also expected when b is large—that is, when increased growth rates lead to rapid increases in juvenile mortality. Such a rapid increase is expected when food availability to juveniles is low, so that even mini-

mal growth rates lead to a nonzero chance of starvation (rates toward the right-hand side of Fig. 5.7). Conversely, if food availability is high, even fairly high growth rates may lead to imperceptible increases in mortality (rates toward the left-hand side of Fig. 5.7). Our review clearly indicated that juvenile primates, being less experienced and less efficient foragers, are vulnerable to temporary reductions in food supply. Their ecological problems are arguably more severe than those faced by many other immature organisms. In many animals, immatures occupy a niche that is distinct from that of adults because they remain in nests until nearly adult size (many mammals and birds), they possess distinct adaptations for food acquisition (many arthropods), or they grow up without parental care and thus without adult constraints (most fish, amphibians, and reptiles). The life-style of most primates (arboreal, wide-ranging, and social) precludes such isolation between the juvenile and adult ecological niches. Thus a young primate is forced to fit partially into an adult ecological niche when it is much smaller and less experienced than an adult. Moreover, living in cohesive groups exacerbates food competition and may reduce effective food availability. Thus various arguments suggest that food availability may be relatively lower for primates than for many other animals.

In all, the typical combination of primate traits, group-living and effective communal vigilance, and the mixing of adults and young in the same group, arguably reduces c/b and explains primates' slow growth. Further predictions can be developed. Primates with distinct juvenile niches (such as nest-dwelling prosimians) should have higher growth rates than predicted for primates of their body mass. Conversely, other mammal species in which juveniles are constrained to conform to the adult niche (e.g., many mobile social carnivores and ungulates) may be expected to have growth rates lower than would similar species with a distinct juvenile niche (assuming overall predation rates are the same).

Growth and Learning: Cause and Effect Revisited

We have presented evidence that adult-like ecological skills are gained well before the end of the juvenile period. However, it could be argued that the relevant constraint is not ecological experience, but social experience. Testing this possibility is outside the scope of this chapter, but is perhaps irrelevant. Even if learning complex social behaviors requires a long developmental period, it does not of itself require a low growth rate, as a juvenile could accelerate growth (but not learning) and then continue to mature behaviorally at a large body size. One might also argue that social competition against adult-sized individuals provides an advantage to remaining small. However, small individuals are also less able to defend themselves against such aggression. For instance, in a study of rank acquisition among female macaques, Datta (1988) found a strong effect of a female's kin support and body size relative to her opponent: larger females were more successful (see also Pereira, Chapter 20, this volume).

If low juvenile growth rates are advantageous, they force long periods before sexual maturity (assuming fixed size at maturity). Although behavioral development could plausibly be fast even though growth is slow, there is no plausible ecological or social factor that would favor a halt in learning or other behavioral development prior to sexual maturity. Thus low growth rates may favor prolonged development by forcing delayed sexual maturation. Given that the effect of delayed maturity on growth rate is merely permissive, while that of growth rate on time to adult size is unavoidable, we prefer the latter interpretation of the relationship between these two variables. However, there is still a third possibility. If the length of the juvenile period is set not by the time needed to reach adult size (as assumed in our model), but by other (say, social) requirements, and high growth rates carry increasing risks of starvation, then natural selection will favor the lowest juvenile growth rate consistent with reaching adult size by the end of the required juvenile period. Distinguishing this alternative from the possibility that low growth rates force delayed maturity will be exceedingly difficult, however.

What other correlated responses to low growth rates should we expect? Protracted learning periods could allow for increased behavioral flexibility and complexity. Such behavioral adaptability would allow more refined social competition and cooperation, thus favoring more complex social systems. Complex social

systems, in turn, appear to favor individuals with larger brains (Clutton-Brock & Harvey 1980). Individuals in such systems might in turn more effectively avoid predation or periods of low food supply, thus setting in motion a selective positive-feedback loop among social complexity, large brain size, and delayed maturity.

Slow Juvenile Growth and Life History

Mammalian life histories vary widely, but since, interestingly, most of the life-history variables are intercorrelated, one can array life histories on a one-dimensional fast–slow continuum: from early-breeding, high-reproductive effort, and early death to late age at first breeding, relatively low reproductive effort, and high maximum life span (Harvey et al. 1989; Charnov 1991). Much of this variability is accounted for by body size, but species of the same size can still vary appreciably in "speed of life." Primate life histories clearly are at the slow end of the continuum (Harvey et al. 1987). Life-history theory thus must answer this crucial question: Why do some species, especially primates, delay reproduction (Promislow & Harvey 1990)?

Theoretical explanations for the slow–fast variation are numerous. One model posits that delayed development is needed to allow for the growth of a large brain, which may be adaptive for either socially or ecologically complex environments (Clutton-Brock & Harvey 1980; Harvey et al. 1987). Within primates, delayed maturity is significantly correlated with adult brain size when the effects of body size are removed (Harvey et al. 1987). If this relationship is causal, then it explains the relatively prolonged juvenility in primates as a consequence of their larger brains compared with those of other mammals (see Martin 1981). Under this "brain growth constraint" model, large adult brain size is adaptive, but has the nonadaptive side effect of a prolonged juvenile period, which, in turn, facilitates extensive learning (Pereira & Altmann 1985). Due to the very high correlations between brain size and body size, it is very difficult to test this model.

Other models assume that interspecific variation in reproductive rates, and thus in other life-history traits, is determined by ecological conditions, either indirectly through the ecological impact on metabolic rates or more directly through the effects of mean population density relative to carrying capacity. The empirical evidence, however, for these models is equivocal, and the theoretical underpinning for the second is dubious (review in Harvey et al. 1989).

A very different set of models bases its predictions for life history on demographic considerations. In particular, optimal life history depends (in a complex interactive fashion) on age-specific mortality rates (Cole 1954; Williams 1966). This hypothesis provides a very good fit to life-history variation among mammals (Promislow & Harvey 1990) and birds (Saether 1988). Variation in primate life histories (Ross 1988) can also be interpreted in this way (Promislow & Harvey 1990). Indeed, since mortality rates tend to scale with body size, this hypothesis may also explain the pervasive body size effect on life history (Pagel & Harvey, Chapter 3, this volume).

The strong association between observed mortality rates and life-history characteristics automatically leads to the next question: What factors determine age-specific mortality schedules? Various possible explanations are currently being considered. One possibility is that an animal facing high overall mortality must mature very fast to maximize the probability that it will survive to reproduce at all (Pagel & Harvey, Chapter 3, this volume). Physiological explanations are also being examined. Currently, the main contender is some form of "optimal life-history" model that assumes that a long juvenile period is adaptive if it improves reproductive value more (through increased breeding success later on) than it decreases it (through juvenile mortality) (Charnov 1990; see also Pagel & Harvey, Chapter 3, this volume). For this to work, we need low juvenile mortality and decreasing costs of reproduction with increasing age. This age effect is often observed among mammals (Promislow & Harvey 1990), but the underlying cause is not really clear. Experience and social position (territory ownership, dominance) are possibilities (cf. Rubenstein, Chapter 4, this volume).

Our model attempted to provide an explanation for the slow growth rate of primate juveniles, which seems to be dictated by the aversion of starvation and predation. The conditions in which slow growth may be favored by selection were found to be low unavoidable juvenile mortality and bad prospects for increased food intake. We concentrated on growth, but slow growth implies slow development and, relative to other animals of the same adult body size, delayed breeding and long life span (see Charnov 1991). Thus this "juvenile risks" model may be relevant to the debate on life-history evolution. Specifically, it suggests that the selection on immature growth rate by unavoidable mortality or temporal variation in food supply determines the length of the juvenile period and, through the as yet incompletely understood but universal correlation with other life-history traits, drives life-history evolution, at least in some taxa.

Is it possible to develop tests that distinguish between the three remaining most plausible models: the "brain growth constraint" model, the "optimal life-history" model, and the "juvenile risks" model? The models differ in what they view as causal variables, and in what particular conditions are needed to produce the observed correlations (Table 5.4). The "brain growth constraint" model sees adult brain size as the primary predictor of delayed maturity. Among primates, age at maturity is significantly related to adult brain size, even after adjusting for the effect of body size. However, since brain size and body size are measured with different degrees of accuracy, this correlation must be interpreted with caution. Furthermore, the data of Harvey et al. (1987) also show that age at maturity is better correlated with neonatal brain size than with adult brain size, which suggests that delayed maturity is not directly linked to the time needed to grow a large brain.

The "optimal life-history" hypothesis sees low adult mortality as the major predictor for delayed maturity. Any correlation between delayed maturity and brain size can hold only if the two are linked to low adult mortality. Finally, the "juvenile risks" hypothesis requires that delayed maturity is a consequence of slow juvenile growth, fostered by the rapid increase in mortality of more rapidly growing individuals and a low juvenile predation rate. To explain the observed correlation between delayed maturation and brain size, the model must assume that the risk of juvenile death influences both these variables.

The last two hypotheses make contrasting predictions about how diet affects mortality rates. Folivorous rodents have relatively fast life histories (Mace 1979), and folivorous primates show a tendency in the same direction. The "op-

Table 5.4. Comparison of Causal Variables and Predictions of Various Hypotheses Predicting Delayed Adulthood in Primates

	"Brain growth constraint"	"Optimal life history"	"Juvenile risks"
Causal variable	Slow growth of large brains	Low adult mortality	High mortality risks for juveniles
Slow growth rate predicted?	Yes	No	Yes
Growth rate affected			
By feeding ecology?	No	No	Yes
By juvenile mortality?	No	Yes	Yes

Note: Predicted effects assume that other variables are held constant.

timal life-history" model requires that rapid maturation of folivores, relative to nonfolivores, be associated with high adult mortality, because of higher variation in food abundance. According to the "juvenile risks" hypothesis, by contrast, folivores can grow more rapidly: increased growth rates are less risky for folivores if their food supply is less seasonal than that of frugivores and insectivores. Translation of phenological data into measures of potential critical shortages has not yet been attempted. However, reproduction in folivorous primates is usually less seasonal than in nonfolivores (C. Hemingway, unpublished). This observation supports the "juvenile risks" model, if it is interpreted as indicative of a less seasonal food supply. However, it can also be interpreted as merely reflecting the point that folivores have faster life histories. Problems such as this illustrate the difficulty of testing between these ideas, and their proper evaluation awaits the development of more refined tests.

ACKNOWLEDGMENTS

This is contribution 836 from the Graduate Program in Ecology and Evolution at the State University of New York at Stony Brook. We thank S. Altmann and M. Pereira for extensive comments.

Part II

GROWING INTO DIFFERENT WORLDS

This part of the volume highlights the diversity of primate life histories and modes of social organization. In the best known species, females reside in their natal groups for life, whereas males disperse around puberty and transfer to other groups (e.g., van Noordwijk et al., Chapter 6). But there are numerous variations on just the theme of sex-biased dispersal. In the population of Hanuman langurs studied by Rajpurohit and Sommer (Chapter 7), for example, males are often forced to disperse as juveniles, some as young as 18 months of age. In the South American muriqui (Strier, Chapter 10), males remain in the natal group and females disperse at puberty. In red howler monkeys, some females remain to breed in the natal group, but others depart as juveniles to join other groups or establish new ones (Crockett & Pope, Chapter 8). In the nongregarious primates (Nash, Chapter 9), both males and females leave their mother and spend most of their lives ranging alone, but males often disperse farther than females. Social organization in the great apes (Watts & Pusey, Chapter 11) includes female dispersal from harem-type groups in mountain gorillas and from fission–fusion communities in common chimpanzees.

Primates born into different "worlds" of social organization can be expected to exhibit diverse behavioral adaptations in development. The chapters in this section attempt to relate juvenile behavior and emerging sex differences to the socioecology and adult reproductive tactics of the species. They are generally concerned with functional aspects of behavior: How do juvenile behavioral adaptations help developing individuals survive the juvenile period and mature into competent adults?

In many ways, these chapters are as striking for the commonalities as for the differences they reveal among the natural histories of juvenile primates. Regardless of major differences in phylogeny and social organization, for example, young juveniles typically maintain close social bonds with their mothers. Also, when a likely father is available, youngsters usually spend much time associating with him (van Noordwijk et al.; Nash; Strier; Watts & Pusey). Van Noordwijk et al. explain the tendency of young juveniles to seek the company of tolerant adults in terms of the need for safety. The juveniles of all species face the common problem of surviving until adulthood because they are typically more vulnerable than adults to predators, accidents, and potential malnourishment. Thus young juveniles of all kinds reside in their natal group and associate with adult kin.

Under some circumstances, however, adults coerce juveniles to adopt hazardous patterns of life history. Given the opportunity, for example, a juvenile male Hanuman langur will remain in his natal group at least until puberty (Rajpurohit & Sommer). But to avoid injury or even death from the aggression of a new adult male, he may be forced to disperse long before then. Crockett and Pope describe a complementary pattern for juvenile female red howler monkeys. Low adult mortality generates competition for female breeding opportunities in their study population, leading some females to leave the safety of their natal group as early as 2 years of age. Mortality rates among such young emigrants are extremely high. Rajpurohit and Sommer report that half of all juvenile male Hanuman langurs die within 6 months of emigration, and Crockett and Pope present similarly high estimates for emigrant juvenile female red howlers.

As juveniles mature, they increasingly show sex differences in behavior that reflect those of the adults of their species. For example, sex differences in foraging behavior characteristic of adults are evident among older juvenile long-tailed macaques (van Noordwijk et al.). Subadult muriquis show the sex-typical patterns of social association characteristic of their patrilocal system, with strong bonds among males and far weaker ones among females (Strier). Old juvenile galagos exhibit sexual behavior in accord with male and female roles (Nash). In chimpanzees, where adult males

often travel together, juvenile and adolescent males increase their association with adult males with age. By contrast, in mountain gorillas, where groups are dominated by one adult male, juvenile male attraction to adult males declines with age (Watts & Pusey). By the end of the juvenile period, adult-like sex and species differences in behavior are generally well developed.

Species members who remain in their natal groups to reproduce have opportunities to form lasting bonds with familiar companions and, especially, to exploit kin support during the transition to adulthood. Emigrants face the challenges of adjusting to unfamiliar areas and social partners and to enhanced conflict with conspecifics. Crockett and Pope (Chapter 8) show that one factor that can contribute to increased risk of mortality for emigrants is dietary deficiency. The emigrant female red howlers also exhibited delayed reproduction in comparison with nonemigrants.

Dispersing individuals often make the best of a difficult situation by associating with kin or other familiar individuals. Whenever possible, juvenile male Hanuman langurs follow their fathers and brothers into all-male groups (Rajpurohit & Sommer). Also, maturing male red howler monkeys form coalitions with their fathers and brothers to defend the natal group or invade neighboring groups (Crockett & Pope). In many of the nongregarious species, dispersing daughters establish territories close to their mothers' (Nash). Overall, maturing individuals able to maintain social ties with kin may best be able to mitigate the costs of emigration and the transition to adulthood.

6

Spatial Position and Behavioral Sex Differences in Juvenile Long-Tailed Macaques

MARIA A. VAN NOORDWIJK, CHARLOTTE K. HEMELRIJK,
LOUIS A. M. HERREMANS, and ELISABETH H. M. STERCK

The most important achievements for a juvenile primate are to survive until adulthood and to develop the behavioral skills required for reproductive adult life. Survival depends on ecological skills, such as avoiding predation and acquiring sufficient food. Juveniles are vulnerable to predation because of both their inexperience and their small size, which leads to a larger number of potential predators. It has been shown that larger groups of primates detect predators at larger distances (van Schaik et al. 1983b), so being present in a large group should provide a juvenile better protection than being in a small group. Few field studies have concentrated on juvenile behavioral ecology, but the available data show that juveniles in several group-living diurnal primates tend to remain in the center of a social group within short distances of other group members (Robinson 1981; van Schaik & van Noordwijk 1986; Pereira 1988a). Juveniles can thus be characterized as risk averse (see Janson and van Schaik, Chapter 5, this volume).

Because a juvenile's survival is in the direct interest of the mother, she could be expected to help her offspring remain in safe spatial positions. However, occupying the center of a large group also entails the disadvantage of increased food competition (Waser 1977a; van Schaik et al. 1983a; Watts 1985b). Providing social protection, high-ranking mothers may be able to provide simultaneously a safe environment and

access to high-quality food patches to their independently moving offspring. The effectiveness of this kind of maternal protection can rarely be assessed in the wild because mortality estimates require large datasets (but see Altmann et al. 1988; Cheney et al. 1988). However, if maternal protection increases a juvenile's chances of survival, we should expect dominance rank to influence both an immature's spatial position in the group and its activity budget.

For juvenile primates, sex differences in social behavior are well documented for many species (reviews in Jolly 1985; Pereira & Altmann 1985; Walters 1987a). These sex differences might be enhanced by differential patterns of social interaction with older conspecifics that reinforce a juvenile's own predispositions. Among adult primates, females and males are known to differ not only in their social relationships (Walters 1987a; Walters & Seyfarth 1987), but also in their ecology (e.g., Terborgh 1983; Fragaszy 1986; Robinson 1986; van Schaik & van Noordwijk 1986, 1989). In several primate species, adult females and males seem to specialize on different food sources requiring different manipulative skills or amount of foraging time. An important question is whether juveniles show sex differences in maintenance activities that reflect their later adult activity profiles, and how and when such sex differences arise.

This study focused on juveniles in a natural

population of long-tailed macaques (*Macaca fascicularis*). In the study population, about 20% of the infants die before they are 1 year old, and of the remaining juveniles another 15–20% do not survive to 4 years of age (van Noordwijk & van Schaik 1987, unpublished data). Although the cause of mortality is often unknown, predation cannot be ignored as a factor in survival. The most important diurnal predators in the study area are found on the ground and in the lower canopy (clouded leopard, tiger, other cats, python). Thus to avoid predation, an immature juvenile should not only be close to other group members, but also avoid the lower canopy layers. Earlier observations revealed that compared with adults, juveniles as a class are more often present in the (safer) main party of the group, tend to have more neighbors, and spend more time foraging and less time resting and eating fruits (van Schaik & van Noordwijk 1986).

This pilot field study focused on variation within the juvenile class. As noted above, we anticipated that maternal dominance status along with juvenile age and sex might be important determinants of juvenile behavioral ecology. Unraveling their relative importance may help us to understand better the major socioecological challenges of primate development.

STUDY SITE AND METHODS

Observations were made in two groups of long-tailed macaques (Table 6.1) living in their natural habitat around the Ketambe Research Station, Gunung Leuser National Park, northern Sumatra, Indonesia (described in van Schaik & Mirmanto 1985). This population has been under observation almost continuously since the end of 1979, and individuals and their relationships are known for several groups (van Noordwijk & van Schaik 1985, 1987, 1988).

In this study we included individuals between 20 weeks and 6 years old (Fig. 6.1). We considered animals as juveniles from the age of 20 weeks, because of their high degree of loco-

Table 6.1. Group Compositions, Including Identities, Ages, and Maternal Dominance Ranks of Focal Subjects

Timsel (1982/1983)			House (1982)			House (1986)		
Focal individuals	Age (years)	Rank	Focal individuals	Age (years)	Rank	Focal individuals	Age (years)	Rank
Females			Males			Females		
Gitta	5	8	Waf	6	1	Trutje	0.9	1
Gerdien	4	1	Saudara	6	6	Grietje	1.2	4
Gombak	3.5	2	Schele	5	4	Doris	1.2	7
Lijs	3	1	Thijs	4	1	Els	1.2	10
Sugar	3	5	Ron	4	11	Males		
Bentik	3	7	Toon	3	1	Karel	2.2	6
Males			Gerard	3	3	Tonnis	2.0	1
Punt	5.5	1	Klaas	3	4	Brio	1.2	3
Bengkok	4.5	2	JP	3	11	Obelix	1.1	9
Klovis	3.5	3	Dirk	2	7	Tarzan	0.6	2
Snowy	3	10	Goof	1	3	Quandi	0.4	8
			Erik	1	10	Wolf	0.4	5

Other group members	Timsel (1982/1983) (N)	House (1982) (N)	House (1986) (N)
(Sub)adult males	8	6	8
Adult females	9	11	13
Juvenile 3–5 years male	Above	Above	6
Juvenile 3–5 years female	Above	0	5
Juvenile 1–2 years male	3	Above + 3	Above
Juvenile 1–2 years female	4	3	Above
Juvenile <1 year male	1	3	Above + 2
Juvenile <1 year female	0	2	3
Total	35	40	48

*a*During middle observation period.

Fig. 6.1. Juvenile long-tailed macaques: 1, 2, and 3 years old.

motor independence: after 20 weeks of age, clinging by immatures to their mothers has decreased to 0% during maternal foraging and about 25% during maternal walking (Karssemeijer et al. 1990). One-year-old juveniles still clung to their mothers occasionally during travel over long distances. Although 20-week-old immatures may still depend appreciably on their mothers for nutrition, they also feed themselves considerably. Our sample included three males that met the age criterion only during the last two observation periods (see below). The oldest individuals sampled were two males about 6 years old (just prior to natal emigration)

and a female approximately 5 years old that had not yet conceived her first offspring.

Focal animal samples of juveniles were made in two periods by different observers: C.K.H. studied 12 1- to 6-year-old juvenile males in House group from August to November 1982, and 10 old juveniles (3–5 years: 4 males and 6 females) in Timsel group from December 1982 through January 1983 (Table 6.1). L.A.M.H. and E.H.M.S. studied 11 young juveniles (0.5–2.5 years: 7 males and 4 females) in House group from July to December 1986. Because most births were confined to the second half of the year (van Schaik & van Noordwijk 1985b), it was easy to assign some old juveniles with unknown birthdates to their age class. These juveniles were also assigned to a mother on the basis of agonistic aiding and grooming (see Walters 1981) observed during several observation periods. Juvenile dominance ranks within age–sex classes corresponded to those of their assigned mothers. (In general, dominance relations closely resembled those found in captivity [Angst 1975; de Waal 1977] and among other macaques [Kawai 1958; review in Walters 1987a], except that males tended to outrank female age peers).

In the study area, long-tailed macaques are highly arboreal and inhabit dense riverine forest. Observation conditions are variable, making it difficult to follow small individuals for long periods of time. For the old juveniles (sampled in 1982/1983), focal animal samples lasted 4–8 minutes and were collected between 6.30 and 12.00 hours. About 9.5 hours of sample were collected for each subject. For the young juveniles (1986), focal samples lasted 2–15 minutes, and data were collected between 6.30 and 13.00 hours evenly spread within three 6-week periods (about 3.3 hours/individual/period). Consecutive focal samples on the same individual were separated by at least 30 minutes.

Focal animal samples were collected in the same way as those on adults in the same population (van Noordwijk & van Schaik 1987, 1988): at 1-minute intervals, activity, height (reliable data were obtained only during part of the 1986 sample), and the presence and identity of neighbors within 5 m were recorded. Activities included locomotion, rest, stand, feed on a clumped food source (i.e., mainly fruit), forage on dispersed food sources, groom, and other social activities (almost exclusively play). Study

groups often split up during the day into two or more parties. So, in addition, presence or absence in the main party was recorded per hour for all focal subjects.

The morning was divided into three time blocks, because activities were not evenly spaced over the observation hours. To obtain an estimate of the time budget of whole mornings, we averaged the percentages of each activity per time block. For the young juvenile sample, each 6-week period was treated as a separate estimate, and the average values of these were used in the analyses.

Because the data were collected in different years, in different seasons, in different groups, and by different observers using slightly different methods, analyses were done separately for each group and observation period. Age and rank effects were analyzed for all three groups. Sex differences, however, could be analyzed only among the old juveniles of Timsel group and among the young juveniles of House group (1986). Throughout, two-tailed nonparametric tests were used following Sokal and Rohlf (1981).

ACTIVITIES

Age and Rank Effects

In the young juvenile sample, percentage of time resting decreased, while time spent feeding on clumped food increased with age (Table 6.2). Also, the youngest males spent more time playing than did the older young juvenile males.

In the older juvenile sample (1982/1983), the oldest males spent more time feeding on fruits than did the younger ones (Table 6.2), but females showed the opposite age effect. Both female and male juveniles decreased the time

Table 6.2. Kendall Correlations Between Activities and Age

	House (1986)[a]		Timsel (1982/1983)		House (1982)	
Activity	Total ($n = 11$)	Males ($n = 7$)	Females ($n = 6$)	Males ($n = 4$)	Males ($n = 12$)	Fisher test
Rest + groom[c]	−0.64**	−0.62	+0.75*	+1.00*	+0.26	
Feed	+0.61**	+0.62	−0.89**	+0.67	+0.47**	**
Forage	+0.17	+0.55	0.0	−0.33	+0.44**	
Locomotion + stand	+0.17	+0.24	+0.45	−0.67	−0.30	
Play	−0.36	−0.71*	−0.30	−1.0*	−0.47**	**

Note: For the two data sets of old male juveniles (Timsel and House [1982]) a Fisher combination test was done. *p < 0.05; **p < 0.01.
[a]No separate correlations for the young juvenile females in House (1986) because they were all of the same age.
[b]Including cling for the young juveniles.

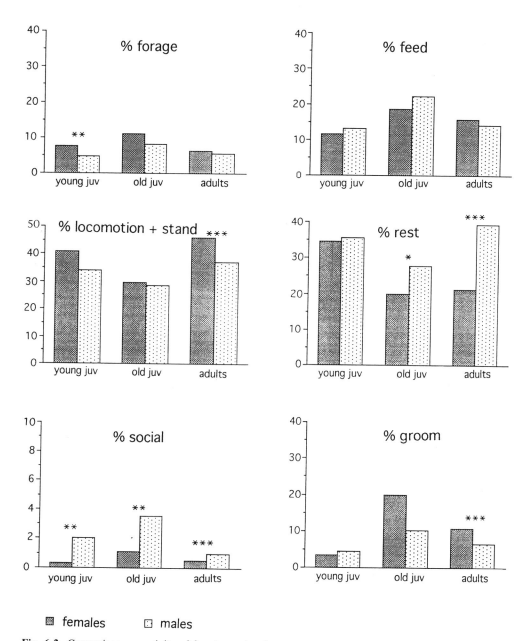

Fig. 6.2. Comparison per activity of females and males among young juveniles, among old juveniles, and among adults (van Schaik & van Noordwijk, unpublished data). Significant sex differences within an age-class are indicated by **$p \leq 0.01$ and ***$p \leq 0.001$. Because samples for different groups were gathered during different studies (see "Study Site and Methods"), comparisons can be made only within age classes; see van Schaik and van Noordwijk (1986) for a broad comparison of all age–sex classes.

spent playing. The oldest juveniles tended to rest more often than younger juveniles in the same group.

Maternal dominance rank hardly affected the time budgets of young or old juveniles.

Sex Differences

Even among the young juveniles, males and females showed some differences in activity budgets that mirrored adult sex differences (Fig. 6.2). Most strikingly, females as young as 60 weeks already spent more time foraging for dispersed food (mostly high-protein food such as insects, larvae, and young liana shoots) than their male peers (on average, 7.6 vs. 4.8% of time; Mann–Whitney $U = 28$, $n,m = 4,7$, $p < 0.01$). Between successive foraging bouts, an individual walks slowly and stands while scanning for another potential food source. So a high percentage of time devoted to foraging will be accompanied by a high percentage of time standing and walking. Indeed, like adult females, the young juvenile females tended to stand and move more time than did young juvenile males (percentage locomotion + stand: $U = 24$, $p < 0.10$). Data gathered from old juveniles showed nonsignificant trends in the same direction (Fig. 6.2). The young juvenile males tended to spend more time feeding on clumped food such as fruit than their female peers ($U = 15$, $n,m = 4,4$, $p < 0.10$, excluding the three youngest males because of the significant age effect). The old juveniles showed a weak trend in the same direction. Like adults, old juvenile males spent more time resting and they tended to spend less time grooming than their female peers (Fig. 6.2).

On average, the four young juvenile females spent 0.3% of their time playing, whereas their seven male peers averaged 2% ($U = 26.5$, $n,m = 4,7$, $p < 0.05$). In another season and social group, old juvenile females spent on average 1.1% playing vs 3.5% by their male peers ($U = 24$, $n,m = 6,4$, $p < 0.01$).

SOCIOSPATIAL POSITION

Height

The danger of predation probably is highest below 10 m height in the trees (van Schaik et al. 1983b). The youngest, more vulnerable, juve-

niles were more often above 10 m than older young juveniles ($\tau = -0.491$, $n = 11$, $p < 0.05$). No rank effect or sex difference was apparent among young juveniles. No reliable height data could be collected for the old juveniles in 1982/1983.

Presence in the Main Party

The size and composition of the main party are strongly influenced by social and seasonal/ environmental factors (van Schaik & van Noordwijk 1986). When the data on young juveniles were collected, House group was larger and clearly less cohesive than when most of the adult and old juvenile data were collected.

Within samples, younger juveniles were more often present in the main party than older ones (young juveniles: $\tau = -0.495$, $n = 11$, $p < 0.05$; old juveniles [House 1982/1983] excluding the two youngest: $\tau = -0.662$, $n = 10$, $p < 0.01$). Among maternal brothers the older ones also were less often present in the main party (all seven brother pairs in House group and in the only brother pair in Timsel). Younger juveniles, it seems, did not yet often risk leaving the main party. Although presence in the main party is correlated with dominance rank for adult females (van Noordwijk & van Schaik 1987), there was no significant effect of maternal rank on the presence of young or old juveniles. Neither was there a sex difference in presence.

Neighbors

Within the young juvenile sample, the youngest subjects were more likely to have at least one neighbor ($\tau = 0.642$, $n = 11$, $p < 0.01$). The presence of the mother within 5 m distance decreased with age among young juveniles ($\tau = -0.532$, $p < 0.05$), but she remained a frequent neighbor even for old juveniles (Fig. 6.3).

Young juvenile males were more likely than were their female peers to have other juveniles, especially males, as neighbors (Mann–Whitney tests: all juvenile neighbors, Fig. 6.3: $U = 27$, $n,m = 7,4$, $p < 0.05$; juvenile male neighbors: $U = 28$, $n,m = 7,4$, $p < 0.01$). Old juveniles appeared to have juvenile neighbors less often than young ones (see also van Schaik & van Noordwijk 1986). Old males more often had juvenile male neighbors than did old females ($U = $

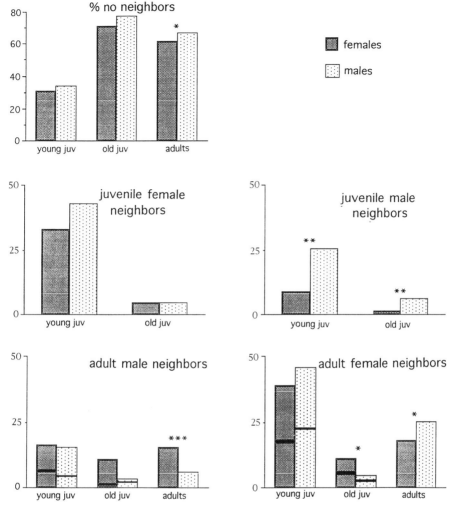

Fig. 6.3. Neighbors: comparison of sexes among young juveniles, among old juveniles, and among adults. Inside the bars for male and female neighbors, the contribution of the leader male and mother is presented under the line. $*p \leq 0.05$; $***p \leq 0.001$.

24, $n,m = 6,4$, $p < 0.01$), who more often had adult female neighbors (Fig. 6.3).

Young juveniles from high-ranking families more frequently had an adult female as a neighbor (including mother: $\tau = -0.727$, $n = 8$, $p < 0.01$; excluding the three youngest ones) and in total more often an adult neighbor ($\tau = -0.509$, $n = 8$, $p < 0.05$). In addition, the leader male (i.e., male dominant when the youngest cohort of infants was conceived: see van Noordwijk & van Schaik 1988) was more often a neighbor for the high-ranking young juveniles (Fig. 6.4; $\tau =$

-0.532, $n = 11$, $p < 0.05$). However, the frequency of having no neighbor at all did not correlate with dominance rank.

DISCUSSION

Juveniles and Safety

The juvenile period constitutes only a part of an individual's potential life span. Since survival probably has the highest priority for a juvenile (Janson & van Schaik, Chapter 5, this vol-

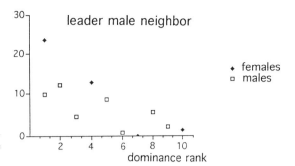

Fig. 6.4. Maternal dominance rank and the frequency with which a young juvenile had the leader male as a neighbor.

ume), we expected juveniles to remain in relatively safe spatial positions. Indeed, younger juveniles avoided the lowest canopy layers, where predation risk is presumably highest (van Schaik et al. 1983b); they spent more time than older juveniles in their group's safer main party, and more often had at least one neighbor.

In the field, it is hard to determine what makes a juvenile leave the main party and join a small subgroup. By leaving the center of the group, an individual reduces direct competition over food items at the cost of increased predation risk. If food competition would be the major factor in leaving the main party we should find a lower presence of low-ranking (unprotected) juveniles, as in adult females (van Noordwijk & van Schaik 1987). However, we did not find a significant effect of (maternal) dominance rank on an individual's presence in the main group. Since some subgroups visit other fruit trees than the main party, food choice might be a motivation to leave the main party (see also van Schaik & van Noordwijk 1986), especially for males that increase their fruit eating with age. Whether an individual forms or joins a subgroup is probably also determined by its history of social relationships in the group and the location of preferred partners (e.g., Fairbanks, Chapter 15, this volume).

A juvenile's safety in the main group or a subgroup is also influenced by the number and identities of its close neighbors. With age, juveniles were less often close to their mothers and less often had any neighbor nearby, suggesting that juveniles gradually accept the risks of utilizing more dangerous locations. The young offspring of high-ranking mothers more often had adult females and the leader male as neighbors, indicating likely safer spatial positions for them: leader males are more likely than other males to

protect infants and young juveniles both in interactions with conspecifics and against predators by alarm calling or even physical protection (van Noordwijk & van Schaik 1988; see also Pereira 1989). So low-ranking juveniles seem to have a slight disadvantage at the most vulnerable phase of life. It is too early, however, to determine the significance of this difference in spatial position for survival.

Sex Differences in Social Development and Ecology

In their future lives, females will stay in their natal group, whereas males will emigrate before fully maturing (van Noordwijk & van Schaik 1985, 1988). All juvenile males in the study population had left their natal group before they were 7 years old, most being 4 or 5 years old. Age at natal emigration was not related to maternal rank or the time spent peripheral to the main party (Ijsseldijk, personal communication; van Noordwijk & van Schaik 1985, unpublished data). Accordingly, there was no significant sex difference in the percentage of time juveniles were present in the main party.

Natal emigration is almost always done with several peers, and such emigrant cohorts are sometimes joined months later by one or more other juveniles from their natal group (see also van Noordwijk & van Schaik 1985). The older peers in emigrant cohorts probably just reached sexual maturity, but are still several years from reaching adult body size (cf. van Schaik et al., in preparation). Followers are often much younger. Long before male juveniles emigrate, they differ from females in their social relationships. Male juveniles concentrated their social activities (play and grooming) and proximity more on other juvenile males with whom they might eventually migrate (see also Fairbanks, Chapter 15, this volume). In contrast, female juveniles concentrated on a female social network and later included adult males, thus establishing relationships for the rest of their lives in the natal group. Pereira (1988a; also Pereira & Altmann 1985) noted a similar sex differentiation in juvenile savannah baboons.

Hiraiwa-Hasegawa (1989) found sex differences among infant and juvenile chimpanzees in frequencies of ant eating and daytime nest making, differences that mirrored pronounced adult sex differences. Likewise, juvenile capuchin monkeys show dramatic sex differences in foraging behavior that mirror adult differences (*C. apella:* Terborgh 1983; *C. olivaceus:* Fragaszy 1986; Robinson 1986; *C. albifrons:* van Schaik & van Noordwijk, unpublished data). In species where adults show only small sex differences in foraging ecology, no sex differences are found among juveniles either (vervets [*Cercopithecus aethiops sabeaus*]: Harrison 1983; baboons [*Papio cynocephalus*]: Post et al. 1980). Long-tailed macaques seem to be intermediate, with moderate adult sex differences that also appear among juveniles: females tended to spend more time foraging, standing, and walking.

The question arises as to whether these early sex differences are functional. For species with sex-specific foraging substrates, early specialization might provide training that facilitates efficient adult foraging. However, Janson and van Schaik (Chapter 5, this volume) argue that juveniles generally acquire these skills relatively fast, so early differentiation between the sexes would seem unnecessary. Moreover, training cannot play a role in long-tailed macaques, because the sexes differ mainly in time spent on different classes of food, without sex-specific techniques. Another possible functional explanation would be that juvenile males and females already differ in nutritional requirements. However, size and growth rates begin to differentiate only after about 5 years (van Schaik et al., in preparation), well after sex-specific foraging patterns are established. The most plausible explanation seems to be that the sexes simply differ in their developmental programs, and that early sex differentiation in foraging behavior is selectively neutral.

The same could be said for the sex differences in social behavior: male long-tailed macaques leave their natal group only after their first 4–6 years of life. However, preliminary results indicate that 1- to 2-year-old females and males groom and play with almost every other age–sex class, but they already concentrate on different classes. The most plausible conclusion again seems to be that an individual has its own femaleness or maleness that shapes its behavior from very early ages, long before sexual maturity and dimorphism are apparent (see also Milton, Chapter 12, this volume; cf. Pereira 1988a and Nakamichi 1989 on social development).

Thus it appears that a sex-specific behavioral repertoire emerges before it gives an individual a clear advantage in survival, growth, or reproduction. It could be argued that at least some aspects of behavioral sex differentiation are nonfunctional results of observational learning from particular classes of neighbors (cf. Pereira 1988a; Edwards, Chapter 22, this volume). Observational learning may well be important to develop or refine a full behavioral repertoire, but the initial choice by the individual to direct its attention to a same-sex conspecific (see Pereira 1988a) remains unexplained. In conclusion, the behavioral transition into an adult of either sex starts earlier than would be required by divergent nutritional or social needs and is unlikely to be fully explained by observational learning.

ACKNOWLEDGMENTS

We sincerely thank the Indonesian Institute of Sciences (L.I.P.I.), the Directorate General of Nature Conservation (P.H.P.A.), and the Universitas Nasional in Jakarta (UNAS) for sponsoring the field work in Indonesia. We thank Mr. Idrusman for his assistance and companionship in the field and Carel van Schaik for stimulating support throughout the project. We thank L. A. Fairbanks, an anonymous reviewer, and especially M. E. Pereira and C. P. van Schaik for extensive comments on the manuscript. This study was financially supported by the Netherlands Foundation for the Advancement of Tropical Research (WOTRO) and the Dobberke Foundation.

7

Juvenile Male Emigration from Natal One-Male Troops in Hanuman Langurs

LAL SINGH RAJPUROHIT and VOLKER SOMMER

The Hanuman langur (*Presbytis entellus* Dufresne 1797) is the best studied and most adaptable South Asian colobine, occurring in a wide range of habitats from the Himalayas and peninsular forests to semiarid woodlands, villages, towns, and cultivated land (Oppenheimer 1977; Roonwal & Mohnot 1977; Vogel 1977). In addition to this remarkable environmental adaptability, the species has a highly variable social organization. The two basic types of social groups are bisexual troops and all-male bands. Troops are matrilineal groups of adult females and offspring with either one adult male (one-male troops, harems) or more than one adult male (multimale troops). The percentage of one-male troops versus multimale troops, and the corresponding number of extratroop band males, varies from site to site (for recent review, see Newton 1988).

Previous field studies documented that non-adult males (at least at one-male sites) transfer regularly from bisexual troops into male bands, whereas females tend to remain throughout their lives in their natal troops (e.g., Sugiyama 1965, 1967; Mohnot 1974, 1978; Hrdy 1977; Moore 1982, 1984a, 1985; Laws & Laws 1984; Mathur & Manohar 1991). Exact ages of emigrants are not known, since these can be gained only from long-term data on known individuals. Moreover, the studies are biased toward females because of the difficulties of following emigrating males. Consequently, in previous investigations of male bands (e.g., Moore 1985), the natal troops of the males as well as their genealogical relationships remained largely unknown.

Thus well-designed theoretical models on the potential influence of nepotism and/or mutualism in the organization of competition and cooperation among langur males (Moore 1985) could not be tested against sufficient data. Moreover, quantitative data on emigration patterns in langurs could help to clarify the persistent debate over the relative importance of intrasexual competition and inbreeding avoidance in producing sex-biased dispersal (for reviews, see Moore 1984a; Moore & Ali 1984; Pusey 1987; Pusey & Packer 1987).

This chapter addresses these problems utilizing data collected during a 12-year study of free-ranging langurs living at Jodhpur, India. Methodological problems are minimized, since (1) the population is geographically isolated (no other langurs within a radius of at least 100 km), and migration to other populations is negligible, allowing reliable total counts of both sexes; (2) the reproductive units are largely harem structured, with a consequently high certainty of paternity; and (3) the population is subject to the only long-term study spanning enough years to allow fairly accurate estimates for crucial reproductive parameters.

STUDY SITE AND METHODS

Study Site

Jodhpur is located in the state of Rajasthan at the eastern edge of the Great Indian desert. The climate is dry, with a maximum temperature of up to 50°C during May and June and a minimum temperature around 0°C during December and January. Jodhpur receives 90% of its scanty rainfall (annual average 360 mm) during the monsoon period (July to September). The town stands on a hilly sandstone plateau covering approximately 85 km², surrounded by flat semidesert. The natural vegetation is open scrub dominated by xerophytic plants such as *Prosopis juliflora, Acacia senegal,* and *Euphorbia caducifolia.* Parks and fields are dependent on irrigation.

The Jodhpur plateau is inhabited by a geographically isolated population of about 1300 Hanuman langurs (*Presbytis entellus entellus*), which has been studied by various Indian and German research workers since 1967. There are no other Hanuman langur groups within a radius of 100 km. The habitat is greatly impacted by humans, particularly because the population of Jodhpur town increased markedly from about 350,000 in 1968 to about 750,000 in 1988. The whole plateau is subject to deforestation, except for a protected reserve of scrub forest with a

total area of about 20 km². The langurs live in the open scrub forest, but several groups include human habitations (roofs, deserted buildings), parks, and field areas in their home ranges. All permanent water sources (tanks, artificial lakes, wells) are man made. The diet consists of items from approximately 190 plant species. Most langurs are fed by local people for religious reasons with wheat preparations, vegetables, and fruits. The degree of provisioning varies considerably between different groups. Some langur groups raid crops. As they are considered sacred, however, the langurs are not hunted. Apart from feral dogs, natural predators are absent. The langurs are easy to observe, as they are not shy and spend most of the daytime on the ground (for details, see Mohnot 1974; Winkler 1981; Sommer 1985; Srivastava 1989).

Study Population

The reproductive units of the Jodhpur langurs are bisexual one-male troops (harems) with a

Fig. 7.1. Juvenile members of a Hanuman langur male band watching a fight between adult males for possession of harem troop Kailana I on September 23, 1982. The life history of three males was followed from birth: M8.2 (second from left), 23.1 month old, transferred recently into this all-male band (AMB) 10, Sidnath. He met two paternal half-brothers: M11.2 and M7.4 (third and fourth from left). The latter had emigrated 3 month before during a previous resident male replacement, being 18.0 and 15.5 months old. (Photo, V. Sommer)

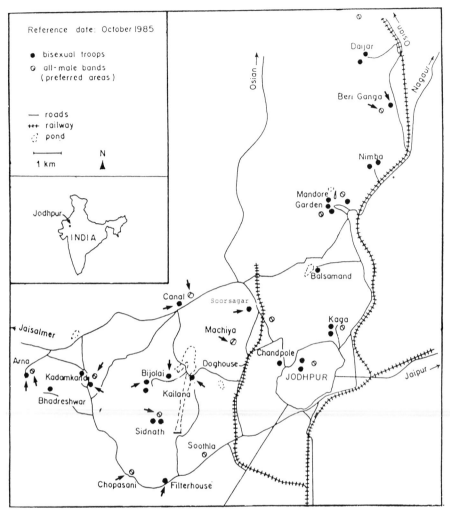

Fig. 7.2. Locations of bisexual troops and all-male bands around Jodhpur. Natal troops of males relevant to the present study, and bands into which they emigrated, are marked by arrows.

single adult breeding male. Each troop occupies its own home range of about 0.5–1.3 km². Troops containing several adult males are rarely found (<10%; see below). Long-term demographic monitoring of known individuals indicates that (with few exceptions) females remain in their natal troop throughout their lives. Males emigrate—usually as juveniles—and join all-male bands (Fig. 7.1), whose moving ranges can be up to 20 km². On average, adult males weigh 18.5 kg and adult females 11.7 kg (for details, see Mohnot 1974; Winkler et al. 1984; Sommer 1985).

Seven censuses were carried out between 1983 and 1986 that covered the entire isolated population. The censuses had a high accuracy because all bisexual troops and male bands were included. These counts revealed population sizes between 1130 and 1298 animals, and male:female sex ratios of 1.1 in infants, 1.5 in juveniles, and 0.2 in subadults and adults (Mohnot et al. 1984; Rajpurohit & Sommer 1991). Because of fusions and fissions, the number of bisexual troops in the population has varied between 27 and 29, with an average troop size of 38.5 (range: 7–93). The number of all-male bands has remained about 13, with an average of 11.8 members (range: 2–47) (Fig. 7.2).

Reproductive Pattern

Jodhpur langurs breed throughout the year, although there is a birth peak in March and a minimum in November. Menstruations are almost always visible. The mean reproductive parameters are age of menarche, 29.0 months; cycle length, 24.1 days; gestation length, 200 days (6.6 months); and interbirth interval, 16.7 months (for details, see Sommer et al., 1992).

All-male bands invade home ranges of bisexual troops in an unpredictable pattern, sometimes resulting in harem holder replacements. Such resident male changes occur during all months of the year. They can be rapid takeovers (occurring in days) or gradual processes (up to several months). During gradual replacements, temporary multimale stages and successions of short tenures (interim residencies) may develop. The social situation stabilizes after a male that is able to defend the troop for a longer period of time gains residency (Vogel & Loch 1984; Sommer 1988; Sommer & Rajpurohit 1989). A new resident may kill the infants sired by his predecessor (e.g., Mohnot 1971; Sommer 1987; Agoramoorthy & Mohnot 1988; for review, see Sommer, in press). Residencies of single adult males in bisexual troops range from just a few days to at least 74.0 months, with an average of 26.5 months (Sommer & Rajpurohit 1989).

Assessments of Paternity

One-male troop types of social structure must not necessarily correspond with a mating system of "harem" polygyny where only the resident adult male has reproductive priority to the troop females. Males other than harem holders might sire infants (1) during multimale stages in connection with resident male changes, (2) if they are able to "sneak" copulations during otherwise stable one-male situations, and (3) if females leave their troop temporarily and copulate with extratroop males.

At Jodhpur, however, the possibility of males other than the current harem holder inseminating troop females is probably negligible for several reasons. First, of a troop's history 95.4% was characterized by stable one-male situations. Periods of social instability with multimale in-

fluxes during resident male changes accounted for only 4.6% of the time (database: 9560 days of observation in three troops between 1977 and 1988: Sommer & Rajpurohit 1989). During multimale stages, females copulated promiscuously (Sommer 1988). The number of infants conceived during stable versus unstable social situations was 95.3 versus 4.7%. However, during unstable situations, a maximum of four (2.7%), but probably only one, infant (0.7%) was fathered by low-ranking band members and not by high-ranking interim residents (Sommer & Rajpurohit 1989). Second, at Jodhpur, most troops are geographically isolated (Fig. 7.2) and dwell in open habitats with good visibility. During foraging, the troops remain cohesive. It is virtually impossible for extratroop males to approach a troop without being noticed by the harem holder, who will then chase the intruders away. It cannot be completely excluded that extratroop males penetrated *unnoticed* into a troop's home range, but the probability is very low; for example, during 288 days of observation in one study troop between 1981 and 1982, a harem holder was never observed to discover an extratroop male *after* the latter had contacted a female. If a harem holder was unable to prevent such contacts, he was likely to be replaced in the near future. Third, during stable one-male situations, single females left their troops for some hours or at the most one night probably less than 5% of all days of observation. Such females rarely copulated with neighboring residents or male band members at the periphery of the troop. Only a fraction of these extratroop copulations could have resulted in conceptions, since the respective females were not always fertile (Sommer, unpublished).

Therefore, the situation at Jodhpur differs crucially from other langur study sites such as Rajaji, India, with much higher multimale influxes (Laws & Laws 1984) or from the situation described for some populations of forest guenons or patas monkeys, where resident males probably sired few offspring born into groups with a basically one-male structure (review in Cords 1987).

Thus at Jodhpur, in the vast majority of cases, the sire of an infant can be inferred fairly accurately by calculating which male resided in a particular troop around the time of conception. Moreover, the cycle of many females was monitored over years, and it was possible to pinpoint the conception because copulations were observed during the conception estrus (Sommer et al. 1992), which further supports assumptions of paternity.

Nevertheless, without direct determination of genetic relationships (e.g., through DNA fingerprinting), assessments of paternity remain a probabilistic approach. Therefore, the following definitions are used.

1. The likely father is termed the *paternal male* in reference to infants that were conceived while a single adult male resided in a constantly monitored troop characterized by a stable one-male situation. If the conception took place during periods when the troop was not monitored, kin relationships were termed *presumed*.
2. Adult male immigrants that established themselves as new harem holders after replacement of the paternal male are termed *replacement male*.

Age Classification

For males, the following categories were applied (detailed physical descriptions are provided by Moore 1984a and Rajpurohit & Sommer 1991): *infant I* (0–5 months), *infant II* (5–15 months), *juvenile* (15 months to about 4 years), *subadult* (about 4–6 years), *young adult* (about 6–8 years), and *adult* (from 7 to 8 years onward).

Study Design

This chapter reviews data collected between 1977 and 1988 by numerous observers that have been published in various sources (Vogel 1979; Winkler 1981; Mohnot et al. 1984; Vogel & Loch 1984; Winkler et al. 1984; Sommer 1985, 1987, 1988; Agoramoorthy 1987; Rajpurohit 1987; Agoramoorthy & Mohnot 1988; Borries 1988; Rajpurohit & Mohnot 1988; Sommer & Rajpurohit 1989; Srivastava 1989; Rajpurohit & Sommer 1991), supplemented by yet unpublished observations made by H. Loch, P. Winkler, and C. Vogel between 1977 and 1980 and by the authors between 1984 and 1988.

The data pool comprises 79 individually known males that survived up to juvenile age (>15.0 month) in nine different natal troops. Juvenile members of male bands with unknown origin were *not* considered. Forty-eight males (60.7%) were born into three focal troops (denominated Kailana I, Kailana II, and Bijolai) whose composition was constantly monitored throughout the study period. All animals in these troops were individually known since 1977 (troops Kailana I and Bijolai) or since 1983 (troop Kailana II). Presumptions on paternity are particularly reliable for infants born into these troops. The members of a fourth troop (Canal) were identified since 1985; its resident male was known since 1983. The remaining individuals were natal to five other troops (Beriganga, Soorsagar, Filterhouse, Kadamkandi E, and Arna), whose demographic development was regularly monitored since 1984; at least the resident males of these troops were identified since 1985.

The nine natal troops experienced 20 harem holder replacements. Usually as a result of such resident male changes, the juvenile males joined seven different all-male bands (AMB 2A, Beriganga; AMB 7, Machiya or Doghouse; AMB 7A, Canal; AMB 10, Sidnath; AMB 11, Chopasani; AMB 12, Kadamkandi; AMB 13, Arna). At the end of the study, 17 individuals were still with their natal troop.

The demographic development of the all-male bands was likewise monitored. Between 1977 and 1980, only a few male band members were known individually. From 1981 onward, the identification has improved a great deal, and by 1985, almost all members of the respective bands, including juveniles, were individually identified.

It was often difficult or impossible to follow individually known males once they left their natal troop, particularly if they joined male bands containing many like-aged individuals. However, even in such cases it was usually possible to determine at least those bands into which they had *not* immigrated, and in this way make sure that records concerning kin relationships of males were not blurred in all bands. Continual identification of male residents that were ousted from their troop was much easier, since all these individuals already had unambiguous marks, such as scars on their ears, faces, or tails, or missing fingers or toes. Identification became impossible for 22 of 62 immature emigrants (35.5%) immediately after they left their natal troop. The other males (64.5%) could be identified for various periods of up to 1 month (9 males), 6 months (12 males), 12 months (5 males), 18 months (2 males), 24 months (3 males), 36 months (6 males), or longer (3 males). Several of these males died in the course of the study.

It should be emphasized that emigrants that could not be further identified were not treated as "dead" individuals, but were omitted from the calculation of survivorship curves. This conservative treatment of the data minimizes biases concerning the estimation of rates of survival after emigration and of further contact of paternal siblings among each other and with the paternal male. However, since the study spanned a 12-year period, the sample sizes were usually still fairly large and seldom fell below 20 individuals.

Three life-history periods of juvenile males were analyzed: (1) *time spent in natal troop* (date of birth, paternity, and time spent with replacement male if paternal male was ousted as harem holder); (2) *separation from natal troop* (age, social situation, mode of separation, transfer into male band, and paternal sibling cohort size); and (3) *fate after separation* (time spent with paternal male, time spent with paternal siblings, and survivorship).

The proportion of individuals varies for which information concerning different variables was available. Selected case studies are provided in the Appendix.

LIFE HISTORIES OF JUVENILE MALES

How to Survive to Juvenile Age

At Jodhpur, langurs breed in one-male troops. A male infant can only grow up to juvenile age (i.e., >15 months) *during* the residency of the paternal male if the latter's tenure exceeds 21.6 months (6.6 months gestation plus 15.0 months). This was true for 35 (55%) of 64 tenures of known length (Sommer & Rajpurohit 1989).

However, infants were sired throughout a given tenure. Thus of 52 male infants born between 1977 and 1988 into four carefully monitored troops, only 16 infants (30.8%) were older than 15 months during the paternal male's ousting. The majority of individuals experienced a harem holder replacement during infancy. This had a strong influence on further life expectancy because at Jodhpur, 35.5% of 110

(male and female) infants present during, or born shortly after, resident male changes fell victim to infanticide committed by immigrating adult males (for review, see Sommer, in press). Most observers agree that this represents a strategy in which immigrating males shorten the period of temporary lactational sterility by killing unweaned offspring sired by their predecessors.

The probability of surviving up to juvenile age depends on when the resident male changes relative to the age of a given infant. For immature male langurs, the various possibilities are summarized in Figure 7.3.

1. *Male infants born within one gestation period after the replacement* (excluding terminated pregnancies: see Agoramoorthy & Mohnot 1988). About two-thirds survived and grew up in their natal troop during the residency of the replacement male (upper-left corner of Fig. 7.3). The remaining one-third of the infants born shortly after the replacement were killed within the next 3 months (lower corner of Fig. 7.3).

2. *Male infants up to 8 months of age already present during resident male changes.* About one-half of these individuals were killed within 1 month after takeover (corner on the lower right side of the x axis in Fig. 7.3). The others survived throughout the residency of the replacement male.

3. *Male infants between 9 and 15 months of age present during resident male changes.* Most of these immatures were not harmed by the new male and remained in the troop until the next harem holder replacement occurred. These semi-independent individuals probably had good chances of surviving because they no longer hampered their mothers' renewed insemination and/or could more efficiently avoid male attacks, or be protected by adult females. On the other hand, the prospects of surviving without maternal support outside the natal troop are probably very low for males younger than 15 months.

4. *Juveniles (age >15 month) present during resident male changes.* Of 27 individuals, only three (aged 15.4, 15.6, and 17.6 months) remained in their natal troop once a new adult male took over, whereas 24 (88.9%) immediately left. All these males probably had a realistic chance of surviving without maternal

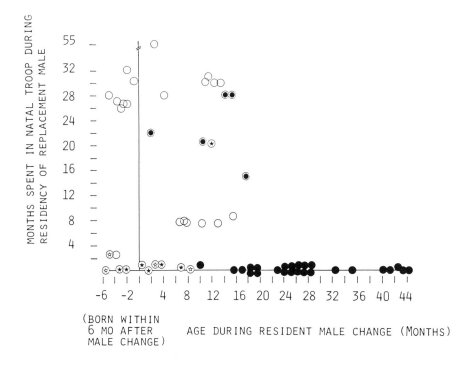

Fig. 7.3. Fate of immature males after the paternal male's replacement by a new resident. *x* axis: age (months) of the respective individuals during the resident replacement; negative numbers indicate that a given individual was sired by the previous resident but born only after the paternal male had been replaced (e.g., −2 means 2 *months* afterward). *y* axis: time (months) a given individual spent with the replacement male. Data are for 51 individually known males born between 1977 and 1988 into troops Kailana I, Kailana II, Bijolai, and Filterhouse.

support. Thus no male older than 18 months remained in his natal troop once a resident male change occurred.

Age of Emigration

The youngest male emigrant was 14.6 months old. He was chased and wounded by a new resident taking over his natal troop in the course of all-male band invasions. The male joined one of the bands, but disappeared after 7 days; he most likely died. The second youngest emigrant (15.5 months) survived for at least 21 months.

Thus males can survive without maternal support at least from an age of 15.5 months onward.

More than half of all emigrations had taken place by the age of 27 months, and 93.1% by 48 months, the end of the juvenile stage (Fig. 7.4). Hence only a few individuals were still with their natal troops as subadults. The oldest individual to leave his troop was a young adult male of about 80 months that succeeded in taking over a neighboring troop 6 months after his emigration. However, not a single case was observed in which a male grew up in his natal troop and took over as a resident.

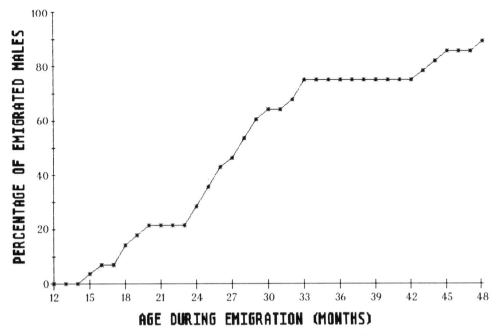

Fig. 7.4. The proportion of males emigrating from their natal troop as a function of age. ($n = 28$ males born into 5 different troops. Cases only with positive evidence for transfer into a male band are included; juveniles who died in their natal troop were not included.)

For a total of 28 males, the average age of emigration from the natal troop was 2.5 years (30.5 ± 14.4 months; upper half of Fig. 7.5). The mean age of juvenile and subadult males that were still residing in their natal troops at the end of the study was 2.8 years (33.7 ± 11.9 months, range 16.0–60.5 months; lower half of Fig. 7.5). It is therefore unlikely that at any given time the average age of potential emigrants will fluctuate much because, with the same probability that some males will grow older in their natal troop (thus increasing the average age of potential emigrants), there will be male infants maturing to juvenile age (which decrease the average age of potential emigrants).

Mode of Separation from the Natal Troop

Males emigrated during all months. The basic social situation within the natal troop had a strong impact on the likelihood of juvenile male emigration ($n = 50$ cases): (1) 76.0% occurred during harem holder replacements (mean age of emigrants: 24.2 months; cases 1–5 and case 7);

(2) 6.0% occurred during harem holder replacements in neighboring troops (mean age of emigrants: 39.3 months); and (3) 18.0% of all males left their natal troop during undisturbed harem holder residencies (mean age of emigrant: 38.9 months; case 1).

Hence the typical situation of emigration is a change of the adult male resident in the natal troop. Those males that left during other situations (2 and 3) were, on average, 1.2 years older, suggesting that maturing males experience conflict with the adult resident once they approach subadult age. Male changes in neighboring troops probably facilitated their departure from their own troop because immatures had the opportunity to come into contact with members of invading male bands.

Males leaving during undisturbed residencies can return to their natal troop. This happened twice: (1) a cohort of three juvenile emigrants returned after 4 months (case 1), and (2) the two youngest emigrants from a trio aged about 48, 60, and 80 months returned after about 6 months. Familiarity with the current resident (usually the paternal male of the emigrants) ap-

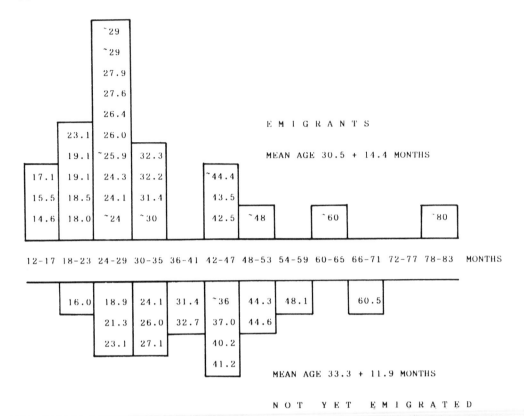

Fig. 7.5. Age of 28 males that emigrated in 1977–1988 from their natal troop compared with 17 males older than 15.0 months who had not emigrated as of November 1988 (birthdates known to the day or month; estimated ages are indicated as ~).

pears to facilitate such returns. A new harem holder, on the other hand, is not likely to tolerate returning emigrants (case 3). In only a single instance did a juvenile fluctuate continually between his natal troop and a male band for a period of several months when an unrelated harem holder was present.

It is difficult to decide if emigrations were the result of "force" by adult males or if they had a "voluntary" component because maturing males preferred to live in a male band anyway. Resident male changes often involved several male bands (Vogel & Loch 1984; Agoramoorthy 1987; Sommer 1988). During such multimale stages, immature natal males preferentially interacted with like-aged members of male bands. Sooner or later, high-ranking males tried to chase rival bands, as well as members of their own band, out of the home range of the contested harem. In the course of these waves of

expulsion, natal males were likewise driven away. However, it could not be decisively determined that they were "singled out" by adult males attempting to install themselves as residents. Moreover, in the odd instance of continual wandering between troop and band (case 2), a voluntary component of the transfer could hardly be denied, since the individual was not attacked by the new resident.

In the majority of cases, however, "force" was the obvious proximate cause for transfers. Among the 50 male emigrants, 53.8% were chased by adults and another 12.8% were chased and wounded (cases 1 and 5); one wounded individual died. In 36.1% of the emigrations, chases were not observed, although most of these individuals followed the former resident after he was forcefully ousted by another male (cases 3 and 4). Of the 23 cases of chasing and/or wounding, 91.3% were commit-

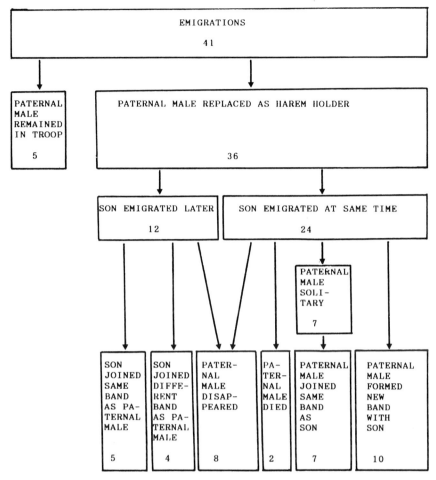

Fig. 7.6. Relation between emigrating males ($n = 41$) and the paternal male. Numbers indicate the number of emigrants for which a certain constellation was true.

ted by nonpaternal males. The remaining two cases occurred when a paternal male chased his subadult sons out of his troop (without wounding them).

Emigrants and the Paternal Male

The future relationship between a son and his father depends primarily on whether they emigrate together from a given troop. In this regard, sufficient information is available concerning 41 cases of immature male emigration (Fig. 7.6): (1) few males (12%) emigrated while the paternal male was still resident; (2) 29% emigrated *after* the paternal male had been replaced (usually during the replacement of his successor;

cases 2 and 4); and (3) 59% emigrated *with* the paternal male during his replacement (cases 3, 5, and 7).

After emigration, sons and paternal males sometimes relocated in the same band and sometimes were separated, since the fate of male residents can be quite variable in the course of their ousting (see review in Rajpurohit & Mohnot 1988):

1. Residents may be killed by invaders or due to accidents. Several males disappeared, some of them after they had been severely wounded by competitors. They most likely died (case 4).

2. Ousted residents regularly rejoined male

bands, often those bands from which they originated before they took over a harem.

3. Some ousted residents tried to regain a residency and spent some time (usually less than a month) solitarily while monitoring their former harem from a distance (case 7). At Jodhpur, however, ousted males were never successful in reestablishing themselves as harem holders. Consequently, solitary males usually also rejoined male bands (case 7).

4. Ousted residents may form a new male band with their sons.

At Jodhpur, only one case of new male band formation has been observed (case 3). It is unknown how stable such bands are. However, newly formed bands are likely to mingle sooner or later with other bands because there are only a few suitable areas where extratroop males can survive. If the paternal male and a son leave the troop during different times, they may (case 3) or may not (case 4) join the same band. In one well-documented case, the paternal male and son recognized each other after weeks of separation (case 6). It is unknown how long this kind of kin recognition could potentially last.

In total, 54% of all emigrants ended up in the same band as the paternal male (Fig. 7.6). However, the actual time spent together varied considerably, because either the paternal male or the son disappeared or died: (1) 4% of the emigrants spent only several days with the paternal male; (2) 44% spent a period of up to 3 months with the paternal male; and (3) 52% spent more than 5 months with the paternal male.

Survivorship After Emigration

It is difficult to quantify if and how fathers provide special care to their sons, although it was observed that paternal males defended them during encounters with rival male bands (case 5). If sons lived with the paternal male, they had a somewhat higher probability to survive during the first 6 months after emigration compared with males that emigrated without paternal males; mortality rates became equal 9 months after emigration (Fig. 7.7). The differences were not statistically significant, not even during those periods with the greatest discrepancy in survival probability [month 1 after emigration: $\chi^2(1) = 0.265$, ns; month 3: $\chi^2(1) =$

0.984, ns; month 6: $\chi^2(1) = 0.119$, ns; 2×2 contingency tables]. However, it has to be considered that the sample size was rather small for statistical treatment. Interestingly, the mean age of those males emigrating with the paternal male was 28.6 ± 8.7 months ($n = 21$). Those surviving without paternal males were, on average, older individuals (37.4 ± 17.9 months; $n = 12$), for whom paternal support might not any longer have been crucial.

Cohorts of Paternal Siblings

Males sired by a particular resident are (at least) paternal half-brothers ($r \geq 0.25$). Most of them might even be more closely related because the females in langur one-male troops have a high degree of interrelatedness (see Seger 1977). Exact genealogical relations via the maternal side, however, could not be assessed. Cases in which full-brothers (same mother and same paternal male) matured together to juvenile age were not observed, although such cases might occur in the Jodhpur population. However, since some infant males survived resident male changes (Fig. 7.3) and matured together with sons sired by the new resident, the juveniles living in a given troop are not necessarily *all* half-brothers.

Consequently, the following figurations were observed.

1. All half-brothers emigrated together into the same band (cases 3 and 7).
2. Half-brothers transferred in different years (case 1 and 2) and joined different male bands (case 4) or relocated in the same band (case 5).
3. Cohorts of emigrants contained males sired by different males (case 1).

The cohort size during emigration (independent of the emigrant's kin relation) could be documented for 56 individuals that left their troops during 19 waves of emigration. Of these males 10.7% emigrated alone, 48.2% in cohorts of 2 to 4 individuals, and 41.1% in cohorts of 6, 7, and 10 individuals, yielding a mean cohort size of 2.9 individuals.

Information concerning paternity of emigrants was available for 44 individuals. The mean cohort size of paternal siblings emigrating together was slightly smaller, 2.4 individuals,

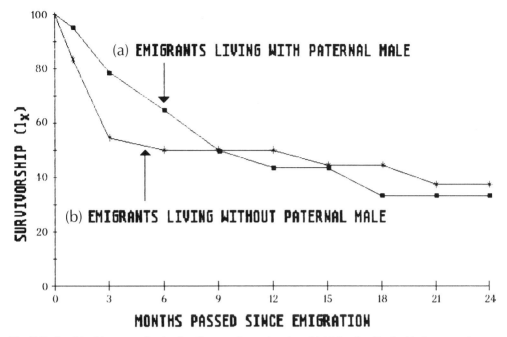

Fig. 7.7. Survivorship curves for the first 2 years after emigration. (a) Males that lived with the paternal male after they left their natal troop are compared with (b) males that lived without the paternal male. Sample of individuals: (a) $n = 12$ in month 1, decreasing to $n = 8$ in month 24; (b) $n = 21$ in month 1, decreasing to $n = 15$ in month 24.

due to the splitting of cohorts sired by the same male. Of all males, 15.9% emigrated without any paternal sibling, 47.7% in cohorts of 2 to 4 siblings, and 54.5% in cohorts of 6 to 10 siblings.

Soon after emigration, the mortality of males rose dramatically, leading to a male:female sex ratio for the subadult and adult age class of 1:4.1 (Rajpurohit & Sommer 1991). As a consequence, cohorts of two, three, or four males may eventually be wiped out or reduced to single individuals. Information in this regard is available for only 30 individuals, since the other males could no longer be identified after their emigration. The proportion of males surviving "alone" (i.e., without paternal siblings) rose steadily as a function of time since emigration (Fig. 7.8). After 26 months, more than half of all male-band members had no siblings in their band. The figure rose to 75% after 30 months, and 38 months after emigration, no paternal sibling was left for those few males that had survived.

DISCUSSION

Emigration: Inbreeding Avoidance or Intrasexual Competition?

In several primate species—for example, toque macaques (Dittus 1979), rhesus macaques (Wilson & Boelkins 1970), and capuchin monkeys (Robinson 1988a)—migrating males face increased risks of mortality from predation, starvation, and hostility from strange conspecifics. Langur males are no exception in this regard because they die significantly more often than females from "external causes" such as accidents, intraspecific killings, and predation (Rajpurohit & Sommer 1991). Why, then, do they leave their natal troop? Two major functional explanations have been proposed for the broad pattern of sex-biased dispersal in animals, particularly in primates (for review, see Pusey 1987; Pusey & Packer 1987).

1. Individuals disperse as a consequence of intrasexual competition for mates and food (e.g., Moore & Ali 1984).

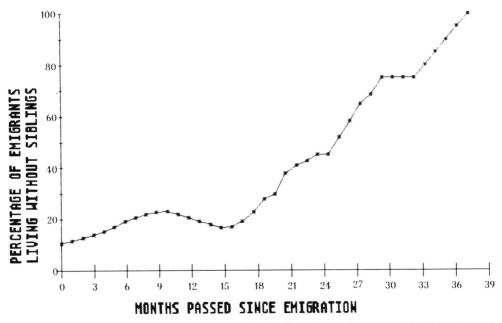

Fig. 7.8. Cumulative proportion of surviving emigrated males (5 months moving averages) that do not live with any paternal sibling plotted against the time passed since emigration. The initial sample of 49 individually known males decreased gradually to 11 in month 18 and 3 in month 38. Thus the proportion of males living "alone" increased as the total number of survivors decreased.

2. Dispersal evolved as a mechanism to avoid inbreeding (e.g., Packer 1979a).

Observations on proximate causes can give clues about ultimate functional reasons. If competition is the primary cause, emigrations should take place as a result of aggression. Inbreeding avoidance as the evolutionary explanation for dispersal would be more consistent with voluntary emigrations and/or as a result of sexual attraction to extragroup mates or as a result of sexual disinterest among natal troop females (for review, see Pusey & Packer 1987).

Male transfer in seasonally breeding species during the mating season seems to be often motivated by sexual attraction (e.g., toque macaques: Dittus 1977; Barbary macaques: Paul & Küster 1985). Jodhpur langurs, on the other hand, breed year-round. Indeed, males emigrated during all months of the year, but this is not consistent with the sexual attraction hypothesis because they joined male bands. Moreover, langur males are capable of copulatory behavior from an age of about 4 years onward. Of 28 emigrants, only 10% were older than 4 years

(Fig. 7.5), making it very unlikely that sexual attraction caused them to leave their natal unit, especially because such males have virtually no chance of successful competition with harem residents for access to females.

Intrasexual competition can more easily explain langur emigrations, including those of males beyond puberty, since at least two of three sexually mature emigrants were chased by the current resident. Intolerance of a new resident male obviously accounts for at least those 76% of emigrations, which occurred in the course of adult male replacements. Two-thirds of all males, including most of those that emigrated during undisturbed harem holder residencies, left their natal troop as a direct consequence of chasing and/or wounding by adult males.

The remaining males were not directly observed to be attacked, but transferred at the time of the paternal male's eviction by a new resident. A single juvenile emigrated rather voluntarily, maintaining constant fluidity between his natal troop and a neighboring male band for a period of 7 months. Although these cases seem to be more consistent with the inbreeding avoidance hypothesis, it has to be considered that all these emigrations forestalled an inevitable eviction, since not a single male in the Jodhpur population was observed to grow up to adulthood in his natal troop.

The juvenile that alternated for months between natal troop and male band was the last offspring of an old female that vigorously protected her son on the slightest indication of agonistic behavior of adult males. Her increased investment and risk taking could well be a result of her low reproductive value, which would favor for increased effort into her last offspring (Trivers 1974). The rather gradual emigrations and the temporary return of emigrants into their natal troop that were observed during this study have been previously described for the Jodhpur site by Mohnot (1978). He reported that adult males consistently harassed juvenile males aged about 14–16 months until they became peripheral and, after 4–6 weeks, joined a neighboring band. (It was not specified whether the harassers were paternal or replacement males.) Sugiyama (1967) described that a replacement male chased and bit all juveniles until they emigrated. Boggess (1982) reported the exclusion of subadult males during the mating season.

At sites with predominantly multimale troops (e.g., Gir Forest: Starin 1978; Kaukori: Jay 1963b; Melemchi: Bishop 1979; Polonnaruwa: Muckenhirn 1972; Simla: Sugiyama 1976a; Rajaji: Laws & Laws 1984) juveniles have not been seen in male bands (for review, see Moore 1985). If paternity certainty is low in multimale troops, the absence of juvenile emigration could be explained "by the reluctance of adult males to inflict the costs of expulsion on potential sons *until* those potential nonsons are old enough to be sexual competitors" (Moore 1985, p. 174). Consequently, it can be expected that males emigrating *newly* into multimale or one-male troops would try to expel not only the sexually mature residents of a troop, but also all juveniles (Moore 1985). This is exactly the pattern found at Jodhpur, a site where the reproductive units are almost exclusively harem structured, with a concomitant high certainty of paternity.

Male-Biased Dispersal in Langurs

Several functional explanations address widespread sex differences in dispersal patterns. Male-biased dispersal is strongly linked to po-

lygyny. It is assumed that intrasexual competition for mates is more intense among males than among females in polygynous species (Trivers 1972), and males may therefore be more likely evicted by rivals (Clutton-Brock & Harvey 1976; Greenwood 1980). Moreover, the sexes are believed to differ concerning the resources that limit their potential lifetime reproductive success. Because of the average low paternal investment, which is also typical for Hanuman langurs (e.g., Hrdy 1977; Vogel 1979; Sommer 1985, 1987), most mammalian males compete for access to females, whereas females do not mainly compete for mates, but for access to other resources such as food (Trivers 1972; Wrangham 1980). Since foraging efficiency probably increases with familiarity of an area (Waser & Jones 1983), females may consequently benefit more than males from remaining in the natal area.

These explanations can well be applied to Hanuman langurs. In the Jodhpur population, female transfer is very rare. During the 12-year study, there have been four observations of single adult females in temporary association with male bands. Three of these females carried infants. At least two of them emigrated after a new, and potentially infanticidal, male entered their troop (Rajpurohit & Sommer, personal observation; Hrdy [1977] and Moore [1985] report similar observations for Abu/Rajasthan). A few other langur females occasionally left their natal troop for a part of the day or at the most one night and copulated with neighboring residents or male band members at the periphery of the troop (Rajpurohit and Sommer, personal observation; Winkler, personal communication). Finally, it was observed once that a young female in estrus left her natal troop, which was still under control of the male that sired her, and copulated with the highest ranking member of a male band living about 600 m away (Sommer, personal observation). However, we have no evidence that such behavior led to permanent female transfer.

The latter case represents the only indication that incest avoidance might be a factor in structuring transfer patterns at Jodhpur. It is not known whether copulations occur between fathers and daughters or if natal males and natal females would copulate, if they would not be prevented by other males. However, the likelihood that this could be reasonably explains why males evicted during undisturbed harem residencies tend to be older males approaching or past puberty. On the other hand, the fact that the timing of juvenile male emigration is much sooner than the age at sexual maturity does not, in itself, mean that competition for mates is not a factor, since the replacement male may find it physically easier to force out potential competitors for mates when they are rather small instead of waiting until they actually are sexual competitors.

Recently, it has been suggested that food competition is a major factor in delayed growth among juvenile primates (Janson & van Schaik, Chapter 5, this volume). In this regard, it should also be considered that competition for food, rather than for mates, could be the principal driving force leading to emigration of immature langur males. Because the average tenure of harem holders is 26.5 months, the adult residents may often no longer be in the troop once the young males would have reached sexual maturity. On the other hand, male juveniles are competitors for food with the male and his future offspring. Female juveniles are also food competitors, but more importantly are prospective mates, especially since menarche in female langurs occurs at an early age (29.0 months). Since gestation is 6.6 months and the average birth interval 16.7 months, the total number of different females a male can inseminate may be very important, since he may not be present long enough to reproduce repeatedly with the same female.

Maximizing breeding success in a short time period is critical for harem holders because, at Jodhpur, no male gained residency in more than one troop (Sommer & Rajpurohit 1989). The optimal strategy for immigrant males may therefore be to kill suckling infants (review in Sommer 1992) in order to shorten the mothers' lactational amenorrhea and to remove food competitors of the male's future offspring as well as to force out other extraneous food competitors—that is, male juveniles.

The two hypotheses (intrasexual competition for mates and food) are not of course mutually exclusive, and it is at present also not possible to dismiss the role of inbreeding avoidance completely.

Male Cooperation: Nepotism or Game Theory?

Dispersal can be divided into *natal emigration* (i.e., emigration from the natal group or range) and *natal transfer* (i.e., emigration from the natal troop followed by immigration into a group containing breeding members of the opposite sex) (Pusey & Packer 1987). In the langurs of Jodhpur, natal emigration occurred at an average age of 2.5 years, but it is unlikely that a male was ever able to immigrate permanently into a bisexual troop before he was at least 7–8 years old. Thus many males live for more than 5 years in male bands. Other males spend their whole postjuvenile life in a male band, because about one-quarter of all adult males probably never gain a residency (Sommer & Rajpurohit 1989).

Secondary transfer ("breeding dispersal," i.e., movement from one bisexual troop to another) was never observed in Jodhpur langurs, since no male gained a residency in more than one troop (Sommer & Rajpurohit 1989). However, ousted male residents regularly rejoined all-male bands, often those bands containing individuals with whom they were familiar before they took over a harem (Rajpurohit & Mohnot 1988).

Since langur males spend a considerable portion of their lives in all-male bands, it is important to quantify the role of kin selection and/or reciprocity (mutualism) among nonrelated individuals in structuring the composition of bands as well as in determining the coalitional behavior of individual males (Moore 1985). The evidence presented in this study revealed great variability of band composition: juveniles may emigrate with or without the paternal male, as a unit of paternal half-siblings, or they may split up and join different bands. The formation of a new band by an ousted resident that is followed by his (presumed) weaned sons has already been documented by Sugiyama (1965), Mohnot (1977), and Hrdy (1977, p. 298). At Jodhpur, 54% of all emigrants ended up at least temporarily in the same band as their father. The presence of paternal males seemed to increase the likelihood of survival during the initial 6 months after emigration (Fig. 7.7), although the difference was not statistically significant. Moreover, about 85% of all males emigrated with at least one paternal sibling.

However, the potential influence of fraternal kin support decreased steadily because of high mortality among emigrants: about 3 years after emigration, there was no surviving male left who still lived with a sibling (Fig. 7.8). Since emigrating males were, on average, 30.5 months old, they were, on average, 68.7 months old when they lost their last paternal sibling. Such males, with an age of 5.7 years, were young adults that influenced the daily activities of male bands (moving pattern, agonistic interactions with rival bands and troops) more strongly than did subadult or juvenile male band members (Moore 1984a). Thus it is not likely that cooperation of close kin plays a crucial role among high-ranking members of male bands.

Langur males often join forces while ousting a current harem resident. A similar cooperation among males has been observed in male lions when they oust other males from residence with a pride of females. However, after a takeover, male lions in the ruling coalition rarely fight openly over access to receptive females, possibly because females copulate hundreds of times per conception and thus the "worth" of exclusive access is devalued. In one case, a coalition was a cluster of brothers, half-brothers, and cousins originating from the same pride, but most coalitions consist entirely of unrelated individuals (Bertram 1978; Bygott et al. 1979; Packer & Pusey 1982).

In Jodhpur langurs, multimale troop structures may develop during resident male changes, but they are usually temporary transitional periods that later become one-male structures again. There have been only three cases observed where troops developed semipermanent multimale structures (Srivastava et al. 1986; Mohnot et al. 1987). Two cases concern residents that tolerated three subadult to young adult males in their troop. These males might have been their own sons. They were lower in rank, and the resident probably prevented them from copulating before they ultimately left the troop. In at least one case, they were chased by the resident. The third case comes closer to the lion pattern, since two males (whose genealogical relationship was unknown) jointly took over a troop and remained together for the next 17 months. Observations were discontinuous, but those copulations that were recorded were not mutually harassed by the two males.

However, in the vast majority of resident male changes, the highest ranking male of the invading band immediately tried to expel his former allies once they had successfully ousted the previous resident (Mohnot 1974; Vogel & Loch 1984; Agoramoorthy 1987; Sommer 1988). This suggests that in lions, the probability that a solitary male will be able to subdue cooperating rivals is smaller than in langurs,

where residents can be in sole control of a troop for several years. Moreover, mortality rates among emigrating male langurs at Jodhpur seem to be even higher than among subadult lions, with the consequence that kin are almost never available for adult coalitions. The mechanisms of cooperation and competition in male bands will be subject of forthcoming papers.

APPENDIX: SELECTED CASE STUDIES

Observers: GA, G. Agoramoorthy; VD, V. K. Dave; HL, Hartmut Loch; LR, Lal Singh Rajpurohit; VS, Volker Sommer; AS, Arun Srivastava; MT, Monica Tobler; PW, Paul Winkler

Case 1 (Troop Bijolai, 1977–1983). M1 was the adult male resident of troop Bijolai from at least February 1977 to September 1980. During his tenure, seven male offspring were sired that survived up to juvenile age. These paternal half-siblings and two more juveniles (which might have also been sired by M1 before February 1977) left their troop during four different occasions. (Most of them transferred into male bands.)

1. One male aged 23.6 months disappeared during the paternal male's residency.
2. Three males (aged 32.2, 31.4, and 26.4 months) left between December 20 and 25, 1979 during the residency of M1, but returned 4 months later.
3. These three males and four additional paternal siblings left the troop during the replacement of M1 in September to October 1980. At least two of them were chased by interim residents.
4. The last son of M1 transferred during the replacement of the successor of M1 and left with two paternally unrelated males during January 1983. (Observers: HL, PW)

Case 2 (Troop Kailana I, 1982–1983). The troop contained three juvenile males, sired by a resident that had been ousted the previous year. During another resident replacement in June 1982, two of these juveniles joined all-male band (AMB) 10, Sidnath (M7.4, 15.5 months; M11.2, 18.0 months). Their elder sibling (M8.2, 20.1 months) remained with the troop. He was the last infant of F8, a female of old physical appearance ranking in the lower third of the dominance hierarchy that acted particularly protective toward her son (e.g., in terms of grooming him, still allowing nipple contact, and defending him during agonistic encounters with troop males). The two other mothers (F7, middle-aged; F11, young, 6 years) showed these interactions less often and weaned their sons earlier. During September 1982, another male change began in troop Kailana I, lasting until January 1983, with 45 males from several different bands involved. In the mornings and evenings, these bands regularly invaded the troop's home range, and various adult males installed themselves as interim residents. The previous troop members M7.4 and M11.2, now members of AMB 10, sometimes interacted with juvenile females. Their mothers, however, chased their juvenile sons as soon as they approached them. Juvenile M8.2, on the other hand, began to move back and forth between his natal troop and AMB 10 (Fig. 7.1). Until his final departure from the natal troop at the end of April 1983, he stayed for hours or periods of up to about 10 days with the male band, depending on the moving pattern of the band. His mother regularly groomed him whenever he expressed signs of fear toward adult males. Even after some months, he seemed undecided whether to stay with his natal troop or the male band; for example, on December 20, 1982, at 17:45, troop KI and the invading AMB 10 separated and moved toward different sleeping sites. Juvenile M8.2 sat in the middle between the two troops. Emitting mild squeals, he alternated in looking in the direction of the troop and the male band. Ultimately, he started gnawing a stick. Shortly thereafter, his preferred playmate, a juvenile male band member, joined him and initiated a rough-and-tumble play. Thereupon, both followed behind the male band. M8.2 survived for at least another 9 years. (Observers: VS, VD, GA, L, LR)

Case 3 (Troop Beriganga, April 26, 1984). The previous resident M103 was expelled by four male invaders. He left the troop with his 10 presumed sons, aged about 20–45 months. His sons were not attacked by the invaders. Paternal male and sons formed a new male band, which was stable for at least 45 days. They stayed about 1 km away from the troop. The new resident, who installed himself on May 5, 1984, did not allow them to come closer. Most juveniles disappeared within the next 7 months. During November 1984 only one juvenile and M103 were seen. (Observer: LR)

Case 4 (Troop Kailana I, December 22, 1985). The previous resident M46 was severely wounded by an immigrating male, who established himself as new resident. When M46 left the troop, two juveniles (M4.6, 27.9 months; M12.2, 26.0 months) followed him without being chased by intruder M119. (These juveniles had been sired during the tenure of the previous resident M11, who, after his ousting in mid-1983, joined AMB 11, Chopasani.) M46 and M12.2 disappeared within 7 days. They probably died, since they were not seen again despite constant monitoring of all surrounding male bands until March 1988. Juvenile 4.6 remained solitary (at least during July 1986); at the latest, by October 1986, he had joined AMB 7, Machiya. This male band did not contain any maternal or paternal half-siblings. (M4.2, a maternal half-brother of M4.6, born 6 years before, was identified in AMB 10, Sidnath, during 1982/1983, but disappeared later.) M4.6 survived until at least June 1990. (Observers: AS, LR, VS)

Case 5 (Troop Filterhouse, March 15, 1985). Resident M125 was replaced and joined AMB 11, Chopasani. He was accompanied by one of his presumed sons, about 26 months old, who was chased and wounded by the new resident. On November 21, 1985, another male change took place in troop Filterhouse. This time, the four remaining presumed sons of former resident M125, aged between 24.3 and about 30 months, were chased out of the troop by the new resident. They joined the paternal male's and paternal siblings' male band. The paternal male remained with his presumed offspring until November 1986. Occasionally, he defended them during encounters with rival male bands. (Observers: LR, VS)

Case 6 (Troop Canal, January 12, 1987). A juvenile male, aged about 15 months, was removed from his natal troop, Canal, after he had fallen into a gorge. He recovered in the camp of the observers and was introduced into AMB 7, Machiya, on February 4, 1987. The band was attacked on the same day by the resident of troop Canal, whose tenure began at the latest in January 1983. The resident chased the young adult and juvenile members of the band (who did not originate from troop Canal). His juvenile son was the only individual that did not run away after the resident had dashed into the male band. The resident's aggressive attacks came to an immediate halt when his son approached him. They obviously recognized each other despite 23 days of separation. The paternal male moved back to his troop. His son, emitting mild squeals, initially tried to follow him, but could not catch up and remained with the band. (Observers: VS, MT)

Case 7 (Troop Arna, December 8, 1984). After his replacement, former resident M129 remained solitary for about 15–20 days. His six presumed sons, aged about 18–30 months, were chased out of the troop by the new resident on December 25, 1984, and transferred into AMB 13 at Arna. M129 joined them later and stayed with them for at least 5 months. (Observer: LSR)

ACKNOWLEDGMENTS

We are gratefully indebted to S. M. Mohnot, Jodhpur, and Christian Vogel, Göttingen, for supervision. L.S.R. was sponsored by grants from Man-and-the-Biosphere Programme, Department of Environment, Government of India, New Delhi, and the German Academic Exchange Service (DAAD), Bonn. V.S. was supported through the Feodor-Lynen-Programme of the Alexander von Humboldt Foundation and the German Research Council (DFG; Grant So-218, 1-3). For helpful comments on the manuscript, we are thankful to Jim Moore, Amy Parish, Eckart Voland, Andreas Paul, and Sarah Blaffer Hrdy.

8

Consequences of Sex Differences in Dispersal for Juvenile Red Howler Monkeys

CAROLYN M. CROCKETT and THERESA R. POPE

Howlers, as a genus, have been relatively well studied with respect to demography and ecology (Crockett & Eisenberg 1987). As for many primate species, the problems faced during the juvenile stage, per se, have not been addressed specifically. In this chapter, we attempt to correct this, focusing on the red howler monkey (*Alouatta seniculus*). The behavioral ecology of the red howler shares many similarities with the other five species (references in Crockett & Eisenberg 1987; new data on *A. caraya*: Rumiz 1990; *A. fusca*: Mendes 1989; and *A. belzebul*: Bonvicino 1989). The most divergent species is the mantled howler (*A. palliata*), and some of its differences and their implications are discussed toward the end of this chapter.

Howlers differ from most polygynous primates in that immatures of both sexes emigrate from their natal troops (Crockett & Eisenberg 1987; see Pusey & Packer 1987). This bisexual emigration has its roots in the juvenile stage, when most natal dispersals occur. We describe the conditions and costs of emigration, and the alternative behavior patterns that juveniles might employ to achieve breeding status and, ultimately, reproductive success as adults.

METHODS AND STUDY AREA

These data were collected on red howler monkeys at Hato Masaguaral, Venezuela (Table 8.1). The study area has been described in nu-

merous publications (e.g., Rudran 1979; Troth 1979; Crockett 1985; Crockett & Rudran 1987a). The concentration of Woodland troops is separated from the Gallery Forest population by about 4 km of open savanna dotted with clumps of shrubs and a few small trees (see Crockett 1985, Fig. 1). The two study populations are part of a larger "metapopulation" (Gilpin 1987) within which some flow of individuals occurs.

Using scars, ear tags, and other characteristics, our individual recognition of troop members was nearly 100% for adults and larger juveniles, and somewhat less for younger animals. In 1981, 153 howlers were captured, weighed, and ear-tagged (Crockett & Pope 1988), including 26 from the Gallery population and 17 of 35 Woodland howlers marked in 1978 (Thorington et al. 1979).

During census contacts, the observer recorded the number of individuals in each age–sex class, all known individual identities, and *ad libitum* behavioral observations. Maternal identities were known for most infants born into troops during our study period. For each troop, a timeline of every individual's presence was constructed, including birth or immigration and disappearance or emigration (Crockett & Rudran 1987a). Few untagged or unscarred juveniles could be identified positively outside the context of their natal troops. Thus many of their fates had to be categorized as "emigrated/disappeared."

Table 8.1. Study Periods, Researchers, and Number of Troops Censused (in Parentheses) per Habitat at Hato Masaguaral

Year	Woodland	Gallery Forest
1976–1978	Rudran (20), Mack, Thorington[a]	Robinson[a] (11)
October 1978; March 1979– February 1981	Crockett[b] (26)	Crockett[b] (17)
August 1979– August 1980	Sekulic[c] (4)	
January 1981– January 1982	Pope[d] (29)	
November– December 1981	Crockett[b] (29)	Crockett[b] (22)
May–August 1982	Saavedra[e] (2)	
February 1983	Crockett[b] (29)	Crockett[b] (21)
February 1984	Crockett[b] (29)	Crockett[b] (25)
February 1984– July 1985	Pope[d] (33)	
March–April 1987		Crockett[f] (31)

[a]Robinson (personal communication, 1978 unpublished report); Mack (1978 unpublished report); Rudran (1979); Thorington et al. (1979); earlier studies listed in Crockett and Eisenberg (1987).
[b]Crockett (1984, 1985, 1987); Crockett and Eisenberg (1987); Crockett and Pope (1988), and others. There were 115 dawn–dusk observations of two troops from March 1979 to February 1981. There were more than 212 howlers in other areas in February 1983.
[c]Sekulic (1982a, 1982b, 1982c, 1983).
[d]Pope (1989, 1990) collected activity budget and feeding behavior data on 19 pairs of extratroop individuals and troop members matched for age and sex (10 pairs of males, 9 pairs of females), and tracked the movements of three female and four male radio-collared emigrant extratroop howlers.
[e]Saavedra (1984, personal communication).
[f]Unpublished data.

Behavioral observations were made during census contacts with many troops and extratroop animals over 6 years and during intensive sampling of selected troops and individuals (Table 8.1). Data analysis methods are described in conjunction with specific results. Since many more Woodland than Gallery animals were tagged, and Pope's (1989, 1990) data came exclusively from the Woodland, some quantitative analyses exclude the Gallery Forest population.

Populations in the two study habitats were growing; however, they differed in mean troop size, adult composition, and density (Crockett & Eisenberg 1987). In 1981 the mean troop size and composition in the Woodland was 10.5 individuals, with 1.6 adult males and 2.9 adult females, compared with 7.7 individuals, 1.2 adult males and 2.5 adult females in the Gallery.

Furthermore, the Gallery population density was less than half that of the Woodland. Perhaps as a consequence, population growth and new troop formation were greater in the Gallery Forest, where new troops outnumbered established troops by 1987 (Crockett 1991). This has implications for the fates of juvenile dispersers of both sexes, who are more likely to find breeding vacancies there.

PREREQUISITES FOR SUCCESSFUL REPRODUCTION

Most juvenile red howlers disperse from their natal troops but must become troop members before reproducing. Howler troops include at least one male and two females that defend a territory and produce offspring. (This chapter is not the forum for addressing the problematical question of whether howlers exhibit "territoriality" as classically defined; see Sekulic 1982b; Crockett & Eisenberg 1987; Whitehead 1989.) Individual troops display strong site fidelity and are found in the same ranging area year after year. Although red howler territories overlap with adjacent troops, they are vigorously defended with howls and chases. Adults and older juveniles of both sexes actively participate in territorial defense (Fig. 8.1).

Appropriation of a defended home range seems crucial for breeding. Dispersing animals must either immigrate into existing troops or form new ones. Extratroop individuals of both sexes constituted 6 to 9% of the approximately 500 animals censused, but they never succeeded in forming new reproductive troops without first establishing a territory. Between 1979 and 1985, 22 extratroop associations persisted, produced offspring, and became new troops (Crockett 1985; Pope 1989). As described later, associations can be ephemeral, and it is difficult to pinpoint how many failed to become troops. Although two new troops included related males, females dispersed singly and were not known to have formed new troops with related females.

Once a female begins reproducing, she gives birth on average every 17 months and is a permanent troop member until her death or the dissolution of the troop; mortality of troop-living adult females is low, about 0.05 deaths per female year (Crockett & Rudran 1987b). For an

Fig. 8.1. Red howler adult female (left), adult male (center), and 4-year-old juvenile male (right) howling at neighbors during an intertroop encounter. About 2 years later, both males invaded a neighboring troop. (Photo, C. M. Crockett)

adult male, in contrast, residence in a troop does not ensure successful reproduction. Paternity exclusion analyses indicate that when more than one adult male is present, only one sires the troop's infants (Pope 1990). About 38% of all troops, 55% of Woodland and 14% of Gallery, contained more than one adult male (Crockett 1985).

DEFINING THE JUVENILE STAGE

Pereira and Altmann (1985) defined the onset of juvenility as the age at which an individual can survive the death of its mother. The juvenile stage for a red howler monkey begins at about 1 year of age when its mother ceases lactation. The age at weaning is usually 10.5 to 14 months (Crockett, unpublished data) and occasionally is as long as 18 months (Mack 1979). Given the low rate of maternal death, only a few cases of known-aged orphans exist. The two youngest who survived were 13 and 15 months old, and all orphaned at 10 months of age or younger died.

The juvenile stage ends when most red howlers are first capable of successful reproduction (Fig. 8.2). Females first gave birth at 4 to 7 years of age (median = 5.17 years), and males first became fathers at 6 to 8 years of age (median = 7 years). Male ages came from troops in which only one father was possible; no evidence of paternity by males living outside of a troop was found by Pope (1990). A male of 5.6 years was suspected to be a father, but his 7.7-year-old brother was also present; thus neither could be included in Figure 8.2. An estimate of the age of sexual maturation can be calculated by subtracting gestation length of 6.3 months (Crockett & Sekulic 1982), or typically 4 years for females and 5 years for males.

Reproductive capacity begins before maximum weight is reached, at least 1 year earlier for females and 2 years for males. The rate of weight gain declines at 4 years of age for both sexes; maximum weight is reached at about 5–6

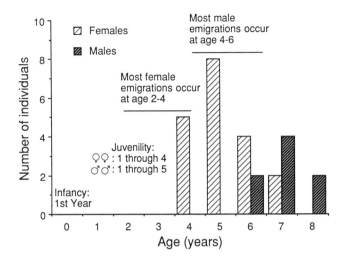

Fig. 8.2. Frequency distribution of age at birth of individual's first offspring. Ages of 5 females and 4 males were estimated from their size at first contact; ages of 14 females and 4 males were known to within a few months or exact month.

years for females and 7–10 years for males (Fig. 8.3). Although we recognize that adulthood is a gradual transition, the ages of 5 for females and 6 for males are convenient to mark the end of the juvenile stage. We are including as juveniles the

adolescent age class, defined elsewhere as "subadult" (36–59 months for females and 48–71 months for males; Crockett & Pope 1988). Thus juvenility lasts for 4 years for red howler females and 5 years for males.

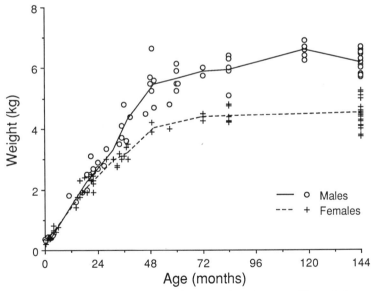

Fig. 8.3. Weights in kg of 145 red howlers captured for ear-tagging. Ages up to 60 months known to ± 1 month; additional points through 84 months (immature when first recognized) estimated to nearest year (or less for 60 months). Males ($n = 6$) plotted at 118 months were considered young adults (~84 months) 34 months previously (Thorington, personal communication; Thorington et al. 1979). Points at 144 months include middle-age, older, and all unknown-aged adults ($n = 15$ males, 26 females). Solid (male) and dashed (female) lines connect mean weights per 12-month period, plotted at mean age for both sexes combined (equally spaced distance curves [e.g., Coelho 1985] would be misleading due to skewed age distribution of the sample within years of age).

SEX DIFFERENCES IN JUVENILE
SURVIVAL AND DISPERSAL

Successful reproduction for a female is not simply a matter of surviving long enough to reach maturity. The single most important factor is whether she emigrates. Red howler males usually emigrate from their natal troops as adolescents or young adults, 4 to 6 years of age, as is typical of polygynous primate species (Pusey & Packer 1987). What is unusual is that most females emigrate singly as juveniles, at age 2 to 4 (Crockett 1984). The youngest female disperser was less than half the age of the youngest verified male emigrant (20 vs. 43 months).

The pattern of emigration and subsequent dispersal appears to have a profound effect on survivorship of the two sexes. Survivorship of 234 red howlers born into the population was analyzed (see Crockett & Rudran 1987b, Fig. 1). Whereas about 56% of males were still present in the population at 7 years of age, only 26% of females were known to have survived to this age. Actual survivorship was probably somewhat higher, especially for females, since some animals that disappeared probably left the study area. Emigrant females traveled over much larger areas than males and moved far from their natal ranges (Pope 1989). However, as a consequence they probably were more exposed to mortality factors such as predation (see Isbell 1990; Isbell et al. 1990). Evidence from extra-troop females within the study population indicates that few surviving emigrants manage to reproduce successfully. A comparison of female and male fates through young adulthood illustrates the different challenges faced by members of the two sexes.

Fates of Juveniles

Females. The most common outcome for females was "disappeared" (Fig. 8.4a). Using an estimate of 21% female infant mortality (Crockett & Rudran 1987b), we calculated that the 62 Woodland juvenile females potentially reaching sexual maturity between January 1981 and July 1985 (Pope 1989) originated from a birth cohort of 78 females. Fifteen of the 17 females that remained in their natal troops at the end of the study period had borne offspring, and the other two eventually may have been re-

cruited as natal breeders ("recruitment" begins with the birth of a female's first infant). Thus 19–22% of Woodland females born became reproductive females in their natal troops.

About 58% of females born in the Woodland emigrated or disappeared as juveniles. Because untagged or unscarred emigrants from study troops cannot be distinguished reliably from new immigrants into the area, the fates of tagged females ($n = 20$) were used to estimate the fates of the unmarked females that disappeared ($n = 25$). For example, one of the 20 (5%) definitely died after emigrating; thus 5% of the population that might have emigrated, or 2.9% of females born, was estimated to have died.

Based on these calculations, about 18% of females born definitely emigrated from their natal troops. One-third of these successfully bred within the Woodland population in a new troop formed with other emigrant males and females. Three females that bred outside their natal ranges did so on average 1.4 km away, or 5.1 home range diameters (Pope 1989). Obviously, this estimated value represents the mean minimum dispersal distance.

No female emigrants succeeded in entering and breeding in a previously established troop, and this is generally rare (Crockett 1984). Breeding vacancies in established troops are nearly always filled by natal females. Two-thirds of the emigrants survived in the area for a while, and then disappeared or, in one case, died. One of the tagged Woodland emigrants was last seen in a Gallery troop, 6 km from her natal troop (Crockett 1985); she had disappeared by 1987, and it is not known whether she successfully bred. The remaining 40% of Woodland females born simply disappeared.

Only seven immature females were tagged in the Gallery, and no marked female emigrants were located. Nine females bred in their natal troops and 15 juvenile females disappeared during the same period. Assuming 21% female infant mortality, 30% of Gallery females born bred natally and 49% emigrated or died as juveniles (Fig. 8.4a). The rate of natal breeding in the Gallery was nearly 10% greater than in the Woodland, but the difference is not significant [chi-square test of independence, $\chi^2(1) = 0.83$, $p > 0.30$, NS, $n = 26$ natal recruits, 60 emigrated or disappeared].

Thus at 40 to 49%, the modal fate of juvenile red howler females is "disappeared." Given the

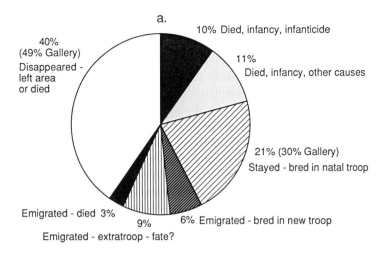

a.

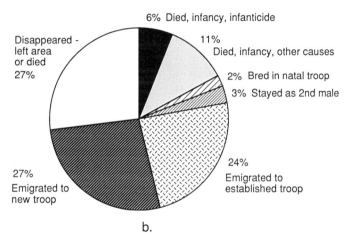

b.

Fig. 8.4. Pie charts depicting fates of (a) 78 females and (b) 76 males through young adulthood. Percentages for infant mortality are based on both habitats. The remaining pie segments are based on Woodland data, with Gallery female percentages in parentheses. Fates of identifiable animals were used to estimate the fates of those who could not be recognized after dispersal.

low mortality of juvenile males in troops and the ranging patterns of extratroop females, described later, we believe that many of these females initially emigrated and rapidly moved out of the study area. A few of them probably succeeded in breeding elsewhere, but an extensive search of areas outside the two study populations in 1983 failed to locate any tagged animals of either sex among the 212 individuals censused (Crockett 1985).

Males. Mortality for males through 7 years is significantly lower than for females (Crockett & Rudran 1987b), and this is reflected in a dramatically different pattern of male fates (Fig. 8.4b, outcomes calculated as for females). Most males were still alive and in the same population as young adults.

Because males do not disperse as far as females, there is greater certainty about their fates through young adulthood. Four males born stayed in their natal troops. Two of six males that reached adulthood prior to 1981 bred in their natal troops; thus an estimated one-third of these four males, or almost 2% of males born, eventually bred natally, and two-thirds (about 3%) stayed as the second adult male, usually with a known or suspected male relative. Approximately 51% of males born emigrated from their natal troops and were relocated within the study area after joining established (24%) or new (27%) troops. None of the male cohort studied were observed to be solitaries.

The mean primary dispersal distance for males was one home range diameter (275 m); for primary and secondary dispersals combined, 71% ($n = 35$) immigrated into an adjacent troop (Pope 1989). Some males travel farther: one Woodland male was located in the Gallery forest about 3 km from his natal troop (Crockett 1985).

The remaining 27% of males in the study cohort simply disappeared, and some may have died prior to dispersal. This figure is far lower than female rates of disappearance.

Data on juvenile male fates in the Gallery are scant because only five males were marked. One ear-tagged male bred in a newly formed troop 600 m from his natal troop (two ranges away); two unmarked but distinctive males bred in troops that newly formed adjacent to the males' natal ranges.

Of course, not all surviving males breed, and males tend to move more than once in their lifetimes. Unlike females, who have a low probability of dying once they reach a reproductive position in a troop, males may incur repeated injuries in fights with other males over breeding status. Three adult males have been found dead after replacement by new breeding males, and many evicted males were wounded prior to disappearing (Rudran 1979; Crockett & Sekulic 1984). Breeding-male tenure length averaged 5 to 7.5 years compared with a maximum reproductive life span estimated to be 20 years (Crockett & Sekulic 1984; Pope 1990). Males moved an average of about three home range diameters during their lives (Pope 1989), suggesting that many of the recognizable males that disappeared probably died.

Costs of Female Emigration

Females either breed in their natal troops or disperse great distances. Emigrant females have larger ranges while extratroop and breed farther from their natal troops than do males. Pope (1989) radio collared three solitary females and four solitary males in the Woodland. During a mean observation period of 3.8 months, the females' mean ranging area (estimated from minimum convex polygons) was larger than the males': 133 ha for females versus 42 ha for males (the three females had ranges larger than three of four males; Mann–Whitney $U = 2$, $n_1 = 3$, $n_2 = 4$, $p = 0.22$, two-tailed, NS). In comparison, Woodland troop home ranges are about 6 ha, so extratroop animals of both sexes have vastly larger ranges. The mean daily minimum distance between radio-tracked locations was significantly greater for females [316 m for females vs. 208 m for males; $F(1) = 5.32$, $p < 0.05$].

Crockett (1984) speculated that emigration was more costly than natal breeding for females, and now we have supporting evidence. Many of the disappeared emigrants must have died. By ranging widely and far from their natal areas, initially surviving emigrants must expend more energy and are exposed to greater risks than nondispersers. Solitary females receive aggression from troop members and suffer a higher rate of injury than natal females (Sekulic 1982a; Crockett & Pope 1988).

Perhaps because they must forage in unfamiliar areas and frequently are displaced by troops, solitary females have diets deficient in some nutrients important to reproduction (Pope 1989). In eight paired activity budget samples (three female pairs and five male pairs), Pope (1989) determined dietary content for solitary animals and age-matched troop members using standard techniques of nutritive component analysis. The foods eaten by troop-living females were significantly higher in crude protein and phosphorus and significantly lower in ash-free neutral detergent fiber (NDFA) than foods of solitary females, solitary males, or troop males (the last three groups did not differ from one another). For example, troop females' diets had 20% protein compared with 13–14% for the solitary females and troop and solitary males. Lower NDFA indicates a higher ratio of nonstructural to structural carbohydrates, and may be responsible for the higher in vitro digestibility found for troop females' diets. During pregnancy, mammals' requirements for protein, phosphorus, calcium, and kilocalories increase dramatically (Lloyd et al. 1978). Even though the female troop members were neither pregnant nor lactating, their diets had levels of nutrients more appropriate for achieving breeding condition than did solitary females' diets.

Exposure to greater risks and dietary deficiencies are not the only penalties paid by dispersing females. They also suffer delayed age at first breeding. The females that succeeded in breeding in new troops did so at a significantly older age at first birth: a median of 6.9 years compared with 5.1 years for natal females (Mann–Whitney $U = 7, n_1 = 3, n_2 = 17, p < 0.05$, two-tailed).

Emigrant females that succeed in finding breeding positions usually do so in newly formed troops. Herein lies another cost. Although newly formed troops in the lower density Gallery Forest were usually successful, the more recently formed troops in the Woodland had a high rate of male replacement, attendant infanticide, and dissolution; for example, in 1984, five new troops formed in the Woodland, infant mortality was 50%, and two of the troops dissolved (Crockett, unpublished data; Pope 1989). Overall, the average reproductive success of females that breed natally is bound to be much higher than that of those that emigrate.

PATTERNS OF JUVENILE BEHAVIOR RELATED TO REPRODUCTIVE COMPETITION

The challenges of survival and reproduction faced by juvenile males and females clearly differ. Given the potential costs of emigration, females should attempt to stay in their natal troops if they can. Despite the apparent benefits of philopatry, however, most females disperse as juveniles. Males, in contrast, usually remain in their natal troops and emigrate as adolescents (subadults) or young adults. In this section, we examine behavior patterns exhibited by each sex, and their potential influence on lifetime reproductive success.

Females

Natal Breeding Related to Female Group Size. For a female, emigration seems to be determined primarily by factors that are unrelated to specific behaviors that she might employ. The percentage of females recruited as natal breeders depends on the number of resident reproductive females (Fig. 8.5). No juvenile females bred natally in troops that already had four adult females, and few did in those with

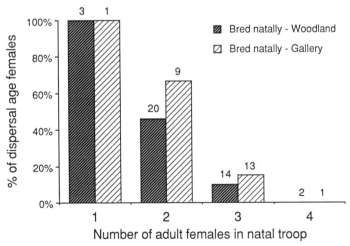

Fig. 8.5. Number of adult females in troops and the percentage of available natal females recruited. Numbers above bars indicate the total number of females reaching dispersal age per female group size. Females that were not recruited natally either emigrated or disappeared (fate unknown) as juveniles.

three. Most natal recruitments were into troops with two adult females. In the few cases where adult female membership dropped to one due to death of the second female, a natal female always was recruited if one of appropriate age was available. Whereas the same pattern was obtained in both habitats, proportionally more females bred natally in two-female troops in the expanding Gallery population [chi-square test of independence, $\chi^2(1) = 1.17, p > 0.20$, NS].

The Importance of Maternal Presence. Maternal presence seems to influence whether a juvenile female stays. In 16 of 17 cases of natal recruitment (14 Woodland and 3 Gallery) for which there was sufficient maternity information, the recruit's mother was also present. In the discrepant case, a Woodland female's mother had died 13 months prior to her recruitment, dropping the number of adult females to one. In a Gallery troop, a juvenile female's mother died when she was only 13 months old, leaving a single adult female of unknown relationship. Although this created a breeding vacancy and she was the only juvenile female present, she apparently was too young to take advantage of the opportunity; a nulliparous female immigrated when the juvenile was 2 years old, and by age 3 the juvenile was gone. This case illustrates the sort of circumstances under which a rare immigration into an established troop might occur.

Multiple recruitments usually involve daughters of the same female. In the Gallery, only one troop had more than one recruit. The mother of the first recruit was unknown; the second and third were thought to be maternal sisters, but maternal identity was not certain. In the Woodland, there were three cases where two maternal sisters were recruited. In one, two daughters of one female were recruited, while three daughters of another disappeared. In another troop, the daughter of a dead female was recruited prior to the recruitment of two younger daughters of a second female. In other cases of multiple recruitment, mothers of the older recruit(s) were unknown. Although these data are limited, they suggest that, over time, one female's daughters are recruited, eventually producing a single matriline. Thus female relatedness varies from troop to troop, being highest in long-established troops and lowest in ones of recent origin composed of unrelated females with no natal recruitment.

The presence of a maturing female's mother may increase her chances for recruitment. Injuries on adult females suggest that they are actively involved in recruitment disputes (Crockett 1984). Furthermore, once one female's daughter is recruited, it may be that the two of them can form a coalition to ensure that the unrelated female's daughters are not allowed to stay. However, a troop that already has three adult females can at most recruit one more, and only 11% of troops consist of four females.

Juvenile females do engage in harassment-type behaviors directed toward the other females and males as well (Crockett 1984). Saavedra (personal communication) observed a juvenile female in each of two troops in mid-1982. Eventually, one female was recruited and the other disappeared. The recruited female was the oldest immature female, and an old female died at the time she reached sexual maturity. The unsuccessful female repeatedly incited one of the adult males into a coalition against an adult female that was not her mother. Although her two older sisters were recruited in 1982 and 1984, there were by then four adult females in the troop and she eventually left. Although behavioral tactics may play a role in who gets to stay, ultimately demographic factors are more important.

Infant Care by Juvenile Females. A juvenile behavior that might influence a female's reproductive success is alloparental care (Fairbanks, Chapter 15, this volume). Considerable interest is shown toward young red howler infants by females of all ages, and much of it is resisted by the infants' mothers. Such resistance has also been reported for mantled howlers (Clarke & Glander 1981; Chapman & Chapman 1986). However, there are cases of red howler behaviors that truly qualify as "caretaking." In these examples, females carry another female's infant for extended periods.

In the best documented case, Crockett observed the adolescent daughter of one Gallery female to carry her infant brother virtually all day when he was about 4 to 6 months old. He was returned to the mother at night and during daytime rest periods when he nursed. During dawn-to-dusk observations, the mother often was completely out of sight of the infant and subadult caretaker.

Similar caretaking behavior was observed during census contacts. Thirteen of 16 cases involved immature females. In 11 of these, a female carried her own mother's infant—that is, her full or half-sibling. In two cases, a female carried the infant of a female that was not her mother, but the infant was probably a half-sibling.

How might caretaking by juvenile females enhance their own reproductive success? The nepotistic pattern suggests that females do not forestall eviction by helping nonrelatives care for their infants. Moreover, 7 of 12 caretakers eventually emigrated. Because many females first give birth in new troops in the absence female relatives, prereproductive caretaking experience could increase primiparous infant survival (Fairbanks 1990).

Interbirth Intervals. Interbirth intervals after the births of females that became natal breeders were significantly longer than those after females that emigrated (19.9 vs. 16.6 months: Crockett & Rudran 1987b). There are no behavioral data to indicate whether these longer birth intervals resulted from behavior on the part of these females that might have suppressed conception of the mother's next infant, or whether the mothers themselves invested more in these females by lactating longer.

Males

Causes of Dispersal. Most natal male dispersals were identified from census data, and the proximate cause of emigration could only be guessed from circumstantial evidence. The following proximate factors were identified: (1) interacting aggressively with resident males (natal/related, nonnatal/unrelated, adult and/or immature); (2) leaving after the natal troop was invaded by one or more new males; (3) joining another troop or extratroop association; and (4) leaving after a period of coming and going that might last for months (often to join another group). Thus some male dispersals are "voluntary" and others are evictions.

Forming Alliances with Other Males. Males have little chance at winning in male–male competition with adult males until they reach adult size at age 5–6 (Fig. 8.3). Males emigrating as juveniles and forming an extratroop association with emigrant females are likely to be replaced by another, larger male. Even a single male of adult size has little chance of successfully invad-

ing a troop and replacing the resident male. Two collaborating males were much more successful in doing so (Crockett & Sekulic 1984). Furthermore, the likelihood of a troop being invaded decreases with the number of adult and subadult males present (Pope & Crockett, unpublished data). Thus juvenile males might increase their inclusive fitness by delaying dispersal and helping prevent their younger siblings from becoming infanticide victims. Indeed, few males emigrated prior to age 5 years and many were fully adult.

Alliances among males may be formed in natal troops or sometimes by apparently unrelated males after dispersal. Whereas females dispersed singly, males sometimes departed in the company of other males. The tenure of two or more related adult and subadult males was significantly longer than when the males of multimale troops were not related (Pope 1990). Associations of male relatives and potential relatives had a mean survivorship of 8.2 years, significantly longer than the mean survivorship of 2.3 years when males were unlikely to be related. The 14 associations in the first category included five father–son pairs, three pairs of full sibs, one pair of half-sibs, one quartet of combined sibs and half-sibs, and four associations of males that emigrated together from the same troop. Unrelated associations included 8 cases of single male troops that were joined by solitary invaders that then coresided with the original resident and 10 cases in which the males came from different troops.

Some examples illustrate the roles that juvenile males play in cooperative associations:

1. Young adult male 7231 invaded an adjacent troop and coresided with long-time resident 7111 for the next 4 years; they fought on numerous occasions. When two of 7111's sons were about 4 years old, the three of them collectively evicted male 7231. Ironically, one of the sons had survived an apparent infanticidal attack by 7231 when he first invaded (Sekulic 1983; Crockett & Sekulic 1984, Case c).

2. A troop was invaded by a coalition of two unknown males. The resident male 911 and his two 3-year-old sons repeatedly tried to chase away the intruders. However, as the three of them pursued one male, the other would approach the females and infants. The juveniles' coalition with the father perhaps forestalled an immediate male replacement but did not prevent

three apparent infanticides (Crockett & Sekulic 1984, Cases 6, 7, and 8). Male 911 and the older son disappeared about 6 months later, apparently evicted by the two invaders.

3. Two brothers dispersed about a year apart but rejoined in a newly formed troop, suggesting that alliances formed during the juvenile period can be reinstated after a considerable period of separation.

4. When possible, males choose relatives over nonrelatives as partners. Male 6312 and three other males invaded a troop in 1977 (Rudran 1979). By mid-1980, only 6312 and natal adult 6331 remained. When, in 1981, 6331 began to spend time away from the troop in a manner typical of young males just prior to emigration, male 6312 actively sought him out in a peripheral part of the troop's home range. Male 6331 lunged at him, but 6312 approached a second time. After a brief head-bobbing session, they play-grappled for several minutes, rested next to each other for an hour, and remained within close proximity to each other for the rest of the day. Young male 6331 remained in the troop for 3.5 more years. The older male was an active rather than a passive participant in maintaining a relationship with the younger male, whose departure would have left male 6312 the only adult male in the troop. However, 1 month after 6312's own son reached adolescence, male 6331 emigrated completely and was observed later as an extratroop solitary.

5. Adult and subadult males in a noninvaded troop (where the males were probably related) simultaneously participated in howling bouts more often than in a troop where one male was an invader; male–male aggression was also higher in the latter troop (Saavedra 1984).

Males are likely to increase their inclusive fitness by forming long-term associations with related males, and some of these cooperative alliances are formed in natal troops by juveniles. We suspect that in some of the troops where we see an older male that is not the breeding male compatibly coresiding with a younger breeding male that the younger male is his son or brother. The older male had some personal reproductive success while younger, and is now contributing to his inclusive fitness.

As for females, however, males are constrained by demographic factors. Since troops are small, there are relatively few potential peers. By chance, there may be many, few, or no juvenile males to form coalitions. As in example 2, a male's sons may be too small to aid in defense, or, as in example 4, allegiances may change from an unrelated male to a maturing son.

Both Sexes: Coping with Extratroop Existence

Dispersing howlers of both sexes face threats to their survival from intraspecific aggression as well as from the environment. Troop adolescents and extratroop individuals have significantly more injuries than expected (Crockett & Pope 1988). Solitary individuals receive aggression from troop members (Sekulic 1982a; Pope 1989). Emigrants must be able to travel alone, sampling new foods in unfamiliar areas. Howlers that disperse from one habitat type to another have to discover new food varieties, since only 38% of plant species eaten are found in both the Woodland and the Gallery Forest (Crockett 1987).

To form a new troop, dispersers must establish affiliative, cooperative relationships with strangers. Pope (1989) observed extratroop individuals for 3- to 4-day periods. These individuals spent about 52% (males) to 62% (females) of the time alone, but the rest of the time in an extratroop association (two or more cotraveling howlers among whom no infants had been born). The typical association was composed of a single male and two females. These mixed-sex groups separated and came together again many times during a single observation day. In a several-day period, a given animal was either alone or a member of one particular association. Visual location of radiocollared animals over 2- to 3-month periods showed that males associated with an average of five different individuals representing 2.75 different associations, whereas females associated, on average, with six different animals representing 3.67 different associations. Since newly formed troops are usually composed of two females, it might be considered advantageous for females to disperse with another female relative. However, only one example of two juvenile females disappearing from the same troop during the same month has been detected, and they were never seen together (Crockett 1984).

Many extratroop howlers, especially males, are no longer juveniles. However, the lifetime reproductive success of these monkeys may depend directly on competitive abilities, foraging skills, and endurance developed while immature.

THE HOWLER PATTERN IN BROADER PERSPECTIVE

For red howler monkeys, several important factors affecting a juvenile's potential reproductive success appear unrelated to particular behavior patterns. For females, the major consideration is how many reproductive females are already in their natal troops. For males, an important element is the presence of related males with which to form alliances. For both sexes, population density and habitat characteristics affect the probability of successful new troop formation and, thus, breeding vacancies that may be obtained without physical challenge. The importance of these variable demographic factors has been emphasized previously (Altmann & Altmann 1979).

A significant sex difference in red howlers is the age by which future reproductive success can be predicted with reasonable certainty. Whereas males may influence their reproductive success through repeated opportunities for male–male competition, expected lifetime reproductive success of females is often determined during the juvenile stage, being sharply reduced if they emigrate.

In many ways, the juvenile experience of red howler males is similar to that of most polygynous primates and other social mammals. There is time for play, which mimics actions used in fighting and develops one's competitive abilities and physical fitness (Bekoff 1988; Fagen, Chapter 13, this volume). In general, red howler males do not leave their natal troops until they approach full size. Some are evicted in conjunction with male invasions, but are not as young as Hanuman langurs (Rajpurohit & Sommer, Chapter 7, this volume). In contrast to most species where male dispersal is the rule, however, a small proportion does not disperse, and some males have even bred in their natal troops with related females; natal breeding also has been reported for *Papio cynocephalus ursinus* (Bulger & Hamilton 1988) and *Alouatta palliata* (M. Clarke, personal communication).

Although much of the evidence is anecdotal, males do seem to solicit long-term alliances with other males. Red howler males have been observed to act cooperatively to invade troops and evict resident males and to defend their own troops from invaders. Especially in the case of troop defense, many males that assist are still juveniles. Some males form long-term relationships with other males that begin during the juvenile stage. Although male kinship is not a prerequisite for cooperation, associations of related males last significantly longer than those of unrelated males.

The role of kin selection in the evolution of red howler society is not fully understood. Howlers form alliances with kin and nonkin alike, as juveniles and adults, although there seems to be a preference for kin when kin are available (also true for male lions [*Panthera leo*]: Packer & Pusey 1982). There are no data to indicate whether howlers are able to recognize kin, and preference for kin could be based entirely on the familiarity of growing up in the same troop (see Walters 1987b).

Coalitions of unrelated males that cooperate to defend or invade troops do not fit a model of reciprocal altruism (e.g., Packer 1977) because multimale troops are usually genetically one-male harems (Pope 1990). More likely, given high postdispersal mortality, male cooperation should be viewed as "helpers at the nest" where nonbreeders' reproductive success is enhanced by being part of a reproductive unit and increasing their chances of obtaining the next available breeding vacancy (Emlen 1984; Wiley & Rabenold 1984). More often, the "vacancy" is created by a direct challenge than a random death. In red howlers, status changes between coresiding males have been documented several times (Crockett & Sekulic 1984). Thus the collaboration of red howler males may fit a coalition game model (Noë 1990).

The life history of female red howlers is unusual among primates, with bimodal outcomes. Some proportion breed in natal troops among kin, whereas the others disperse, usually far away, and attempt to find breeding opportunities elsewhere. The evidence points to very high mortality among juvenile females, especially postdispersal.

The factors associated with female dispersal in red howlers have been discussed previously (Crockett 1984; Crockett & Eisenberg 1987; Pope 1989). The constraints of foraging and the associated pattern of exploitation characteristic of this species involve site-dependent spacing and territorial defence. Since it is critical to reproduction that females maximize food intake, and since diet analysis suggests that resources important to female reproduction are limited, it is in females' interests to limit group size. It appears that the number of reproductive positions in red howler troops is limited (to our knowledge, no troops with five or more reproductive females have been unequivocally reported). Observed and inferred aggression among females, reflecting female–female reproductive competition (see also Cheney & Seyfarth 1987), is the likely mediating factor leading to the emigration of some females. Clearly, then, juvenile females, as well as males, should engage in activities to improve their competitive abilities.

Dispersal is not always disadvantageous for females. Howlers are thought to be colonizing species (Eisenberg 1979). One could envision ecological circumstances where new habitat would become available and some colonizing dispersers might realize high reproductive success. Within the protected ranch boundaries, improved habitat has made it possible for emigrants to form new troops, some of which have been as successful as established troops in terms of birthrate and infant survival. However, howler habitats beyond the ranch boundaries are shrinking, not expanding.

Contrasts with Mantled Howlers

The general pattern of some females emigrating and breeding outside their natal troops seems characteristic of the genus *Alouatta*. Female dispersal has been observed directly or inferred from the sightings of solitary females in other howlers (*A. palliata, A. caraya, A. pigra*), and new troop formation has been documented for *A. palliata* and *A. caraya* (Crockett & Eisenberg 1987; Rumiz 1990). In the mantled howler (*A. palliata*), most juvenile females are evicted at age 2–3 years by female nonrelatives, a pattern reminiscent of red howlers; males, too, emigrate primarily through aggression by unrelated males but are only 1–2 years old, much younger than red howler male dispersers (Clarke & Glander 1984; Clarke & Zucker 1989; Clarke 1990; M. Clarke, personal communication). In contrast to our finding that female emigrants

first breed at an older age than natal breeders, Clarke and Zucker (1989) report that nongroup juveniles of both sexes mature at an earlier age than do troop members, as indicated by development of adult genitalia (testes descent occurs at about 1 month of age in red howlers compared with about 3 years of age in mantled howlers: Glander 1980; Crockett & Eisenberg 1987).

Another problem arises in that the explanation of limited breeding positions that seems to hold for red howlers and the other howlers with small troop size cannot apply directly to mantled howlers (elaborated in Crockett & Eisenberg 1987). In that species, there may be as many as 14 adult females in a troop, and immigration into established troops is rather common. In general, howlers pose some problems for theories predicting female relationships (van Schaik 1989). However, in all howlers studied, female–female competition seems to be an important factor affecting female reproductive success.

Clearly, an important direction for future research is to understand the causes of the somewhat different pattern of juvenile experiences in red versus mantled howlers. Although all howlers seem to share the same dietary niche, being selective folivore-frugivores (Crockett 1987), mantled howlers differ from red howlers in several respects: (1) by possibly being less territorial, although playback experiments indicate behavioral flexibility in spacing patterns with some territorial defense (Whitehead 1989); (2) by adjusting troop size by forming subgroupings when resources are clumped in low density (Chapman 1990) rather than by limiting female group size; and (3) by showing the least sexual dimorphism in adult body weight of all howlers, perhaps reflecting reduced male–male competition (Crockett & Eisenberg 1987). Given these differences, we cannot generalize the juvenile pattern from red howlers to all howlers without qualification.

Comparisons with Other Species

This pattern of female dispersal, and its consequences for juveniles, is not unique to howlers. Strier (1990) suggests that female dispersal is so common among New World primates that it may represent the ancestral form. It appears that in spider monkeys (*Ateles paniscus*) and muriquis (*Brachyteles arachnoides*), females disperse, males do not disperse, and coalitions are formed among related males (Strier 1990; Symington 1990; Strier, Chapter 10, and Milton, Chapter 12, this volume). Red howlers show yet another New World variant in having bisexual emigration but with elements reminiscent of their ateline relatives.

Callithrichids also demonstrate bisexual dispersal, but their smaller body size, greater vulnerability to predation, and essentially monogamous breeding system create different selection pressures on dispersal patterns (Goldizen 1987). Some Neotropical primates show dispersal patterns and resulting experiences of juveniles that are more similar to Old World Cercopithecines (e.g., wedge-capped capuchins: O'Brien & Robinson, Chapter 14, this volume).

Bisexual emigration and female transfer are common in gorillas (Watts 1990a; Watts & Pusey, Chapter 11, this volume). Mountain gorilla socioecology bears some resemblance to that of red howlers: females appear to leave groups that get too large, because of feeding competition; however, females with young must be associated with an adult male to defend against infanticide. Bisexual emigration might also occur in some other primate species that live in small, polygynous groups. One candidate is the De-Brazza's guenon (*Cercopithecus neglectus*), which lives in small groups variously described as monogamous or polygynous (Cords 1987; see Kirkevold & Crockett 1987).

Interesting parallels between red howlers and ringtailed lemurs (*Lemur catta*) are emerging (Vick & Pereira 1989; Pereira, Chapter 20, this volume); all females were thought to be philopatric until recent observations showed that some are forced to emigrate. Over the years, one matriline has been evicted in several groups. Forced female dispersal is suspected to result when group sizes grow too large due to low mortality in expanding populations.

Insights may be gained from comparisons with nonprimates. For example, bisexual dispersal, immigration, and new group formation occur in the dwarf mongoose (*Helogale parvula*) (Rood 1983). Like red howlers, dwarf mongoose females that attained breeding status remained in the same home range throughout their lives. Breeding in new troops led to earlier breeding than staying in one's natal group (this may also be true for red howler males, but not females). Related and unrelated individuals of both sexes exhibit alloparental helping behavior.

Predation is suspected to be common, and unrelated helpers may be next in line for breeding vacancies. In the vicuña (*Vicugna vicugna*), limits in group size seem to be correlated with food, but it appears to be the breeding males that reduce group size by expelling juveniles of both sexes (Franklin 1983).

FUTURE RESEARCH DIRECTIONS

We have introduced a picture of red howler juvenility that is intriguing and provocative. Red howler males have experiences reminiscent of those of many species, and their behavior generally makes sense. We would like to know more about male behavior involved in alliance formation. The early emigration of mantled howler males, as described previously, begs explanation. Indeed, there are several contrasts between mantled and red howlers that need to be investigated.

For females, one must ask the question: If it is obviously so advantageous to breed natally, why disperse? The proximate answer usually must be that forced eviction makes it impossible to stay. However, we have few direct observations and mostly circumstantial evidence from injuries indicating how this happens. The number of cases suggesting that red howler troops develop toward one matriline over time is small, and more data would strengthen this generalization.

We also would like to learn more about how patterns described in this chapter may vary as a function of different stages of population growth and density. Although not statistically significant, female recruitment, especially into two-female troops, was higher in the Gallery than the Woodland. This is somewhat counterintuitive, since there seemed to be more opportunity for successful new troop formation in the Gallery. Data on females' older age at first reproduction in new troops comes from the saturated Woodland habitat. Perhaps some of the females founding the many new troops that formed in the Gallery were younger. Thus under circumstances of population expansion into previously unavailable habitat (e.g., when flooding levels drop due to climatic changes or habitat alteration), dispersing females might have higher reproductive success than if they had not emigrated. The pattern of emigrating females dispersing far away could result from two, not mutually exclusive, factors: (1) new troops might have to form primarily on the margins of a concentration of troops, and/or (2) females may be avoiding inbreeding, since males usually disperse only one or two home range diameters away.

Janson and van Schaik (Chapter 5, this volume) suggest that delayed maturation is related to reducing risks of starvation and other costs of rapid growth, and that when predation risk is low, juveniles should prefer smaller groups. This seems inconsistent with the different modal age at emigration of red howler males and females relative to their size, since the larger male sex stays longer than females. It is also inconsistent with the poorer diets found for both sexes while solitary (but consistent with their having to learn to exploit new food sources). It may be that red howler troop sizes are so small that they never reach the levels of detrimental intratroop feeding competition, but this would be contrary to our explanation of female group size being related to limiting natal female recruitment. However, mantled howlers may be responding differently, and the more rapid maturation reported by Clarke and Zucker (1989) perhaps is related to the nutritional advantage of being solitary in a habitat where predation on solitaries is negligible.

ACKNOWLEDGMENTS

This research was funded by the Smithsonian Institution International Sciences Program, Friends of the National Zoo, National Geographic Society, the Harry Frank Guggenheim Foundation, and the Katharine Ordway Chair. We thank all those listed in Table 8.1 and our previous publications, and the T. Blohm family for allowing us to work at Hato Masaguaral. This manuscript was improved by helpful comments from M. Pereira, L. Fairbanks, M. Clarke, V. Sommer, and reviewer 2.

9

Juveniles in Nongregarious Primates

LEANNE T. NASH

The nongregarious, or "solitary," primates include the diurnal orangutans and most of the nocturnal primates (Table 9.1). "Solitary" denotes the opposite of "gregarious," not of "social" (Charles-Dominique 1978; Bearder 1987). Although, in these species, most foraging is done apart from conspecific adults, individuals maintain social networks by vocalizations, scent marking, sleeping associations, and occasional interactions during the active period (for specific examples, see Table 9.1). The nongregarious species show considerable diversity of diets, predator avoidance methods, and patterns of sociality as revealed by ranging and dispersal patterns (Bearder 1987; Richard 1987; Table 9.1).

Variability in these species' ecological and, especially, social worlds (e.g., the degree to which animals interact both during the active period and in sleeping groups) has not been fully appreciated. Further, their patterns of sociality, including the social worlds of juveniles, form a bridge to semigregarious ("fission–fusion") modes of primate sociality (e.g., *Pan* and *Ateles:* Symington 1990; ruffed lemurs: Pereira et al. 1988; Vick & Pereira 1989). Bearder's categorization of variation in sociality (1987; Table 9.1), based primarily on range overlap, also involves these patterns of interaction. In contrast to most gregarious fission–fusion primate species, whose sociality depends primarily on food patch size (e.g., Symington 1990), fission–fusion variation in the nongregarious species relates to both food supply and protected sleeping sites. The latter may involve more predictable variation in social interactions than where food supply alone influences the occurrence of groupings.

Until recently, most information on juvenile primates came from the semiterrestrial cercopithecine monkeys, especially baboons, rhesus and Japanese macaques, and vervets (Pereira & Altmann 1985; Caine 1986; Walters 1987a). These species' societies are relatively homogeneous, showing female philopatry and male dispersal, dominance rank among females highly dependent on alliances (especially with kin), male dominance over females, sexual dimorphism, aggressive male competition for access to mates, and little male care of young. Among these species, the juvenile period is risky. Young juveniles appear to seek companions that can provide protection from risk. Among the survivors, important sexual diethisms develop (Pereira & Altmann 1985; Caine 1986; Walters 1987a; van Noordwijk et al., Chapter 6, and Fairbanks, Chapter 15, this volume). Old juvenile females, compared with males, play less, play less aggressively, show greater interest in interactions with infants, become increasingly more enmeshed in affiliative relationships with other females (especially within their own matriline), and appear to work to form alliances with both closely related and higher ranking females. Old juvenile males play roughly with each other, practicing the fighting skills, but attend less to forming alliances with females. For species where sex-biased dispersal is reversed, limited data suggest that juvenile sex differences are reversed (Walter 1987a).

Few of the solitary species have been subject to longitudinal studies of known individuals (Table 9.1). Quantitative data on juvenile social interactions or demography (e.g., sex differences in mortality rates) are rare (but see

119

Table 9.1. The Solitary Primate

Species[a]	Body weight[b]	Weaning[c]	Mature[d]	Breeding season[e]	Litter size[f]	Social system[g]
Galagos and Lorises						
Galagoides demidoff[h,j]	0.07	40–50	8–10	N	1(w2)	I
Galagoides thomasi	0.10			?	?	?
Galagoides zanzibaricus[h,j]	0.13/0.16			Y2	1(c2)	III
Galagoides alleni[h,k]	0.31		8–10	N	1(w2)	I
Galago moholi[h,l]	0.16–0.19/0.19–0.21	70–100[h]	7–9	Y2	w2	I
Galago gallarum	?			?	?	?
Galago senegalensis[m]	0.20/0.23	70–100[h]	11–13	Y2	1(c2)	I?
Galago matschiei	0.21			?	1	?
Galago elegantulus[n]	0.29		c.10	?	?	I?
Otolemur garnettii[h,o]	0.72/0.82	140[h]	Over 12	Y	1(c2)	I
Otolemur crassicaudatus[h,p]	1.25/1.50	70–134[h]	15	Y	w2(w3)	I
Arctocebus calabarensis[q]	0.31/0.32	100–130	9–10	N	1	II?
Perodicticus potto[h,r]	1.08/1.02	120–180	c.18	Y	1	II
Loris tardigradus[s]	0.26/0.29	169[h]	11	Y	1(c2)	II?
Nycticebus pygmaeus	?			?	?	?
Nycticebus coucang[t]	1.20/1.30	180	c.18	Y	1(c2)	II?
Tarsiers						
Tarsius bancanus[u]	0.11/0.12	60–90+		Y	1	II or I?
Tarsius syrichta	0.12/0.13			?	1	?
Tarsius spectrum[v]	0.20/0.20			Y	1	III
Lemurs						
Microcebus murinus[w]	0.06	c.60	7–10	Y2	2(c3)	IV?
Microcebus rufus[x]	0.05			Y2	?	IV?
Mirza coquereli[y]	0.30	86		Y	2	III or IV
Cheirogaleus medius[z]	0.18	61	c.12	Y	2(c4)	?
Cheirogaleus major	0.40	70		Y	2(c3)	?
Phaner furcifer[aa]	0.04/0.04			Y	1	III
Lepilemur mustelinus[bb]	0.64–0.75	135	18	Y	1	I or II?
Daubentonia madagascariensis[cc]	2.80		24+	Y	?	?

(continued)

Clark 1985). The juvenile period's bounds (i.e., weaning and reproductive maturation) are not known for many species and are based on limited samples for most others (Table 9.1). However, juveniles in nongregarious species face the same categories of problems that confront gregarious primates: they must survive the juvenile period, develop skills and relationships for later use, and manage social conflict. This review will focus on what is known about juveniles of solitary species, with special focus on sex differences in adult and juvenile worlds, as compared among nongregarious species and with the cercopithecine model. As with gregarious species, it is assumed that, in general, food is the resource most influencing female reproductive success, but access to females is the resource most influencing male reproductive success (Wrangham 1979a).

DIVERSITY AMONG THE SOLITARY SPECIES AND COMPARISONS WITH GREGARIOUS SPECIES

Feeding

Although sociality among solitary species seems to be limited by the distribution of food into small patches, these species encounter most of the feeding problems that gregarious species do, with the exception of status limiting access to foods in face-to-face encounters. We know little about the role of competitive abilities in the acquisition of preferred feeding territories. Orangutan foraging requires tracking the distribution and complex fruiting cycles of many, often rare, tree species, learning habitual foraging pathways, and processing foods requiring considerable strength or manipulative skills

Table 9.1. (*Continued*)

Species[a]	Body weight[b]	Weaning[c]	Mature[d]	Breeding season[e]	Litter size[f]	Social system[g]
Apes						
Pongo pygmaeus[dd]	37.0/69.0	2.5–3	8–12	N	1	I

[a]Taxonomic sources: Olson (1979) and Nash et al. (1989) for galagos; Tattersall (1982) for Madagascar prosimians; Niemitz (1984a) for tarsiers; Napier and Napier (1967) for others. Sources, including background information on ecology and social systems, are given for species in notes *j–dd*. See also Tattersall (1982) for lemurs.

[b]In kilograms; female–male if different and population differences indicated by "–" (Napier and Napier 1967; Tattersall 1982; Harvey et al. 1987; Nash et al. 1989). These figures are only approximate, ignore the marked seasonal weight variations among cheirogaleids, and include data from both wild and captive animals, favoring the former where possible.

[c]Weaning age in days, except for *Pongo*.

[d]Age of sexual maturity in months, except for *Pongo*, in years.

[e]Breeding seasonality: N, none; Y, yes; Y2, yes, two per year; ?, unknown. For summaries see Tattersall (1982), Bearder (1987), and Richard (1987); otherwise see citations for each species.

[f]Litter size: modal(maximum), maximum omitted if same as mode; c, data from captive colonies, w, data from wild. For summaries, see Tattersall (1982), Bearder (1987), Richard (1987), and Nash et al. (1989); otherwise, see citations for each species.

[g]Social system type following Bearder (1987). Type I: male ranges larger than females'; matriarchies present. Type II: male ranges larger than females'; matriarchies absent. Type III: male and female ranges coincide—nongregarious. Type IV: male and female ranges largely separate from one another. Alternative patterns and/or "?" listed where information is unclear or contradictory.

[h]Fieldwork of at least 12 months monitoring individuals of one population. Weaning based on manually expressing milk; otherwise based on behavioral observations, which are systematically shorter (Izard 1987). For useful, but somewhat outdated, reviews of prosimian development, maternal behavior, and reproduction, see Doyle (1974, 1979), Klopfer and Boskoff (1979), and van Horn and Eaton (1979).

[i]Charles-Dominique (1977a); Charles-Dominique and Bearder (1979).

[j]Nash (1983); Harcourt and Nash (1986); Harcourt and Bearder (1989).

[k]Charles-Dominique (1977a, 1977b).

[l]Sauer and Sauer (1963); Doyle et al. (1969); Bearder and Doyle (1974); Kingdon (1974); Doyle (1979); Charles-Dominique and Bearder (1979); Bearder and Martin (1980); Harcourt (1986); Izard and Simons (1986); Bearder (1987); Rasmussen and Izard (1988); Harcourt and Bearder (1989).

[m]Izard and Nash (1988); Nash and Whitten (1989); Zimmerman (1989a); Nash (personal observation).

[n]Charles-Dominique (1977a).

[o]Ehrlich (1977); Nash (1983); Nash and Harcourt (1986); Izard (1987); Nash (personal observation); Ehrlich and Macbride (1990).

[p]Clark (1978a, 1978b, 1985); Harcourt (1986); Bearder (1987); Izard (1987); Rasmussen and Izard (1988).

[q]Charles-Dominique (1977a).

[r]Blackwell and Menzies (1968); Jewell and Oates (1969); Charles-Dominique (1974, 1977a); Oates (1984).

[s]Petter and Hladik (1970); Izard and Rasmussen (1985); Rasmussen and Izard (1988).

[t]Ehrlich (1974); Izard et al. (1988); Rasmussen and Izard (1988); Ehrlich and Macbride (1989); Zimmerman (1989b).

[u]Fogden (1974); Niemetz (1974, 1984b, 1984c, 1984d); Wright et al. (1986); Crompton and Andau (1987).

[v]MacKinnon and MacKinnon (1980); Niemetz (1984b, 1984c, 1984d).

[w]Petter-Rousseaux (1964); Martin (1972, 1973); Hladik et al. (1980); Pages-Feuillade (1988).

[x]Martin (1972).

[y]Pages (1980, 1982, 1983); Hladik et al. 1980).

[z]Hladik et al. (1980); Foerg (1982).

[aa]Petter et al. (1975); Charles-Dominque and Petter (1980); Hladik et al. (1980).

[bb]Hladik and Charles-Dominique (1974); Russell (1977).

[cc]Petter and Petter (1967); Petter (1977); Winn (1989).

[dd]MacKinnon (1971); Rodman (1973, 1977, 1979, 1988); Horr (1977); Rijksen (1978); Galdikas (1979, 1981, 1982, 1984, 1985a, 1985b, 1985c, 1988); Schürmann (1982); Sugardjito (1983); Mitani (1985, 1989); Schürmann and van Hooff (1986); Rodman and Mitani (1987); Sugardjito et al. (1987); Nadler (1988); te Boekhorst et al. (1990).

(Schürmann & van Hooff 1986; Rodman & Mitani 1987; Rodman 1988; te Boekhoerst et al. 1990). Diets of nocturnal prosimians vary in the proportions of insects, fruits, and gums included, but all species must learn appropriate foods (e.g., avoidance of toxic insects or secondary compounds), to acquire hidden foods (e.g., tapping, listening and gnawing of aye-ayes), and their locations (e.g., gum sources visited on regular routes) (Charles-Dominique 1977a; Ganzhorn 1988, 1989; Iwano & Iwakawa 1988; Nash & Whitten 1989; but see Milton, Chapter 12, this volume). Among the lorisines, specialization on noxious insects may be related to lower metabolic rates and slower development (Rasmussen 1986; Rasmussen & Izard 1988).

Limited data suggest that learning foods' identity, location, and processing develop during the juvenile period. Both solitary play (exploration) and observation of tolerant older animals (e.g., kin, including potential fathers) appear to be involved. Choices of companions are more limited than in gregarious species. Among orangutans, solitary play is more common than social play and contains much manipulative and locomotor content (MacKinnon 1971), which may enhance food procurement skills. Age–sex differences in diets and activity budgets (Rodman 1977; Rijksen 1978; Galdikas & Teleki 1981; Galdikas 1988; Mitani 1989) suggest that male and female immatures face different tasks in the development of foraging strategies (see van Noordwijk et al., Chapter 6, this volume). Juvenile orangutans are rarely alone; they travel, rest, and feed with their mothers (Horr 1977; Rijksen 1978; Rodman 1979; Galdikas 1985a), who may share food with them (Horr 1977). When alone, juvenile females use the same routes and feed in the same trees as they do when following the mother (Rodman 1973), suggesting acquisition of traditional foraging routes. How males acquire sex-specific foraging skills and ranges is not clear.

Among the potto and all galago species studied, juveniles begin foraging by following their mother or other adults, including their mother's male associates (Charles-Dominique 1977a; Bearder & Martin 1980; Clark 1985). In most galago species, juveniles take food from adults (Sauer & Sauer 1963; Ehrlich & Macbride 1989; Nash 1990), leading to what Charles-Dominique (1977a) has called "dietary conditioning." Charles-Dominique (1977a) describes the fatal results of hand-reared young *Galagoides demidoff* eating toxic insects that wild animals were not seen to eat. Captive juvenile *Galago senegalensis* were reluctant to take cicadas, until an adult took one; then the young immediately tried to take food from the adult (Nash, personal observation). Juvenile *Lepilemur* also share the mother's range (Russell 1977) and presumably learn feeding patterns from her. These patterns are characteristic of "socially dependent" food learning mechanisms (Whitehead 1986; cf. Milton, Chapter 12, this volume). Unique among primates, *Cheirogaleus medius* undergoes marked torpor during the dry season. Young animals feed longer into the dry season than do adults (Hladik et al. 1980). It is not known if they have already learned the adult diet by the time adults enter torpor.

Play elements may enhance locomotion involved in food-getting abilities (Crompton 1983). Juvenile *Galagoides alleni* show the same extreme preference for leaping between vertical supports in play that characterizes adult locomotion (Charles-Dominique 1977a). The tapping, rolling, and gnawing of objects by aye-ayes in play (Winn 1989) is similar to adult foraging activities. However, some feeding may not be learned. Niemetz (1984b) has argued that tarsiers capture some prey, such as poisonous snakes, instinctively, since there is no room for trial and error!

Predation

Although food distribution patterns are probably the main factor constraining sociality in solitary primates, and often lead to territoriality (Wrangham & Rubenstein 1986), smaller, nocturnal species may also be solitary to avoid predators (Clutton-Brock & Harvey 1977; Charles-Dominique 1978). The largest nocturnal prosimians (e.g., *Otolemur crassicaudatus:* Clark 1985) may have more complex social relations, in part, because size releases them from predation pressure experienced by smaller sympatric species (e.g., *Galago moholi:* Charles-Dominique & Bearder 1979). However, even orangutans are not immune from predators (e.g., tigers, wild pigs: Rijksen 1978; Galdikas 1988). In all species, smaller size and lack of experience, especially when not foraging with an adult, may make juveniles more liable to predation, although no mortality data are available to test this. Compared with gregarious species, solitary species give fewer warning calls and engage in less direct defense from others. Sex-biased dispersal patterns (see below) may also produce higher average risks for males but more variable risks for females.

The slow-moving lorisines reduce sociality to a minimum, apparently to increase crypsis and better avoid predators (Charles-Dominique 1977a). Some may produce glandular secretions that are toxic (Alterman 1990; Alterman & Hale 1991). Pottos butt at threats with their scapular shield, and, if this fails, they drop to the ground. Infant pottos, while riding on their mother's back, may playfully butt at leaves and branches (Charles-Dominique 1977a). It is not known if the longer periods of immaturity in lorisines, as compared with similarly sized galagines, relate, as in other mammals, to lowered mortality risks (Harvey 1990).

Galagos and tarsiers appear to avoid predation by vigilance and quick escape via leaping. Galagos will also produce warning calls (Charles-Dominique 1977a; Niemetz 1984d; Nash, personal observation). Locomotor play, as in the preference for leaping between vertical supports shown by young *G. alleni,* may aid the development of species typical escape patterns. Juvenile solitary locomotor play—involving leaping, hanging, twisting, and running—is seen in captive *G. senegalensis* (Nash, personal observation), and wild and captive *O. garnettii* and *O. crassicaudatus* (Welker et al. 1982; Nash, personal observation; Clark, personal communication), wild *Lepilemur* (Russell 1977), and captive *Mirza coquereli* (Pages 1982, 1983).

Dispersal and Territory Acquisition

Male-Biased Dispersal. In those species where dispersal patterns are known, the males disperse more often, farther, and, in some cases, earlier than females; in many species, adult males have larger ranges than females (Rodman 1973: Charles-Dominique 1977a; Bearder & Martin 1980; Galdikas 1984, 1985c; Clark 1985; Harcourt & Nash 1986; Nash & Harcourt 1986; Schürmann & van Hooff 1986). However, among these species some females also disperse, so female dispersal options may be more variable than for males (Clark 1985; Bearder 1987; te Boekhoerst et al. 1990). Thus the risks associated with dispersal (need to learn new ranges to avoid predation and malnourishment), as in cercopithecines, usually fall more on males. Unlike in cercopithecines, dispersal options and thus risks are more variable for females. Both sexes need to develop the agonistic skills to compete for territories, so fewer sex differences in rates of play might be expected, but sexes may differ in preferred play partners. Juvenile females, more than males, may need to keep options open by maintaining affiliations among female kin.

Among galagos and orangutans, dispersal comes after the juvenile period. In galagos, juvenile males may expand their ranges beyond their mothers' by following adult or subadult males, who tolerate them (Charles-Dominique 1977a; Bearder & Martin 1980; Clark 1985), but otherwise there are few data on juvenile sex differences in ranging patterns (*G. demidoff:* Charles-Dominique 1977a; *G. moholi:* Bearder & Martin 1980; Bearder 1987; *O. crassicaudatus:* Clark 1985; *G. zanzibaricus:* Harcourt & Nash 1986; *O. garnettii:* Nash & Harcourt 1986). Juvenile orangutans rarely travel alone and generally coordinate their activities

with their mothers (Horr 1977; Rijksen 1978; Rodman 1979; Galdikas 1985a), but juvenile males may travel apart from their mothers at an earlier age than females (Horr 1977). Occasionally, mothers may disperse, leaving ranges to daughters (pottos: Charles-Dominique 1977a; galagos: Bearder 1987), further varying options open to maturing females.

Short-term studies of Malagasy prosimians suggest similarities to galagos and orangutans (*Microcebus murinus:* Martin 1972, 1973; Pages-Feuillade 1988; *Lepilemur:* Russell 1977). Since female *M. murinus* can have two litters of twins within 60 days and several adult females may nest together (Petter-Rousseaux 1964; Martin 1973), multiple siblings and other young could develop together. Pages-Feuillade (1988) speculates that coresident males that shared their range with several females during the mating season could have been brothers, although no dispersal data confirm this possibility. *Mirza coquereli* juveniles continue to sleep with the mother, but forage alone (Pages 1980); subsequent dispersal is unknown.

Tarsius spectrum juveniles forage in part, but not all, of their parents' range (MacKinnon & MacKinnon 1980). Second-year males establish ranges away from the natal range while females continue to range within their parents' range. Dispersal of young has not been recorded in *T. bancanus,* but presumably both sexes disperse because juveniles do not appear in adult ranges (Niemetz 1984b, 1984c).

Matriarchies and Local Resource Competition. Thus, at dispersal, both sexes must establish territories, but these dispersal patterns generate sets of females with overlapping or contiguous ranges that are likely to comprise matriarchies. They also generate local resource competition (LRC), which, in turn, may lead to male-biased sex ratios (Clark 1978b). Mothers, sisters, aunts, and nieces may compete for ranges. Within a species, females' social conditions, and thus degree of LRC and litter sex ratios, may vary (Perret 1990). Clark (1985) documents apparently idiosyncratic differences in affiliative and antagonistic relationships between different mother–daughter pairs within the same family. Competition is mediated by direct aggression, especially between nonkin, or by physiological suppression. Agonistic interactions disrupt reproductive cycling in subordinate *M. murinus* females (Perret 1986), and such reproductive suppression may occur widely across solitary species (see also Pereira, Chapter 20, this volume).

Competition Between Males for Mates

Body Size and Reproductive Suppression. As in most mammals, males must develop sexual assertiveness to acquire mates. Male competition for mates is partly mediated by body size in orangutans and some galagos, where larger males tolerate smaller adults and subadults in their ranges, but dominate them in efforts to control access to sexually receptive females. In *G. moholi* and *G. demidoff,* for example, A males are large, vocal, and dominant, and have a characteristic odor, whereas B males are smaller, quiet, and subordinate, and lack the odor (Charles-Dominique 1977a; Bearder 1987; cf. Rodman & Mitani 1987). Similarly sized males, in contrast, can be fiercely aggressive to each other. In *Microcebus,* and perhaps other Malagasy prosimians, subordinate males are instead found in ranges peripheral to those of females and dominant males (Martin 1972, 1973; Pages-Feuillade 1988). A smaller but sexually mature "vagabond" class of males, as in *G. moholi* or *G. demidoff,* may also exist in *T. bancanus* (Fogden 1974).

Testosterone and, in orangutans, secondary sexual feature development (cheek flanges, beards, gular pouch, and characteristic odor) are suppressed in younger male orangutans and *Microcebus* by the presence of a dominant adult male (Kingsley 1982, 1988). In *M. murinus,* the presence or even scent of a dominant or larger male reduces sexual development and levels of testosterone in other males (Perret 1977, 1982a; Petter-Rousseaux & Picon 1981; Perret & Schilling 1987). By dispersing, young male orangutans and *Microcebus* may escape suppression of sexual development by adult males. Hormonal studies confirming suppression have not been done in other species (e.g., galagos), but scent marking in all prosimians could potentially play a parallel role (e.g., Kappeler 1990a).

The wrestling social play of young galagos (Ehrlich 1977; Doyle 1979; Ehrlich & Macbride 1990; Nash, personal observation) may help develop skills needed for intrasexual combat. Similar grappling play has been documented in *T. spectrum* (MacKinnon & MacKinnon 1980),

Lepilemur (Russell 1977), *Mirza* (Pages 1982, 1983), and *Loris* and *Nycticebus* (Rasmussen 1986; Ehrlich & Macbride 1989). When orangutans contact peers, juveniles and adolescents are sociable and ready to play, although they rarely play with the mother (Horr 1977; Rijksen 1978). When fruiting distributions allow it, immatures may become more sociable than adults. This may enhance acquisition of social skills through play (Sugardjito et al. 1987). Locomotor skills developed in solitary play, are probably also important in chases, both agonistic and sexual. Although the sexes might be expected to benefit equally by developing social skills in play, information comparing play of the two sexes is unavailable (Fagan, Chapter 13, this volume).

Female Choice. In many species of nocturnal primates, males' courtship includes chasing the female before she will allow him to mount her (Petter-Rousseaux, 1964; Doyle et al. 1967; Charles-Dominique, 1977a; Nash, personal observation). An estrous female's scent can attract several males simultaneously, some from several ranges away (e.g., Bearder & Martin 1980). Even among seasonal breeders, each female may be receptive on different days (Pereira 1991; Clark, personal communication), facilitating choice. Female orangutans show highly proceptive behavior, rather than broadcast scent. Extreme dimorphism allows subadult male orangutans, who appear unpreferred, to force matings on females. Female protests bring resident adult males, who dominate the smaller males (Galdikas 1984, 1985a, 1985c; Mitani 1985). In all cases, females may be choosing stronger, more vigorous males (see Small 1989).

The locomotor play described earlier may facilitate both male and female sexual behavior. Some juvenile sexuality has been noted among these species. Juvenile female orangutans, for example, sexually solicit subadult males, and adolescent females solicit adult males (Galdikas 1979). The rapidly maturing and seasonally breeding prosimians may show less immature sexual practice than anthropoids. However, juvenile *Loris* mount their mothers (Rassmussen 1986), and juvenile *Mirza* show gestures of courtship (mounts, neck-bites, thrusting) toward the mother (Pages 1982, 1983; cf. sniffing in *G. senegalensis,* which is discussed below).

Infant Care. Females might choose mates on the basis of some form of male parental investment. For example, captive *Cheirogaleus* males interact more with females and offspring than do *Microcebus* (Pereira, personal communication). Males in most solitary primates, however, show little care of young, so females must provide all care. Among nocturnal prosimians, opportunities for juvenile females to interact with infants vary depending on maturation rates and seasonality of reproduction. Juvenile galago females (see below) may show more interest in young than do male peers (see Clark 1985). Juvenile *T. spectrum* may play with infants, but sex differences have not been noted (MacKinnon & MacKinnon 1980). Prior experience with young may be of limited importance in the development of mothering skills in most solitary primates (Capitanio 1986), whereas experience is important in the orangutan (Maple 1980) and, of course, Cercopithecines (Capitanio 1986).

Life Histories, Sexual Dimorphism, and Female Dominance. The most notable contrasts between orangutans and the nocturnal prosimians are in the greatly extended period of development and the extreme sexual dimorphism of orangutans. Both of these differences relate in part to body size and phylogeny (Clutton-Brock 1985; Harvey & Clutton-Brock 1985; Harvey et al. 1987; Pagel & Harvey, Chapter 3, this volume) and so cannot be neatly disentangled among the solitary species. Small body size may constrain the degree of sexual dimorphism possible among all non-human primates (Kappeler 1990b). However, some nocturnal prosimians do show significant size dimorphism (e.g., Nash et al. 1989; Kappeler 1990b). Many prosimians are seasonal breeders, which may limit the "potential for polygamy" (Emlen & Oring 1977) and thus one advantage of large male size, although asynchronous estrus during breeding seasons complicates the issue (Pereira 1991). There has been no entirely adequate test of whether degree of dimorphism covaries with degree of polygamy among solitary prosimians (but see Kappeler 1990b). Complicating the dimorphism issue is that among some, but not all, Madagascar prosimians, females may be both dominant to (in feeding contexts) and larger than males (Jolly 1984; Richard 1987; Pereira et al. 1990; Young

et al. 1990). Documentation of such female priority in nocturnal prosimians is weak (Charles-Dominique & Petter 1980; Perret 1982b; Pages-Feuillade 1988), but at least two species, *M. murinus* and *Phaner furcifer,* may have females that are larger than males (Harvey et al. 1987; Kappeler 1990b). Whether such species differences relate to differences among juveniles is, so far, unknown (see Pereira, Chapter 20, this volume).

CAPTIVE JUVENILE *GALAGO SENEGALENSIS BRACCATUS:* AGE AND SEX DIFFERENCES

Although not known well in the wild (Kingdon 1974), *Galago senegalensis braccatus* (Fig. 9.1) appears similar in behavior to the well-known *G. moholi* from South Africa (previously, also named *G. senegalensis*) (Nash et al. 1989), whose social system is characteristic of many solitary primates (Bearder 1987). Preliminary data on sex differences in juvenile development of captive *G. senegalensis* may provide clues to the significance of the juvenile period in this species and others with similar social systems.

Methods

Subjects. All subjects were captive born. Most births were singletons. Among subjects, all siblings were of different ages. Weaning in *G. senegalensis braccatus* occurs at about 10 weeks (Nash 1990), and a mother may conceive on a postpartum estrus, giving birth when the previous infant is 22–26 weeks old. Animals of both sexes reach sexual maturity at about 1 year of age (Table 9.1). Consequently, the young juvenile age class is defined here as 11–26 weeks of age, and the old juvenile age class as 27–52 weeks of age. Data are based on a mixed cross-sectional, longitudinal sample representing, at any one age period, four to seven young juvenile males, one to five young juvenile females, four to seven old juvenile males, and one to two old juvenile females. The numbers of subjects sampled anytime during an age class were seven young juvenile females, eight young juvenile males, two old juvenile females, and seven old juvenile males. For comparison, three females and four males aged 2.5 to 3 years old were

Fig. 9.1. Adult female *Galago senegalensis braccatus*. (Photo, R. A. Barnes)

sampled for adult behavior. This age was chosen to maximize the number of subjects that had been observed at the same age in adulthood.

Subjects lived in social groups of varying composition. All groups were generated by the breeding of their founding pairs. There were, at most, two breeding females, mother and daughter, simultaneously in a group. Consequently some subjects had age-peers, others had older or younger immature cagemates (or both), and a few had immature cagemates of all three categories. Although maximum group size occasionally reached seven, groups more commonly consisted of three to six individuals of varying ages. Cages were either 2.4 m² or 2.4 × 1.2 m, both 2.4 m high, and equipped with mul-

tiple perches, dividing panels, ledges, and nestboxes. Animals were maintained on a reversed light cycle (13L:11D) and fed daily at the start of the night light condition (see Nash & Flinn 1978).

Procedures. Focal animal samples, during which each behavior was treated as an event for which rates/hour of observation were calculated, were taken during the first and last 3 hours of the waking cycle, when the animals were most active. For most subjects less than 23 weeks old, three 10-minute samples were taken during each target hour, for a total of 180 minutes/week/subject. For older animals, one 10-minute sample was taken during each of the six target hours. Samples were smaller for a few animals, but always proportionate to these protocols.

Data were tallied in 2-week age blocks for animals under 26 weeks of age and 4-week blocks for animals 27 to 52 weeks of age. Because not all subjects were present at all ages, it was not possible to perform an analysis for sex differences in each age class with the biweekly or monthly data points as repeated measures. Analysis was further complicated by small sample size, especially for old juveniles and adults. Consequently, sex differences were tested using the van der Waerden test (Conover 1980) on each subject's median score across its age blocks comprising each age class. The alpha level for a significant difference was set at $p < 0.05$, and $p < 0.20$ to indicate a "trend." Trends clearly need confirmation in a larger sample. To examine partner preferences for each behavior, the percentage of all bouts of the behavior by each subject with each age–sex category of partner was computed. The median percentage, for all subjects with that partner category available, is presented.

Results

Grooming. "Grooming" was performed, as in all prosimians, primarily with the procumbent lower incisors. Among adults, although males showed higher medians for both receiving and performing grooming, sex differences were not significant (Fig. 9.2). Similarly, no marked sex difference in grooming were shown in either age class of juveniles.

Mothers were prominent grooming partners for all juveniles (Fig. 9.3). Females also often shared grooming with older immature males and sisters. Young juvenile males also often groomed with other older females, perhaps reflecting the development of sexual behavior. Young females expressed interest in male groupmates and in kin, with whom they may share a range in adulthood. Males and younger groupmates were the prominent partners of old females, suggesting that sexual and parental activities were developing. Older males gave and received grooming relatively equally to all nonmaternal partners.

Social Sniffing. "Social sniffing" involved any apparent nasal exploration of a partner, except for sniffing of genitals. Among adults, there was a trend for females to receive more than males, but there was no clear sex difference in performing the behavior (Fig. 9.4a and b). Among juveniles, no marked sex differences in receiving social sniffing appeared until late in the first year, when, approaching the adult rate, a trend emerged for females to receive more. Young juveniles of both sexes received less sniffing than adults. Juvenile females sniffed others more than adult females. Juvenile males generally sniffed others less than adult males.

Females tended to sniff with their mothers more than males (Fig. 9.5a). Among young juveniles, males tended to sniff adult females more than they sniffed adult males, while females reversed this pattern, suggesting a sexual aspect to the action.

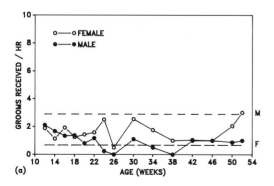 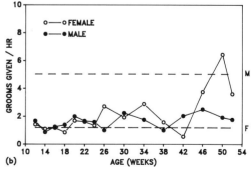

Fig. 9.2. Median rates per hour of (a) receiving grooming and (b) giving grooming. Adult levels are indicated by horizontal short dashed lines ending in M for males and horizontal long dashed lines ending in F for females.

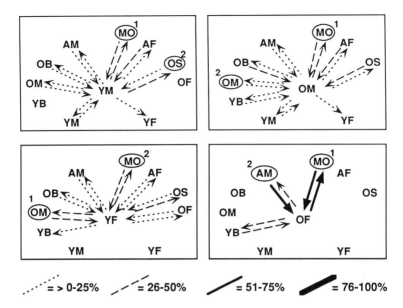

Fig. 9.3. Median percent of grooming bouts to and from each partner type by subjects. YM, young juvenile males; OM, old juvenile males; YF, young juvenile females; OF, old juvenile females. Partner types (females on right of panels, males on left of panels) with top ranking partners *to whom* behavior was given circled and numbered by rank order (1: highest): AM, adult male; OB, older brother; OM, older nonsib, nonadult male; YB, younger brother; YM, younger nonsib male; MO, mother; AF, adult female not the mother; OS, older sister; OF, older nonsib, nonadult female; YF, younger nonsib female. Percentage of bouts indicated by width of arrows as in legend; blank if partner was available but median equalled zero. The median was taken across all subjects at that age that had the partner type available. The figures thus represent a "composite individual" as if all partner types had been available for that sex for that age.

Genital Sniffing. Among adults, as with social sniffs, there was a trend for females to receive more genital sniffing than males, but was no clear sex difference in performing genital sniffs (Fig. 9.4c and d). These differences appeared among juveniles in both age classes and were especially marked from week 40 on.

Young and old juvenile males generally sniffed females' genitals most, but received genital sniffs most from other males (Fig. 9.5b). For females, the same was true, suggesting that males develop into sexual initiators, while females monitor the sexual states of other females in their range.

Chasing. "Chasing" is a prominent part of courtship, but also can be aggressive, especially between same-sex animals. Males rarely chased females aggressively, and females almost never chased males aggressively. Male vocalizations mark most courtship chasing, whereas calls

were usually not given by aggressors. Animals aggressively chased usually vocalized. The behaviors were rare in young juveniles (Fig. 9.6a and b). By the end of late juvenile period, the female may experience her first estrous period and be chased at or above adult female rates. Although the differences in medians were as expected if most chasing was courtship, sex differences in neither adults nor juveniles were significant.

Young and old male juveniles primarily chased older females (Fig. 9.7a). There is no incest avoidance: males will chase and copulate with their own mother and with sisters in captivity (cf. Pereira & Weiss 1991). In the wild, they would disperse. Although the limited demographic data available from wild galagos suggest that fathers are not likely to be present by the time daughters mature (*G. demidoff:* Charles-Dominique 1977a), the extensive chasing during courtship might allow females to

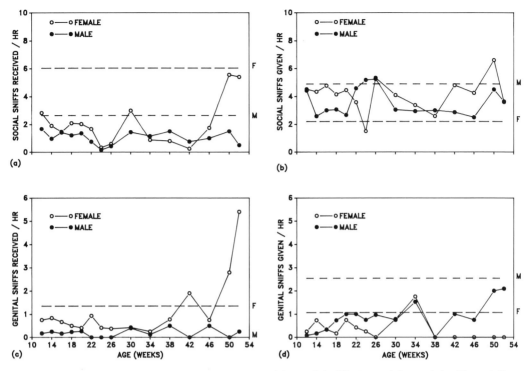

Fig. 9.4. Median rates of (a) receiving social sniffs, (b) giving social sniffs, (c) receiving genital sniffs, and (d) giving genital sniffs. Conventions as in Figure 9.2.

avoid mating with possible fathers. Young and old male juveniles were primarily chased by older males, but among old juvenile males, younger animals also chased them often. This may be play, along with efforts by youngsters to take food from elders (Nash 1990). Female juveniles most often chased their mothers, possibly to take food. Female juveniles were chased most by adult males or brothers.

Play. "Play" involves wrestling, play biting, and grappling (often hanging upside-down from branches, face to face). Young juvenile males showed significantly more play than their female peers (Fig. 9.6c). Sex differences in adults were not significant, but in the same direction. Juveniles play more than adults, and by the end of year one, play rates dropped to adult levels.

Play showed a very different pattern from chasing in partner relationships. Juvenile males in both age classes play most with other males,

usually older ones (Fig. 9.7b). Females play most with their mothers, other females, and, for younger juveniles, older brothers. Old juvenile females also played often with young brothers.

Supplants. "Supplants" were scored when one animal left immediately on another's approach, but only when the supplanted animal appeared to see who was approaching and was not simply startled when another suddenly leaped into close proximity. Among adults, there was a trend for males to get supplanted more than females, but both sexes supplanted others equally often (Fig. 9.8c and d). The opposite tends to occur in juveniles.

Juvenile males supplanted and were supplanted mostly by other males (Fig. 9.9a). Juvenile females most frequently supplanted adult males and other older males. Females could supplant older sisters more than older females and were supplanted by older females more than by sisters.

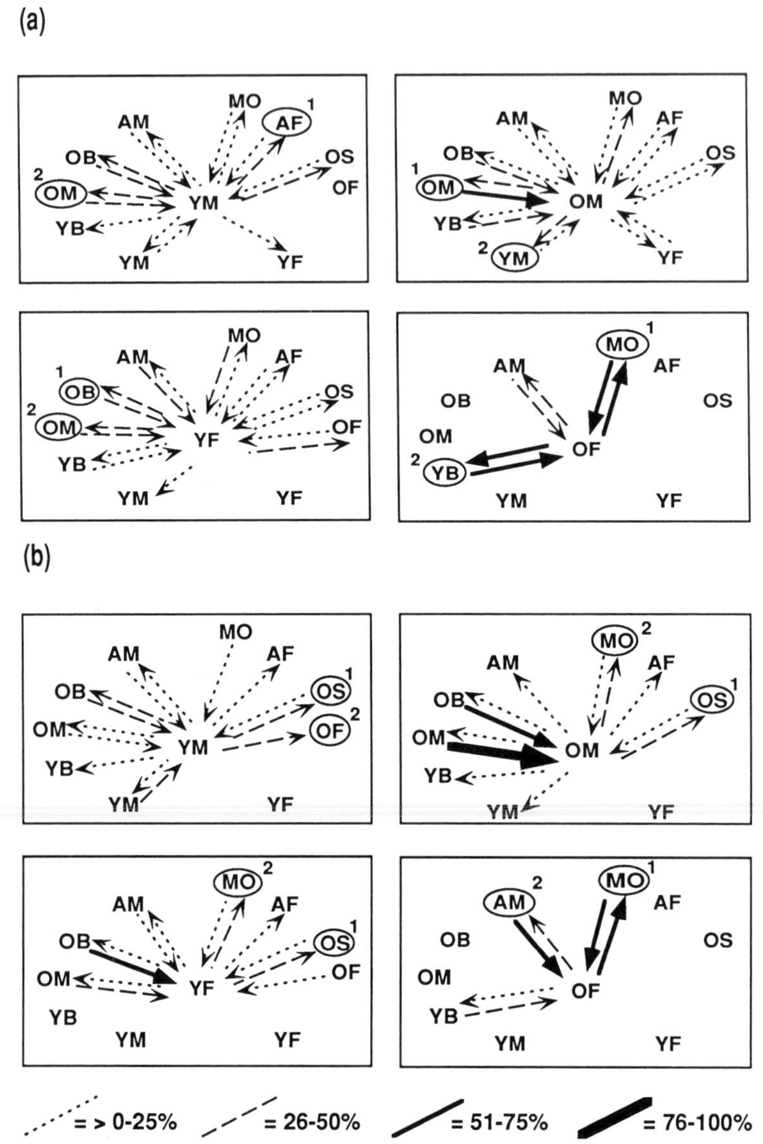

(a)

(b)

Fig. 9.5. Median percentage of bouts to or from each partner type by subjects of (a) social sniffing and (b) sniffs genitals. Conventions as in Figure 9.3.

Aggression. "Aggression" involves grappling, biting, boxing, and other potentially harmful contacts. No juvenile or adult sex differences were evident in giving or receiving aggression (Fig. 9.8a and b).

Females often aggressively rebuffed sexual advances by males (Fig. 9.9b). Aggression was rarer for males, but the most severe aggression ever seen was between maturing males, including some brothers (Fig. 9.9b). Such aggression was not always from older to younger. Also, the oldest adult male was usually not involved in such fights, suggesting his recognition as an A male. The other context in which severe aggres-

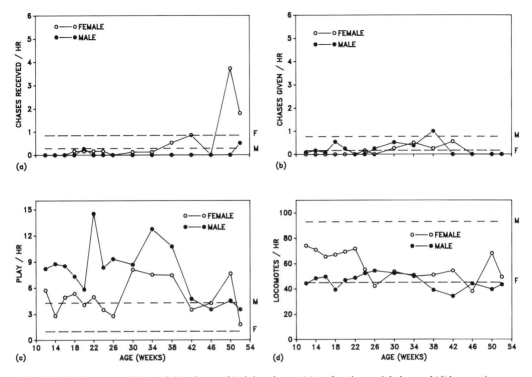

Fig. 9.6. Median rates of (a) receiving chases, (b) giving chases, (c) performing social play, and (d) locomoting. Conventions as in Figure 9.2.

sion was observed was by periparturient females, typically directed at males. The most frequent target was the oldest male, but occasionally immatures, including previous offspring, were targets. Both forms of severe aggression began suddenly and rarely stopped until the target or aggressor was removed. In both temporal patterning and targets, these modes of aggression resembled the two modes of targeting aggression reported for lemurs (Vick & Pereira 1989; Pereira & Weiss 1991).

Locomotion and Visual Scanning. Both locomoting (Fig. 9.6d) and scanning measured general levels of activity. Both showed a trend toward being more frequent in adult males than adult females. During the late juvenile period, there are few consistent differences and rates for both sexes in both behaviors approximated those of adult females. Males had not yet reached adult male rates.

Urine Wash. This behavior involved urinating on the hand and wiping the ipsilateral foot and then usually repeating the movements on the contralateral side of the body. Adults of both sexes urine wash at the same rates. There are two main hypotheses concerning the function of urine washing. Charles-Dominique (1977b) has suggested that for *G. alleni* it is a territorial marker. Harcourt (1981), using data from *G. moholi,* suggested that the behavior serves to facilitate grip. These functional explanations are neither mutually exclusive nor exhaustive. Another possibility is that other social information is communicated—for example, identity and, for females, reproductive condition. In our colony, animals of both sexes can distinguish the sex and estrous state of urine markers (Carpetis & Nash 1983).

Relative to the grip hypothesis, it is notable that urine washing began almost as early as locomotion outside the nestbox. Like adults, young

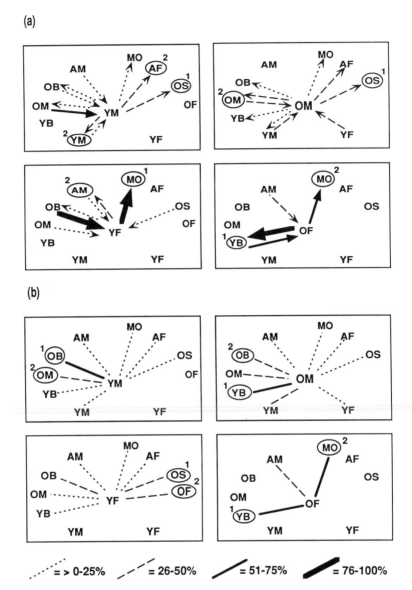

Fig. 9.7. Median percentage of bouts to or from each partner type by subjects of (a) chasing and (b) play (actor/receiver roles not distinguished). Conventions as in Figure 9.3.

juveniles showed no sex differences, and they urine washed less than adults. During the late juvenile period, females' rates significantly exceeded those of males, and early in the period they exceeded adult rates, before falling back to below adult rates.

DISCUSSION

Diversity Among the Solitary Primates

As emphasized earlier, diverse forms of sociality are found among the so-called solitary primates (Bearder 1987). The "rules" (Clutton-

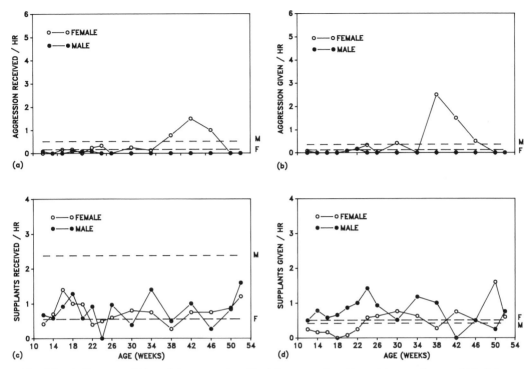

Fig. 9.8. Median rates of (a) receiving aggression, (b) giving aggression, (c) receiving supplants, and (d) giving supplants. Conventions as in Figure 9.2.

Brock & Harvey 1976) governing sociality are likely to be the same as in gregarious species, so nongregarious species should be included in developing socioecological and sociobiological hypotheses (Clark 1985). Although better information (e.g., genetic data) is needed on the extent to which ranging and sleeping arrangements reflect mating systems, the variety of ranging and sleeping patterns imply forms resembling one-male harem groups, age-graded one-male groups with varying rates of male turnover (e.g., *G. demidoff, G. moholi,* orangutans), monogamy (e.g., *Phaner furcifer, G. zanzibaricus,* possibly *T. bancanus*), and multimale promiscuous systems (e.g., *M. murinus, O. crassicaudatus*). Considerations of fission–fusion sociality in primates should therefore include nongregarious primates (cf. Pereira et al. 1988; Vick & Pereira 1989) (Table 9.1).

Solitary species also vary in the nature and amount of contacts (vocal, olfactory, grooming,

resting, playing) maintained during the active period. For example, in this study no adult or juvenile sex differences in urine washing were found for *G. s. braccatus,* whereas Doyle (1979) reported that both adult and juvenile males of the closely related *G. moholi* urine washed more than females. Male *G. moholi* frequently urine wash directly on female groupmates (Doyle 1979). This is less consistently seen in *G. s. braccatus* (Nash, personal observation). Clark (1982a, 1982b) pointed out that species may differ in the degree to which urine is used as a signal. Sex differences may also relate to the nature of the signal (e.g., lower sex differences in territorial marks, greater sex differences in signals involved in male mating competition). Such difference may be reflected in juvenile behavior.

It has been documented recently that the radiation of galago species is much wider than previously recognized (Nash et al. 1989); recent

(a)

(b)

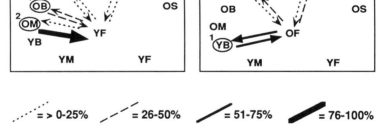

$\cdots$ = > 0-25% $\diagup$ = 26-50% $\diagup$ = 51-75% $\blacksquare$ = 76-100%

Fig. 9.9. Median percent of bouts to or from each partner type by subjects of (a) supplants and (b) aggression. Conventions as in Figure 9.3.

fieldwork likely will reveal the existence of additional species (Bearder, personal communication). Diversity in adult social relations, diet, body size, and other life-history parameters (e.g., age at maturity, litter size, rate of reproduction) all should influence juvenile behavioral tactics. Extending comparisons beyond galagos to the other solitary primates adds to diversity in metabolic rates, diet, life histories, patterns of sexual dimorphism, female feeding priority, seasonality of reproduction, and the extent of local resource competition with related sex ratio bias. As with galagos, research on other genera (e.g., *Tarsius:* Niemitz 1991) is likely to reveal greater species diversity than is currently recognized. Consequently, although too little is currently known to make specific predictions about species differences in juvenile behavior, the species diversity among solitary primates offers important opportunities to test hypotheses developed from research on gregarious primates.

The results from juvenile *G. senegalensis* suggest that sexes may not differ in the rates of affiliation expressed by grooming, in contrast to cercopithecines, but partner preferences may relate to patterns of dispersal. Since dispersal is both risky and critical to reproductive success, much better data are needed about intra- and interspecific variability. Those available are based on so few individuals that it is difficult to distinguish species differences from sampling error. Dispersal patterns also determine the likelihood that kin of either sex may associate during adulthood (Pusey & Packer 1987). This, in turn, affects our expectations about behavioral sex differences and the extent to which juveniles may prefer or need to avoid kin as they age.

Although female philopatry appears to be the rule (Jolly 1984; Clark 1985, Bearder 1987), variation in the size and functioning of "neighborhoods" of overlapping or contiguous ranges of matrilineally related animals is not known. As indicated for *Microcebus*, where more than one litter can be produced in a year, it may also be possible for males to disperse with brothers and other male kin. *Microcebus* and *G. moholi* may show more potential for male affiliation than *G. senegalensis*, which probably has one offspring per year. *O. crassicaudatus*, which bears twins or triplets once per year, presents yet a different set of possibilities for kin affiliation.

The Problems of Being a Juvenile

The primary survival challenges for juveniles of each sex appear to be learning appropriate foods, developing locomotor skills, and avoiding potential dangers. Especially among relatively monomorphic species, adult females may require more protein (for reproduction), and adult males may require more energy (for patrolling larger ranges). If so, juveniles of each sex might benefit by perfecting somewhat different foraging tasks (Cords 1986; van Noordwijk et al., Chapter 6, this volume). Documentation

of feeding differences between the sexes is needed for animals of all age classes. Whereas gregarious species engage in more social than solitary play (Walters 1987a), in nongregarious species, skills developed through solitary play (locomotion, exploration, manipulation) may be more important than those developed in social play. However, differences in degree or style of "solitariness" might be reflected in juvenile play budgets.

Male dispersal and female philopatry would suggest that young males would benefit from learning about a larger range than females (although in captive *G. senegalensis*, adult males' higher activity levels apparently develop after the juvenile period, perhaps at the time dispersal might occur). As seen in a few species, juvenile males more than females may follow animals other than the mother while foraging. If so, then males should select kin (e.g., brothers, probable fathers), when available, to increase chances for tolerance. In galagos, A males (probable fathers) should be more attractive to or tolerant of juveniles than B males.

Among seasonally breeding species, competition among immatures, especially littermates, likely intensifies when they disperse in search of territories. It would be interesting to know if contacts made during the juvenile period allow individuals to assess the competitive abilities of different neighbors and so influence direction of dispersal. However, as kin, same-sex littermates can be expected to show a complex balance of cooperation and competition (Clark & Ehlinger 1987). Clark (1985; Clark & Ehlinger 1987) documents the apparent "idiosyncratic" individual differences in temperament and social relationships of twin *O. crassicaudatus*. Whether such individual differences relate to the balance of cooperation and competition among juvenile kin or are adaptive has yet to be determined (Clark 1991). The development of matrilineally based rank relationships is not important to females of solitary species, as it is to cercopithecine monkey females, and they do not expend the considerable effort cercopithecine monkey females do in establishing matrilineal dominance status (Chapais & Gauthier, Chapter 17, and de Waal, Chapter 18, this volume). Similarly, males among solitary species do not form alliances with others. Thus little of the coalition formation apparent among juveniles in gregarious species is to be expected among juveniles of solitary species.

Anticipating Adulthood

Available data suggest that adults of each sex often maintain territories to exclude most same-sex conspecifics. Males must also directly exclude mating competitors. As in the captive juvenile *G. senegalensis,* male juveniles may play more than females, but the differences are not as dramatic as in Cercopithecines (Fairbanks, Chapter 15, this volume). Each prefer like-sex play partners, which may relate to practice for later competitive interactions. Also, since play may also function to develop affiliative bonds, and since males all disperse but female dispersal may be more varied, detailed studies of social play might reveal that males focus more on rougher play-fighting than females and that female play patterns and play partners are more varied than those of males. The difficulty in testing this will be to control for variations in partner availability (Altmann & Altmann 1979). Where same-sex partners are unavailable, kin might be preferred, because tolerant relationships should reduce chances for play-fighting to become truly aggressive. The supplanting data on captive juvenile *G. senegalensis* also suggested that sisters may be more tolerant of each other than they are of other females of similar ages.

The data presented on *G. senegalensis* suggest that juveniles show patterns of sociosexual behaviors that are sex appropriate (e.g., sniffing, genital sniffing). Where olfaction communicates sex, reproductive condition, and probably individual identity (Clark 1982a, 1982b; Carpetis & Nash 1983), experience with a wide variety of partners' scents may be important. The data presented here also suggest that older males may monitor the development of younger males through olfaction.

Unlike orangutans, solitary prosimian juvenile females may not require experience with infants to be adequate mothers. Nevertheless, since males among those species exhibit no evidence of infant care, it is expected that interest in infants should be higher in females than males. This sex difference might be seen only in attention directed to quite young infants because older infants are able to play and may make appropriate play partners for juvenile males and females.

CONCLUSION

For the solitary primates, almost nothing is known about modal patterns or variation in juvenile dispersal and philopatry, in juvenile social interactions with peers, or in juvenile social interactions with older individuals. The similarities and differences among solitary primates, and between them and the gregarious species, suggest that related variations in juvenile experience should occur. Only quantitative, longitudinal studies of identified animals in natural or naturalistic captive environments will reveal these patterns and allow the expectations made here to be refined adequately and tested.

For the nocturnal prosimians, new technologies, such as radio-tracking using solar charged receivers, make data collection potentially more efficient than in the past. However, research on solitary species will always be more time-intensive than research on gregarious ones. Study of nocturnal species will require consideration of communication along sensory channels virtually inaccessible to human observers (e.g., odors, ultrasound). The nocturnal species, especially, require many difficult life-style adjustments of the fieldworker. Balancing those costs are the benefits of studying shorter lived species: it should be possible to gather data on lifetime residence patterns and reproductive success in fewer years than with most gregarious species. Such studies would form an important bridge between information on solitary nonprimate mammals and gregarious primates that should inform models on the evolution of complex sociality in the gregarious primates (Waser & Jones 1983). Finally, there is the urgency shared with their gregarious relatives that many of the solitary species are endangered or threatened.

ACKNOWLEDGMENTS

The *Galago senegalensis braccatus* colony is supported by the Department of Anthropology, Arizona State University. Preparation of this manuscript was assisted by a grant from the College of Liberal Arts and Sciences, ASU. I thank R.

C. Williams for assistance with the figures, J. Douglas for assistance with figures and bibliographical work, and K. Smith for assistance with sources in French. I am greatly indebted to the many students over the years who contributed to the behavioral data set on *Galago senegalensis* *braccatus* behavior and to R. A. Barnes for expertly caring for and photographing the animals. At various stages, the manuscript was improved by comments from L. Smith, M. Matevia, S. K. Bearder, A. Ehrlich, L. Fairbanks, and, especially A. Clark and M. Pereira.

10

Growing Up in a Patrifocal Society: Sex Differences in the Spatial Relations of Immature Muriquis

KAREN B. STRIER

The muriqui, or woolly spider monkey (*Brachyteles arachnoides*), is a monotypic member of the Atelinae. It is endemic to the Atlantic coastal forest of southeastern Brazil, where severe deforestation has reduced its habitat to less than 5% of its original extent (Fonseca 1985). The total population of muriquis is estimated at fewer than 500 individuals, distributed among less than a dozen isolated forest tracts (Mittermeier et al. 1987).

Systematic data on muriqui behavioral ecology are available from only three of these sites. Muriquis are reported to have a polygamous mating system, in which both sexes mate with multiple partners (Milton 1985; Strier 1987a), but their grouping patterns appear to be variable. At Fazenda Montes Claros and Fazenda Esmeralda, muriquis were observed in cohesive, multimale, multifemale groups (Strier 1987b, 1989; Lemos de Sa 1988), whereas at Fazenda Barreiro Rico, a group of females and immatures had fluid associations, and heterosexual associations between adults occurred only when a female was sexually receptive (Milton 1984b).

Like that in its closest relatives, spider monkeys (Symington 1988, 1990), woolly monkeys (Nishimura 1990), and howler monkeys (Crockett & Eisenberg 1987; Crockett & Pope, Chapter 8, this volume), dispersal in muriquis appears to be female-biased (Strier 1987c, 1991). Long-term data on muriqui dispersal come from a single site (Fazenda Montes Claros) with only two groups, and thus it is likely that sex differences in genetic relatedness are lower than they would be in a larger population with multiple dispersal options. Nonetheless, sex differences in the social relations of adult muriquis are similar to those described in other patrifocal species, such as spider monkeys (Ahumada 1989; Symington 1990) and chimpanzees (Halperin 1979; Nishida 1979; Goodall 1986), and differ from those in matrifocal species, such as baboons and macaques (see review in Walters & Seyfarth 1987).

Both adult male and adult female muriquis spend more time in proximity to same-sex adults than expected by chance, but adult males spend a greater proportion of their time in proximity to one another than do females (Strier 1990, 1992a). Muriquis do not groom one another, but males are more likely to participate in polyadic embraces than females. Female muriquis, like female spider monkeys and chimpanzees, may avoid intragroup contests over food by avoiding feeding in close proximity to one another. In muriquis, however, there is no evidence of agonistically mediated dominance relationships among or between male and female group members.

CORRELATES OF SEX-BIASED DISPERSAL AND RESIDENCY

Strong affiliative relations among male kin in other patrifocal species and among female kin in matrifocal species are expressed by their frequent spatial associations and grooming bouts. In both cases, affiliative relations among related individuals have been attributed to the importance of cooperation in intergroup contests over access to critical resources. Female philopatry and strong female relations are believed to occur when the advantages of cooperative intergroup defense of preferred food patches outweigh the costs of intragroup feeding competition (Wrangham 1980; van Schaik 1989). Similarly, male philopatry and strong male relations are believed to occur when males benefit by cooperatively excluding unrelated males from associating with community females (Wrangham 1979b).

The social relations of immature primates also reflect sex-biased dispersal and residency patterns. As Walters (1987a) has noted, the natal group is the social unit for both sexes through puberty, but it is the future social unit only for the sex that remains through adulthood. Among matrifocal species, the strongest affiliative associations are those among females and between females and their offspring, whereas male associations are less frequent and less predictable. Thus females in matrifocal groups mature into their female kin and social networks, whereas associations between males and their female kin tend to grow weaker over time. The converse situation occurs in patrifocal species, with males associating increasingly with adult male kin as they mature. In both cases, however, juveniles have strong associations with same-sex age peers.

Elsewhere (Strier 1990, 1992a, 1993) I have discussed the significance of sex differences in adult muriqui social relations and dispersal patterns. Here I examine preliminary data on the social relations of immature muriquis inhabiting the 800-ha forest at Fazenda Montes Claros in Minas Gerais, Brazil (19° 50′ S, 41° 50′ W). In general, immature muriquis are expected to exhibit sex differences in their social relations corresponding to those observed among adults in this patrifocal society, and to associate preferentially with same-sex age peers.

EXAMINING SOCIAL RELATIONS IN IMMATURE MURIQUIS

The social repertory of muriquis includes sexual inspections, copulations, embraces, and play, but muriquis devote less than 2% of their time to these activities (Strier 1987b). Nevertheless, spatial relations correlate well with other indices of social relations in other primates, and may provide the most reliable means of assessing sex differences in immature muriqui social relations.

General Patterns of Development

During their first year of life, muriqui infants are almost fully dependent on their mothers. They cling to their mothers' bellies or sides for their first 3–6 months. By 6 months of age, they begin to ride jockey-style on their mothers' backs with their tails wrapped around those of their mothers for added support (Strier 1986, 1992b). Mothers may park their infants on thin branches while they feed, but infants are always retrieved before the mothers move off.

By 1 year of age, muriquis begin to feed and travel with increasing independence, and for this reason I consider 12 months to be the cut-off between infant and juvenile age classes (Fig. 10.1). Weaning occurs at 18–24 months, and interbirth intervals have averaged 33.8 months (Strier 1991).

Muriquis at Fazenda Montes Claros are born with light fur and dark faces. Fur color begins to darken and facial pigmentation begins to appear by approximately 3–4 years of age. By the time they are 5 years old, males and females are still slightly smaller than adults in body size, but their genitalia visibly resemble adult proportions. Menses was observed in one 5-year-old female, suggesting that the onset of puberty in female muriquis may occur at this age. However, no external signs of puberty were observed in any of the 5- to 7-year-old subadults described below. Subadult females between 5 and 6.5 years of age migrate from their natal group. By contrast, no male migrations have been documented during the decade that this group has been monitored (Strier 1991, 1992b).

All subadult and adult muriquis at this site are easily identified by their distinct facial markings. Juveniles can sometimes be recognized by

Fig. 10.1. Juvenile muriqui (*Brachyteles arachnoides*).

more subtle differences, such as muzzle or brow shape, prior to facial pigmentation. However, better lighting and less obstructed views were necessary to identify the young juveniles during the study period described here.

Methods

Data on immature muriquis were obtained during a systematic study of muriqui behavioral ecology from June 1983 through July 1984. At the onset of this period, the group consisted of 23 individuals, including six adult males, eight adult females, two subadult males, two juvenile males, four juvenile females, and one newborn female. The six juveniles were estimated to be between 15 and 18 months at this time; the subadults were estimated at 5–6 years of age. Subsequent observations of individuals of known age have confirmed these estimates (Strier 1991).

By July 1984, the group had increased to 26 members due to the immigrations of two subadult females and one birth during the last month of the study. Because there was only one infant in the group during the first 13 months, only data on the six juveniles and four subadults are considered below.

It must be emphasized that research during this period in 1983–1984 did not focus exclusively on immatures. This, in addition to the difficulties of reliably identifying juveniles, contributes to the small sample sizes for this age class. Only observations of recognized juveniles are included in the following analyses, but the results must be regarded as tentative until more detailed data become available (A. O. Rimoli, unpublished).

Data Collection. Instantaneous scan samples (Altmann 1974) were conducted at 15-minute intervals whenever any members of the study group were visible. Data on spatial relations included the proximity between each scanned individual and its nearest neighbors, and the identities of its nearest neighbors. Proximity was distinguished by three categories: in contact, within a 1-m radius, and within a 5-m radius. If the nearest neighbor was greater than 5 m away, it was scored as "out of proximity" and no nearest neighbors were recorded.

Whenever possible, the identities of all nearest neighbors in the closest proximity category were recorded as individuals. However, if poor visibility precluded individual identifications, the most precise age–sex classes were noted. Data on nearest neighbors are included only when the identities or age–sex classes could be determined. Nearest neighbors were often, but not always, reciprocal. Thus if individual A was less than 1 m from individual B, and individual C was less than 5 m from both A and B, B would be scored as A's nearest neighbor, A would be scored as B's nearest neighbor, and both A and B would be scored as C's nearest neighbors.

Data Analysis. A total of 4334 scan samples, representing 16,847 individual activity records, were conducted (Strier 1987b). Although 3164 of these records involved independent immatures, only 1868 were of positively identified individuals (see above). Sample sizes varied between individuals due to differences in their visibility, and between age–sex classes due to differences in their representation in the group.

A total of 637 observations of identified juveniles were obtained during scan samples, with 449 involving juvenile females and 188 involving juvenile males. Roughly one-third of these observations were made when mothers were carrying their daughters (32.07%) or sons (35.64%). Only those observations of spatial relations in which juveniles were observed independently (420) are considered below. A total of 1231 observations of subadult spatial relations were obtained during the scan samples, with 881 involving subadult males and 350 involving subadult females.

MURIQUI SOCIAL DEVELOPMENT

Spatial Relations of Immature Muriquis

Juveniles were observed in proximity to other group members more often than were subadults (Fig. 10.2). However, mothers accounted for the majority of nearest neighbors of both juvenile males (73%) and juvenile females (73%). When mothers were excluded from these comparisons, there were no significant differences in the frequencies that juveniles and subadults were observed in proximity to other group members [$\chi^2(1) = 0.265, p > 0.05$].

Nonmother adults were the most frequent nearest neighbors of both juveniles and subadults (Fig. 10.3). Adults accounted for roughly 85% of the nearest neighbors of both immature age classes, whereas subadults accounted for only 12–15% of their nearest neighbors. Juveniles were underrepresented (2%) as nearest neighbors to both juvenile and subadult age classes.

For both juvenile and subadult age classes, females were observed in proximity to other group members less often than males (Fig. 10.4). These findings are consistent with sex differences observed in the spatial relations of adult muriquis (Strier 1992a, 1993). To the extent that spatial relations reflect the strength of social relations, these findings are also consistent with predictions that female muriquis have weaker social relations than males.

The fact that both subadult females were recent immigrants to the group might explain why they were observed in proximity proportionately less often than juvenile females. However, natal juvenile females also exhibited weaker spatial relations than juvenile males (Fig. 10.4), suggesting that recent membership in an unfamiliar group is not the only cause of these sex differences. Rather, it appears that sex differences in the strength of muriqui spatial relations emerge at an early age.

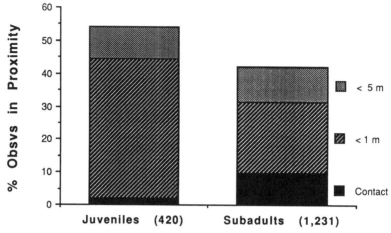

Fig. 10.2. Proportion of observations of independent juveniles and subadults in each proximity category. The total number of observations for each age class is given in parentheses. Juveniles were observed in proximity to other group members significantly more often than subadults [$\chi^2(1) = 78.76, p < 0.001$].

Ontogeny of Female Relations

The nearest neighbors of immature females changed substantially with age (Fig. 10.5). Juvenile females were observed in proximity to adult males proportionately more often, whereas subadult females were observed in proximity to adult females more often. Juvenile females associated with nonmother adult females, subadult males, and subadult females in proportion

to the representation of these age–sex classes in the group. While subadult females were never among one anothers' nearest neighbors, they were observed in proximity to subadult males more often than expected by chance [$\chi^2(1) = 6.79, p < 0.01$]. Identified juveniles were underrepresented among the nearest neighbors of both juvenile (1.67%) and subadult females (7.58%).

Juvenile females were not expected to exhibit

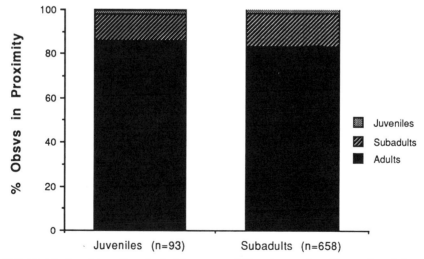

Fig. 10.3. Distribution of juvenile and subadult nearest neighbors by age class. The number of observations of juveniles and subadults is shown in parentheses. Only nonmothers are included as nearest neighbors here.

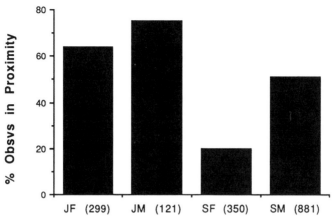

Fig. 10.4. Sex differences in the proportion of observations immatures were observed in proximity. JF, juvenile females; JM, juvenile males; SF, subadult females; SM, subadult males. The number of observations for each age–sex class is given in parentheses. Juveniles were observed in proximity significantly more often than same-sexed subadults [JF–SF: $\chi^2(1) = 130.99, p < 0.001$; JM–SM: $\chi^2(1) = 25.16, p < 0.001$]. Males were observed in proximity more often than females for both juvenile [$\chi^2(1) = 5.01, p < 0.05$] and subadult [$\chi^2(1) = 100.37, p < 0.001$] age classes.

strong spatial relations with nonmother adult females because female-biased dispersal tends to preclude the development of strong social relations among females. It is not clear, however, why juvenile females had such strong spatial associations with adult males. It is possible that juvenile females associated with adult males who were their fathers, older brothers, or other male kin, but if this were the case, it seems unusual that subadult males would not also be strongly represented among their nearest neighbors. Knowledge of at least extended matrilineal kinship prior to the onset of the study is needed before the relatedness between these juvenile females and their male nearest neighbors can be evaluated.

In muriquis, as in many other primates with sex-biased dispersal, immature females appear to leave their natal group voluntarily (see Pusey & Packer 1987). There was no evidence that adult females (or other group members) behaved agonistically toward immature natal females prior to their dispersal. However, all the natal females that have dispersed from this group did so following agonistic intergroup encounters in which adults of both sexes, as well as subadult males, participated.

Similarly, the eight females that have migrated into the study group appeared within a few days of different intergroup encounters.

Their immigration attempts followed a similar pattern in which the female shadowed the group for days and was threatened and chased by resident adult females whenever she tried to approach the group. After repeated efforts spanning several days, the threats toward immigrant females finally ceased (Strier 1992b). Although newly immigrant females remained peripheral (as is evident from their weak spatial relationships; Fig. 10.4), they were observed in proximity to subadult males more often than expected.

It is possible that subadult males assist these females in their efforts to immigrate in some subtle and as yet undetectable ways. It is also possible that these heterosexual relationships persist into adulthood (Strier et al. 1993). For example, the frequent spatial associations that adult females had with particular adult males (Strier 1992a) might reflect long-term relations that were established when these females presumably immigrated.

The agonistic behavior that resident adult females displayed toward subadult females attempting to immigrate may be attributed to the fact that adult females regard these younger females as potential competitors. However, the absence of overt agonistic contests over access to food suggests that intragroup contest competition was minimal and of the scramble type

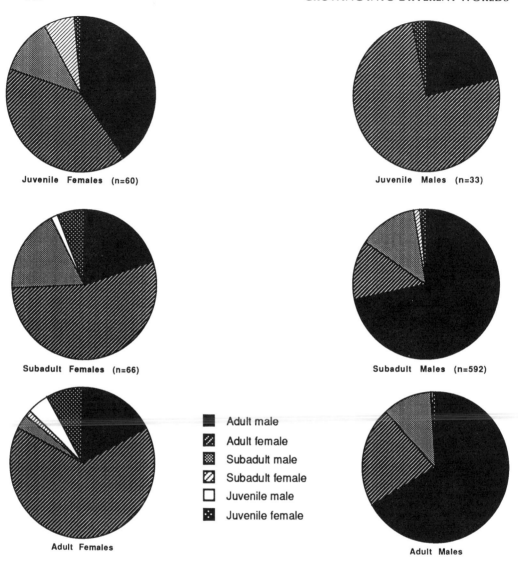

Fig. 10.5. Distribution of nearest neighbors across age–sex classes. Juvenile nearest neighbors are based on nonmothers only. The number of observations for each age–sex class is given in parentheses (for information about adults, see Strier 1992a). It is important to note that the availability of possible nearest neighbors differs across age–sex classes due to the exclusion of mothers among juvenile nearest neighbors and to the fact that individuals cannot be their own nearest neighbors.

(van Schaik 1989), and that females could avoid direct contests by avoiding feeding near one another (Strier 1990). Similarly, males do not appear to provide any parental assistance, and females' abilities to choose their mates appear to be unrestricted (Milton 1985; Strier 1987a, 1992a, 1992b). Nonetheless, without independent measures of ideal group size, which may differ in different habitats, it is difficult to evaluate the costs of additional females to resident females.

Indeed, resident females may eventually tolerate immigrant females because the effort involved in trying to repel them outweighs the costs of any increased competition. Both of the subadult females who immigrated during this study period have now reproduced in the group, and are frequently in proximity to the same females that once threatened their immigration attempts (Strier 1992b).

Ontogeny of Male Relations

The strong spatial relations exhibited by juvenile and subadult males (Fig. 10.4) are consistent with the strong spatial relations between adult male muriquis and with those exhibited by males in other patrifocal species. However, the distribution of immature male spatial relations, like that of immature females, appears to undergo a fundamental developmental shift prior to subadulthood (Fig. 10.5).

Nonmother adult females accounted for a strikingly high proportion of juvenile male nearest neighbors. The nearest neighbors of subadult males, by contrast, directly paralleled those of their adult counterparts (see Strier 1992a, 1993). Subadult males associated much more frequently with adult males than with adult females. The two subadult males were also observed in proximity to each other, but were rarely observed among the nearest neighbors of juvenile males. Identified juveniles were underrepresented among the nearest neighbors of both juvenile (3.03%) and subadult (1.18%) males.

Although juvenile males have strong social relations with adult females, it is clear that by subadulthood, male muriquis begin to seek affiliations with adult males, as has been reported for immature male chimpanzees (Pusey 1983; Watts & Pusey, Chapter 11, this volume). Elsewhere (Strier 1993) I present data demonstrating that subadult males were responsible for maintaining spatial proximity with adult males. It is possible that the two juvenile males in this study had not reached the age when juveniles begin to shift their associations toward older males. If females do not disperse repeatedly, however, these early associations with nonmother females might be equally important to their future social and perhaps reproductive relations.

Muriquis are large-bodied primates (Rosenberger & Strier 1989), and they appear to have correspondingly long life spans. Both of the two juvenile males described in this study are now fully adult males, and all the adult females are still present and reproductively active in the group. There is no reason why nonmother adult

females might not be future mates for juvenile males.

The near absence of overt competition between males, together with their strong affiliative relations with subadult males, suggests that cooperation among male muriquis may be more advantageous than competition (see Strier 1992a). Male muriquis cooperate with one another in intergroup encounters with other groups of males, and there is recent evidence suggesting that the size of male units is an important determinant in these contests among males (Strier et al. 1993), as has been reported in chimpanzees (Wrangham 1979a). Juvenile male muriquis do not participate in intergroup encounters unless they are being carried by their mothers. Subadult males never initiated threats and chases in encounters with nongroup males, but they did join in. Consequently, immature males may become increasingly important allies to adult males as they mature (Strier 1993).

FUTURE QUESTIONS

The preliminary results to emerge from this study raise more questions than they answer. While the sex differences observed in immature muriqui spatial relations are consistent with those in other patrifocal species, the development of these social relations and their significance require further study. For example, both immature females and immature males shift their spatial associations from opposite to same-sex adults between the early juvenile and subadult period, from roughly 2.5 to 5 years of age. Although there were no juveniles in this age span in the group at the time these data were collected, this is clearly an important age group for future research.

Similarly, the reasons for the disproportionately high representation of adult males among the nearest neighbors of juvenile females are not clear. It does not seem likely that adult males protect juvenile females from harassment because no such harassment has been observed. It is possible that juvenile females that succeed in establishing strong affiliations with adult males are more likely to remain in their natal group. However, there were no obvious differences in the nearest neighbors of the one juvenile female that has reproduced in her natal group and those of the other three juvenile females from her co-

Table 10.1. Distribution of Juvenile Nearest Neighbors

Juvenile	Adult males							Adult females									Subadult males		Subadult females		Juveniles	N
	SC	CL	MR	IV	MK	PR	AM	SY	AR	BS	RO	NY	DD	MO	LS	AF	SO	CY	CH	BL		
Females																						
SU	0.9	0.9	0.9	0.9				59			0.9				2.8	11	1.9		2.8	0.9	17	107
AD	8								88						4							25
BR	4.4		1.5	4.4	1.5		2.9			66		1.5		1.5	2.2			4.4			4.4	68
PE	6.7		2.2	6.7			2.2				76				2.2	5.9	2.2	4.4			2.2	45
Males																						
NI		2.9		1			4			3.9		69	5.9	7.8	4.9	1			2.9		2.9	102
DI	1		4										80	4						8		25

Note: Values represent the percentage of all nearest neighbors observed for each juvenile. Totals exceed those given in text because unidentified adult males (AM), adult females (AF), and juveniles (JV) are included. All other initials refer to identified individuals. Values in boldface indicate proximity between each juvenile and its mother. BR was the only juvenile female that has since reproduced in her natal group. See text for discussion.

hort that subsequently migrated (Table 10.1). Obviously a much larger number of observations on recognized juvenile females is needed to evaluate this possibility with more confidence.

The absence of strong associations among the six individuals composing the juvenile cohort is striking by comparison with other primates. Determining whether juveniles actively avoid proximity with one another requires further study.

Ongoing research at Fazenda Montes Claros will permit us to address these questions as systematic data on juvenile behavioral development are compiled (e.g., A. O. Rimoli, unpublished). Nonetheless, comparative data on other muriqui populations are urgently needed before results from this study can be broadly interpreted. As mentioned earlier, female-biased dispersal in a population with only two groups has obvious consequences for the genetic composition of the groups. Whether female-biased dispersal characterizes other muriqui populations is still unknown, although suggestive evidence of an immature female crossing a pasture to another forest patch has been reported from another site (Lemos de Sa 1988).

Understanding the ecological determinants of female dispersal in muriquis, and the implications for the development of muriqui social relations, will require long-term data on groups such as that at Fazenda Montes Claros, where individual histories are known, and comparative data from larger populations where a wider range of dispersal options exist.

ACKNOWLEDGMENTS

The Brazilian government and CNPq provided permission to conduct this research in Brazil. Professor Celio Valle provided generous sponsorship. Financial support was provided by NSF Grants BNS 8305322, BNS 8619442, and BNS 8959298, the Fulbright Foundation, Grant 213 from the Joseph Henry Fund of NAS, Sigma Xi, the L. S. B. Leakey Foundation, and the World Wildlife Fund. E. Veado, F. Mendes, J. Rimoli, and A. O. Rimoli contributed to the long-term demographic records. I thank M. Pereira, L. Fairbanks, and M. Hiraiwa-Hasegawa for inviting me to participate in their symposium at the XIIIth IPS Congress, where a version of this chapter was presented, and for the comments that they, C. Snowdon, and the anonymous reviewers provided on an earlier version of this manuscript.

11

Behavior of Juvenile and Adolescent Great Apes

DAVID P. WATTS and ANNE E. PUSEY

The great apes—chimpanzees (*Pan troglodytes*), pygmy chimpanzees (*Pan paniscus*), gorillas (*Gorilla gorilla*), and orangutans (*Pongo pygmaeus*)—are the largest nonhuman primates and have the longest periods of immaturity. Their ecology and social systems vary markedly. Comparative study of this group could thus address several major questions raised in this volume. First, does the length of immaturity in these species scale with adult body size, and, if not, can the variation be explained by differences in ecology, social system, and demography (see Pagel & Harvey, Chapter 3, and Janson & van Schaik, Chapter 5, this volume)? Second, to what extent is the behavior of immature great apes geared to their current survival, and to what extent to learning behavior that will contribute to their reproductive success as adults (Bekoff 1977; Pereira & Altmann 1985; Pereira 1988a)? Immatures need to develop foraging skills to survive, for example, but they may sacrifice foraging efficiency for opportunities to watch and interact with other group members and thereby gain social expertise. Third, how much do sex and species differences in the behavior of immatures resemble those of adults, and to what extent can such differences be explained by the need to seek particular learning opportunities relevant to adult behavior (e.g., observation of and interaction with infants and mother–infant dyads for juvenile females) (Pereira 1988a; Edwards, Chapter 23, this volume)?

Unfortunately, relevant data are limited by the difficulties of observing these long-lived and slow-developing animals. In this chapter, we first review information on great ape ecology to provide a context for understanding challenges that immatures face and review the physical characteristics and development schedules of each species. We then review data on the behavior of immatures, focusing on chimpanzees and mountain gorillas, and present considerable new data for mountain gorillas. Finally, we discuss the data in light of the three questions above and raise questions for future research.

TERMINOLOGY AND DATA BASE

Juveniles are prepubertal animals that can survive the death of adult caretakers (Pereira, Chapter 2, this volume). Adolescents are postpubertal individuals not yet fertile (Pereira & Altmann 1985). Labial swellings indicate the onset of puberty in females in some species, and menarche, which occurs somewhat later, has been recorded in all species (Table 11.1). Pubertal onset in males is more difficult to determine, and fertility onset is unknown. In species with conspicuous testicles, we can record acceleration in testicular growth. Gorillas lack conspicuous testicles, and the first complete copulation is probably the best indicator of the onset of adolescence (Watts 1990a, 1991a; Table 11.1).

In this chapter, we call mountain gorillas in their fourth year (i.e., weanlings) "young juve-

Table 11.1. Developmental Parameters in Wild Great Apes

Parameter	Age (years) or interval in			
	Chimpanzees	Pygmy chimpanzees	Gorillas	Orangutans
Weaning	5 (1)		3–4 (2)	
First labial swelling	8 (3–5)	c ~ 7 (6)	6–7 (7,8)	
First adult-size sexual swelling	10 (3–5)			
Age at menarche	11 (3,4)	c ~ 9 (6)	7–8 (7,8)	
Female age at first birth	14–15 (3–5)	c ~ 13–15 (9)	10–12 (7,8)	15 (10)
Start testes growth acceleration	9 (3)	c ~ 9 (9)		8–9 (10)
First ejaculation	9 (3,5)		~ 9 (8)	
Adult-size scrotum	12–13 (3,4)			13–15 (10)
Male growth complete	14–17 (3,4)		16	19–20[a] (10)
Mean adult body weight (kg)				
Males	40[b] (11,12) 60[c] (13)	45 (13)	159–175 (13)	80–91 (14)
Females	32[b] (11,12) 47[c] (13)	33 (13)	72–98 (13)	33–45 (14)

[a]Maturation possibly delayed in some populations (te Boekhoerst et al. 1990).
[b]*Pan troglodytes schweinfurthii.*
[c]*P. t. troglodytes.*
Sources (given in parentheses): (1) Pusey (1983); (2) Fossey (1979); (3) Pusey (1990); (4) Goodall (1986); (5) Nishida et al. (1990); (6) Kano (1989); (7) Harcourt et al. (1980); (8) Watts (1991); (9) Kuroda (1989); (10) Galdikas (1985a); (11) Wrangham and Smuts (1980); (12) Uehara and Nishida (1987); (13) Jungers and Susman (1984); (14) Markham and Groves (1990).

niles" (YJ) (Fig. 11.1). Males between 4 years and age at first copulation, and females between 4 years and age at first labial swelling are "old juveniles" (JM/F). Males between age at first copulation and the time when they first receive wounds from adult males (~8–10 years) are young adolescents (YA); from 11 to 13 years, they are old adolescents (OA). Old adolescents are more likely to disperse and to be wounded by silverbacks in intragroup aggression than young adolescents. Males are fully grown at 15–16 years, but can sometimes dominate older males and mate exclusively with fertile females at 14 years (Watts, personal observation) and should then be called "silverbacks." The commonly used term "blackback" (e.g., Schaller 1963) is roughly synonomous with "adolescent," but lacks precise definition. Females between age at first labial swelling and at fertility (~7–9 years) are adolescents (AF).

Researchers commonly divide postpubertal male orangutans into adolescents, subadults, and adults, without clear distinctions between the first two classes (MacKinnon 1974, 1979; Rijksen 1978; Galdikas 1985a; te Boekhoerst et al. 1990). We use the terminology of the original authors.

Most information below comes from published field studies. We augment these with data on mountain gorillas from the long-term records of the Karisoke Research Centre in Rwanda and from fieldwork by D. P. Watts. The records give data on demography and life-history events, sociosexual behavior and allogrooming (all-occurrences sampling), and proximity to adult males (scan sampling; for definitions, see Altmann 1974). Most fieldwork was focused on adults, but data include 310 hours of focal samples on two young juvenile males, two adolescent females, and two old adolescent males in Karisoke Group 4 in 1978; some focal data, plus all occurrence and scan data on juveniles and adolescents in Group 5 and Group Nk in 1984–1985; all-occurrence and scan data on all juveniles and adolescents in Group 5 and Group Bm, 75 hours of focal samples on two adolescent females and 153 hours on three adolescent males in Group 5, and 114 hours on two adolescent males in Group Bm in 1986–1987; and 60 hours of focal samples on three adolescent females in the same two groups in 1991.

Focal data included affiliative and agonistic interactions (e.g., allogrooming, play, displays), sociosexual behavior, and, in 1984–1985 and 1986–1987, proximity within 5 and 2 m (recorded at 5-minute intervals) and approaches to within 2 m.

Statistical tests are two-tailed and follow Siegel and Castellan (1988) and Sokal and Rohlf (1981).

Fig. 11.1. Young juvenile male mountain gorilla (*Gorilla gorilla beringei*).

GREAT APE SOCIOECOLOGY

Interspecific variation in behavioral ecology means that immatures of the four species face different challenges. It also produces important variation in the constancy with which particular social partners are close enough for interaction and observation and the number of particular dyads an individual is typically exposed to, as either participant or observer.

Chimpanzees, pygmy chimpanzees, and orangutans have highly eclectic but largely frugivorous diets (Rodman 1977; Wrangham 1977; Rijksen 1978; Badrian & Malenky 1984; Good-

all 1986; Galdikas 1988; Ghiglieri 1987). Gorillas are more folivorous, although the extent of folivory and frugivory varies among subspecies (Watts 1990b; Williamson et al. 1990).

Chimpanzees and pygmy chimpanzees live in permanent, multimale, multifemale social groups (communities) of up to 100 or more individuals and have fission–fusion social systems. Feeding and travel party size depend largely on food patch size and dispersion, but also on social factors like the presence of estrous females. Female chimpanzees are less gregarious, probably because their food often is in smaller patches (Wrangham 1977, 1979b; Wrangham & Smuts 1980; Badrian & Badrian 1984; Goodall 1986;

Ghiglieri 1987; White & Wrangham 1988; White 1989). Anestrous female chimpanzees are alone with offspring in distinct but overlapping core areas up to 70% of the time; estrous females and adult males are more gregarious and travel more widely (Wrangham & Smuts 1980; Hasegawa 1990).

A chimpanzee community contains a core of related males that jointly patrol its boundaries. Social bonds are usually stronger between males than between females, although mother–daughter bonds may be very strong, and male–male alliances are important in male mating strategies (Nishida 1979, 1983; Tutin 1979; Wrangham 1979a; de Waal 1982; Hasegawa & Hiraiwa-Hasegawa 1983, 1990; Goodall 1986). Bonds seem to be weaker among males in pygmy chimpanzees and stronger among females, and high-ranking males may be better able to control access to estrous females than their chimpanzee counterparts (Kano 1980; Kuroda 1980; Badrian & Badrian 1984; White 1988, 1989, in press; White & Burgman 1990; but see Furuichi 1989).

Orangutans are semisolitary: the only permanent associations are those of females with dependent young. Independent juveniles and adolescents are more gregarious than adults. Paterns of food dispersion, foraging energetics, and feeding competition are thought to constrain gregariousness. Contact between adult females is only intermittent, and they have mostly neutral to antagonistic relationships. Adult males compete intensively for mating opportunities, and "residents" establish home ranges that overlap those of several females (Rodman 1973, 1979; MacKinnon 1974; Rijksen 1978; Galdikas 1985a; Sugardjito et al. 1987).

Gorillas live in permanent, cohesive groups of 1–14 females, one or more males, and their offspring. Their terrestrial food supply is more densely distributed and constantly available than those of other apes. This limits feeding competition and allows more constant association (Wrangham 1979a; Stewart & Harcourt 1987; Watts 1990b). Male–female social bonds are strong, but those between females are weak except those between mothers and daughters or sisters resident in the same group (Harcourt 1979a, 1979b; Watts 1991b, 1992, unpublished data).

Patterns of dispersal vary. Natal transfer by females is common in gorillas and chimpanzees, but some females (about 40% in mountain gorillas) reproduce in natal groups or communities. Secondary transfer is also common in gorillas (Harcourt 1978; Pusey 1979, 1980; Wrangham 1979a; Yamagiwa 1983; Goodall 1986; Stewart & Harcourt 1987; Nishida et al. 1990; Watts 1990a, 1990b, 1991a).

Male chimpanzees are philopatric, but about 50% of male mountain gorillas emigrate near physical maturity, become solitary, and try to attract females, whereas others remain in their natal groups as adults. Gorilla males in different groups compete intensely for mates. Males in multimale groups sometimes form coalitions against extragroup males, but compete aggressively with one another (Harcourt 1978; Wrangham 1979a; Harcourt & Stewart 1981; Goodall 1986; Stewart & Harcourt 1987; Watts 1990a, 1990b, 1991a).

Limited data suggest that although all orangutans disperse from their mothers' ranges, females settle nearby, whereas males disperse more widely (Horr 1977; Galdikas 1979).

GROWTH AND MATURATION

All great ape species show sexual dimorphism in adult body size and sexual bimaturism (sex differences in developmental schedules), with males reaching puberty and full size later, and becoming larger, than females (Table 11.1). Dimorphism and bimaturism are greater in gorillas and orangutans, probably because one-on-one combat is more important in male mating competition, than in either chimpanzee species (Jungers & Susman 1984; Shea 1985, 1986; Rodman & Mitani 1987; Uehara & Nishida 1987; Markham & Groves 1990). Male maturational delays induced by proximity to adult males may extend bimaturism in orangutans (Kingsley 1982; te Boekhoerst et al. 1990).

The great apes do not exhibit the general positive correlation between the duration of immaturity and adult body size (Pagel & Harvey, Chapter 3, this volume). Gorillas grow to be much larger than chimpanzees and orangutans in a similar time, perhaps shorter for females (Shea 1983, 1985, 1986; Table 11.1). Also, gorilla females reach sexual maturity earlier than females of the other species and have shorter interbirth intervals (Goodall 1986; Stewart & Harcourt 1987; Galdikas & Wood 1990; Watts

1992a; Table 11.1). We lack good data on relative life spans, but captive data suggest that orangutans and chimpanzees live longer than gorillas (Harvey & Clutton-Brock 1985).

RESULTS OF FIELD STUDIES

Proximity and Association Patterns of Immatures

With Adult Females. For several years after weaning, juvenile male and female chimpanzees, orangutans, and gorillas associate almost constantly with their mothers, who tend to be their most frequent neighbors. Proximity then declines with age. The decline is abrupt in chimpanzees when males reach puberty and when females start mating with adult males, and for some orangutans after the birth of a new sibling (MacKinnon 1974; Horr 1977; Rijksen 1978; Pusey 1983, 1990; Galdikas 1985b; Hayaki 1988). It is more gradual in gorillas (Fig. 11.2).

Two of four juvenile female mountain gorillas and the one male with data available had higher proximity scores with their mothers than with any other individual at all ages. Two other females each had one higher score (resting or feeding within 2 m) with silverbacks than with their mothers for 1 year.

Immature male chimpanzees show more interest than females in associating with unrelated adult females, particularly estrous females. The time that juvenile males spend in parties with estrous females depends mostly on maternal sociability. Once they have left their mothers, adolescent males spend more time than either juvenile males or juvenile and adolescent females with estrous females and with nonfamily members in general (Pusey 1983, 1990; Kawanaka 1984; Goodall 1986; Hayaki 1988).

Immature male gorillas hardly need alter their spatial patterns to associate with females other than mothers, but tend to do so more than females as they approach adulthood. Juvenile and adolescent females, and juvenile and young adolescent males, spent significantly more time near related females (including mothers) than nonrelatives (Table 11.2). In 1986–1987, two immature females approached related females significantly more often than nonrelatives (Wilcoxon–Mann–Whitney tests; JF Jz: $W_x = 30$, $m,n = 3,8, p < 0.01$; AF Mg: $W_x = 38, m,n = $

4,7, $p < 0.005$). So did two young adolescent males (YA Ca: $W_x = 44$, $m,n = 5,7, p < 0.05$; YA Sh: $W_x = 42$, $m,n = 4,8, p < 0.05$), but the older had his highest proximity scores with an unrelated sexual partner. OA Pb associated more than younger males with nonrelatives (Table 11.2) and approached them and relatives equally often ($W_x = 39$, $m,n = 4,7$, ns). He associated and interacted little with his mother (Fig. 11.2) (Watts 1990a, 1992). (In 1991, OA Ca, then the same age as Pb in 1987, had a much higher rate of interaction with his mother; Watts, unpublished data.) Group Bm had two old adolescent males, two adult females, and four juvenile females in 1986–1987. OA Ts spent more time within 5 m of one unrelated adult than did all females other than her juvenile daughter, and as much or more near the second adult (a possible relative, but with whom he had not resided for most of his life) as all group members other than her juvenile daughter. He copulated with both adult females (Watts 1990a). OA Ha, conversely, although attracted to adult females, was socially and spatially peripheralized by the silverback; before his emigration in 1987, he had by far the lowest proximity scores with both adult females.

With Adult Males. There are sex differences in association between immatures and adult males in chimpanzees and gorillas, and species contrasts in the pattern of differences. By age 5–6 years, male chimpanzees have great interest in joining parties that contain adult males, but often have difficulty in persuading their mothers to accompany them (Pusey 1983). Adolescent males spend more time than juveniles in parties with adult males (Kawanaka 1984; Goodall 1986; Hayaki 1988; Pusey 1990), but are less often in close proximity to adults when in such parties than are juvenile males (Hayaki 1988; Pusey 1990).

Anestrous adolescent female chimpanzees in their natal communities spend less time than estrous peers or male peers near adult males. Recent adolescent immigrants, whether estrous or not, associate with adult males relatively often; they may thereby avoid harassment from resident females (Pusey 1980, 1990; Wrangham & Smuts 1980; Goodall 1986; Hayaki 1988; Nishida 1989).

In contrast to chimpanzees, adolescent female gorillas associate more than adolescent

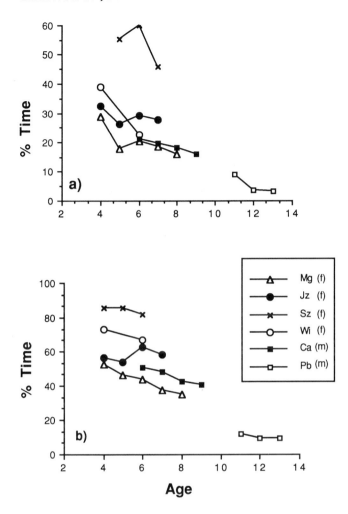

Fig. 11.2. Proximity of juvenile and adolescent mountain gorillas to their mothers: (a) time within 5 m while resting; (b) time within 5 m while feeding. The pattern for time spent within 2 m (values not shown) was very similar.

males with adult males. At age 3–4, female mountain gorillas spent a median 38.6% of feeding time (range 20.4–55.3%) and 60.9% of resting time (34.9–83.3%) within 5 m of their groups' silverback or the older of two silverbacks (the more likely father). Male values were 39.5% (30.8–51.8%) for feeding and 65.7% (64.2–66.1%) for resting (Fig. 11.3). Proximity gradually declined with age, with no significant sex difference through age 8 (Mann–Whitney tests, ns). By then, female proximity reached adult levels (Harcourt 1979; Watts 1992), but male values declined through age 13 (Fig. 11.3). The lowest scores were for OA Ha with an unrelated silverback, but by age 13, males (e.g., OA Pb in Group 5) can spend less than 5% of feeding time and 10% of resting time within 5

m of related silverbacks with whom they compete for mates.

Grooming

Juveniles usually receive most of their grooming from their mothers (Horr 1977; Rijksen 1978; Goodall 1986; Nishida 1988; Pusey 1990). Eleven of 16 juvenile gorillas received more grooming bouts per hour from mothers than from any other partners (Table 11.3). Juveniles received a median of 83% of their total grooming from mothers, and 12 of 14 got more grooming from mothers than from any other partners (Table 11.4; durational data were not always congruent with bout rates; for example, compare grooming by JM Bo in Tables 11.3 and 11.4).

Table 11.2. Results of Wilcoxon–Mann–Whitney Tests (W_x values; Siegel & Castellan 1988) Comparing Proximity of Juveniles and Adolescents to Related and Unrelated Adult

Animal	Sex	Age	m,n	Proximity category	
				Feed, 5 m	Rest, 5 m
Jz	F	4–5	5,5	15***	15***
Jz	F	5–6	3,7	27**	26*
Mg	F	5–6	5,5	15***	15***
Mg	F	6–7	3,7	27**	27**
Mu	F	7–8	4,5	30**	30**
Pp	F	8–9	5,5	15**	15**
Ca	M	6–7	5,6	15***	15***
Ca	M	8–9	5,7	50****	50****
Sh	M	7–8	5,6	15***	15***
Sh	M	9–10	5,7	49***	43
					$p = 0.053$
Pb	M	10–11	5,6	17**	17**
Pb	M	12–13	5,7	35	36
				ns	ns

Note: In significant differences, more time was spent with relatives. Data for 5 m only shown; data for 2 m follow the same pattern.
  *$p < 0.05$.
 **$p < 0.01$.
***$p < 0.005$.
****$p < 0.001$.

Juvenile chimpanzees gave most of their grooming to mothers and siblings. Gombe females gave more than 90% of their grooming to mothers, and males gave 67% to mothers and siblings. Both males and females developed balanced reciprocal grooming relationships with mothers (Pusey 1983, 1990). Many immature gorillas groom their mothers less extensively and exclusively. Juveniles gave a median of only 28.2% of total grooming to mothers (Table 11.4). Only four of nine females and two of six males groomed mothers more often than any other partner (Table 11.3). In eight of nine mother–juvenile daughter dyads, mothers did more grooming (usually much more) than daughters (Table 11.3). All juvenile females groomed mothers less, absolutely and as a percentage of all grooming given, than mothers groomed them (Table 11.4). Mothers groomed five of six juvenile males more often than the reverse (Table 11.3), and all sons received more grooming than they gave (Table 11.4).

Many old juvenile gorillas groom silverbacks

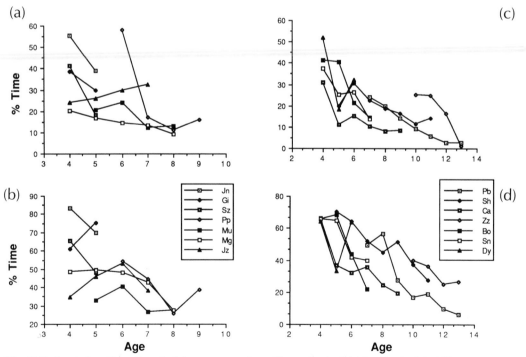

Fig. 11.3. Proximity of juvenile and adolescent mountain gorillas to the single/older silverback in their groups: (a) females, time within 5 m while feeding; (b) females, time within 5 m while resting; (c) males, time within 5 m while feeding; (d) males, time within 5 m while resting.

Table 11.3. Juvenile and Adolescent Mountain Gorilla Grooming with Mothers and Silverbacks

Individual	Age (years)	Bouts/hour from mother	Mother top partner?	Bouts/hour to mother	Mother top partner?	Bouts/hour from silverback	Bouts/hour to silverback
Females							
Wil[a]	3	0.080 (90.9)	Yes	0.002 (80.0)	Yes	0	0
Jn1[b]	4	0.074 (88.9)	Yes	0.008 (60.0)	Yes	0	0.037 (71.8)
Wi2[b]	4	0.068 (56.5)	Yes	0.022 (61.6)	Yes	0.021 (26.9)	0
Jn2[c,d]	5					0.003 (9.1)	0.003 (15.4)
Jzl	5	0.015 (15.1)	No	0.002 (33.3)	No	0	0
Szl	5	0.149 (100)	Yes	0.005 (22.2)	No	0	0.011 (78.8)
Mul[e]	5	— (62.3)	Yes	— (21.1)	No	0	— (48.3)
Jz2	6	0.026 (27.5)	No	0.003 (24.0)	No	0	0.008 (66.7)
Sz2[c,d]	6	0.136 (89.4)	Yes	0.015 (71.5)	Yes	0.009 (5.8)	0
Gil	6	0.067 (93.6)	Yes	0.002 (1.8)	No	0	0.069 (78.9)
Pol[e]	6	(12.0)	No	(28.6)	No	0	(14.3)
Mg1	6	0.002 (14.3)	No	0.002 (14.3)	No	0	0.007 (50.0)
Mg2	7	0.003 (19.4)	No	0.008 (17.5)	Yes	0	0.007 (13.8)
Cl	7	0.002 (4.3)	No	0.016 (14.5)	No	0	0
Mu2	7	0.011 (83.3)	Yes	0.001 (2.8)	No	0	0.026 (66.7)
Gi2[c,d]	7					0.015 (50.0)	0.003 (4.3)
Au[d]	7					0	0
Po2	8	0.001 (33.3)	No	0.001 (1.9)	No	0	0.040 (68.5)
Males							
Kw	3	0.135 (84.4)	Yes	0	No	0.004 (2.7)	0
Ts	4	0.012 (8.7)	No	0	No	0.033 (24.7)	0
Sh1[d,e]	5					(8.9)	(95.3)
Dy	6	0.020 (75.0)	Yes	0.010 (33.3)	No	0	0.081 (88.3)
Bo	7	0.030 (80.0)	Yes	0.016 (25.0)	No	0	0.008 (13.8)
Sn	7	0.013 (36.0)	No	0.033 (44.0)	Yes	0	0.015 (20.0)
Cal	7	0.013 (70.6)	Yes	0.010 (50.0)	Yes	0	0.009 (45.8)
Sh2[d]	8					0	0.060 (90.3)
Pb1[d,e]	8					0	(88.1)
Ca2	9	0.035 (76.8)	Yes	0.002 (8.0)	No	0	0.002 (8.0)

(*continued*)

Table 11.3. (*Continued*)

Individual	Age (years)	Bouts/hour from mother	Mother top partner?	Bouts/hour to mother	Mother top partner?	Bouts/hour from silverback	Bouts/hour to silverback
Sh2[d]	10					0	0.012 (37.8)
Tg[d]	10					0.014 (100)	0
Pb2	11	0	No	0	No	0	0.004 (100)
Bm[c,d]	12					0	0
Pb2	13	0	No	0.001 (0.4)	No	0	0

Note: Values are grooming rates, measured as bouts per hour, with percentage of total bouts given in parentheses.
[a]Numbers indicate same individual in different study periods.
[b]Mother emigrated or reimmigrated during study; only data while mother was present used to determine if mother was top partner.
[c]Silverback not a relative.
[d]Mother not present in group.
[e]No sample times available for calculation of rates; only percentage of bouts given.

extensively: a median of 47.1% of their total grooming went to the only or older silverbacks in their groups (Table 11.4). The highest values were for maternal orphans grooming fathers, but 7 of 11 others groomed related silverback more than mothers (Tables 11.3 and 11.4). In Group Nk, the silverback had significantly more grooming bouts with juveniles than with females [*G* test; $G(1) = 100.23$, $p < 0.0001$]. Juvenile–silverback grooming was 40% of the group total (excluding mothers grooming infants), more than between any other age–sex classes. Almost all silverback–juvenile grooming was always unidirectional (Table 11.3).

Adolescent female chimpanzees mostly groom their mothers, and have balanced reciprocal grooming relationships with them (Nishida 1988; Pusey 1990). Not all gorilla females do, nor are mothers necessarily their top grooming partners. Mothers groomed more in two of four dyads, and daughters groomed more in the other two (although grooming was rare for AF Po2; Tables 11.3 and 11.4). Only one female groomed her mother more often than any other partner, and only one was groomed most often by her mother (Table 11.3). Two were groomed longest by mothers of all partners, but none groomed mothers longest (Table 11.4). Two groomed silverbacks much more than their mothers (Tables 11.3 and 11.4).

Adolescent male chimpanzees groom adult males and full-sized females more than adolescent females do at all ages, although they are not fully integrated into male grooming networks until adulthood (Goodall 1986; Pusey 1990;

Takahata 1990). Gorillas show similar sex differences with females, but opposite ones with males. Five of seven male gorillas, but only two of nine females, groomed unrelated adult females. At the extreme, OA Pb groomed nearly twice as much as anyone else in Group 5 (excluding mothers grooming infants); 75% of this went to unrelated adult females that were also his sexual partners (Watts 1992). Males groom less than adolescent females with adult males (Tables 11.3 and 11.4), however, and silverbacks in the same group almost never groom each other. Silverback–adolescent grooming was unidirectional (Table 11.3).

There are too few data to examine reciprocity in mother–adolescent son grooming in gorillas.

Affiliative bonds exist between some adolescent female orangutans at Tanjung Puting; grooming occurred only in these dyads and between mothers and offspring (Galdikas 1985b).

Agonistic Behavior

Juvenile chimpanzees and gorillas are subordinate to adults (Harcourt & Stewart 1987, 1989; Pusey 1990), except that juvenile male chimpanzees at Gombe neither submitted to nor received submission from adult females (Pusey 1990). Adolescent female gorillas submit to adults in decided interactions (Watts 1985b; Harcourt & Stewart 1987, 1989), but many are undecided and adolescents show some aggression to adults (Watts, unpublished data). Daughters of high-ranking female chimpanzees may give more aggression to females in their

Table 11.4. Juvenile and Adolescent Grooming with Mothers and Their Groups' Single or Older Silverbacks[a]

	Age (years)	Minutes grooming received	Minutes grooming given	From mother	Mother top partner?	To mother	Mother top partner?	From SB	To SB
Females									
Juveniles									
Wil[a]	3	26	30	92.3	Yes	87.5	Yes	0	0
Jn	4	107	156	20.7[b]	Yes	3.9	No	0	92.3
Wi2	5	188	84	51.1	Yes	45.8	Yes	15.2[d]	0
Jn2	5	73	30	[b]				9.6	6.8
Jz1	5	453	3	11.9	No	100	Yes	0	0
Sz1	5	592	85	100	Yes	28.2	No	0	65.9
Jz2	6	688	159	27.5	No	11.3	No	0	42.1
Sz2	6	320	47	88.0	Yes	53.4	No	5.9[d]	0
Gi	6	262	623	95.0	Yes	0.6	No	0	91.5
Mg1	6	66	20	45.5	No	15.0	No	0	55.0
Adolescents									
Mg2	7	73	237	34.3	Yes	18.2	No	0	16.2
Cl	7	24	88	8.3	No	22.7	No	0	0
Mu2	7	51	328	88.2	Yes	1.0	No	0	82.6
Gi2	7	47	266	[c]				0[d]	6.7[c]
Au	7	55	165	[c]				0	8.5
Po2	8	3	357	33.3	No	0.8	No	0	82.6
Males									
Juveniles									
Kw	3	186	5	90.0	Yes	0	No	5.7	0
Ts	4	345	15	37.1	No	100	Yes	19.4	0
Dy	6	238	445	90.4[b]	Yes	1	No	0	95.1
Bo	7	170	228	84.4	Yes	21.5	No	0	25.0
Sn	7	139	332	25.2	No	44.0	Yes	0	22.0
Cal	7	103	85	81.6	Yes	50.6	Yes	0	47.1
Adolescents									
Sh1	8	4	315	[c]				0	96.8
Ca2	9	365	78	86.9	Yes	5.1	No	0	47.1
Sh2	10	30	150	[c]				0	39.3
Pb1	11	2	12	0	No	0	No	0	100
Tg	11	19	8	[c]				94.7	0
Bm	12	1	1	[c]				0[d]	0
Pb2	13	92	995	0	No	0	No	0	0

Note: Values are percentage of total duration of grooming received or given.
[a]"1," "2," and "3" refer to the same individual in different study periods.
[b]Mother emigrated during study period; value is percentage while she was present.
[c]Mother not present.
[d]Silverback not a relative.

natal communities than do those of low-ranking females (Goodall 1986; Pusey 1990). All adolescent immigrant female chimpanzees, and some gorillas, are frequently harassed by resident females (Pusey 1980, 1990; Nishida 1989; Watts 1992). As juvenile female orangutans mature to adulthood, their relationships with adult females and peers become less tolerant (Horr 1977; Rijksen 1978; Galdikas 1985b).

Male chimpanzees and gorillas make protracted challenges of adolescent and adult females before achieving dominance over them (first over adolescents). In chimpanzees, this starts when males are half grown and smaller than the females, who initially ignore, threaten, or attack them (Pusey 1990). Similar age-related differences in female responses to male displays illustrate the process in gorillas. Young and old adolescent males displayed at females from 0.05 to 0.58 times per male–female pair hour. Rates varied little among YA Sh, YA Ca, and OA Pb in Group 5 [$n = 118$ displays; $\chi^2(2) = 0.31$, ns] and between two Group 4 males one year apart in age ($n = 38$ displays, $\chi^2 = 0.95$, ns). Females responded more aggressively to young adolescents than to OA Pb in Group 5, however [$\chi^2(1) = 5.21$, $p < 0.05$], and gave submission significantly more often to OA Pb's

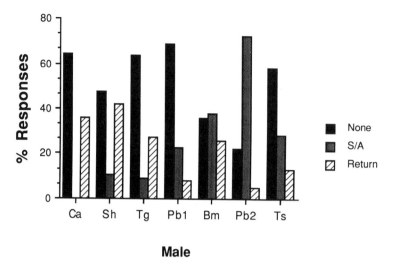

Male

Fig. 11.4. Female responses to aggression by adolescent males in Group 5. None, no overt response; S/A, submissive or appeasement behavior; Return, aggression returned. Male ages given in Table 11.4.

displays $[\chi^2(1) = 33.51, p < 0.0001$; Fig. 11.4]. Still, they ignored more displays from OA Pb than from the silverback and were not aggressive to the silverback (Watts 1992).

The highest adolescent display rates (OA Ha: 0.14/pair hour; OA Ts: 0.46/pair hour; YA Pb: 0.58/pair hour) were in groups with recent female immigrants. OA Pb displayed more in 1984–1985, when Group 5 had five recent immigrants, than in 1986–1987, when it had none $[X^2(1) = 52.03, p < 0.0001]$. He displayed more often at the immigrants than at six long-term residents [76 of 109 displays at immigrants; $X^2(1) = 33.51, p < 0.0001$]. OA Ts also displayed more at a recent immigrant (44 of 60 displays) than at a longer term resident in Group Bm $[X^2(1) = 21.99, p < 0.0001]$.

The rate of adult–adolescent male aggression increases as adolescents mature in gorillas (Fig. 11.5) and chimpanzees. This is associated with decreased affiliation and proximity in gorillas, but adolescent chimpanzees seek proximity with adult males despite a rise in aggression and tension as they work their way into the male dominance hierarchy and grooming and alliance networks (Goodall 1986; Pusey 1990; Takahata 1990). Adult males orangutans tolerate juvenile and adolescent males while feeding (although the immatures are wary), but not subadults, especially near estrous females (MacKinnon 1974, 1979; Horr 1977; Rijksen 1978; Galdikas 1985a, 1985b; Mitani 1985).

Relatedness; adolescent age; age disparity; the presence, number, and identity of estrous females; and the history of the males' social relationship can influence aggression rates among male gorillas (Harcourt & Stewart 1981; Watts 1990a, 1991a, 1991b). For example, Pb received far more aggression at age 12–13 than at age 10–11 (Fig. 11.5). He often avoided or was supplanted by the silverback, or ceased his activities and watched warily when the silverback approached. They were paternal half-brothers that had an affiliative relationship when both were immature, but tension between them increased as their sexual rivalry grew. Serious silverback aggression toward old adolescents commonly involves females. For example, OA Ts and a silverback fought several times over access to estrous females and when Ts supported a female at whom the silverback was displaying.

The severity of aggression also increases in gorillas, especially between males, as immatures grow older. Silverback aggression to juveniles is rare; for example, there was none between the silverback and two juvenile males in Group 4. Adults rarely, if ever, wound young juveniles in their own groups. Adult females sometimes wound old juveniles; they wound young adolescent males more often and wound adolescent females at rates equal to those between adults (Table 11.5; Watts, unpublished data). Silverbacks rarely wound females. Adult

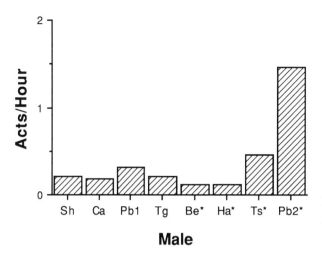

Fig. 11.5. Rate at which adolescent males received aggression from silverbacks. Male ages given in Table 11.4. *old adolescent.

females do not wound old adolescent males, who are larger than they are, but silverbacks do, sometimes at rates much higher than those among females (Table 11.5).

Agonistic Support

Adult female mountain gorillas and chimpanzees typically support immature close relatives against larger or more distantly related opponents (Goodall 1986; Harcourt & Stewart 1987, 1989). Female transfer limits opportunities for persistent alliances between close female relatives (Harcourt & Stewart 1987, 1989), but these occur when relatives reside together as adults (Watts 1992, unpublished data).

Table 11.5. Rates at Which Immature Mountain Gorillas Were Wounded by Adults, Measured as Number of Wounds Inflicted per Female (or Silverback) per Day

Immature class	n	From adult females	From silverbacks
Young juveniles	5	0	0
		(0–0)	(0–0)
Old juvenile females	7	0.001	0
		(0–0.002)	
Old juvenile males	5	0.002	0
		(0–0.004)	(0–0)
Adolescent females	2	0.004	0
		(0.003–0.005)	(0–0)
Young adolescent males	2	0.004	0
		(0.003–0.005)	(0–0)
Old adolescent males	4	0	0.034
		(0–0)	(0.10–0.044)

Note: Values are medians, with ranges in parentheses.

Maternal support may help some young female orangutans and chimpanzees establish home ranges or core areas, and support from high-ranking mothers can help nondispersing female chimpanzees acquire high rank (Rijksen 1978; Goodall 1986). Female pygmy chimpanzees form alliances more commonly than females of other species and may use them to defend access to food (White & Wrangham 1988; White, in press); some alliances may develop from close associations between immigrant adolescents and resident adults (Furuichi 1989). Furuichi (1989) and Kuroda (1989) state that female pygmy chimpanzees help sons to rise in rank, but give no data.

Adult male gorillas and chimpanzees protect recent adolescent and adult female immigrants from harassment by residents (Pusey 1980, 1990; Wrangham & Smuts 1980; Goodall 1986; Nishida 1989; Watts 1992). Silverbacks also support offspring against larger opponents, but otherwise mostly end conflicts without supporting either opponent (Harcourt & Stewart 1987, 1989; Watts 1992, unpublished data).

Adolescent male and female orangutans at Tanjung Puting sometimes cooperate to supplant adult females from food trees (Galdikas 1987).

Some adolescent male chimpanzees develop alliances with particular adult males that help them to rise in rank (Pusey 1983, 1990; Goodall 1986). Male gorillas do not develop alliance networks like those of chimpanzees, but adolescents in multimale groups sometimes gain support from closely related silverbacks (fathers or full brothers) against less closely related silverbacks (Watts, unpublished data).

Sex Differences in the Effects of Relatedness in Gorillas

Like adult females (Harcourt 1979a; Stewart & Harcourt 1987; Watts, unpublished), immature female mountain gorillas have more affiliative interactions, and fewer aggressive ones, with female relatives than with nonrelatives. Four females in Group 5 all had more affiliative than aggressive interactions with all female relatives ($n = 4$–5). Three had more affiliative than aggressive interactions with two nonrelatives each; otherwise, all had more aggressive than affiliative interactions with nonrelatives ($n = 5$–7). All had more aggressive interactions with nonrelatives (AF Mu: $W_x = 15$, $m,n = 5,5$, $p < 0.01$; AF Mg: $W_x = 10$, $m,n = 4,7$, $p < 0.01$; AF Ma, $W_x = 11$, $m,n = 4,7$, $p < 0.01$; JF Jz: $W_x = 10$, $m,n = 4,7$, $p < 0.01$), but more affiliative interactions with relatives (Mu: $W_x = 40$, $m,n = 5,5$, $p < 0.01$; Mg: $W_x = 38$, $m,n = 4,7$, $p < 0.01$; Jz: $W_x = 38$, $m,n = 4,7$, $p < 0.01$; AF Ma: $W_x = 13$, $m,n = 4,7$, $p < 0.05$).

Adolescent males in the same group showed a different pattern. They usually had higher rates of aggression than affiliation regardless of female relatedness (8 of 13 related and 19 of 23 unrelated dyads). YA Ca had more affiliative interactions with relatives ($W_x = 45$, $m,n = 5,7$, $p < 0.05$); otherwise, rates of affiliation and aggression with relatives and nonrelatives were equal (aggression: OA Pb: $W_x = 42$, $m,n = 5,7$, ns; YA Sh: $W_x = 29$, $m,n = 4,8$, ns; YA Ca: $W_x = 31$, $m,n = 5,7$, ns; affiliation: Pb: $W_x = 33$, $m,n = 5,7$, ns; Sh: $W_x = 36$, $m,n = 5,7$, ns).

Play

It is difficult to disentangle possible sex and species differences in play frequency and possible partner preferences from small sample effects and from demographic constraints on partner availability and presence. Sex differences in chimpanzees have not been firmly documented (Pusey 1983, 1990; Hayaki 1985a). Male gorillas and orangutans seem to play more than females, as expected in highly dimorphic species if play promotes fighting ability (Fagen 1981, Chapter 13, this volume). Four adolescent female gorillas played as much as or less than

two old adolescent males and less than three young adolescents older than the females (Fig. 11.6). There are no quantitative data on rough-and-tumble play. Immature orangutans in temporary social groups at Ketambe played up to 10% of the time; play among males was more common, and average bout length longer, than male–female or female–female play (Rijksen 1978). Also, adolescent and subadult males sometimes engage in considerable rough play; at least at Tanjung Puting, this has not been observed among adolescent females (Rijksen 1978; Galdikas 1985b). Rijksen suggests that rehabilitant males at Ketambe used rough play to test each other's physical ability and aggressive tolerance and to establish and maintain rank.

Immatures seem to prefer play partners of similar size and age when they are available. Most immature chimpanzees at Gombe played most often with siblings, but played at higher rates with age peers when they were present (Pusey 1990). Adolescent female orangutans at Ketambe solicited play with males more than with females, but males responded more to other males (Rijksen 1978). Juvenile and adolescent gorillas played mostly with age–sex peers (Table 11.6), with a few exceptions. Two Group 4 adolescent females played more with males (especially juveniles) than with each other. A Group 5 adolescent played little with juvenile males and more with infant males, however, and as much as expected with her female peer (Table 11.6). OA Pb, who had no peers, played mostly with two young adolescent males (Table 11.6). They were equally responsible for initiating play bouts with each other [$n = 108$; $X^2(1) = 0.93$, ns] and with him [YA Sh: $n = 27$; $X^2(1) = 1.81$, ns; YA Ca: $n = 14$, $X^2(1) = 1.14$, ns]. In Group Bm, OA Ts played exclusively with four immature females and not at all with OA Ha, an unrelated peer with whom he had played regularly when both were younger. Play between male gorillas usually stops as they become mating rivals. OA Ts stopped playing with his group's silverback (a nonrelative), for example, and OA Pb no longer played with a silverback half-brother (both silverbacks played with other individuals).

Young males should play more than females with larger partners if they have more need to develop motor skills through play (Fagen, Chapter 13, this volume). Several observations

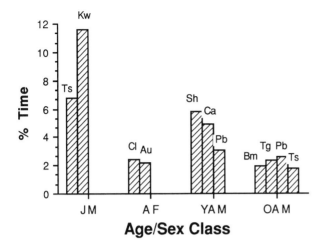

Fig. 11.6. Percentage of time spent playing by immature mountain gorillas. JM, juvenile male; AF, adolescent females; YAM, young adolescent male; OAM, old adolescent male.

fit this prediction. Immature male chimpanzees play more than immature females with adult males (Pusey 1990). Group 4 juvenile males played more than adolescent females with adolescent males, YA Ca and YA Sh in Group 5 played more with OA Pb than adolescent females played with any adolescent male (Table 11.6), and juvenile male orangutans at Ketambe were more interested than juvenile females in play with adolescent males (Rijksen 1978).

Also, infant male gorillas played strikingly often with adolescents of both sexes in Group 5 (Table 11.6); one adolescent played more with a male infant than with his adolescent peer. Three of eight infant males accounted for most of the play, however, and the difference may have been among individuals rather than between the sexes.

During travel or social play, juvenile chimpanzees often use several motor elements that

Table 11.6. Distribution of Play Among Age–Sex Classes in Two Mountain Gorilla Groups

I. Group 4, 1978: Percentage of Play by Members of Column class that Was with Row Class

Class[a]	Juv	Adol F	OAdol M	Adult F	SB
Juv ($n = 2$)	66.0	81.9	55.1	100	
Adol F ($n = 2$)	18.8	6.2	25.3	0	
OAdol M ($n = 2$)	15.1	11.1	19.6	0	
Adult F ($n = 3$)	1.0	0	0	0	
SB ($n = 1$)	0	0	0	0	

II. Group 5, 1986–1987 and 1991: Percentage of Play by Three Adolescent Males and Two Adolescent Females with Members of Different Age–Sex Classes

Males, 1986–1987 (age in years)	Inf M ($n = 5$)	Inf F ($n = 2$)	Juv M ($n = 0$)	Juv F ($n = 2$)	Adol F ($n = 1$)	YAdol M ($n = 2$)	OAdol M ($n = 1$)	Adult F ($n = 11$)	SB ($n = 1$)
Cantsbee (9–10)	49.4	13.9	0.2	1.9	32.4	0.7	1.6	0	
Shinda (10–11)	44.9	7.2	0.9	4.2	27.5	4.3	5.1	6	
Pablo (12–13)	6.5	0.9	0	3.2	81.7		6.4	0	

Females, 1991 (age in years)	Inf M ($n = 3$)	Inf F ($n = 2$)	Juv M ($n = 6$)	Juv F ($n = 1$)	Adol F ($n = 1$)	YAdol M ($n = 0$)	OAdol M ($n = 1$)	Adult F ($n = 10$)	SB ($n = 3$)
Mahane (6)	45.8	2.1	3.1	9.7	9.4	0	29.2	0	
Umwana (6)	17.7	5.9	0	11.8	64.7	0	0	0	

[a]Inf, infant; Juv, juvenile; Adol, adolescent; YAdol, young adolescent; OAdol, old adolescent; SB, silverback; M, male; F, female.

appear in adult charging displays, such as stamping or slapping the ground. Juvenile males do so much more often than juvenile females, which parallels the adult sex difference in display frequency (Pusey 1990). Immature gorillas also commonly use display elements in play, but there are no data on possible sex differences.

Sexual Behavior

Male chimpanzees show elements of sexual behavior by 1 year. They show copulation sequences complete except for ejaculation by age 4 (Pusey 1990), earlier than gorillas (7–8 years); they are also likely to copulate at much higher frequencies than gorillas as adults and more often in the presence of other males. Immature males of all species probably get few fertile matings. Adult male chimpanzees tolerate copulations, and interference in their own copulations, by adolescents less than by juveniles, and courting or copulating adolescents are warier than are juveniles. Copulation rates with estrous females are high for juveniles, then decline through adolescence. Immature males copulate mostly with adolescent females, nulliparous adults, and parous adults not near peak estrus. They do so mostly in opportunistic situations and do not form consorts successfully before late adolescence (McGinnis 1979; Tutin & McGinnis 1981; Hasegawa & Hiraiwa-Hasegawa 1983, 1990; Hayaki 1985a; Takasaki 1985; Goodall 1986; Pusey 1990). Silverbacks tolerate copulations between their daughters and adolescents, but often interfere when adolescents copulate with the silverbacks' own sexual partners. Adolescents sneak some copulations with these females. Old adolescents sometimes interfere in copulations by young adolescents (Watts 1990a, 1991a). Adolescent and subadult male orangutans cannot compete aggressively with fully grown adults and often copulate forcibly. This seems a maturation-dependent tactic in their case, but adult males in some populations also frequently copulate forcibly (MacKinnon 1974, 1979; Rijksen 1978; Galdikas 1981, 1985a; Schurmann 1982; Mitani 1985; Rodman & Mitani 1987).

Pubertal and adolescent female chimpanzees and orangutans copulate with immature males before adult males show interest in them (Goodall 1968, 1986; Galdikas 1981, 1985a; Schurmann 1982; Pusey 1990). Silverbacks show in-

terest in adolescent relatives other than daughters and in adolescent immigrants (Harcourt et al. 1981; Watts 1990a, 1991a). Adolescent female mountain gorillas copulate with adolescent and adult males in their natal groups; some subsequently mate there with relatives other than fathers (Stewart & Harcourt 1987; Watts 1990a, 1990b, 1991a).

Alloparenting

Data on alloparenting in apes are still insufficient to reveal clear age or sex differences. Adolescent and nulliparous adult female chimpanzees were more active alloparents than members of other age–sex classes at Mahale (Nishida 1983), but late adolescents seldom interacted with infants at Gombe (Pusey 1990). At both sites, siblings received most alloparenting and other friendly interactions. There was no overall sex difference in rates of interaction with infants at Gombe, but immature females interacted with infant siblings at higher rates and with other infants at lower rates than did males (Pusey 1990). The amount of alloparenting received from nulliparous females at Mahale was inversely correlated with infant age (Nishida 1983).

Immature female mountain gorillas seem particularly interested in infants. Two in Group 5 more often peered at and touched infants less than 3 months old while mothers were holding the infants than two young adolescent males did [$n = 43$ interactions; $X^2(1) = 8.31, p < 0.01$]. The females rested from 1.2 to 2.7 times as much within 5 m of five unrelated adult females when the adults had infants less than 8 months old than when they did not have infants, a significant increase in both cases (Wilcoxon matched pairs tests, $T+ = 15, p < 0.05$). The immature females made the majority of approaches in all dyads. The males each rested closer to only three of the same five females when they had infants, and the increases were less than those for immature females. Males sometimes carry unrelated infants, however.

Orphans

In keeping with their earlier weaning, mountain gorillas orphaned by maternal death or emigration between ages 3 and 5 may have better survival prospects than same-aged chimpanzee orphans. Chimpanzee orphans 3–4 years old rarely survive (Goodall 1971, 1986; Pusey 1983). Eight of nine Karisoke gorillas that lost mothers or both parents between the ages of 3 and 5 years (including two of three 3 year olds) have survived more than 5 years (one died of a poaching injury) (Fossey 1983). Orphaned chimpanzees typically have close adult associates, most often older siblings (Goodall 1971, 1986; Pusey 1983; Nishida et al. 1990). Juvenile gorillas that have lost mothers, but not fathers, associate closely with fathers, who share their night nests with the orphans and may groom them extensively (Fossey 1979). In 1984–1985, for example, a 3-year-old female increased time spent within 5 m of her father from 21.0 to 74.3% during feeding and 43.3 to 85.5% during resting, and a 4-year-old male from 21.6 to 36.3% during feeding and 38.9 to 70.1% during resting. Orphans with neither parent have other close associates. For example, two Group 4 juvenile males slept with an older paternal half-brother at night; he groomed both extensively, as did another, unrelated adolescent male (Watts, unpublished data). One orphan maintained a long-term bond with the unrelated male and stayed in his group as a silverback.

Orphans seem depressed and may temporarily stop playing (for up to 1 year in gorillas). Chimpanzee survivors may have long-term deficits in tool use and social behavior (Pusey 1983). Such deficits may be less likely in gorillas, given more constant availability of other social partners and paternal care.

Behavior in Intergroup Encounters

Chimpanzee boundary patrols are primarily a male activity. Participation increases with age to a peak among prime adults. Adolescents accompanying adults on silent patrols are touched or threatened if they vocalize and gradually learn to maintain silence and vigilance (Goodall et al. 1979; Goodall 1986).

When their groups encounter another or a lone silverback, juvenile and adolescent male gorillas often display at the opposing side. Silverbacks mostly ignore displays of juveniles and young adolescents and are more likely to ignore old adolescents than silverbacks. Old adolescents sometimes briefly follow opposing groups or stay near them as their own groups move off; this may be important practice

(Crockett & Pope, Chapter 8, this volume), but exposes them to some risk of attack. Female juveniles and adolescents display less than male peers, but usually express interest and often remain near displaying males.

Ontogeny of Foraging Behavior

Both gorillas and chimpanzees acquire the basic adult feeding repertoire by the end of infancy, and, at least for gorillas, the diversity of weanling diets is comparable to that of adult diets (Watts 1985b; Hiraiwa-Hasegawa 1990). The adult sex difference in time devoted to feeding on ants is evident in juvenile chimpanzees (Hiraiwa-Hasegawa 1989). Mountain gorillas do not show obvious sex differences in diet.

The proportion of time that mountain gorilla juveniles spend feeding is comparable to adult values (Watts 1988), but juveniles feed more slowly than adults (Watts, unpublished data). Because juveniles travel in cohesive groups, their habitat use patterns are determined by those of adults. Adolescent females that transfer enter unfamiliar areas, but do so with groups familiar with those areas. Once chimpanzees and orangutans start to move independently of mothers, they have no comparable, constant social companions that can serve as sources of ecological information. Whether it is more difficult for them than gorillas to acquire such knowledge is a crucial question with regard to hypotheses about developmental speed, but one that cannot be answered at present.

DISCUSSION

Developmental Schedules

Gorillas grow faster, reach sexual maturity faster relative to body size, have shorter interbirth intervals, and may have shorter life spans than chimpanzees and orangutans. They have higher infant survival than at least chimpanzees (Goodall 1983; Watts 1989; Nishida et al. 1990) and low juvenile mortality (unpublished Karisoke records), but comparisons of this measure or of adult mortality are not available. Several possible explanations for differences in the speed of life may apply to the great apes. First, gorillas have smaller brains than the others relative to body size; this is consistent with the general primate trend for folivores to have relatively smaller brains than frugivores (Clutton-Brock & Harvey 1980). Across primate subfamilies, age at maturity is highly correlated with adult brain size once the body size effect is removed; this suggests that postnatal brain development, learning skills, and social maturity are intimately linked with physical maturity (Harvey & Clutton-Brock 1985). Relatively large brains may be associated with relatively complex social life (Humphrey 1976), but the solitary nature of large-brained orangutans makes it hard to apply this argument to the great apes. Alternatively, orangutans and chimpanzees may need relatively larger brains than gorillas because their diets are more diverse and their food supplies less evenly distributed in space and time. Such conditions require the storage and processing of much ecological information (Clutton-Brock & Harvey 1980; Milton 1981b).

Second, Pagel and Harvey (Chapter 3, this volume) argue that the general relationship between large body size and slow development arises because large size reduces adult mortality rates, partly through decreasing predation risk, and that low adult mortality automatically leads to delayed maturation. They suggest that after removal of body size effects, remaining variation in developmental speed is due to variation in the time that juveniles need to acquire skills relevant to survival and reproduction. On these grounds, we would expect gorillas to have relatively high adult mortality, perhaps because terrestriality carries high predation risks, or juvenile gorillas to benefit less than juvenile chimpanzees and orangutans from time to learn and to develop skills. As for the brain size argument, social skills would probably benefit only chimpanzees, but they and orangutans may take longer than gorillas to acquire foraging skills. The limited data on the ontogeny of foraging are insufficient to resolve this argument.

Finally, Janson and van Schaik (Chapter 5, this volume) argue that juveniles of many primate species grow relatively slowly because individuals growing as fast as physiologically possible would face excessive mortality risk during times of low food abundance. Mortality would come either from starvation, induced by competition for food with larger and more capable adults, or from predation, more frequent because of the need to spend more time feeding and/or to reduce vigilance while feeding faster. For this hypothesis to explain variation in devel-

opment rates among the great apes, either gorillas would have to face such high predation rates that they could not afford to delay maturity, or chimpanzees and orangutans would have to be more susceptible to starvation than gorillas. Orangutans and chimpanzees face more temporal variation in food abundance than do gorillas, but the costs discussed by Janson and van Schaik may not apply to them with much force because they usually do not feed in groups and they may suffer less predation than smaller primates.

Current Survival versus Future Gains

Postweaning relationships with mothers are close and prolonged in all great apes. In the more solitary orangutans and chimpanzees, mothers are the individuals best able to provide protection and information about food location. In these species and gorillas, dispersal and adult association patterns mean that most immatures of both sexes eventually spend little time with their mothers. Affiliative behavior with mothers, such as grooming, may therefore reflect current needs, not a need to cultivate future support. It is thus interesting that grooming reciprocity with mothers is similar for female and male juvenile chimpanzees. This is unlike baboons, in whom reciprocity is higher with daughters, who are philopatric and rely on maternal support for acquiring adult dominance rank more than sons do (Pereira & Altmann 1985; Pereira 1988b). Immature gorillas (except adolescent males) may groom their mothers relatively less than chimpanzees do, and tend to groom silverbacks extensively, partly because they rely less on mothers and more on males for support. Whether individual females consistently groom more with daughters than sons, as in vervets (Fairbanks & McGuire 1985), is unknown; they may usually not do so because daughters usually do not become their mothers' adult allies.

In gorillas, attraction to silverbacks presumably lessens vulnerability, as demonstrated by silverback attention to orphans, while also placing juveniles in a position to observe adult males and to develop skills in interacting with them.

Besides allowing opportunities for play and sociosexual behavior, relatively frequent association among immature orangutans may increase safety and, sometimes, foraging efficiency. Smaller size means more vulnerability to predators (Galdikas 1988), and independent immatures, especially females, may face pressure to form groups whenever ecological conditions permit. Grouping is less ecologically costly for them than for same-sex adults because they are smaller, and it allows them to supplant adult females from food sources on occasion and may allow males to benefit from using adolescent females' familiarity with their foraging areas (Galdikas 1988).

Some patterns of association and behavior seem to have immediate costs, but presumably lead to increased proficiency in adult behavior. Play, with its energy costs and risks, is a good example (Fagen, Chapter 13, this volume). So, too, is the emergence of a sex difference in chimpanzee association patterns. Females forage with mothers until they start mating and are less gregarious than male peers; in this respect, they are following, rather than practicing for, an adult strategy that increases foraging efficiency by reducing competition (van Noordwijk et al., Chapter 6, this volume). By associating more with estrous females and adult males, pubertal males probably face increased feeding competition at a time when they would benefit from efficient foraging, but they gain opportunities to watch and interact with adults that they do not have while with their mothers.

The benefits of other behavior are more difficult to categorize. For example, young male chimpanzees and gorillas start to challenge females for dominance when smaller than the females (considerably so in chimpanzees). Males thereby appear to risk attack by a female or her supporters. Such challenges can immediately help rapidly growing males if they increase access to monopolizable food patches (as in baboons: Pereira 1988b, 1992). This may occur in chimpanzees, but most mountain gorilla food sources are not clumped and monopolizable and contest competition is infrequent (Watts 1985a). Early dominance over females may be important mostly because it increases future mating opportunities (Pusey 1990) and/or is relatively safe practice of behavior shown in future aggressive competition with other males (Pereira 1988b). In this regard, it is interesting that silverbacks as well as adolescents display particularly often at females that have recently joined their groups (Watts 1992, unpublished data). Displays may give the females, who are un-familiar with the males, information about male quality as mates and may be attempts to induce the females to mate with the males either immediately or in the future.

Sex and Species Differences in Adults and Immatures

Immature gorillas and chimpanzees show many adult-like sex differences. For example, whereas both sexes have strong bonds with their mothers and adult sisters as long as they reside together, males cultivate relationships with unrelated adult females more actively than do females, through association, grooming, aggression, and sexual behavior. The adults are potential mates for chimpanzees and, at least for a limited time, gorillas. Even gorilla males that subsequently emigrate can practice behavior involved in mate attraction and retention by interacting with all adult females in their natal groups. At least in chimpanzees, there is also a sex difference in display frequency that foreshadows the adult sex difference.

Social play is thought to enhance social and agonistic skills and is predicted and often found to occur at higher rates in males in species where fighting ability strongly influences male reproductive success (Owens 1975; Symons 1978; Fagen 1981; Pereira & Altmann 1985). Compared with female peers, immature male gorillas and orangutans seem to play more, with a wider range of partners, and proportionately more with available same-size or larger partners. Similar sex differences occur in many other primates (e.g., yellow baboons: Pereira & Altmann 1985), but are less evident in chimpanzees. This may be because chimpanzees are less sexually dimorphic and male reproductive success depends less on male intrinsic power (Fagen, Chapter 13, this volume). Alternatively, it could be an artifact of small sample size and/or due to demographic constraints on partner availability. We should use care in invoking such constraints, though: if play is more important to males, this should be evident from comparison of males and females that have the same partners available.

The relationships of immatures with adult males show particularly strong adult-like species differences. Only in chimpanzees do adult males have strong cooperative social bonds marked by extensive grooming and frequent al-

liance formation. Conversely, adult female gorillas generally have closer affiliative relationships with adult males than do adult female
chimpanzees. As they mature, immature male
chimpanzees associate and interact increasingly
often with adult males, both affiliatively and agonistically, and they gradually join the male-
bonded social networks of their natal communities. Immature females show increases only
when in estrous or, temporarily, after transfer to
a new community. In contrast, juvenile male
gorillas initially maintain close spatial association with silverbacks and often interact affiliatively with them, but as males grow older, proximity and affiliation decline and the frequency
and intensity of aggression from silverbacks
increases. This pattern commonly precedes adolescent male emigration in male-dispersal species (e.g., redtail monkeys: Struhsaker & Leland 1979; yellow baboons: Pereira 1988a).
Spatial association also declines between immature female gorillas and silverbacks as the
females grow older, but this stops at adolescence and females maintain high rates of affiliative interaction with silverbacks.

Directions for Future Research

Clearly we need more data on immature pygmy
chimpanzees and orangutans, and sample sizes
on chimpanzees and gorillas are much smaller
than those on species like baboons. Apparent
differences in adult associations and behavior
between the two chimpanzee species raise important questions about the extent of affiliative
behavior between adults and same-sex immatures and possible support of sons by mothers in
pygmy chimpanzees. If affiliation and cooperation among adult males is less developed in pygmy chimpanzees, for example, there should be
relatively less sex difference in association and
affiliation among immatures in this species. Do
immature male orangutans make more effort
than females to watch adult males, as in chimpanzees? Immature males sometimes shadow
consorting adults (Galdikas 1985a), although
they may in fact be attracted to the females. For
that matter, do immature male gorillas selectively monitor adult males in their groups, as
appears to happen in baboons (Pereira 1988a)?

An intriguing and particularly important
question to investigate in the great apes is the
extent to which individual differences in the experience of immatures lead to individual dif-

ferences in adults. Chimpanzees show great individual differences in behavior and personality,
reflected, for example, in variation in male persistence in striving for dominance and success in
forming alliances and in the adequacy of maternal behavior (Goodall 1986, 1989a). Social experience of immatures must be more influenced
by mothers in chimpanzees than in group-living
species in which there are always other social
companions available. Individual differences in
maternal sociability and the presence or absence
of mothers or infant and adult siblings probably
contribute to differences in adult chimpanzee
behavior.

Dispersal patterns in gorillas and chimpanzees are less regular than those in many other
primates. The variation may result from variation in the presence or characteristics of kin, in
the affiliative relationships of immature individuals with relatives, and in the amount of intrasexual competition (Harcourt & Stewart
1981; Goodall 1983; Pusey 1983, 1990; Stewart
& Harcourt 1987; Watts 1990a, 1990b), but
we cannot yet test proposed explanations adequately. Competition among immature peers,
possibly for opportunities to remain in natal
groups or communities, deserves more study.
Variation in competition for male tolerance, in
availability of potential mates, and in relatedness between males could help explain the large
individual differences in the behavior of immature gorillas. Silverbacks tolerate sons and full
brothers more than paternal half-brothers. Juvenile males that have no male peers, but have
natal females available as potential mates (see
Watts 1990a, 1990b), may not need to groom
silverbacks, whereas those with relatively many
peers (e.g., juvenile males in Group Nk) may be
highly likely to do so.

All these questions require continued long-
term studies, but all study of great apes in the
wild has great urgency because of the animals'
endangered status. Conservation issues are beyond the scope of this chapter, but we can have
no illusions about meeting research goals without solutions to the enormous socioeconomic
problems underlying the destruction of great ape
habitats and populations.

ACKNOWLEDGMENTS

D.P.W. thanks L'Office Rwandaise du Tourisme et
des Parcs Nationaux for permission to work in
Rwanda, the Zaire Office of Parks for allowing

Karisoke personnel to work in Le Parc National des Virungas, and the late D. Fossey for permission to work at Karisoke in 1978–1979 and 1984–1985. A. E. P. thanks the Tanzanian National Parks for permission to study at Gombe and is grateful to J. Goodall for help and encouragement at all stages of the study. A. E. P. was supported by grants from The W.T. Grant Foundation, the Harry Frank Guggenheim Foundation, The Eppley Foundation, The L. S. B. Leakey Foundation, and the American Philosophical Society. D. P. W. was supported by NIMH Grant 5T32 MH 15181-03; by grants from The L. S. B. Leakey Foundation, The Chicago Zoological Society, The World Wildlife Fund U.S., The Wildlife Preservation Trust International, and The Eppley Foundation; and by The Digit Fund. A. H. Harcourt, M. E. Pereira, and B. Smuts provided constructive criticism of the manuscript, and M. E. Pereira provided encouragement at all stages of writing. Finally, we thank all who have contributed to the long-term records at Gombe and Karisoke and particularly acknowledge the invaluable contribution and friendship of Tanzanian and Rwandan research assistants.

Part III

DEVELOPING SKILLS AND RELATIONSHIPS FOR LATER USE

Whether the selected aspect of protracted development in primates is expanded reproductive careers, reduced juvenile mortality, or both, individuals should benefit by using their juvenile time "wisely." The juvenile years provide opportunities to observe and to practice behavior and to develop long-term social alliances, all of which could potentially enhance individuals' competitive capacities in relation to their peers. The chapters of this part focus more specifically on the question of how juvenile behavior and social tactics may promote adult fitness.

In research on these group-living species showing delayed maturity, large brains, and long lives, there has been, perhaps, an overemphasis on the importance of adult role models; too much of behavioral development is attributed to social learning. In the opening chapter here, Milton exploits a quasi-natural experiment to show that adult role models are not always necessary for normal development. When spider monkeys were reintroduced to Barro Colorado Island, the sole survivors were immatures. These individuals formed a group, located appropriate food sources, and developed species-typical behavior in the absence of

adults. Through exploration and trial-and-error learning, presumably in conjunction with evolved perceptual and dietary predispositions, these monkeys selected a diet of ripe fruit that differed from those of other monkey species on the island but resembled those of spider monkeys at other locations. In addition, they developed the fission–fusion pattern of association and sex differences in day ranging that are characteristic of their species.

To say that juveniles are not required to observe adults to develop species-typical behavior, however, is not to say that experience does not influence development. Just as the quality of the juvenile diet will affect the rate of growth, adult stature, and ultimate success of a primate (e.g., S. Altmann 1991), so too should early opportunities to increase knowledge and to practice behavior affect adult behavioral flexibility and competence. Juveniles spend much time and considerable energy on activities that would appear to have relevance for later life.

Play is one of the striking ways in which juvenile primates spend their time and energy. In Chapter 13, Fagen reviews the evidence on the form and function of primate play and argues that in addition to the likely functions of physical training and motor skill development, play provides the young primate opportunities to develop the cognitive skills involved in competition, cooperation, interpretation of behavior, and individual creativity. He raises the issue of trade-offs in development that involve behavior by considering sex differences in the decline of play across the juvenile years. The juvenile must balance the changing risks and benefits of play with allocation of energy toward growth and maturation. Behavioral expression at any age depends not only on the costs and benefits of a given behavior, but also on individuals' conflicting immediate demands and future constraints and opportunities.

The three remaining chapters of this part explore the tactics that juvenile monkeys use in forming social relationships with other group members and examine the long-term consequences of these relations for the juveniles and the adults. Juveniles appear to recognize and respond to the attributes of others, and they associate preferentially with those that are likely to share beneficial relationships with them. O'Brien and Robinson (Chapter 14) investigate social relations among female peers, Fairbanks (Chapter 15) examines aspects of relations with peers and with adults, and Horrocks and Hunte (Chapter 16) focus on juvenile relations with adult males. These works all find affiliative associations to be nonrandom—strongly influenced by age, kinship, sex, and dominance rank—and they all find juveniles playing active roles in initiating and maintaining preferred social relations that are likely to influence their long-term fitness.

In studying a free-ranging one-male group of vervet monkeys, Horrocks and Hunte (Chapter 16) found that juveniles interact with adult males in ways that might influence the outcome of male–male competition and thereby help determine which male would be available in the future. Juveniles were supportive of paternal males and aggressive to challenging males when residents were challenged by potential immigrants. But also, the quality of a relationship was influenced by the age, sex, and dominance rank of the juvenile. Paternal males and young juveniles of either sex were mutually supportive and affiliative, whereas old juvenile females were more likely to show interest in challenging males. In addition, behavioral asymmetries between juveniles created by matrilineal rank were offset to some degree by disproportionate attention from paternal males to the offspring of low-ranking mothers.

Chapters 14 and 15 contribute to the small but growing database on longitudinal social relations in primates. In O'Brien and Robinson's study of free-ranging capuchins and in Fairbanks's work with captive vervet monkeys, social relations established by juveniles are shown to persist into adulthood. O'Brien and Robinson reveal remarkable consistency in patterns of grooming, support, and aggression by female wedge-capped capuchins studied as juveniles and again 10 years later as the adult members of their group. Advantages established during the juvenile years appear to have been maintained throughout the 10-year interval.

Fairbanks's research on partner preferences emphasizes that not all juvenile social relationships are designed for longevity. While the proximity relations that juvenile vervets established with adults persisted over time, relationships with play partners and with infants did not. The results are consistent with the hypothesis that juvenile choices of partners for play and allomaternal care function to develop skills to be applied during adulthood but not to establish long-term social bonds. Whether the tactic of a particular activity is to estab- lish or reinforce social relations or to prac- tice behavior, however, the juvenile years clearly are important in determining indi- vidual differences likely to influence adult fitness.

12

Diet and Social Organization of a Free-Ranging Spider Monkey Population: The Development of Species-Typical Behavior in the Absence of Adults

KATHARINE MILTON

Consistencies in the behavior of particular organisms from generation to generation are generally described as species-typical. In higher organisms such as primates, many such behavioral consistencies are regarded as acquired through the result of social learning, often through exposure to the behavior of older and/or more experienced conspecifics. Although the capacity for learning is based on genetically acquired learning mechanisms, learned behavior, as such, cannot be inherited but must be learned anew by members of successive generations. There may, however, be innate attributes that predispose the members of particular species to learn certain behaviors far more readily than others (Galef 1976; Cambefort 1981; Pereira 1988a; Guilford 1990). Further, responses that appear innate or unlearned are not necessarily immune to the effects of experience (Guilford 1990). Each individual also has its own unique genetic inheritance that establishes the potential range of behavior that may develop (e.g., Baldwin & Baldwin 1979). Thus in speaking of species-typical behavior, particularly with respect to primates where social learning appears to play a major role, it is often difficult to determine whether an animal is behaving in a certain way because observation of or interaction with others has influenced its behavior or whether

such behavior results largely from some innate predisposition to respond in a particular way (Bateson 1976; Galef 1988; Pereira 1988a).

Primates clearly show behavioral flexibility; indeed, some regard primates as having the highest level of behavioral plasticity (Bernstein & Smith 1979) and/or the greatest learning capacity of all mammals (Cambefort 1981). There are abundant accounts in the literature attesting to the capacity for intra- and interspecific behavioral plasticity in a wide array of primate species (Rowell 1967; Fairbanks 1975; Baldwin & Baldwin 1979; Cambefort 1981). Yet when we look at the behavior of primates in the natural environment, behavioral plasticity—at least in terms of most daily activities—is not something that strikes the observer. Rather, species-typical behaviors are manifested with monotonous regularity and from study site to study site, with minor modifications often attributed to features of particular environments, we find that baboons act like baboons, howler monkeys like howler monkeys, chimpanzees like chimpanzees, and so on. This statement is not meant to ignore the fact that behavioral differences exist between individuals in the same social unit and between primate groups of the same species or to question the obvious evolutionary importance of behavioral novelty (e.g., Mayr

1963; Bateson 1976, 1988; Wcislo 1989; West-Eberhard 1989). Rather, it is an attempt to focus attention for a change not on this "potential for innovation"—which certainly must be a universal trait of all living organisms and not just primates—but on the monotony of much primate behavior and on factors that may be involved in its ontogeny. Because of the overwhelming importance of social learning to humans, we readily view our closest living relatives as similarly dependent on learned behavior and attribute much of this learning to direct observation and imitation of the behavior of other, older group mates (e.g., Cambefort 1981, p. 244). It is my contention, in fact, that the acquisition of many species-typical behaviors in nonhuman primates does not requires social learning through observation of more experienced individuals. I base this contention on observations of the behavior of a free-ranging spider monkey (*Ateles geoffroyi*) population in Panama that developed initially from a small group of immatures, raised in the absence of conspecific adults. These young monkeys nonetheless managed to become completely self-sufficient nutritionally and, as adults, displayed a diet, a pattern of social organization, and many sex-typical behaviors indistinguishable from those of wild spider monkey populations elsewhere. Without exception, all field studies of wild spider monkeys have shown that the annual diet consists largely of ripe fruits (Klein & Klein 1977; van Roosmalen 1985; Chapman 1987, 1988a, 1988b; Symington 1988). Without exception, spider monkeys are reported to show a fission–fusion pattern of social organization with strong male bonding and separate foraging by all-male groups (Eisenberg 1976; van Roosmalen 1985). Males are reported to travel farther during the day than females, and immatures and females are reported to have apparent core areas (van Roosmalen 1985; Symington 1988; Chapman 1988a). Individual females and their dependent offspring are particularly likely to be found in these core areas during times of year when ripe fruit is scarce. All these behavioral patterns, as discussed below, are also characteristic of the Panamanian spider monkey population. This example suggests that for some primate species, social learning resulting from the observation and/or imitation of the behavior of older, more experienced individuals may not be required for the development of many species-typical social and foraging behav-iors (for similar views, see, for example, Fairbanks 1975; Galef 1976, 1988; Cheney & Seyfarth 1990; Visalberghi & Fragaszy 1990a).

Galef (1988) provides an excellent review of the evidence for imitative learning in animals and differentiates among the many terms used to describe learning in a social context. In this chapter, I am not concerned with the occurrence or nonoccurrence of true imitative behavior in primates (i.e., "purposeful, goal-directed copying of the behavior of one animal by another," Galef 1988), nor can I shed much light on the degree of importance of social enhancement or eventual social transmission (sensu Galef 1988) in the development of the species-typical behavior that took place among these young spider monkeys. What my work does address is the question of whether these young spider monkeys required exposure to the behavior of older, more experienced conspecifics to develop normal, species-typical and sex-typical behaviors. The answer suggested by the data is that they did not.

HISTORY OF SPIDER MONKEYS ON BARRO COLORADO ISLAND

Study Site

Barro Colorado Island (BCI) is a 1600-km^2 nature preserve, located in Lake Gatun, the principal water supply for the Panama Canal in the Republic of Panama. A complete description of the history, physical attributes, climate, flora, and fauna of the island can be found in the literature (Milton 1980; Leigh et al. 1982).

Introduction of the Founder Population

Spider monkeys once occurred naturally in the vicinity of BCI. However, prior to 1923 when the island was declared a nature reserve, spider monkeys had been exterminated by hunters. Between 1959 and 1961, various spider monkeys were purchased in the central market in Panama City by the Smithsonian Tropical Research Institute and released onto BCI in hopes they would survive and repopulate the forest (Eisenberg & Kuehn 1966; Dare 1975; Eisenberg 1976). Capture sites of these monkeys are unknown, but presumably the monkeys were captured in forests southeast of Panama City, many dozens, if not hundreds of kilometers from BCI.

Since infant spider monkeys typically can be captured only if the mother is shot and falls from the tree with the infant, it is likely that all spider monkeys released onto BCI were small infants (less than 2 years old) at the time of capture. Spider monkeys infants have an unusually protracted period of maternal dependence (Eisenberg 1976; Milton 1981a; van Roosmalen 1985). The first year is spent largely on the mother, who continues to carry the infant extensively during its second year as well; spider monkey infants nurse until more than 2 years of age (Milton 1981a; Symington 1988). A juvenile spider monkey continues to associate and sleep with the mother for some months after a new sibling is born and does not begin to move about independently until around 3 years of age (K. Milton, personal observation; see also Ahumada 1989). Sexual maturity for captive male and female spider monkeys is placed at 4.75 and 5 years, respectively (Eisenberg 1976); wild males show some correlates of sexual maturity between the fourth and fifth year, whereas sexual maturity for wild females is estimated to occur between 6.5 and 7.6 years of age (Milton 1981a).

Records apparently were not kept on details of the reintroduction process, and no exact figure can be provided on either the number of spider monkeys released or their respective ages. Estimates of ages of the released monkeys is further complicated by the fact that in the early 1960s, the unusually slow maturation rate of spider monkeys was not fully appreciated, leading to the possibility for considerable error in age estimates. Dare (1975) estimates that as many as 20 spider monkeys may ultimately have been released on BCI, ranging in age from "very young" (presumably indicating an animal around 12 to 18 months of age) to "adult," which could mean anything. Some monkeys were released near the laboratory buildings, and others were released some distance away on the north side of the island. What is certain is that most of these individuals did not survive the introduction and that no individual described as "adult" survived it (Dare 1975). In 1964, Eisenberg (1976) noted that a total of five spider monkeys, four females and one male, composed the BCI spider monkey population. At introduction, the four females were estimated to range in age from perhaps 1.5 to 3 years and the male was estimated to be perhaps 4 to 5 years old

(Dare 1975). On release, the five young monkeys, although free to range in the forest, were provisioned near the laboratory clearing until they appeared old enough to forage independently. After provisioning was ended, the monkeys continued to solicit food whenever they were near the laboratory clearing. At that time (mid-1960s), there were few visitors to BCI and it is likely that supplementary foods were sporadic and unreliable (see Fairbanks 1975). As the objective was to produce a totally self-sufficient, free-ranging spider monkey population, feeding spider monkeys was not encouraged once provisioning was ended.

In March 1974, when I began work on BCI, the population of spider monkeys had grown to 12 individuals. This number included the five founders and seven F_1 offspring. The first infants apparently were born to the founder population in 1966 (Eisenberg & Kuehn 1966; Dare 1975). Estimates and records of births between 1966–1974 and 1974–1980 can be found in Eisenberg (1976) and Milton (1981a), respectively. Between 1974 and 1977, I concentrated on the dietary ecology of howler monkeys on BCI, but in 1978, to complement the data on the howler diet, I began similar field observations on the dietary ecology and social organization of the BCI *Ateles* population. Information compiled as a result of this work is presented below.

Collection of Feeding and Ranging Data

To examine the food choices and ranging behavior of the BCI spider monkey population, data were collected between October 1978 and June 1979, with several additional days of data compiled in 1980–1981 (Table 12.1). During most of these observations, the population was composed of three of the original founding females and nine of their offspring whose ages ranged from 13 years to newborn. Prior to these systematic observations, however, I already had made 4 years of intermittent observations on this population when all five founders were still present.

Two techniques of data collection were employed: instantaneous scan samples and focal animal samples (Altmann 1974). During instantaneous scan samples, the activity of each spider monkey visible to the observer was recorded at 5-minute intervals from 0600 to 1900 hours for 2 or more consecutive or closely spaced days per month. In focal animal sam-

Table 12.1. Sample Data on BCI Spider Monkeys

Date	Principal activities (%)			Travel (m)	Number/type of food species[a]		Type of sample	Hours of sample
	Rest	Travel	Feed					
1978								
Oct. 13	60.2	11.2	27.0	900	5	1F,4L	Scan	10.5
Oct. 15	69.5	8.3	15.8	900	4	3F,2L	Scan	9.0
Oct. 30	63.6	16.7	17.4	550	7	3F,4L	Scan	12.0
Oct. 31	62.5	12.5	22.3	460	8	5F,3L	Scan	12.0
Nov. 01	64.2	9.2	24.3	650	8	4F,4L	Scan	11.0
Nov. 02	56.9	14.6	22.2	600	10	4F,6L	Scan	12.0
Nov. 03	60.0	11.9	23.6	520	9	6F,3L	Focal	12.0
Nov. 03	66.0	11.1	17.3	700	11	5F,5L,3Fl	Focal	12.0
Dec. 09	50.9	12.9	32.4	745	6	2F,4L	Scan	9.0
Dec. 20	43.7	13.0	26.8	940	6	2F,4L	Scan	10.5
Dec. 21	42.9	22.2	24.3	1300	6	2F,4L	Scan	11.0
1979								
Jan. 21	46.1	21.8	31.5	1600	6	4F,1L,1Fl	Scan	13.0
Jan. 25	47.6	17.6	33.3	1220	10	5F,3L,2Fl	Focal	10.0
Jan. 27	48.8	17.3	31.8	1300	9	5F,4L	Focal	13.0
Feb. 25	39.5	14.5	39.9	1000	11	7F,5L	Focal	9.5
Feb. 27	57.6	9.9	25.8	700	5	2F,3L	Focal	11.0
May 02	38.2	26.0	29.1	2400	16	14F,2L,1A	Scan	12.0
May 03	44.6	28.7	21.0	1900	9	6F,3L	Scan	12.0
May 04	23.1	33.7	33.0	2100	20	13F,6L,1Fl,2A	Scan	12.0
May 05	41.6	28.0	22.3	1300	11	9F,3L	Scan	12.0
Jun. 02	41.0	22.8	28.9	2400	8	5F,3L	Focal	12.0
Jun. 03	44.1	33.3	15.7	1600	13	9F,4L	Focal	12.0
Jun. 04	52.7	26.2	13.2	1700	8	6F,2L	Focal	12.0
Jun. 05	42.5	25.5	23.3	1900	11	7F,5L	Focal	10.0
1980								
Jan. 17	49.5	21.4	25.0	1600	10	5F,4L,1Fl	Scan	11.5
1981								
Nov. 12	50.0	16.7	29.9	800	9	3F,6L	Scan	12.0
Nov. 13	45.2	19.1	34.4	900	5	2F,3L	Scan	12.0

[a]F, fruits; L, leaves; Fl, flowers; A, animal matter.

ples, an adult female was selected at 0600 hours and followed for the entire day; every 5 minutes, the activity being performed by the focal individual was recorded. Dates of sample days and the number of hours of observation per sample are listed in Table 12.1. Only days in which 9 or more hours of observation were compiled were used in calculations. Activities were divided into five types: *resting, traveling, feeding, moving,* and *other,* as defined in Milton (1980). Most food species were already known; samples of unknown foods were collected whenever possible and identified in the BCI herbarium.

To compile ranging data, on each sample day a record was kept of the travel movements of the monkey(s). As spider monkeys tended to travel over sections of the island that I knew, approximate locations could be estimated at any time. In addition, spider monkeys occasionally passed over trails marked at 100-m intervals; by noting

these trail markers, precise locations for various points on the travel route at specific times of day could also be recorded. At the end of each sample day, I estimated the number of meters traveled, and the identity, order, and approximate or exact locations of all food trees visited.

To examine the foraging behavior of spider monkeys in relation to seasonal changes in rainfall and fruit abundance, the annual cycle was divided into three 4-month seasons: a transition-dry season, December–March, an early rainy season, May–July; and a late rainy season, August–November. Feeding and ranging data for each sample day were placed in one of these three categories and tested using ANOVA.

Social Behavior

During sample days, social behaviors were noted ad lib. Data were kept on group composition for 18 of the 27 sample days. Although in this study social behavior was observed qualitatively, more recently (1989–1990), quantitative data on some social parameters of the BCI spider monkey population have been compiled by J. Ahumada, many of which lend support to descriptions in this account.

BEHAVIOR, SOCIAL ORGANIZATION, AND DEMOGRAPHICS

Activity Budgets

A total of 27 complete days ($\geq$ 9 hours per sample day; total = 307 hours) of observation, covering portions of 7 months and all three seasons, were ultimately compiled on general activities (Table 12.2). Consolidation of these data permitted calculation of an average annual activity budget (presented as $\bar{X} \pm 1$ SD here and elsewhere in text) as follows: resting = 50.1 $\pm$ 10.5%, traveling = 18.7 $\pm$ 7.4%, feeding = 25.6 $\pm$ 6.5%, and moving = 3.3 $\pm$ 1.9%. A negligible amount of time ($\sim$ 2%) was devoted to "other" activities.

Feeding Behavior

Over the total sample, spider monkeys were observed eating foods from a minimum of 100 plant species (85 identified as to species and $\geq$ 15 more unidentified). An average of 8.9 $\pm$ 3.5

Table 12.2. Comparative Data on *Ateles* spp.

	Activities (%)			Annual feeding time devoted to fruit (%)
	Rest	Travel	Feed	
Klein and Klein (1977)	63	15 (moving)	22	83[a]
van Roosmalen (1985)				83 (annual fruit intake)
Symington (1988)	45	26	29	75
Chapman (1987, 1988a)	<49	23	28	78
Eisenberg and Kuehn (1966)[b]	40	25 (moving)	26	
Milton (this study)	50	19	26	70

[a]Carpenter gives 90% fruit/nuts, food items consumed in 48-day sample; Hladik gives 83% fruit by weight. Both estimates cited in Klein and Klein (1977).
[b]Eisenberg and Kuehn, several weeks of field observations in 1964.

food species were eaten per day; 5.2 $\pm$ 3.2 fruit species and 3.7 $\pm$ 1.3 leaf species. On average, 70 $\pm$ 20.6% of daily feeding time was devoted to fruit eating versus 26.4 $\pm$ 16.7% devoted to leaf eating. Flowers and flower buds were occasionally eaten, as were, on two occasions, insects or insect exudate and, on one occasion, what appeared to be frog eggs. All three feeding observations involving animal matter occurred in May 1979.

Fruit Eating. Daily feeding tended to be heavily concentrated on fruits (Table 12.2). With one exception, all fruits eaten appeared ripe. The only unripe fruits eaten were those of an understory palm. These fruits were still soft, and monkeys were able to crush them, apparently to obtain liquid and soft matter inside the immature husk.

There were clear differences between months and between seasons in terms of the number of fruit species eaten. During the earlier portion of the rainy season on some sample days, spider monkeys took food from 13 to 20 fruit species (Table 12.2). Fewer fruit species were eaten in other months, but, on occasion, the spider monkeys visited four, eight, or more individuals of a single fruiting species in one day, taking some ripe fruits from each tree visited (see also Chapman 1988b). This was observed, for example, in the case of *Dipteryx panamensis, Hieronyma laxiflora, Spondias mombin, Virola surinamensis,* and *Quararibea asterolepis.*

ANOVA results showed that significantly more fruit species were eaten during the early rainy season than in either the dry or the late rainy season ($p = 0.006$, 2 df); the number of fruit species eaten per day in the dry and late rainy seasons did not differ significantly. ANOVA results showed no difference between the dry and early rainy seasons regarding the percentage of daily feeding time devoted to fruit, although significantly more time was devoted to fruit in both these seasons than in the late rainy season ($p = 0.003$, 2 df).

Leaf Eating. Spider monkeys were more restricted in their choice of leaf species, feeding on no more than six leaf species in any single day. With one exception, *Ficus yoponensis,* only young leaves were eaten; mature leaves of *F. yoponensis* have a chemical composition similar to that of young leaves (Milton 1979).

By far the single most important leaf species for spider monkeys was *Poulsenia armata* (Moraceae), which was eaten in five of the seven sample months. Larger individuals of *P. armata* produce some new leaves more or less continuously and flush large new leaf crops two or more times per year (Milton 1991). Other species with young leaves used as food included *Brosimum alicastrum, Platypodium elegans, Terminalia* sp., *Eugenia* sp., and *Inga goldmanii*, as well as a number of vine species, including various members of the Bignoniaceae. Petioles and young leaf blades of *Philodendron* sp. and *Monstera* sp. were also eaten (see also van Roosmalen 1985).

Daily Path Length

Mean daily path length averaged 1203 ± 543 m with a range of 460 to 2400 m ($n = 32$ day ranges). ANOVA results indicate that daily path length was significantly longer in the early rainy season than either the dry or late rainy season; daily path length was significantly longer in the dry season than the late rainy season ($p = 0.0001$, 2 df). Thus monkeys traveled longer distances during the season when the greatest number of fruit species were eaten (Table 12.2).

Male spider monkeys tended to have longer day ranges than females (see also van Roosmalen 1985, pp. 184–185). The 12 individuals might start the day as a fairly cohesive unit but within a few minutes to an hour, the four adult males in the troop often moved away from the females and immatures and were not seen again until late afternoon or the next morning. At times, it was clear from the direction of their travel that they were returning from a considerable distance relative to the area occupied for the day by the adult females and younger animals (see also Symington 1988; Chapman 1988a). Other scientists working on BCI during this study reported their daily sightings of spider monkeys to me and it was not unusual for someone to see the adult males 800 m or more from the females I had been simultaneously accompanying. As there were only 12 spider monkeys on the island during most of this study, it was obvious when males had departed from the area I was in with females and immatures. As a general rule, females and younger animals tended to forage in areas closer to the laboratory clearing than males. On some occasions, all 12 animals foraged for long distances as a fairly cohesive unit. Females and younger animals, however, were never observed to forage further from the laboratory clearing area in a given day than males (see also Symington 1988).

Social Organization

Spider monkeys on BCI showed a fission–fusion pattern of social organization and parties of many sizes and many permutations were observed. The 12 spider monkeys generally slept within a few to a few hundred meters of one another. Once animals woke, they began to call and travel; the entire group then often came together and traveled as a unit, occasionally remaining in loose association for the full day. More typically, however, the group would fission within the hour into smaller units. During the mid-to-late rainy season, when ripe fruit is extremely scarce in the BCI forest, adult females foraged in what appeared to be individual core areas, accompanied by their immature offspring (see also van Roosmalen 1985 and Symington 1988 on *Ateles paniscus*). During this same time of year, however, on occasion, the 12 spider monkeys also came together and foraged and/or rested as a unit.

Correlations between patch size and party size (e.g., Symington 1988) could not be examined for my study, as consistent data were not compiled. My 18 days of association records suggested that females showed a tendency to forage individually, accompanied only by their immature offspring, during the transition between the wet and dry seasons (when fruit is in shortest supply), that various females and immatures showed a tendency to associate during the dry season, and that all members of the community tended to associate somewhat more during the early portion of the rainy season (when fruit production is high in the BCI forest). More data are required, however, to confirm these apparent trends. Overall, the 18 days of association records show that adult males foraged in association with females and immatures approximately one-third (33%, 6 of 18 days) of the time. This percentage is similar to that obtained in 1964 for the pattern of association of the single male founder with the four founding females (Eisenberg & Kuehn 1966, p. 54).

Male Bonding

During the study, the 3.5-year-old son of one adult female left his mother and younger sibling and began to associate, travel, and forage with the four adult males. Similar behavior was observed in 1980–81 and in 1989 with three other young males; each was estimated to be around 3–3.5 years of age. None of the three different mothers (two founders and one F_1 daughter) was observed to be aggressive toward her son; rather, sons appeared to leave their mothers voluntarily to associate with older males.

Population Changes

In the approximately 30 years since spider monkeys were introduced onto BCI, although the population has moved into its third generation, its size has increased only slowly. Data from the BCI record book and Dare's (1975) reconstruction of the population's probable geneology indicate that the first eight surviving infants were male, generating a sex ratio that minimized the population's potential for growth. Further, one of the founding females produced no offspring between 1974 and her death in 1989 and may always have been infertile. In 1977, one founding female and her female infant were removed from the island because the mother had become aggressive toward humans.

At a minimum estimate, *excluding* the five founders and one female infant removed from the island, at least eight male spider monkeys and possibly two females have died. Most male spider monkeys on BCI disappear during subadult and/or early to middle adulthood. In contrast, there is no record of any adult female (adult = more than 5 years) dying of anything other than apparent old age. The three founding females that died on the island were estimated to be between 27 and 31 years old at the time of death, and the one founder male died at an estimated age of 23. Two founder females continued to produce offspring until the approximate time of their respective deaths (as noted, the other appeared to be infertile). As of 1990, the spider monkey population on BCI was composed of 16 individuals that formed a single social unit. All five of the original founders were dead, and the population had moved into the F_2 generation.

DISCUSSION

Results show that the spider monkey population on Barro Colorado Island has a diet and pattern of social organization essentially similar to those of spider monkeys elsewhere. Yet this population was founded from a group of young monkeys that were not exposed during most of their development to the behavior of adult conspecifics. What factors can be suggested to account for acquisition of the behaviors observed?

Development of the Species-Typical Dietary Pattern

A diet composed almost exclusively of ripe fruits is highly specialized relative to the diets of most other anthropoids. The strong focus of these monkeys on ripe fruits raises the question of how this dietary pattern developed. The literature suggests that young primates acquire food habits initially from association with their mothers and later from other members of their social unit (Rhine & Westlund 1978; Cambefort 1981; Watts 1985a; Whitehead 1986). Watts (1985a) stated that observational learning is largely responsible for the transmission of food preferences in mountain gorillas. Similarly, Whitehead (1986) reported that infant howler monkeys must see the mother or another individual eat a leaf before they will sample it. Cambefort (1981) suggested that observational learning in a social setting is of general importance in the acquisition of new feeding habits by primates but pointed out possible species differences in the acquisition of new food habits. Naive individuals of some species (e.g., mandrills) apparently can learn about a food through observation of the feeding of others, whereas in other species (e.g., savanna baboons, vervets) naive individuals must try the new food themselves (Cambefort 1981). Fairbanks (1975) reached similar conclusions in her comparison of pigtail macaques and spider monkeys (see also Zahorik & Houpt 1981; Box 1984). Most recently, Pereira (1988a) noted that social learning is implicated in the acquisition of food preferences by young primates but cautioned that unequivocal demonstration of its fundamental importance was not yet accomplished. Work on the acquisition of foraging skills in capuchin monkeys and chimpanzees shows that when placed in appropriate environments (e.g., those containing potential foods and potential tools for obtaining them), some individuals of both species acquire the ability to use tools, whereas others never do in spite of repeated observations of tool-using behavior by conspecifics over periods of months or years (Boesch & Boesch 1990; Visalberghi 1990; Visalberghi & Fragaszy 1990a). The acquisition of such extractive skills in capuchin monkeys is attributed to a tendency in this genus for monkeys to persistently manipulate and/or strike objects together in combination with trial-and-error learning (Visalberghi 1990; Visalberghi & Fragaszy 1990a). Anderson (1990, p. 144) suggests that some degree of social facilitation may occasionally also play a role here.

In the case of the BCI spider monkeys, there were no older spider monkeys on the island for the five founders to observe while growing up. The most abundant primate species on BCI is the mantled howler monkey (*Alouatta palliata*) (Milton 1980). Unlike spider monkeys, howler monkeys eat considerable foliage at all times of year. Clearly, the young spider monkeys did not adapt the diet of howlers either as a result of direct observation or simply from taking advantage of an available food source. Nor does the dietary pattern of these spider monkeys resemble those of any other primates on BCI (i.e., tamarins, capuchin monkeys, and night monkeys), each of which feeds primarily on considerable animal matter, primarily insects, as well as ripe fruits.

As infants, prior to capture, the five founders doubtless were exposed to foods their mothers were eating, and these foods were almost certainly ripe fruits. Thus these young monkeys may have been conditioned by early experience to associate ripe fruits with feeding, although they were not old enough themselves to eat fruit in appreciable amounts. On BCI, the provisioned diet of the five young spider monkeys included various domesticated fruits (F. Bocanegra, personal communication). Thus in exploring the forest, even if food habits were totally acquired through independent sampling and trial-and-error learning, it would not be surprising for the monkeys to focus on fruit. However, the spider monkeys not only selected ripe fruit as their principal dietary component, but also evinced the ability to discriminate between ripe-

fruit species in terms of quality. On occasion, for example, spider monkeys on BCI were seen traveling through fruiting fig (*Ficus* spp.) trees loaded with pungent ripe fruit without stopping to feed on a single fruit, fruit avidly fed on in the same day by howler monkeys. Rather, the spider monkeys traveled on to more distant ripe fruit of other species that were higher in soluble carbohydrates and/or crude fat (K. Milton, unpublished data). When these higher quality ripe fruits were seasonally unavailable, spider monkeys ate fig fruits. In such cases, however, the monkeys often did not swallow the fruit pulp but dropped much of it to the ground after chewing, sparing the digestive tract from processing this fibrous low-quality material (for similar behavior by *Ateles belzebuth* in Colombia, see Klein & Klein 1977). It seems unlikely that such complex, fruit-specific feeding patterns were learned by the five founders prior to capture.

Work on terrestrial herbivores has shown that they are capable of selecting the "best" foods without previous exposure (Zahorik & Houpt 1981). Lemurs born in captivity showed the ability to utilize visual and chemosensory cues to discriminate the relative palatability of different novel food items (Glander & Rabin 1983). The feeding behavior of the BCI spider monkeys suggests that some chemical constituents of fruits interact with chemosensory receptors to permit them to gauge fruit quality and modify their feeding behavior in response to this information (see also Whitehead 1986). It also suggests that these receptors are particularly sensitive to fruits and relatively unresponsive to young leaves or insects, as, otherwise, one might expect spider monkeys to include more of these palatable and nutritious items in the diet.

There may be less risk to frugivores in trying new items of diet than is the case for folivores (Milton 1981b; Whitehead 1986). Ripe-fruit pulp and arils are generally palatable and serve to attract seed-dispersal agents, whereas leaves are often protected by potentially harmful chemical compounds (Milton 1980). Further, ripe fruits are far more patchily distributed in tropical forests than are young leaves (Milton 1980). Thus it might be advantageous for frugivores to sample new fruits whenever they are encountered in the forest, since the risks are low and the benefit could be considerable. In addition, as suggested by Fairbanks (1975) and Cambefort (1981), there may be a relationship between the pattern of social organization of a species and the manner in which it acquires new food habits (see also Caine & Mitchell 1980; Zahorik & Houpt 1981). If so, we should expect spider monkeys to be particularly adept at learning new food habits through trial and error since their usual mode of foraging should often place them in situations where few, if any, other adult members of their species are present when novel, palatable fruits are encountered.

Studies of young primates, however, suggest that they independently experiment with novel foods to some degree regardless of whether their species tends to focus feeding on leaves or fruits (Itani 1958; Cambefort 1981; Watts 1985a; Whitehead 1986). In addition, it is generally younger primates, particularly juveniles, that are most adventurous in trying new foods (Cambefort 1981). In combination, these observations suggest that in the absence of more experienced conspecifics, young primates of many species, when placed in appropriate environments, may be able to develop a species-typical dietary profile.

Development of Species-Typical Social Behaviors

Trial-and-error learning, coupled with species-specific chemosensory receptors and an expanding pool of shared information, may explain the observed food choices of BCI spider monkeys. Such an explanation, however, does not suffice to explain their pattern of social organization. Fission–fusion sociality of the Ateline sort is rare in primates (van Roosmalen 1985), and no other primate species on BCI shows such a pattern. To explain its occurrence in this population, one must postulate either that very young monkeys observed this pattern prior to capture and remembered it as adults or that a predilection to behave in this manner is somehow inherent. The fact that juvenile males in the F_1 generation, without models, began to associate frequently with one another and the single founding male likewise suggest some predisposition to such behavior. Eisenberg (1976) noted that subadult and adult male spider monkeys give increasing attention to juvenile males, perhaps facilitating the breaking of the bond with the mother. In this respect, it is note-

worthy that in 1964, the founding male prior to the birth and maturation of his sons did not associate with the founding females on a continuous daily basis, even though at that time he had no other male companions (Eisenberg & Kuehn 1966). Eisenberg and Kuehn (1966, p. 54) note that for 19 of their 26 days of observation, the founding male "could be found feeding, moving and resting alone." Given the social nature of spider monkeys, this seems noteworthy and suggests a strong tendency for male disassociation from females during late adolescence and adulthood. The only other explanation—that the founder observed this male behavior prior to capture and remembered it, adopted it, and taught it to his maturing sons—seems extremely unlikely. The foraging range of females relative to males and their use of semisolitary core areas likewise indicate either remarkably retentive memories for sex-typical behavior or an inherent tendency to behave in species- and sex-typical ways at maturity. Likewise, the female founders raised offspring without prior experience with infant care.

CONCLUSION

These data suggest that many species-typical behaviors in primates, both foraging and social, can develop in the absence of opportunities to observe and interact with more experienced individuals. This is not to deny the considerable capacity for phenotypic plasticity in primates or the importance of social learning, but to point out that assumptions of social learning, social imitation, social facilitation, and/or social transmission with respect to the acquisition of a particular species-typical behavior may not always be justified.

ACKNOWLEDGMENTS

Data analysis was facilitated by National Science Foundation Grant 85-12634 to K. M. I was invited to contribute this paper to the volume in 1990 while I was a visitor at the Duke University Center for Tropical Conservation, carrying out research under sponsorship of the John Simon Guggenheim Memorial Foundation. I thank these organizations for their support of my research.

13

Primate Juveniles and Primate Play

ROBERT FAGEN

Play is improvised performance, with variations, of skilled motor and communicative actions in a context separate from the environment in which behavior including these actions proximately increases reproductive success. Chasing, wrestling, and energetic body movements accompanied by context-specific signals distinct from those of predator avoidance and defense or intraspecific fighting all constitute play in animals (Fig. 13.1). Play is considered important for development of skilled movement, for communication, and for social relationships. Play always involves elements of interpretation of an environment that is essentially ambiguous (Fagen 1993). Over the past decade, play attracted increasing attention (Fagen 1981, 1990; Bekoff 1984; Smith 1984; Caro 1988). Recent analyses span ontogeny (Chalmers 1980b), structure (Pellis 1981, 1988), cognition and motivation (Mitchell 1990; Jolly 1991), function (Pellis 1988), and evolution (Ghiselin 1974, 1982; Chiszar 1985; Heinrich 1989). Sophisticated experimental paradigms clearly demonstrate a role for play in development of defensive aggression in laboratory rats (Potegal & Einon 1989; Einon & Potegal 1991). Studies of improvisation in avian song ontogeny (Marler & Peters 1982) and of human parent–offspring play (Als et al. 1979; Stern et al. 1983; Lamb 1984; Dienske 1986; Hinde 1987) additionally indicate potential biological importance of playful and play-like activity.

Elaborate social and solitary play characterize all primate species, from tarsiers (*Tarsius spectrum*) (MacKinnon & MacKinnon 1980) and mouse lemurs (*Mirza coquereli*) (Pages 1983) to orangutans (*Pongo pygmaeus*) (e.g., Rijksen 1978; Galdikas 1979, 1985a, 1985b, 1990; Maple 1980, 1982; Becker 1984) and bonobos (*Pan paniscus*) (e.g., Kuroda 1980; Becker 1984; de Waal 1986a, 1988; Kano 1990). We lack detailed studies of play in most primate species.

POSSIBLE HETEROGENEITY AND LEVELS OF ANALYSIS

Continuing debate (Fagen 1981, pp. 54, 474–475; Smith 1984) reveals need for analysis of play at all levels. In addition to the traditional "four questions" of ethology, analyses of play now consider the kinds of purposive questions appropriate to animals with active, integrative, and interpretive intelligence (Purton 1978; Mitchell 1990). Distinct forms of play become increasingly integrated as brain size increases relative to body size across taxa. For example, infant and juvenile great apes, like carnivores (and perhaps unlike many other terrestrial mammals), make play signals during solo play (Goodall 1968, p. 181; Lethmate 1976, 1977; Fagen 1981, p. 49; Becker 1982, p. 41; Chevalier-Skolnikoff 1982, pp. 309, 323; Becker 1984, pp. 121, 139, 142; Goodall 1986, p. 560, 1989a, p. 19, 1989b; van Hooff 1989, p. 133). Solo and social play of these species may well differ in ontogeny, control mechanisms, and function, but the animals' own natural categories (as imperfectly indicated by play signals) appear to be inclusive.

Fig. 13.1. Juvenile ringtailed lemurs (*Lemur catta*) at play. (Photo, D. Haring)

In this chapter, the form of play being discussed will be stated where such qualification is needed. Other statements made without explicit qualification are intended to apply to all forms of play.

Relationships between forms of play are almost surely biological dependent variables shaped, at least in part, by the animal's own cognitive categories. Play features prominently in the discourse of signing common chimpanzees (*Pan troglodytes*) (Terrace 1979; Fouts 1989; Gardner & Gardner 1989). It would be interesting to use two-way communication to

investigate ways in which trained, sign-using nonhuman primates learn to categorize play. Chimpanzee Nim's expressive vocabulary included signs for the play categories "tickle," "play" (a general sign invented by Nim and used with human adults, children, and living animals but not with pictures of animals or inanimate objects), "jump" (to solicit a jump-chase game), and "happy" (used by Nim when excited, as while tickling). Play ranked high in Nim's signed discourse. His most frequent two- and three-sign combinations were "play me" and "play me Nim," with "tickle me" in third place on the two-sign list and "tickle me Nim" in fourth place on the three-sign list (Terrace 1979, p. 212).

RISKS OF PRIMATE PLAY

Preadult behavior affects primate survival (Konner 1977a; Pereira & Altmann 1985; Caine 1986; Walters 1987a). New evidence for risks of play in immature primates strongly supports this view. A male infant howler monkey (*Alouatta palliata*) fell into a river and drowned during a playbout with two juveniles (Clarke & Glander 1984). Infant and juvenile common chimpanzees fell most frequently during play (Goodall 1986, pp. 98, 100). An infant mountain gorilla (*Gorilla gorilla*) fell into a tree fork, hung helplessly by her neck, and almost suffocated before her mother saved her (Fossey 1983, p. 88). Vervet monkey (*Cercopithecus aethiops*) play groups form at some distance from adults, and members of these subgroups are those most often caught by predatory baboons (*Papio cynocephalus*) in Amboseli National Park (Hausfater 1976). Predation accounts for 15% of all vervet mortality annually in Amboseli, and 63% of these deaths are of infants or juveniles (Cheney & Wrangham 1987). Playing vervets are vulnerable to predators of at least four species (Cheney & Wrangham 1987). Vigorous play of subadult Himalayan langurs (*Presbytis entellus*) was not often seen in the trees around Melemchi, Nepal, but "as a result of one playful wrestle between two subadults, one of them fell thirty feet into a thicket. He emerged looking very sheepish" (Bishop & Bishop 1978, p. 105).

Baldwin (1986) suggests that risk of falling could influence evolution of play in arboreal primates. Unlike falling, predation risk is important whether primates play on the ground or in trees, although predators (carnivores vs. raptors) may differ in the two cases. Biben et al. (1989) observed young squirrel monkeys (*Saimiri boliviensis*) at play. They experimentally demonstrated a compensatory increase in adult vigilance during play and a function for play vocalizations as an auditory cue to alert adults to playing young. Adult primates are attentive to young at play (Fagen 1981, pp. 393–394; Strum 1985) and are known to rescue young from dangers encountered in play (Kum-

mer 1968). Strum (1984) asserts that "the injury rate in baboon play is high enough to make it likely that at some time during its 'play' life, each young Pumphouse baboon will sustain an injury."

Risk in Human Play

Estimating risks of juvenile play requires very large samples, and nonhuman primate evidence is just beginning to be accumulated. But epidemiological studies of injuries to human children yield samples much larger than any now available from nonhuman primates. For example, Sacks et al. (1989) monitored 5300 children aged less than 1 year to 12 years for 12 months. Elardo et al. (1987) followed 133 children aged 2–80 months for 42 months. These studies fully confirm the hypothesis, suggested by research on nonhuman primates, that falls in play are a major source of juvenile injury (Gratz 1979; Garretson & Gallagher 1985; Elardo et al. 1987; Sacks et al. 1989). Playgrounds are a major site of injury and playground falls a major cause of injury (Sacks et al. 1989). Preschool-age children (0–5 years) are one and one-half to two times more likely than older children to suffer fall injuries (Guyer & Gallagher 1985). Among preschool-age children, three factors associated with frequent falls are sex (male), activity level (high), and quality of play (rough) (Bergner et al. 1971). In two studies, 3- and 4-year-old boys had twice as many accidents as girls during uncontrolled free play (Atkinson 1963; Federer & Dawe 1964). Boys suffer more injuries than girls (Rivara et al. 1982). Other studies report no sex differences in childhood injuries (Bittner & Delissovoy 1964; Solomons 1982) or, interestingly, no overall sex differences, but a bimodal pattern among individual girls, who tended to be either accident prone or accident free (Fuller 1948; Elardo et al. 1987). Results to date focus on day-care-center children because they are most accessible for study and because of social concerns. Yet the frequency of accidental falls by preschoolers is even higher in the home than at day care (Guyer & Gallagher 1985).

Play-related risks other than falls include aspiration of toys (Mofenson & Greensher 1985) and other toy-related injuries (Greensher & Mofenson 1985). Other risks to child safety are also play related (Garfield 1983).

Costs of falls and fall-related injuries for a

human child in Western industrialized society range from embarrassment and time spent away from school, home, and friends to medical risks that may be mitigated by availability of pediatric care. Under the socioeconomic conditions of other industrialized nations, or in agricultural or hunter-gather societies, these social costs and more will continue to be present, but the costs of play-related injury in terms of immediate survival and as routes for infection are surely much greater than in an American preschool for middle-class children.

Questions Suggested by Risks of Play. The classic ecological argument for benefits of play emerges with renewed force from evidence of risk. By this argument, only if benefits of play exceed its costs will play persist in evolutionary time. Evidence of costs is undeniable, and yet no benefit hypothesized to date (Baldwin & Baldwin 1977; Poirier et al. 1978; Fagen 1981) has been demonstrated for nonhuman primates. It is conceivable that play might not benefit primates at all. If the ecological argument is incomplete, as some critics suggest, then, risks notwithstanding, play could persist. For example, play could be a byproduct of selection for another character, such as relative brain size, or for some larger phenotypic unit that inexorably produces play as a harmful byproduct in early ontogeny. Play could also persist in the face of natural selection if particular forms of cultural selection operated (see the section "Cooperation, Cognition, and Culture in Play"). Like the ecological argument, which remains to be substantiated by evidence of benefits, these alternative, pluralistic arguments are weak because they, too, lack the empirical evidence they require and even, in some cases, concrete mechanisms. Still, play as a whole and differences in play should not be viewed as adaptations a priori (Fagen 1981, p. 482). If play, like the player's dangerously large, heavy head, is an unavoidable but dangerous consequence of a large brain or of a sophisticated adulthood, questions about the possible adaptiveness of play become less compelling.

The key evidence against these arguments is that play rates drop precipitously when animals are under stress (Fagen 1981, p. 370). Play may even disappear completely at such times. This evidence indicates that mechanisms exist to suppress play under a variety of different environmental circumstances. Therefore, if play were merely a harmful byproduct of some other character, these same mechanisms would be available to block its expression whenever and wherever it proved harmful.

Rates of play track relevant environmental variation with impressive precision. Responses to food availability are the classic example of such presumably adaptive tracking. Fagen (1981, p. 370) lists 15 studies in which decreases in food produced decreases in play. In Amboseli National Park, declining food availability reduced play to trace levels in juvenile vervet monkeys (Lee 1984). Weaning leads to a temporary decrease in play of young savanna baboons (Nicolson 1982). Experimental studies of declines in play in response to food decreases or food unpredictability reveal behavioral mechanisms linking food and play in primates (Rosenblum et al. 1969; Zimmermann et al. 1975a, 1975b; Baldwin & Baldwin 1976; Zimmermann et al. 1976; Rosenblum & Sunderland 1982; Rosenblum & Paully 1984; Boccia et al. 1988) If nonhuman primates must reduce their rates of play in the face of food shortage, food unpredictability, or other stressors, they invariably do so—large brains, inherent playfulness, and complex adulthood notwithstanding. Since animals readily stop playing for extended periods, it seems likely that any form of play lacking direct benefits would be eliminated by natural selection over evolutionary time.

AGE DIFFERENCES IN PLAY

Average rates of play decline consistently in nonhuman primates from late infancy or early juvenility through adolescence to adulthood (Hinde & Spencer-Booth 1967; Goodall 1968; Chivers 1974; Mori 1974; Owens 1975; Leresche 1976; Bramblett 1978; Symons 1978; Levy 1979; Chalmers 1980b; Chulee 1981; Kraemer et al. 1982; Becker 1984, p. 96; Pereira 1984; Hayaki 1985a; Koyama 1985; Bramblett & Coelho 1987; Kondo-Ikemura 1988). This trend holds for total play, for social play, and for nonsocial play (Fig. 13.2). High rates of play characterize late infancy and early juvenility, not juvenility per se, in nonhuman primates.

Regrettably, different forms of play have not always been recorded separately, and studies of infants tended to occur apart from studies of

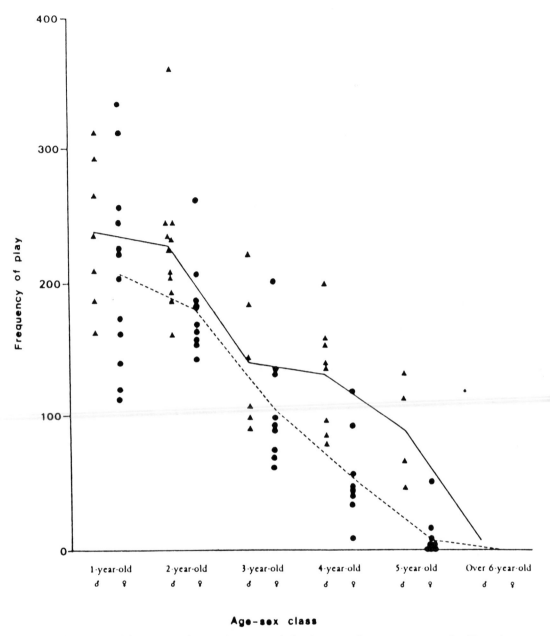

Fig. 13.2. Individual frequency and mean frequency of play in young Japanese macaques. In this study, individual differences, not sex differences, accounted for most observed variation in juvenile play frequencies (Koyama 1985, p. 393).

juveniles, even in the same social group. Further age-structured analyses of quantities and qualities of primate play across the life span are needed to clarify current understanding of primate play and of the ontogeny and phylogeny of primate behavior.

Ontogenetic Decreases in Play

Proximate Causes of Developmental Decreases in Play. Levy (1979), Caine (1986), and Pusey (1990) discuss possible proximate causes for decreases in play with age. Their conclusions differ. Caine (1986) sees male play declining due to intermale competition and heightened aggressiveness. Pusey (1990) found that both sexes of the common chimpanzee at Gombe Stream tended to become more aggressive in play as they reached adolescence. She also observed a possible decrease in adolescent responsiveness to play partners. Adolescents appeared more frequently to ignore invitations to play or to play only briefly.

Levy (1979) discusses ontogenetic declines in play of rhesus macaques (*Macaca mulatta*). She makes three important points.

1. An ontogenetic decline in play characterizes the entire juvenile period, not simply its end.
2. Proximate and ultimate causes of declines in play are necessarily distinct.
3. Reasons for declines in female play may differ from reasons for declines in male play.

Levy argues that old juvenile male rhesus monkeys begin to assert their developing rank in play (see Symons 1978). They monopolize opportunities to perform in particular roles and to make particular moves such as mock attacks. As a result, partners are difficult to attract and play encounters difficult to maintain. At the same time, sexual maturation and subsequent rank reversals further interfere with play opportunities for play. Social behavior of old juvenile females becomes increasingly similar to that of adult females, even before conception, and these increasing social priorities likely leave less time for play (see Fairbanks 1990). Although overall social motivation remains high, it appears to include decreasing play tendencies. A subsidiary factor, shared with males, is avoidance of play due to the inability to accept mock attacks from subordinates or to apparent anxiety over play

partners that insist on playing solely the role of attacker. Levy found no evidence in rhesus monkeys to support the conventional wisdom (e.g., Hall & DeVore 1965; Baldwin 1969) that social play of nonhuman primates becomes increasingly rough with increasing age. However, some aspects of play increasingly resemble aggressive fighting. Levy's key finding, however, involves indications of the monkeys' own perception of behavioral events that a human would perceive as play opportunities. Older juvenile rhesus monkeys, more often than young ones, seem relatively tense and fearful, or simply uninterested, when approached by a like-aged partner soliciting play.

Levy's observations suggest changes in the style of play, increased inhibition of rough play with age, and decreased speed and forcefulness of actions. She found that play developed increasing complexity in older animals due to differentiation of roles, resulting in asymmetric forms of play. Play of like-aged dyads changed with age. Younger animals in like-aged dyads wrestled, mutually pulling and pushing with close physical contact. Wrestling occurred in mutual upright and in stand-up/belly-up positions. These younger animals also chased with role reversal. Wrestling and chasing patterns were also present in play of older animals in like-aged dyads, but these patterns were less frequent relative to other, asymmetric patterns: jump onto, grasp and pull, mouth, lunge at, dodge aside, and crouch. Younger animals seemed to favor play patterns that had fixed contact qualities. They would wrestle with close physical contact, or they would chase each other. Older animals were more likely to integrate noncontact and contact play at each instant of a continuing interaction, rapidly alternating contact and noncontact behaviors while in close physical proximity. Older juveniles were likely to direct mock attacks at each other. The attacker approached its partner directly, grasping, hitting, mouthing, chasing, and gaining a superior wrestling position. The animal attacked would back away, flee, dodge, and/or allow itself to be mouthed or climbed on. Younger juveniles play more mutualistically with face-to-face wrestling and grappling, chases in which the roles of pursuer and pursued reverse, and up/down wrestling with frequent role reversal. This evidence, like the filmed play sequences in Symons and Bishop (1977; see also Symons

1978 and Koford 1966), indicates a change in choreography, rather than a change in roughness. Roles in asymmetric play of like-sexed, like-aged dyads tend to reverse less often with increasing male age, but not with increasing female age. By age 3, the higher ranked member of a like-aged male dyad would jump onto, lunge at, and otherwise physically assert himself in play 1.5 times as often as would his lower ranked partner. This is not a large difference, but it contrasts with the virtual equality of rates for such assertive behaviors in younger like-aged male play dyads.

Gonadectomy and prenatal hormone exposure can alter juvenile primate play patterns as well as many other features of behavior (Goy & Resko 1972; Goy & McEwen 1980; Loy et al. 1984). Castration postpones the decline in play in old juvenile male rhesus monkeys, but does not prevent it entirely (Loy et al. 1984). Female rhesus monkeys treated prenatally with gonadal hormones play to a later age than do intact males when they are housed together (Goy & Resko 1972). Prenatal exposure has no effect on play of juvenile females in all-female groups (Goy & McEwen 1980), but critical prenatal periods exist for effects of exposure to androgen (Goy et al. 1988). In both sexes, the decline in play with age seems to reflect both hormone-dependent and hormone-independent factors.

Possible Ultimate Causation of Developmental Decreases in Play.

Juvenile primates maintain themselves, grow, and enter social relationships. As they do so, play risks and values may change with age. An age-structured fitness model (Fagen 1981) helps clarify thought about age-specific play schedules in relation to these risks and values. The model is an idealized statement of alternative assumptions about costs and benefits of play. In a simplified version, I evaluate the view that play requires so little time and energy (see Martin & Caro 1985) that its only immediate effect is to decrease survival to some degree—for example, in falls or through increased conspicuousness and decreased alertness to predators. I further assume that benefits of a given amount of play at any age are equal (Chalmers 1987). Analysis shows that general age-specific mortality factors whose effects are independent of amount of play will not affect relative amounts of play at different ages. The model makes it possible to understand how older infants and young juveniles can play more than older juveniles, even though the older ages are clearly "safer" in terms of overall survivorship (e.g., Amboseli baboons: Altmann et al. 1985; Cayo Santiago rhesus monkeys: Dunbar 1987; Janson & van Schaik, Chapter 5, this volume). This result holds whether play has short-term effects on juvenile survival or has effects delayed until adulthood.

A Fitness Model for Prereproductive Play in the Life History.

Assume (Fagen 1981) that life histories with play will evolve to maximize the discrete Malthusian parameter λ, defined by the equation (e.g., Dunbar 1987)

$$F(\lambda) \equiv \sum_{x=0}^{w} l_x b_x \lambda^{-(x+1)} = 1$$

Let w be the maximum age, a the age at first reproduction, and z_x the fraction of resources at age x devoted to play ($0 \leq z_x < 1, 0 \leq x < a$). (If total resources available vary with age, then two ages x, y with $z_x = z_y$ could use different absolute amounts of time or of energy in play.) Survivorship from age x to age $x + 1$ is modeled by the relationship $p_x = s_x f_x(z_x)$. Function $f_x(z_x)$ describes the way in which play at age x decreases survival from age x to age $x + 1$. Its value is 1 when $z_x = 0$ and 0 when $z_x = 1$. It is a decreasing function of z_x. Benefits of play are modeled by a function $g(z_0, z_1, \ldots, z_{a-1})$ that enters the demographic equation multiplicatively: $l_x b_x = g(z_0, z_1, \ldots, z_{a-1}) l'_x b'_x$ for all $x \geq a$, where l'_x and b'_x give age-specific survival and reproduction in the absence of any play and do not depend on resources invested in play. Function g is an increasing function of each of its arguments and has value 1 when all of its arguments are 0. An optimal age-dependent schedule of investment in play satisfies (Schaffer 1979; Fagen 1981)

$$\partial\lambda/\partial z_x = 0, \text{ for all } x$$

By implicit differentiation, this condition is equivalent to

$$-(\partial F/\partial z_x)/(\partial F/\partial \lambda) = 0, \text{ for all } x$$

The denominator of the above expression is nonzero and can be eliminated from the equation, yielding the equation

$$-\frac{1}{F_x}\frac{\partial F_x}{\partial z_x}l_a \sum_{x=a}^{w} \frac{l'_x}{l_a} b'_x \left[g(z_0,z_1,...,z_{a-1}) + \frac{\partial g}{\partial z_x} (z_0,z_1,...,z_{a-1}) \right] \lambda^{-(x+1)} = 0$$

The quantity l'_x/l_a depends only on adult survival probabilities, not on the s_x for $x < a$ and not on the z_x. The only quantity in the equation that depends on prereproductive s_x is l_a. If we divide both sides of the equation by the product of all the prereproductive s_x values, we obtain an equation for z_x that does not depend on any of the prereproductive s_x. This familiar result from life-history theory is here shown to be valid for play.

What can make play more common at one age than at another? If the model's assumptions hold, there is only one possible explanation. This explanation would be that *play at different ages is differentially risky*. Thus, for example, if a given amount of play is more dangerous for younger animals, they should play less. Human toddlers and infant great apes appear to be at more risk from play than are juvenile conspecifics (see the section "Risks of Primate Play"). But, contrary to the prediction that follows from this explanation, juveniles actually play less frequently than infants. Although the question of age-specific play risks will need systematic investigation before firm conclusions can be drawn, the model's assumptions appear to be incorrect, given current evidence.

If the model's assumptions are incorrect, then other possibilities exist. Let us systematically explore the possibilities that arise if we contradict each of the model assumptions in turn.

1. Suppose, contrary to the assumption of no time, energy, or nutrient cost, play actually takes significant amounts of resources. Then we would expect that play tends to decrease survival and/or growth, and predict that an age class (or sex) having access to more or richer resources (e.g., milk) may well play more often.

2. Suppose, contrary to the assumption of equal benefits at all ages, that a given amount of play at one age actually has a greater effect on subsequent survivorship or on reproduction than a given amount of play at another age. Then we would predict that if all other factors were equal, play would occur most frequently at the ages where it was most effective. When trade-offs occur, the picture

becomes more complex, but precise quantitative predictions of age-specific play schedules are still possible given sufficient information (Fagen 1980, 1981).

Knowledge of age-specific play schedules, combined with use of a simple fitness model, allows some fairly profound inferences about play. Students of play have long sought to demonstrate that play has major impacts on time and energy budgets and/or that sensitive periods for play exist in ontogeny. Such conclusions now appear to be within relatively easy reach.

SEX DIFFERENCES

Studies of several cercopithecine species and a few other nonhuman primate taxa reveal sex differences in juvenile social play (Aldis 1975; Baldwin & Baldwin 1977; Caine & Mitchell 1979; Mitchell 1979; Beatty 1984; Meaney et al. 1985; Walters 1987a; Meaney 1988; Watts & Pusey, Chapter 11, this volume). Juvenile males tend to play-fight more frequently than juvenile females in some but not all cercopithecines, notably savanna baboons (Owens 1975; Pereira 1984) and rhesus macaques (Symons 1978; Levy 1979). Other primates in which juvenile males play fight more than juvenile females include hamadryas baboons (Kummer 1968; Aldis 1975; Leresche 1976), several other macaque species (Caine & Mitchell 1979), and squirrel monkeys (*Saimiri* sp.) (Baldwin 1971; Biben 1986). Juvenile and subadult female red colobus (*Colobus badius*) of Abuko Nature Reserve, Gambia, play more frequently than juvenile and subadult males (Starin 1990). Both sexes of juvenile common chimpanzees (Hayaki 1985a; Green 1989) and bonobos (Becker 1984) appear to play socially at equal frequencies. Strong individual differences and small sample sizes characterize the data for both *Pan* species.

Orangutan play, like that of cercopithecines, exhibits sex differences. Adult orangutans are highly sexually dimorphic. An adult male may be over twice as heavy as an adult female (Rodman & Mitani 1987). Available evidence suggests that male–male competition for sexual access to females was important in orangutan evolution, and that aggressive encounters between adult males may result in wounds and broken fingers, contrasting with relatively amicable interactions among the more tolerant (but occasionally pugnacious) adult females (Rodman & Mitani 1987). In confinement, orangutans in multiage social groups play together readily (Edwards 1982; Maple 1982; Becker 1984; Voigt 1984; Poole 1987; Tobach et al. 1989). Social play duration of juvenile males is almost double that of juvenile females (Becker 1984, p. 101), same-sex social play duration of immature males is double that of females (Rijksen 1978, p. 254), and females "tend to withdraw from play" (Maple 1982). Maple (1980) and Galdikas (1985a) also found sex differences in orangutan play. But do these differences, described mainly from artificial social groups, matter in the wild? Opportunities for play with other immatures are limited for young of both sexes, especially for infants and juveniles. A young orangutan can play socially with its mother and sibling, with other immatures encountered during social contacts between families, as an adolescent and subadult (Galdikas 1985b; Rijksen 1978), and even during interactions between its mother and a subadult male (Galdikas 1990). Does the semisolitary existence of wild orangutans provide sufficient social opportunities for males to express their greater playfulness and thereby to reap the putative benefits of enhanced fighting skill? Orangutan social play is also a component of adult "cooperative mating" sequences (Rijksen 1978, p. 273). Social play may be important for affiliative relationships in orangutans, as well as for male practice of fighting skill. A sex difference expressed only in confined groups could simply be a nonadaptive consequence of some other characteristic—perhaps itself adaptive—of young male orangutans.

Where higher male frequencies of social play occur, they characterize juvenility, not social play per se. Sex differences in infant play are nonexistent (or were too small for samples to detect) in two troops of savanna baboons (Hendy 1986). Infant female Japanese macaques (*Macaca fuscata*) play socially more often than infant males (Hayaki 1983; Mori 1974), but old juvenile males play more often than old juvenile females (Hayaki 1983; Koyama 1985; Eaton et al. 1986; Nakamichi 1989). Even 1 year olds showed no sex difference in play frequencies (Koyama 1985). Confined, provisioned vervet monkeys follow the conventional cercopithecine pattern in which male juveniles play more than female juveniles (Fedigan 1972; Bramblett 1978; Raleigh et al. 1979), but such differences are not evident in infancy (Lee 1984) and fail to attain statistical significance under harsh conditions in the field (Lee 1983a). Little is known about possible sex differences in nonsocial or adult play in any nonhuman primate. Available information (e.g., Goodall 1968) suggests that adults, like infants, may show no sex differences in social play or that females may even be more playful than males in adulthood.

Massive overlap between the distributions of male and female juvenile play rates is evident in each of two Japanese macaque groups (Koyama 1985; Imakawa 1990). Many females in each group played more than most like-aged males, and this pattern may be common (see, e.g., Cheney 1978a; Lee 1984). Until females are on the verge of adulthood, sex differences in play rates are evident only as statistical abstractions from large samples. The play experience of most males will differ sharply from that of the nonexistent "average juvenile male," and precisely the same statement holds for females, many of whom will experience more social play than the "average juvenile male."

Individual variations in food consumption inevitably contribute to overall error variance in studies of sex differences in play. In an unknown proportion of cases, uncontrolled and unmeasured variation in amount and quality of food consumed may confound existing conclusions about sex differences in juvenile primate play. Few, if any, studies of sex differences in play control or correct for sex differences in food-related variables. Yet evidence that animals at a higher nutritional level play more is incontrovertible. Stuart Altmann (1991, personal communication) estimated total milk consumption and measured protein availability relative to maintenance requirement in 11 immature savanna baboons, along with play bouts and minutes per day in nine of them (Table 13.1). The two food measures were significantly correlated ($r = 0.80$, $p < 0.01$), as were the two play measures ($r = 0.86$, $p < 0.01$). Both play measures varied significantly with milk consumption (bouts: $r = 0.78$, $p = 0.014$; minutes, $r = 0.70$, $p = 0.036$) and with protein (bouts: $r =$

Table 13.1. Play and Diet in Immature Savanna Baboons (Stuart Altmann)

Individual	PDR	Milk consumption (kj/day)	Bouts play/day	Minutes play/day
Alice (f)	0.655	711	12.6	1.3
Bristle (m)	0.728	1240	28.3	3.4
Dottie (f)	0.743	1300		
Eno (f)	0.721	1355	12.5	0.75
Fred (m)	0.632	658	25.4	1.8
Hans (m)	0.770	1970	35.9	3.2
Ozzie (m)	0.720	1313	22.8	4.1
Pedro (m)	0.530	765	6.1	0.5
Pooh (f)	0.500	374	0.0	0.0
Striper (f)	0.671	427		
Summer (f)	0.750	1540	32.3	3.2

Note: PDR (protein deviation from requirement) measures food protein available, expressed as a fraction of the minimum maintenance requirement, after this maintenance requirement is met. Play of Dottie and Striper, pilot subjects in a larger nutrition study, was not sampled systematically.

0.84, $p = 0.01$; minutes: $r = 0.79$, $p = 0.011$). Males consumed (nonsignificantly) more milk, had (nonsignificantly) more excess protein, and played (nonsignificantly) more than females. Food consumption, measured jointly by milk and protein, explained 72% of the variance in bouts and 62% of the variance in minutes. These small sex differences in the play variables could result simply from sex differences in the food variables.

Can allocation of food resources explain both the decline in play with age in both sexes and the earlier decline in females? As immatures grow, their nutritive demands for maintenance increase. Altmann (1980) and Dunbar and Dunbar (1988) based models of maternal time budgets and lactation on precisely this assumption. Given a fixed nutritive input and increasing demands for maintenance, it is difficult to imagine how play could do anything but decrease over the juvenile period.

Added to maintenance are the costs of growth per se and of reproduction. Food demands of young males are expected to increase at or slightly before their adolescent growth spurt. Food demands of young females will increase when they become pregnant. Although the male adolescent growth spurt and first female reproduction may occur at roughly the same age (e.g., in savanna baboons), overall demands on time, energy, and nutrients may differ greatly between the sexes at this age, with the females on a much tighter resource budget. Life-history models (Fagen 1981) with these features predict decreases in play a year or two before life-history periods involving major resource demands. If pregnancy and motherhood demand more of a relatively inexperienced female's energy and time than adolescence requires of males, then juvenile female play should decline in frequency earlier than does that of juvenile males.

Maternal investment in offspring can vary with offspring sex (Pusey 1983; Silk 1987; Hrdy 1989) and should directly affect sex-specific rates of play. Research on sex differences in juvenile primate play will require detailed age-specific and sex-specific data on food consumption and nutrition in order to achieve its goal of evolutionary insight.

Sex differences in juvenile play are interesting, potentially valuable tools for testing hypotheses on primate behavior. In analyzing such differences, it is important to remember that differences in group means mask individual variability, and that nutrition, demography, age, and probably other factors, many poorly understood, can enhance, remove, or even reverse familiar patterns of sex differences in juvenile primate play (e.g., Cheney 1978a; Lee 1984; Goldfoot & Neff 1985).

Evolutionary hypotheses about sex differences in juvenile play of certain species tend to focus concentration on males, with the unfortunate consequence that female play seems attenuated or incomplete. But females are not incomplete, attenuated males. Why do females play? What is distinctive and important about female play in a given species? Features of female play deserve documentation for their own sake. In a recent study of ringtailed lemurs (*Lemur catta*), for example, Pereira predicted that juvenile females would not differ from males in rate or style of play. In contrast to savanna baboons and rhesus macaques, females and males of this species benefit about equally from potentially damaging solo acts of aggression as adults (Richard 1987; Vick & Pereira, 1989; Pereira & Weiss, 1991). Neither Gould's (1990) study of free-ranging *Lemur catta* in the wild nor Pereira's (personal communication) of animals in a large, forested enclosure found gross sex differences in juvenile lemur social play. Pereira's detailed observations showed that eight distinct types of sex difference previously demonstrated in play of juvenile baboons at Amboseli National Park (Pereira 1984;

Pereira & Altmann 1985) were absent from the play of his ringtailed lemur juveniles. Box (1975), Cleveland and Snowdon (1984), and Stevenson and Poole (1982) also found no sex differences in juvenile marmoset play. Further quantitative study of lemurids, callitrichids, and cebids may reveal additional primates whose juvenile females and juvenile males play at similar frequencies and in similar ways.

PLAY COMPANIONS AND PARTNER PREFERENCES

A juvenile's play companions and play tendencies interact so as to determine both quality and quantity of play experience. The experience of a juvenile male in a social group where infants are the only other immatures will differ substantially from that of an infant female in a group offering only old juvenile male partners.

Young cercopithecines (e.g., Berman 1978; Cheney 1978a; Symons 1978; Caine & Mitchell 1979; Levy 1979; Colvin & Tissier 1985; Koyama 1985; Janus 1989) and perhaps other young nonhuman primates (reviews: Fagen 1981, pp. 401–403, 434–435; Walters 1987a) appear to prefer certain classes of play partners. Age, size, sex, kinship, and maternal dominance can all be important, where choices exist. In an especially thorough study of play-partner preferences, Pereira (1984) demonstrated (in addition to the expected age and sex preferences) sex differences in the number of frequent play partners and in rates of play initiation with older group members, but not in rates of play initiation with infants, among juveniles in two groups of savanna baboons.

COOPERATION, COGNITION, AND CULTURE IN PLAY

Levels of cooperation in play of juvenile primates may exceed those predicted by simple evolutionary arguments such as those of Konner (1975), who viewed primate play in terms of conflicts of individual interest. He argued that each animal might have its own preferred play style as a result of age, sex, dominance, early experience, or individuality, and that conflict would therefore be expected between any two potential partners. Implications of these ideas

(Fagen 1981, p. 388; Fagen 1984; Fagen 1986) led me to question how, in a world of selfish juveniles, any two individuals could ever agree to play at all, a problem that I termed "Harlequin's Dilemma."

1. Perhaps individuals' needs are really all about the same when they are young. Only as they get older and develop as progressively more differentiated individuals do such conflicts become substantive enough to depress the frequency of play and ultimately remove it almost entirely from the behavioral repertoire. But infants and juveniles are also individuals, and adults play, so this explanation seems implausible.

2. Potential play partners who encounter each other repeatedly are more likely to cooperate in play because repeated encounters facilitate cooperation (Axelrod & Hamilton 1981).

Cooperation in play seems to occur much more readily than either of these cautious scenarios would predict. The natural history of primate play offers examples of play between strangers belonging to different primate social groups and even to different species (e.g., Goodall 1989b). A young langur (*Presbytis entellus*) named Miro, the only immature in his troop, had play opportunities only during intertroop encounters. At these times, Miro, usually timid and dependent, assertively sought out playmates of any age in the alien troop (Hrdy 1977, p. 295). Juvenile savanna baboons (Ransom 1981, p. 257; Strum 1987, pp. 93, 108) and rhesus macaques (Hausfater 1972) also play with strangers from other troops. Juvenile orangutans occasionally leave their social unit to play with peers (MacKinnon 1978, p. 116; Galdikas 1979).

Human–domestic dog play varies with familiarity of partners, but also can occur cooperatively between strangers who apparently depend on widespread play scripts to stabilize interaction (Mitchell & Thompson 1990). These and Biben's (1989) results indicate that stability of social play between unfamiliar partners stems partly from cultural knowledge and partly from individual intelligence and imagination. Moreover, distinct but equally attractive forms of stable social play are available and used by players in accordance with individual or joint preferences. Cooperative social play seems to thrive in nature for all of these reasons.

Chase:hit–bite:wrestle ratios are simplified measures of the multifactorial structural complexities of play, but they may well reflect individuals' willingness to cooperate in social play. In theory (Fagen 1984, 1986), animals more familiar with one another should use contact play relatively more often and noncontact play relatively less often, because their chances of cooperation are higher and chances for physical injury correspondingly smaller. (The relatively higher risks of wild chases in play above the ground lead to the prediction that active arboreal play should involve more noncontact play, especially chasing, when partners are familiar than when they have relatively little previous contact.)

Other factors also influence qualities of social play in nature. For example, Colvin and Tissier (1985) investigated relative amounts of three kinds of social play, differing in amount of physical contact, among 3-year-old male rhesus monkeys. High rates of hit-and-bite were associated with high rates of approach–withdrawal among sibling (mixed-age) dyads, whereas rates of these two play forms varied inversely for relationships between affiliative peers.

Juvenile primates appear adept at cooperatively solving the problem of playing with partners whose play interests differ from their own (Biben 1989). Failure to cooperate prevents individuals from playing well together and can ultimately lead to a decline in play, if cooperative pairings cannot be established. Why juvenile primates become less willing to cooperate in play, rather than more adept, with increasing age and experience is a puzzle that highlights central issues in the ontogeny of primate play.

How do food variations affect possible sex differences and other qualities of primate play, including cooperation? Zimmermann et al. (1975b) found that when food availability was experimentally reduced, play of infant and juvenile rhesus monkeys changed both in quantity and in quality. In the low-protein group, play solicitation resulted in aggressive responses. Studies of free-ranging common chimpanzees and savanna baboons (Pereira, personal communication) furnish near-perfect parallels to these results. Merlin, an infant male at Gombe Stream, was cared for by his sister Miff after his mother's death, but "became lethargic,"

"played less" (Goodall 1986, pp. 74, 101), and met other chimpanzees' play solicitations with aggression (Goodall 1971, p. 227). Play of low-protein subjects (Zimmermann et al. 1975b) was relatively low in reciprocal biting, mouthing, and chewing. Decreasing reciprocity with age also characterizes play of juvenile rhesus monkeys (Levy 1979). Play reductions due to low nutrition and age may share the associated characteristic of reduce reciprocity. In both cases, the proximate mechanism may be an initial decrease in available food energy, leading to later decreases in the probability of repeated play encounters.

To rhesus monkeys on low-protein diets, to high-risk chimpanzee and baboon infants, and perhaps even to children, play is in the eye of the perceiver. A given individual's behavior, labeled "play" by a human observer, may in fact be an ambiguous social initiation serving to elicit information about a conspecific's basic emotional disposition in relation to its social world. Of course, such dispositions are themselves the results of self-assessment. Pellegrini (1989) found from interviews that sociometrically popular and rejected elementary-school children had strikingly different views of rough-and-tumble play. Popular children viewed rough-and-tumble play as affiliative and "fun," whereas the rejected children saw rough-and-tumble play and aggression as interrelated and serving similar functions. Factor analysis of ethological observations of play in the two groups revealed behavioral differences as well. Rough-and-tumble play among the rejected children included "aggressive" behavior absent from the play among the popular children, including "hit-closed hand" and "hit-at," and play among rejected children was relatively likely to escalate to actual aggression.

Social play appears to have two very different faces. To the fortunate and fit, it brings further motor competence and social affiliation; to those lacking confidence or competence, it is a macabre dance through which existing social and physical asymmetries are reinforced. An existing probability distribution of health and nutritional level in a population of juveniles will be further shaped and propagated by this means. Where this prior distribution is roughly symmetric with a relatively narrow spread, benefits of play will be distributed very widely among individuals, further reinforcing symmetry and equality of future opportunity. But where the prior distribution is skewed, with a long right-hand tail, the benefits of play will be restricted to the small number of high-quality individuals in the population. Transforming effects of interpretive play are not, however, merely restricted to this simple dynamical dichotomy. Indeed, for particular individual juveniles at risk, play can be profoundly restorative and does not inevitably amplify preexisting differences in physical and social competence (Werner 1989, and see below).

Play embodies the full sophistication of primate social communication (Altmann 1988) and leaves its participants to work out their own affective response. As Bishop and Bishop (1978, p. 12) argue, primates "can profit from ambiguities in their actions." One key to understanding the role of play in development of primate social cognition may be to recognize that social stimuli, the intentions of other animals, and environments are often inherently ambiguous, so that the ability to actively and even creatively interpret these ambiguous features becomes a central issue in the evolution of intelligent behavior (Fagen 1992).

Individual creativity, interpretation, improvisation, and cultural transmission (Fagen 1986, 1992) are essentials of juvenile primate play. Playing primates create individual, novel play signals, movements, objects, and games (Goodall 1973, 1989a, p. 17; Eaton 1976; Candland et al. 1978 and refs. cited; Chevalier-Skolnikoff 1982; Fossey 1983, pp. 187, 232; Huffman 1984; Kummer & Goodall 1985, p. 209; de Waal 1986a). Talapoins (*Cercopithecus talapoin*) and bonobos (*Pan paniscus*), for example, invented games in which animals closed or covered their eyes before moving and interacting socially (Gautier-Hion 1971; de Waal 1986a). Novel behaviors transmitted in play can become culturally established (Goodall 1973), and especially creative players have characteristic behavioral profiles (Goodall 1989a). Individual play styles are almost surely transmitted in given families by cultural means (e.g., chimpanzee Flo dangling Flint from her foot in play in a characteristic and unique way, and Flo's daughter Fifi playfully dangling her own infants in the same way: Goodall 1990, p. 39, photo following p. 52). Consequences of play-driven cultural inheritance for behavioral evolution can be complex and nonintuitive (Fagen

1981, 1992). Cultural transmission in the presence of natural selection may be important for maintaining behavioral diversity in natural populations (Fagen 1981; Findlay et al. 1989a, 1989b).

GENERAL DISCUSSION AND CONCLUSIONS

Hypothesized benefits of social play include physical training and motor skill development, facilitation of communication and interaction leading to social bonding, and perhaps more general cognitive effects such as development of interpretive ability (Fagen 1981, 1992; Suomi 1982; Bekoff 1984). In savanna baboons, play is highly positively correlated with dietary variables (Table 13.1), as is fitness (Altmann 1991). Primatologists and, presumably, baboons could therefore use an individual's juvenile play prospectively to measure its lifetime reproductive success, with considerably greater ease than measurement of dietary variables would entail. It is not known whether baboons (or other primates) actually make use of this source of information about their companions' future reproductive success.

Body targets of play movements and the occurrence of tickling (at least in great apes) need to be adequately considered in any analysis of possible benefits. Pellis (1988) demonstrated for several rodent species that play-fighting targets (body parts a partner attempts to contact and control) are not, in general, those expected from aggressive practice but are drawn partly, or even largely, from amicable behavior including sexual behavior. Owens (1975) described play targets in juvenile savanna baboons. These targets may, in theory, differ in play-wrestling and in tickling. Detailed studies of play targets in a number of primate species will be needed to support further the hypothesis that aggressive practice is a major benefit of juvenile primate play. Tickling in juvenile primate play (Andrew 1963; Aldis 1975, pp. 26–27, 247–249; Terrace 1979; Fagen 1981, pp. 103, 107, 109, 325; Chevalier-Skolnikoff 1982; Becker 1984; Rijt-Plooij & Plooij 1987; Fouts 1989; Gardner & Gardner 1989; van Hooff 1989) is well known but seldom discussed (except by sign language-trained great apes). It is difficult to argue that tickling yields any functional benefits hypothesized for play. Why does tickling occur primarily, perhaps even exclusively, in great apes and humans, and indeed why should this phylogenetic oddity exist at all? Great apes certainly tickle each other in play, but whether tickling occurs regularly in other nonhuman primates is unknown (M. Pereira, personal communication). Nor is it known whether tickles, or tickling attempts, occur when members of ticklish (e.g., common chimpanzee) and putatively nonticklish species (e.g., chimpanzee–baboon play at Gombe Stream) play together.

Generalized functional capability, a monkey's "basic stance toward the environment, its characteristic approach to the world" (Mason 1979b) seems global and vague. But experiments with playful dogs as surrogate mothers for infant monkeys, versus mobile cloth mothers, suggest that dyadic responsiveness, and very possibly play itself, is important for normal social development. Dogs were much more effective than passive mechanical "mothers," even when the mechanical devices had moving parts or were programmed to move actively. The dogs treated the young monkeys as playmates and companions, rather than with indifference. Play captures many of the elements of companionship, important in primate and in social canid development, and was one major experiential difference between the monkeys raised by dogs and the monkeys raised by mechanical surrogates.

Longitudinal studies of human development on the Hawaiian island of Kauai (Werner & Smith 1977, 1982; Werner 1985, 1989) link individual playfulness to resilience under stress. Some especially playful children who were at risk because of medical problems or their home situation proved to thrive over the long term despite the odds. Whether play is a consequence of the same behavioral and social mechanisms that protect children at risk or, alternatively, is an actual cause of this sparing remains to be investigated systematically.

In reorganizing a young primate's world and expanding its boundaries (Simpson 1978; Altmann 1980), juvenility can frighten as well as attract. The fun of play can shape a young primate's developing interpretations of a challenging new social and physical world in positive directions. But if fear of such novelty, itself highly adaptive at times (Hinde 1987), overwhelms fun, as discussed earlier in a social con-

text, development may be slowed or diverted into different pathways. A difficult physical or cognitive skill that requires mastery of previously frightening situations can best develop in an initial context of play (e.g., the role of play in the early training of young dancers: Daniel 1978; Kuklin 1989). Play was an important part of Nim's sign-language training (Terrace 1979, p. 81). Clearly, learning sign language is more foreign to an infant chimpanzee than is learning dance to a young child, and lessons often became boring or frustrating. Play can facilitate learning of skilled actions when boredom, fear, frustration, or self-consciousness could have devastating effects on developing skill. And, following mastery, the improvisational richness of play can keep skills sharp and make performances compelling, at least if the following example from human culture is any indication.

Oxnard (1984, pp. 333–334) cites classical ballet as a pinnacle of human creativity, "far removed from . . . animal behaviours." Yet systematic inquiry into development of dancers, or for that matter into human motor skill development generally (e.g., Thelen 1989; Thelen & Fogel 1989), may yield important insights on primate behavior. On stage, skilled performing artists may improvise. They liken themselves to sparring partners (Suzanne Farrell and Jacques d'Amboise; Farrell & Bentley 1990, p. 154; Cheryl Yeager and Julio Bocca; Anon. 1991). The result: "A wild display of flamboyant yet elegant fun" (Farrell & Bentley 1990, p. 153). (If a better general definition of play than this exists, I have yet to find it.)

Why invest scientific effort in the study of play, when two dancer–writers can casually define this behavior with verve, elegance, and clarity? But this is exactly my point: that play interests so many different sorts of interesting people. The study of play joins worlds as separate, narrow, and cloistered as classical ballet and experimental science, links disparate minds, and thereby sows seeds of intellectual revolutions that may lie decades or even centuries in the future.

Students of play explore unusual issues, even traversing the borderland of territory that some might call spiritual. Access to these regions ought to confer a certain humility on primate researchers. It demands that they critically reexamine their goals as scientists and the relationship of these goals to the larger issue of their own humanity. I think that this challenge explains why so many scientists, and perhaps more than a few nonscientists, seem to feel threatened by play and the study of play. Clarity of thought about primate behavior in the broad context furnished by play research can illuminate human nature along perspectives already being pursued with great vigor by courageous innovators in the humanities (Paglia 1991) and in the performing arts (Kirkland & Lawrence 1987, 1990).

Studies of play in juvenile primates reveal a unique complex motor behavior with social, cognitive, expressive, interpretive, and improvisational components. Play bears eloquent witness to the behavioral sophistication and cognitive elaborateness of juvenile primates across the entire order, from common chimpanzees to ringtailed lemurs. Play is physically risky but persists in evolutionary time, for reasons that remain the subjects of energetic scholarly investigation. As a cultural and interpretive activity through which juvenile primates actively define both their moves (physical, social) and the spaces in which these moves occur, play potentially challenges conventional views of primate evolution and human nature, including ideas about gender and cooperation. Play is one reason that scientists and others find juvenile primates both so fascinating to watch and so richly productive of new scholarly ideas and empirical relationships.

ACKNOWLEDGMENTS

I thank Michael Pereira, Stuart Altmann, and Michael Potegal for references and unpublished material. Thanks also to Stuart Altmann, the editors and referees for their comments. Figure 13.2 appears with the kind permission of *Primates* and the cooperation of the author, Dr. Koyama. To collect literature for this review I used Primate Information Center bibliographies (University of Washington, Seattle).

14

Stability of Social Relationships in Female Wedge-Capped Capuchin Monkeys

TIMOTHY G. O'BRIEN and JOHN G. ROBINSON

It has been postulated, but rarely demonstrated, that social relationships developed among juvenile primates persist into adulthood (Sade 1972b; Bramblett et al. 1982; Hausfater et al. 1982; Fairbanks & McGuire 1986). Because juvenile primates must be incorporated into the social network of the adults, it is likely that assimilation proceeds via the development of social relationships among peers and with adults (Cheney 1978b; Walters 1987a). Development of social relationships among juveniles and between juveniles and adults has been well documented for a variety of cercopithecine primates (Cheney 1977, 1978b; Walters 1980; Horrocks & Hunte 1983a; Pereira 1988b), but few studies have documented the persistence of these relationships over time (but see Hausfater et al. 1982; Samuels et al. 1987; Fairbanks & McGuire 1986; Fairbanks, Chapter 15, this volume). No data are available for neotropical primates.

In this chapter, we describe patterns of social interaction between juvenile females in a group of capuchin monkeys (*Cebus olivaceus*), a neotropical, forest-dwelling primate, and describe patterns of social interaction 10 years later between the same individuals as adults. We first compare patterns of aggression, grooming, and spatial association to test the hypothesis that social relationships developed as juveniles persist into adulthood. We then present data that indicate that (1) rank relations among juvenile females are good predictors of the pattern of

aggression and grooming both as juveniles and later as adults, and (2) rank relations among juvenile females are predicted by the relative frequency that adults join coalitions with these juveniles.

DEMOGRAPHY, SOCIAL BEHAVIOR, AND SOCIAL RELATIONSHIPS

The development of female juvenile social relationships is influenced by three important sociodemographic traits: kinship, age, and social dominance. Relationships with kin versus non-kin, peers versus older females, and high- versus low-ranking females are associated with particular suites of social interaction that serve to define the relationships. Because social interactions may function in the acquisition of resources (often through the establishment of social bonds: McKenna 1982) and social partners may vary in quality as a function of age, rank, and relatedness, patterns of social interaction should reflect the relative desirability of partners (Seyfarth 1977).

Social relationships involving a juvenile and related females tend to be affiliative. Juvenile females maintain close grooming relationships with and spatial proximity to closely related females, irrespective of age (Cheney 1977, 1978b; Walters & Seyfarth 1987). Juvenile females handle sibling infants more than unrelated infants, in part due to tolerance of the mother toward her older offspring (Cheney 1978b; Altmann 1980; O'Brien 1990). Juvenile females play more with sibling infants and related peers, if available (Cheney 1978b; Lee 1983a; Walters 1987a). Related females are more likely to join a juvenile in coalitions against unrelated females and are less likely to be aggressive toward a closely related juvenile (Cheney 1977; Massey 1977; Walters 1980; Bernstein & Ehardt 1985a; Pereira 1988b).

Social interactions of juvenile females are affected by the relative ages of the participants. Juveniles females are unlikely to have closely related females as peers (Altmann & Altmann 1979). Given the interbirth interval and survival schedule characteristic of wedge-capped capuchin monkeys, a juvenile female has approximately a 27% chance of having a female juvenile sibling (T. G. O'Brien, unpublished data). Because similar-aged interactants are likely to be unrelated and have similar resource requirements and social objectives, we might expect juvenile social interactions to be competitive rather than affiliative. Cheney (1977, 1978b) found that juvenile female baboons fought among themselves more frequently than with juvenile males because they competed for access to infants. Cheney also found that patterns of alliance formation among juvenile baboons were consistent with a model that assumed social and resource acquisition benefits for participants. When playing with other juveniles, females preferred same-age peers of either sex rather than other juvenile females (Cheney 1978b).

Social relationships between female juveniles and adults may be structured by the relative benefits to each participant (Cheney 1978b; O'Brien 1990). Among unrelated individuals, juvenile females initiate affiliative interactions with adults more often than adults reciprocate (Cheney 1978b; O'Brien 1990). Patterns of aiding during aggression are more variable and involve an assessment of dominance ranks of interactants and risk of involvement (Walters 1980; Cheney 1983; Bernstein & Ehardt 1985a). The asymmetry in attractiveness as measured by affiliative actions received suggests that although juvenile female may be attracted to adults, the reverse may not be true. Among close kin, juveniles groom their mothers more than they receive grooming (Cheney 1978b; O'Brien 1990), but mothers aid juveniles during

aggression more than vice versa (Bernstein & Ehardt 1985a). Even within a matriline, however, daughters are not equal, and a mother may direct affiliative actions and agonistic support preferentially toward one daughter over another (Schulman & Chapais 1980; Bernstein & Ehardt 1985a).

Finally, social relationships can be affected by the relative dominance rank of interactants or their relatives (yellow baboons: Walters 1980; Pereira 1988b; Japanese macaque: Kawamura 1965; Kurland 1977; rhesus macaque: Kaufmann 1967; Sade 1972b; long-tailed macaque: Angst 1975; vervets: Cheney 1983; Horrocks & Hunte 1983a). Among cercopithecine primates, a pattern emerges that emphasizes the role of aggression and agonistic support from adult females in the development of social dominance of juvenile females (Berman 1980; Walters 1980; Berstein & Ehardt 1985a). Daughters usually assume a rank immediately below that of their mother (Hausfater 1975; Kurland 1977), and adult female interventions in the disputes of offspring appear to reinforce juvenile ranks that parallel the ranks of mothers (Sade 1972b; Cheney 1977; Walters 1980, 1987a). Agonistic social interactions involving juveniles, therefore, tend to be directed at lower ranking individuals and affiliative social interactions involve high-ranking females (Cheney 1977, 1978b, 1983; Walters 1980, 1987a; Pereira 1992).

THE WEDGE-CAPPED CAPUCHIN MONKEY

The wedge-capped capuchin monkey (*Cebus olivaceus*) (Fig. 14.1) is a medium-size (2.5–4.0 kg), omnivorous neotropical primate that occurs in northern South America from Colombia to Guyana. It inhabits a variety of habitats, but is most commonly found in dry deciduous and gallery forests of savannah habitats (llanos). Wedge-capped capuchin monkeys live in female-bonded groups in which male dispersal may occur as early as 2 years old (Robinson 1988a, 1988b). Males exhibit dominance hierarchies based on size, and females exhibit dominance hierarchies based on matrilineal inheritance (Robinson 1981). Groups are typically composed of adult females and their offspring, one breeding adult male, and several nonbreeding subadult and (occasionally) adult males.

Fig. 14.1. Juvenile female wedge-capped capuchin monkey feeding on *Cephalotes* ants in a twig.

Group size is bimodally distributed, with small and large groups averaging 11 and 26 individuals, respectively (Robinson 1988a). The annual ranges of groups may overlap entirely, but groups normally avoid one another.

Births occur throughout the year, but most births coincide with the onset of the rainy season (May–June) (Robinson 1986). Infant care is exclusively maternal during the first 2–3 months of life. Allomaternal care complements maternal care for an additional 6 months (O'Brien 1990). By the end of the first year, infants are relatively independent of their mothers but continue to suckle occasionally for another year. The juvenile stage lasts until the sixth year for

both sexes (Robinson 1988a). Mortality rates are highest for infants; 18.2 and 21.6% of female and male infants, respectively, die in the first year of life. By age 2, however, mortality rates have declined approximately by half and continue to decline throughout the juvenile stage (Robinson 1988a). Age at first reproduction is sex specific; females breed at 6 years old. No male has obtained a breeding position in a group before the age of 12; males, therefore, are not considered adults until 12 years of age.

The major social and ecological challenge to female wedge-capped capuchin monkeys is thought to be the reduction of competition for food (Srikosamatara 1987). Feeding competition is mediated at the intergroup level by avoiding contact with other groups and by priority of access to resources based on group size (Robinson 1986; Srikosamatara 1987). Within groups, access to resources is mitigated, in part, by a female dominance hierarchy and by avoidance of other group members (O'Brien 1990). Female reproductive success is affected primarily by group size; females in large groups have higher fecundity and higher expected lifetime reproductive success than females in small groups (Robinson & O'Brien, unpublished data). Within groups, there are no correlates of female reproductive success. Specifically, there is no significant correlation between dominance rank and reproductive success; similar results are reported for vervets (Cheney et al. 1988) and savannah baboons (Altmann et al. 1988). High-ranking females may, however, have an earlier age of first reproduction and a slightly shorter interbirth interval (Robinson 1988b). Lifetime reproductive success of high-ranking females, therefore, may be greater than that of low-ranking females.

The wedge-capped capuchin monkey is an excellent subject for comparison with cercopithicine primates. The genus *Cebus* shares many social and demographic features of paleotropical primates and has been compared socioecologically with the genus *Cercopithecus* (Robinson & Janson 1987). Although there have been many studies of social relationships in paleotropical primates, the majority of these have been on terrestrial cercopithecines, specifically macaques, baboons, and vervets. The ecology of forest-dwelling cercopithecines is very different, however, and we know little about their social relationships. The same is true for forest-dwelling colobines and almost all neotropical primates. Our study is one of a few longitudinal field studies of the demography and social behavior of neotropical primates. The comparison with previous research, therefore, is illustrative both from a terrestrial–arboreal perspective and from a paleotropical–neotropical perspective.

STUDY SITE AND METHODS

Study Site

Our study site is located on Fundo Pecuario Masaguaral, a 5000-ha cattle ranch that has been managed as a wildlife refuge since 1944.

The ranch is located at 8° 34' N, 67° 35' W in the state of Guarico, Venezuela, 50 km south of the town of Calabozo. The area is characterized by savannah plains, with gallery and dry deciduous forests bordering rivers and streams. The vegetation of the ranch has been described in detail by Troth (1979) and Robinson (1986). The climate is strongly seasonal, with a 4-month dry season (December–March), a 6-month wet season (May–October), and two transitional months (April and November). Annual average precipitation is 1450 mm. Temperature and humidity are high; dry-season temperatures range from 30° to 40°C, and average relative humidity exceeds 80% during the wet season (Robinson 1986).

Wedge-capped capuchin monkeys inhabit the dry deciduous and gallery forests in the eastern part of the ranch. Our study area is an approximately 500-ha forest bordered by cattle pasture to the north and south, savannah and pasture to the west, and the Caño Caracol (a stream that forms a natural boundary for capuchin groups' home ranges) to the east. A grid of trails allows access to all parts of the study area. The forest canopy is 15–35 m in height, and many tree species lose their leaves during the dry season. Tree species are patchily distributed and bear fruit seasonally (Robinson 1986). Thus fruit availability varies in both space and time.

Data Collection and Analysis

We compared patterns of association and social interaction for a single cohort of seven female wedge-capped capuchin monkeys during two time periods, 1977–1978 and 1986–1988. The subjects were six juvenile females and one nulliparous, first-year adult. The females were not closely related, with the exception of one presumed aunt–niece dyad (AM–WH). The females were observed as juveniles in 1977–1978 and as adults in 1986–1988. All were members of Main group, whose size increased steadily from 23 individuals in 1977 to 38 individuals in 1985. In 1985, Main group fissioned, and the new fission group (called Splinter group) moved 7 km to an area uninhabited by other capuchin groups. Two study females left Main group as part of Splinter group.

Data were collected for 14 months in 1977–1978 (hereafter called the 77 data set). Daily behavioral observations began between 0545 and 0615 hours and lasted for 12 hours. Observations consisted of a 20-minute scan sample (Altmann 1974; Struhsaker 1975) each half-hour, and *ad libitum* sampling of social interactions that were not recorded during systematic sampling (for details, see Robinson 1981, 1986). Data were collected for 15 months in 1986–1988 (hereafter called the 87 data set). A 20-minute scan sample was collected each hour in the same manner as the 77 data set. Additionally, a 20-minute focal sample of a female was conducted each hour. Social interactions not recorded during systematic sampling were recorded *ad libitum* (for details of sampling, see O'Brien 1990).

Caution is required when making comparisons that involve data collected under different sampling regimes. Unequal sampling effort may introduce biases when comparing within and between studies (Altmann & Altmann 1977). In the present study, sample sizes were large and uniform within and between studies. Preliminary analyses (*G*-tests of heterogeneity: Sokal & Rohlf 1981) of datasets indicated that (1) time budgets developed from 77 and 87 scan samples were not significantly different; (2) time budgets developed from the 87 scan and focal samples were not significantly different; and (3) distributions of social interaction in the 87 focal and opportunistic samples were not significantly different. We therefore considered the two data sets comparable.

We examined four types of social interactions:

1. Nearest neighbor association. The frequency that two females were in closer proximity to each other (less than 10 m) than to other group members.
2. Coalition formation. The frequency that females joined or were joined by other females during an agonistic interaction involving another group member.
3. Grooming sessions. The frequency that females engaged in grooming and the relative contribution by each female as determined by frequency of grooms during a session.
4. Aggression. The frequency that females were involved in supplantation or fights. Fight involved vocalizations, lunges, chases, pulling, or biting that elicited submissive response or retaliatory aggression by the target female.

We initially constructed sociograms of interactions to illustrate the relative magnitude of social interactions among females. We then calculated a similarity index (Dice 1945) for each dyad for each type of social interaction (except coalitions). The similarity index measured the frequency that a dyad engaged in a behavior divided by the sum of the frequencies that each individual of the dyad engaged in the behavior. For example, individuals AA and BB groom each other 30 times. AA grooms a total of 70 times with all members of the cohort, and BB grooms a total of 50 times. The similarity value is $30/(70 + 50) = 0.25$. Using similarity indices, we constructed single linkage cluster analyses (SLCA: Morgan et al. 1976) that group females according to the degree that they engage in a particular behavior. The dyad with the maximum similarity value forms the seed of a cluster, and other females are added to the cluster according to declining scores. The SCLAs illustrate the order of relationships between females in the 77 and 87 data. Finally, we used Spearman's rank correlation analysis to determine if patterns of social interactions in the two data sets were significantly correlated (Sokal & Rohlf 1981).

LONGITUDINAL CONSISTENCY OF SOCIAL RELATIONSHIPS

Of the juvenile females studied in 1977, seven were still alive in 1987 (Fig. 14.2). Of the seven, parentage was known for three. HI, the highest ranking female in the group in 1977, was the mother of WH. BE, a middle-ranking female, was the mother of RO; and BU, a low-ranking female, was the mother of CR.

We divided the seven juvenile females in the cohort of interest into young (2 years old), middle (3–4 years old), and old (5–6 years old) peer groups based on ages of individuals in 1977 (Fig. 14.2) and assigned dominance ranks based on the outcome of aggressive dyadic interactions at that time (see Robinson 1981).

Dominance Rank

Individuals were ranked by the relative frequency that they chased, threatened, and displaced one another (Robinson 1981). The dominance rank within the cohort of interest was generally consistent between 1977 and 1987 (Fig. 14.2).

PEER GROUP	MAIN 1977	1987 MAIN	SPLN

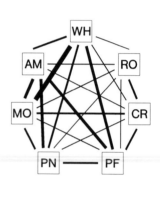

MIDDLE	WH ——————— WH	
OLD	AM ——————— AM	
OLD	MO ——————— MO	
OLD	PN	PN
MIDDLE	PF ⤬ CR	RO
YOUNG	CR ⤬ PF	
YOUNG	RO	

Fig. 14.2. Juvenile peer group classification and dominance rank position of female wedge-capped capuchin monkeys as juveniles and as adults. Females are ordered in descending rank.

WH, the highest ranking juvenile in 1977, was the highest ranking adult female in 1987. As adults, females AM and MO also retained relative rank positions occupied in 1977. In 1977, PF was dominant to CR, but as adults their relative rank had changed. This change was not unexpected. PF and CR exhibited a transitional relationship during the 1977 study. PF was orphaned as a 1 year old at the beginning of the 1977 season, and appeared to have difficulty maintaining her rank position. She was involved in aggression twice as often as expected based on a random distribution of aggressive interactions, and was often the target of aggression by other females. Females PN and RO left Main group as adults during a group fission in 1985, but retained their relative rank positions in their new group, known as Splinter group.

Aggressive Interactions

The pattern of aggressive interaction among individuals was predictable from the dominance rank and was generally consistent across years (Fig. 14.3). In 1977, aggressive interactions were concentrated among the old and middle juvenile peer groups. The two highest ranking juveniles, WH and AM, were each involved in 30% of aggressive interactions. Females MO, PN, and PF each were involved in, on average, 21.1% of interactions, and the lowest ranking and youngest juveniles were involved in, on average, 16.3% of interactions. Patterns of aggression in adults were similar to the juvenile pattern and significantly correlated ($r_s = 0.67, p = 0.036$). Females PN and RO continued to be

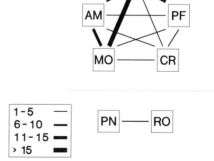

Fig. 14.3. Sociogram of aggressive interactions among juvenile (top) and adult (bottom) female wedge-capped capuchin monkeys. Females are ordered by descending rank counterclockwise. The split in the adult sociogram indicates the fission into two groups.

1977

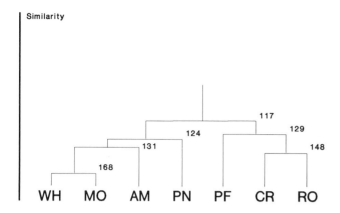

1987

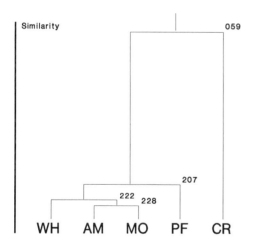

Fig. 14.4. Single linkage cluster analysis of similarity values for aggression in 1977–1978 and 1986–1988 for Main group. Females add to clusters in order of declining rate of interaction.

involved in few aggressive intercations after leaving Main group. An SLCA illustrates (Fig. 14.4) that the pattern of aggressive interactions in 1977 is associated with both rank and age: the cluster of WH, MO, AM, and PN is high- and middle-ranking, old and middle-age animals that are involved in many aggressive interactions; the cluster of PF, CR, and RO is young and middle to low-ranking animals that are involved in fewer interactions. In 1987, there was just one cluster of adults, a high-ranking, high-aggression group comprising WH and AM (frequently the aggressors) and MO and PF (the primary recipients of aggression). CR continues

to be involved in few aggressive interactions as an adult.

Nearest Neighbor Associations

The pattern of nearest neighbor associations among juveniles was based on age: animals in the same peer age group were more closely associated with one another. The SCLA (Fig. 14.5) indicates that WH, in the middle peer group, tended to associate with the young peer group of CR and RO. PF, the other animal in the middle group, tended to associate with MO and PN, two animals in the old group. AM, the oldest juve-

1977

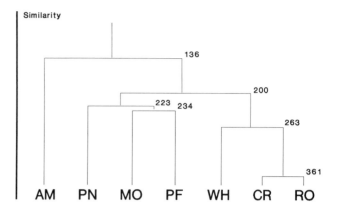

1987

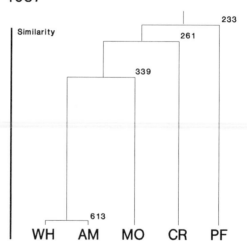

Fig. 14.5. Single linkage cluster analysis of nearest neighbor similarities. Females add to clusters in order of declining similarity values.

nile, was weakly associated with both clusters. Among adults it was based on relative rank: animals of adjacent rank were more consistently nearest neighbors. Accordingly, there was no significant correlation between similarity values of association for the two data sets (r_s = 0.22, p > 0.05). Patterns of association were not constant over time.

Females spent more time with one another as juveniles than they did as adults (Fig. 14.6). In 1977, females spent an average of 36.8% (SE = 3.8%) of their time with other juveniles and 17.7% (SE = 2.2%) alone (no neighbors within 10 m). In 1987, females spent an average of 19.0% (SE = 2.4%) of their time in the company of their juvenile peers and 36.7% (SE = 2.1%) alone. These differences are significant (Wilcoxon signed rank test, two-tailed: in association, p = 0.031; alone, p = 0.016) and are especially pronounced for lower ranking females and the high-ranking female WH. WH's association with other study females declined in the 87 data, as it did for lower ranking females, but without the concomitant increase

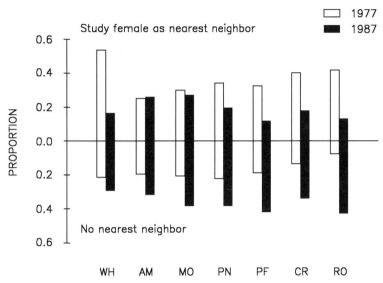

Fig. 14.6. Proportion of time that female spent as nearest neighbor of another study female and time spent without a nearest neighbor in 1977–1978 and 1986–1988.

in time spent alone. The difference was due to WH spending more time with the offspring of AM, the other high-ranking female (O'Brien 1990).

Grooming Relationships

The pattern of grooming among juvenile females in 1977 was related to dominance rank (Fig. 14.7). The four highest ranking females are involved in grooming sessions three times as often as other juveniles. Grooming sessions were usually asymmetric; dyads averaged 1.03 (SE = 0.13) grooms/grooming session, indicating that most grooming sessions did not involve reciprocal grooming. The higher ranking female groomed more frequently than the lower ranking female. Among dyads that engaged in at least three grooming sessions over the course of the study, the higher ranking juvenile groomed an average of 2.8 (SE = 0.46) times as often as she received grooming.

The pattern of grooming among the same individuals as adults in 1987 continues to be related to dominance rank. Grooming relationships remained concentrated among the higher ranking females in Main group, but specific rela-

tionships changed. WH and AM developed a strong grooming relationship that was approximately reciprocal (groom ratio = 1.5), whereas WH and MO lost the strong grooming relationship apparent in the 77 data. Grooming sessions were more reciprocal in the 87 data, averaging 2.5 (SE = 0.59) grooms/grooming session, but the higher ranking female continued to groom more than the low-ranking female [2.2 (SE = 0.24) grooms given per groom received for dyads with at least three grooming sessions]. An SLCA (Fig. 14.8) illustrates the changes in clustering of grooming partners; as adults, WH and AM formed a close relationship at the expense of MO. These three females formed the primary grooming cluster in both data sets. Relationships with other females changed little from the juvenile period to adulthood. Grooming similarity indices were marginally correlated between the two data sets (r_s = 0.61, p = 0.06). The frequencies of grooming interactions in 1977 were correlated with nearest neighbor associations in 1987 (r_s = 0.80, $p < 0.01$) and with grooming frequencies in 1987 (r_s = 0.72, p = 0.02). This suggests that once grooming relationships were established, they persisted and were expressed in other measures of association.

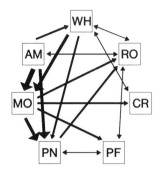

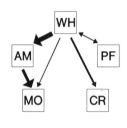

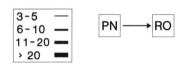

Fig. 14.7. Sociogram of grooming relationships among juvenile (top) and adult (bottom) female wedge-capped capuchin monkeys. Females are ordered as in Figure 14.3. The arrows indicate the dominant direction of grooming.

Acquisition and Maintenance of Dominance Rank

The results indicate that the dominance rank of juvenile females is a good predictor of patterns of aggression and grooming both as juveniles and later as adults. Accordingly, it is important to explore the determinants of this dominance rank.

The dominant adult female of Main group (HI) joined coalitions with juveniles more than any other adult female (85 of 101 recorded supports; Fig. 14.9a). HI supported juveniles 1.5 times as often as juveniles supported her, and her support was directed preferentially toward related, high-ranking juveniles, WH and AM (36 of 51 supports). The dominant female participated against juveniles in 21% of coalitions.

Other mothers (BU and BE) rarely joined daughters in coalitions. Adult females joined coalitions against juveniles only occasionally (19 of 101; Fig. 14.9b); most coalitions were against subadult males. Juveniles joined coalitions against juveniles 28 times (Fig. 14.9c); the high-ranking juvenile female WH was involved in 61% of these coalitions. Juveniles usually joined in coalitions against lower ranking juveniles with one exception, and coalitions were directed toward older juveniles more often than the youngest peer group.

DISCUSSION

Our study demonstrates that social relationships formed as juveniles persist into adulthood. Dominance rank is consistent across years. The patterns of aggression and grooming that define the juvenile relationships also define the adult relationships. The patterns of both aggression and grooming are related to the dominance rank of individuals. The patterns of spatial associations differ between adults and juveniles. They are primarily determined by age in juveniles and by dominance rank in adults.

The importance of dominance rank as a predictor of both juvenile and adult social relationships suggested that we examine how juvenile rank is acquired and maintained. Adult females intervened in aggressive interactions involving juveniles, but interventions by females other than the dominant females were rare.

The dominant female, furthermore, preferentially joined coalitions with high-ranking juveniles. The pattern of coalition support is similar in quality to that reported for paleotropic primates (Bernstein & Ehardt 1985a; Walters 1987a); some adults intervene on behalf of juveniles, and this intervention may aid in the development of rank. That the primary beneficiaries of coalitions with the dominant female were her sister and daughter suggests the importance of adult kin in the development of juvenile dominance rank (Kurland 1977; Massey 1977; Walters 1980).

Why are adult females other than the dominant female not more important in coalitionary support of their juvenile daughters? A likely explanation concerns the spatial distribution of a capuchin group. Robinson (1981) illustrates that independent juveniles cluster centrally as a ju-

1977

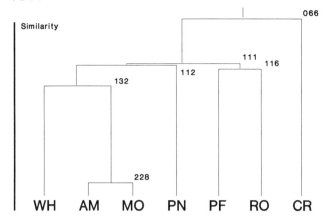

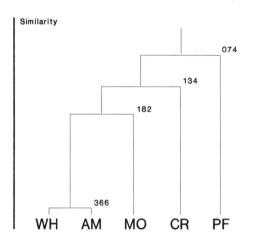

Fig. 14.8. Single linkage cluster analysis of similarity values for grooming in 1977–1978 and 1986–1988 for Main group. Females add to clusters in order of declining rate of interaction.

venile group during daily activities rather than in kin groups with mothers. Because the dominant female occupies a central position and other adult females are peripheral, the dominant female frequently is often the only female nearby when a juvenile is in need of coalitionary support. Additionally, if juveniles are in conflict with another juvenile, a lower ranking juvenile usually does not receive support because capuchins tend to support high-ranking rather than low-ranking females during female–female aggression (Robinson 1981). As has been reported for other primates, coalitions rarely form when

low-ranking females challenge higher ranking females (de Waal 1977; Bernstein & Ehardt 1985a).

Juveniles females are buffered from aggression by subadult males through coalition support and indifference of males toward pre-reproductive females (O'Brien 1991), but they must cope with female–female aggression. Frequency of aggression with other group females (including adult) was high for juveniles in the 77 data (Robinson 1981), especially for WH, AM, MO, and PF. Ratio of wins to losses declined with rank among juveniles, but declined pre-

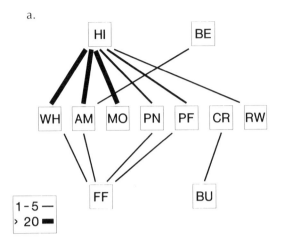

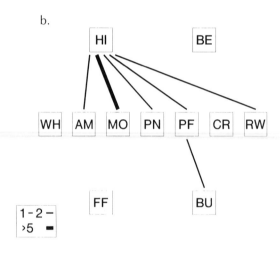

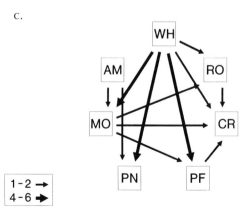

Fig. 14.9. Sociogram of coalitions among female wedge-capped capuchin monkeys: (a) adults and juveniles in coalitions. Adults are in top and bottom row; (b) adults in coalition against juveniles; (c) juveniles in coalition against juveniles. Arrows indicate direction of aggression.

cipitously for females ranking below MO (win/loss = 9.6, 3.4, 1.1, 0.31, 0.25, 0.22, 0.21). The general correspondence between ability to win in agonistic encounters with the decline in frequency of involvement in aggression indicates that, as juveniles, females are beginning to avoid others that may be aggressive toward them.

Close affiliative relationships develop among high-ranking juveniles, but these relationships generally exclude low-ranking females. Although higher ranking juveniles frequently are aggressive to one another, they are also affiliative—grooming and supporting one another during aggression. The lack of coalitionary support for low-ranking females and low participation in grooming sessions, coupled with the high rate of aggression, result in the peripheralization of these females and their decreased involvement from the social network of the group. As juveniles, the low-ranking females exhibit grooming patterns that continue into adulthood (O'Brien 1990); they are poor grooming partners that tend to not initiate grooming, to not reciprocate grooming, and to terminate the grooming session. The low frequency of involvement in grooming and the degree of asymmetry in grooming ratios suggest that low-ranking juveniles do not view higher ranking juveniles as accessible partners (Seyfarth 1977). Juveniles may learn early in life that it is risky or unrewarding to approach high-ranking females. As adults, high-ranking animals approach, groom, and support one another preferentially (O'Brien 1990); low-ranking females are excluded.

In addition to dominance, what are other determinants of social relationships? Kinship plays a role in the development of social relationships among juveniles. The females that were related (WH and AM) displayed low levels of aggression toward each other, and both benefited from coalitionary support of the dominant female. These females, however, did not exhibit the close affiliative relationship that they shared as adults. The benefit of kinship appeared to be a reduction in aggression rather than increased affiliation. This is consistent with results of O'Brien (1990) in which juvenile relationships with siblings and mothers were more neutral than affiliative. Of 13 dyads involving a 0.5 degree of relatedness, 4 failed to form a grooming relationship and only 3 formed close grooming relationships.

Given the long interbirth intervals between surviving offspring (over 24 months), the chances of two juvenile females being siblings are low. Consequently, kinship effects on juvenile relationships are more likely exerted through the support (or lack of support) that a juvenile receives from adult relatives and are reflected by the juvenile assuming a rank commensurate with its matriline. If juveniles are related, they tend to be adjacent in rank. The relationship of siblings or more distant kin, therefore, may be indistinguishable from a relationship between unrelated females of adjacent rank except for a reduction in aggression.

The normal pattern of social development therefore appears to reflect the dominance hierarchy that develops within a group. Deviations from this pattern may arise, however, as a result of development of affiliative or tolerant relationships among females of disparate rank. High-ranking WH developed a close allomaternal relationship with low-ranking CR that involved a high degree of carrying and babysitting relative to other juveniles in the group (Robinson, unpublished data). That this affiliative relationship continued into the juvenile stage is evident from the low rate of aggression that CR received compared with other juveniles. As an adult, CR ranked second as a grooming partner for WH and was least frequently a victim of aggression from WH. Although anecdotal, this example suggests a mechanism whereby females may develop preferences for low-ranking infants that translate into the development of an affiliative or a tolerant juvenile relationship that persists into adulthood.

Our results indicate that female wedge-capped capuchin monkeys share some features of social development with paleotropical cercopithecine primates. As reported for cercopithecines, dominance rank appears to be the primary determinant of the structure of social relationships among juveniles. Dominance rank among juveniles and patterns of social interaction appear to be stable through time for capuchin monkeys, except as modified by demographic events. Although the data are not adequate to address the effect of relatedness among juveniles, anecdotal data and demographic constraints indicate that kinship between juveniles may not be an important factor in the development of relationships. Presence of adult kin may benefit high-ranking juveniles, but agonistic support from adults does not appear important in the development of social relationships among most juveniles. More data are required to determine the exact role of kinship in the acquisition of dominance rank and the way in which the availability of sibling juveniles might affect the role of dominance in the development of juvenile social relationships.

ACKNOWLEDGMENTS

Over the years of this study many people have contributed in ways too numerous to record. We would especially like to thank Thomas Blohm for his support and permission to work on his ranch; John Eisenberg, who gave us the opportunity to begin this study; Rudy Rudran, who has supported this study during the lean times; and a large number of colleagues and students who carried out the demographic surveys. Improvements in the manuscript resulted from reviews by Robert Seyfarth and Margaret Kinnaird. Financial support was provided by National Geographic Grants 2152-80 and 3584-87, funds provided by the Friends of the National Zoo, and NSF Grants BSR 830035 and BNS 8718377.

15

Juvenile Vervet Monkeys: Establishing Relationships and Practicing Skills for the Future

LYNN A. FAIRBANKS

To an evolutionary biologist, social relationships are viewed as tactical arrangements between individuals, each seeking to maximize the benefit–cost ratio of cooperative acts over time. Juvenile primates grow up in a social context and form specific types of social relationships with the individuals around them. To make effective partner choices, a juvenile must make judgments about the relative value of different individuals as social partners for specific activities (Kummer 1978).

It is my intent in this chapter to demonstrate that one of the key objectives of behavior during the juvenile period is to form beneficial social relationships that are likely to influence an individual's competitive ability as an adult. Data from a longitudinal study of captive vervet monkeys will provide evidence that juveniles discriminate among adult group members based on their expected value as alliance partners, and that they are motivated to form long-term relationships with high-value partners that, particularly for females, are likely to persist into adulthood.

The chapter will contrast the selection of adult proximity partners with the choice of partners for play and for allomaternal caregiving. It will argue that partners for play and for allomaternal care are selected for their appropriateness for the activity in question: juveniles chose play partners that are comparable in size and strength and motivation to play, and juvenile allomothers chose infants based on their age and accessibility. Although play and allomaternal care may allow juveniles to practice behaviors that have later consequences, the relationships formed with play partners, and between allomaternal caregivers and infants, do not persist as social bonds into adulthood.

SELECTING PRESENT AND FUTURE ALLIANCE PARTNERS

An alliance partner is one that assists in agonistic conflicts with other group members. If we use Kummer's (1978) scheme of evaluating partners based on their qualities, tendencies, and availability, we would expect that a juvenile would prefer an alliance partner with the qualities of physical strength and social power, the tendency to respond and come to the assistance of the juvenile when needed, and the likelihood to be available both now and in the future.

The quality most often associated with effectiveness as an alliance partner is dominance rank. Seyfarth (1977) recognized the fact that females in several cercopithecine societies tended to be attracted to higher ranking females and argued that high-ranking females are more attractive because they are more effective allies in intragroup competition (Walters 1980; Cheney 1983; Hunte & Horrocks 1987). Competition for access to and fear of aggression from high-ranking females may influence the outcome of female relationships in different social systems, but when given the opportunity female cercopithecines do appear to prefer to associate with higher ranking than with lower ranking partners (Cheney 1978b; Fairbanks 1980; Seyfarth 1980; Silk 1982; de Waal & Lutrell 1986; Zucker 1987; Harcourt 1989; Cheney & Seyfarth 1990).

The attribute most often associated with the tendency to provide assistance in agonistic encounters is kinship. Ever since the pioneering work of the early Japanese primatologists in discovering the importance of matrilineal kinship in rank inheritance, numerous studies of cercopithecines have verified that individuals are more likely to protect, defend, and behave altruistically toward kin than toward nonkin (Cheney 1977; Kaplan 1977; Kurland 1977; Massey 1977; Walters 1980; Bernstein & Ehardt 1985a; Hunte & Horrocks 1987; Chapais 1988a; Chapais & Gauthier, Chapter 17, this volume).

All else being equal, a juvenile should invest more energy in long-term social relationships with individuals that are likely to be available in the future. In matrifocal societies with male emigration, availability is influenced by age and sex. Juvenile females will continue to live and associate with the same adult females, but their relationships with males will be comparatively short-lived. Virtually all juvenile males will eventually emigrate from their natal troop. The limited data that exist following males after emigration indicate that males are likely to move with natal male kin into groups that have already received other males from their natal group (Meikle & Vessey 1981; Cheney & Seyfarth 1983; Crockett & Pope, Chapter 8, this volume; Rajpurohit & Sommer, Chapter 7, this volume). As a result, juvenile males will be more likely to encounter natal group males later in life than natal group females. To the extent that future availability is a factor in partner value, juvenile females should invest more energy than juvenile males in establishing relationships with adults in the natal group, and juvenile females should prefer female and juvenile males should prefer male partners.

SELECTING PLAY PARTNERS

Juveniles should attempt to form different types of relationships to serve different functions. Characteristics that make an individual a good alliance partner would be expected to be different from those that would make an individual a good play partner (Fagen 1981; Janus 1989).

There has been considerable speculation about the benefits of juvenile play, but most agree that two of the most likely functions of social play for primates are to promote physical fitness and to develop fighting skills (Smith 1978b; Symons 1978; Fagen 1981; Chalmers 1984; see also Nash, Chapter 9, Watts & Pusey, Chapter 11, and Fagen, Chapter 13, this volume). An ideal partner for play would therefore be someone similar in size, strength, and motivation to play (Owens 1975; Fagen 1981; Pereira & Altmann 1985). Individuals of the same age and sex would be most likely to satisfy this decision rule.

It has also been suggested that in addition to the functions mentioned above, play serves to facilitate the development of affiliative relationships between individuals (Poirier & Smith 1974; Cheney 1981a; Lee 1983a; Nakamichi 1989). Juvenile vervet monkeys that play together also tend to groom and sit together (Lee 1983a). Nakamichi (1989) claims that the close associations formed during play by juvenile male Japanese macaques are continued over time as foraging and proximity relationships.

SELECTING INFANTS FOR ALLOMATERNAL CARE

As with play, there has been considerable discussion about the benefits of allomaternal care to the participants involved (Hrdy 1976). The differences of opinions that have arisen on this topic can be partially attributed to the diversity in the expression in this behavior across species (e.g., McKenna 1979; Cleveland & Snowdon 1984). Infant caretaking by juvenile females is a commonly observed activity among vervet monkeys, and most investigators agree that in this species, infant caretaking is likely to help juvenile females practice their mothering skills before the birth of their own first infant (Lancaster 1971; Struhsaker 1971; Fairbanks 1990).

If a juvenile female is motivated to hold and carry infants as practice for mothering her own first infant, then she should be attracted to younger rather than older infants (Hrdy 1976; Scollay & deBold 1980), and she should select infants that are available for caretaking—those whose mothers are most likely to permit the juvenile to carry off the infant without resistance. There is evidence that both of these factors influence the attractiveness of infants for allomaternal care in vervet monkeys. Younger infants are more attractive to caretakers than older infants (Lancaster 1971; Struhsaker 1971; Johnson et al. 1980; Fairbanks 1990). Accessibility of infants is influenced by kinship and by dominance rank; for example, mothers are more likely to release their infants to older sisters, and high-ranking mothers are less protective and more willing to allow caretakers to hold and carry their infants than are low-ranking mothers (Johnson et al. 1980; Lee 1989; Fairbanks 1990).

Several investigators have argued that close bonds are formed between allomothers and infants that persist over time and may facilitate adoption of the infant if the mother dies (Hrdy 1976; Quiatt 1979). In most of the case histories that describe persistent relationships of infants with their caregivers, however, the caregivers were adults, not juveniles (e.g., Gouzoules 1984; Chism 1986), and other studies have noted that prior caretaking does not predict who will adopt an orphaned infant (Berman 1982; Dolhinow & Krusko 1984).

PARTNER PREFERENCES OF JUVENILE VERVET MONKEYS

This chapter will present data on the social behavior of captive vervet monkeys to address the following questions:

1. Do juveniles seek proximity with adults who would be expected to have high value as allies, based on rank, kinship and sex?
2. Do the proximity relationships with adults that are established at 2 years of age persist into adulthood?
3. How do the differing needs and opportunities of juvenile males and females influence their partner choices?

4. Do the attributes that govern selection of adult proximity partners differ from those that influence the selection of play partners and partners for allomaternal care?
5. Do the social relationships that juveniles form with frequent play partners, or with the infants they caretake most often, persist into adulthood?

It should be noted that this chapter is about attributes of individuals that influence attraction, not about agonistic alliances per se. The first hypothesis being tested is whether juveniles are attracted to adults that would be expected to be high-value alliance partners, based on their qualities and tendencies. The degree to which the adults actually provide support to the juveniles is a separate question and is not being directly assessed here.

THE VERVET COLONY AND RESEARCH METHODS

The Vervet Colony

In 1975 a colony of vervet monkeys captured from St. Kitts, West Indies, was established at the Nonhuman Primate Research Facility on the grounds of the Sepulveda Veterans Administration Medical Center, Sepulveda, California. The north section of the facility has been reserved for longitudinal studies of social behavior and development (Fairbanks 1988a, 1988b, 1989; Fairbanks & McGuire 1988).

Vervet monkeys are members of the subfamily Cercopithecinae and are often grouped with macaques and baboons in discussions of social organization. Like other cercopithecines, vervet females generally remain in their natal group for life (Horrocks 1986; Cheney et al. 1988). Females can be ranked in a dominance hierarchy based on the outcome of agonistic interactions. Members of the same matriline tend to associate together and typically share the same rank relative to other matrilines (Horrocks & Hunte 1983a). Males emigrate from the natal troop at puberty. They are likely to leave with male kin and to enter groups that have already received other males from the same group (Cheney & Seyfarth 1983).

Social groups in the vervet colony have been managed with the object of maintaining natural group composition as much as is possible in captivity. Females are kept in their natal group; males are removed at puberty, and immigrant adult males are replaced at 3–5 year intervals. To accommodate the population growth, the original group was split into two in 1977, and into four in 1986.

Subjects

The focal subjects for the analyses described here are 56 2-year-old juvenile vervet monkeys, born in the colony from 1982 through 1986 (Fig. 15.1). At 2 years of age, individuals are well beyond nutritional dependency on the mother, but they are still immature physically and socially. Females are capable of breeding at 3 but do not reach adult size until 4 years of age (Fairbanks & McGuire 1984). Males develop their full adult size, musculature, and dentition by the age of 5.

The 2 year olds lived in groups that contained an average of 6.8 adult females (age 4+), 2.7 adult males (age 4+), 9.8 other juveniles (age 1–3), and 4.8 infants.

Data Collection and Analysis

The behavior of the 56 2-year-old subjects was observed for an average of 166 5-minute focal animal samples and 498 independent scan samples each, spaced evenly throughout the year. (For details of the data collection and sampling system, see Fairbanks & McGuire 1985; Fairbanks 1990.)

The attraction of juveniles to adult group members was evaluated by measuring the frequency per hour of approaches observed during focal animal sampling. An approach was recorded whenever the focal juvenile moved from beyond 1 m to within 1 m of the recipient and sat down. This proximity-seeking behavior was used as an indicator of attraction on the assumption that a more attractive individual would be approached more often than a less attractive individual. The definition was intended to exclude approaches that were either accidental or hostile.

Play partners were determined by measuring the frequency per hour that play was initiated or received by the focal juveniles with each other group member. Play was coded as wrestle play

Fig. 15.1. Two-year-old male vervet monkey.

or chase play, and each was recorded once per bout of continuous activity. Wrestle and chase play have been combined for this analysis.

Allomaternal caretaking was coded when the juvenile subject touched, held, carried, inspected, sniffed, or groomed an infant less than 6 months of age during focal animal sampling. Caretaking was coded only once per bout of continuous activity, even if more than one of the above behaviors were observed.

Spatial relationships were derived from scan samples. Proximity refers to the percentage of scan samples that individuals were observed within 1 m of one another.

Dominance rank was determined separately for adult males and adult females. Female rank was categorized as high, middle, or low based on the outcome of agonistic behavior among females in focal animal and target behavior samples. In each of the captive groups, one male was clearly differentiable from all other males as the highest ranking, based on the outcome of agonistic behavior among males. The remaining males were designated as low ranking.

Kinship was determined by matrilineal relationships. During the data collection period reported here, only three of the 2-year-old subjects had a male that was likely to be their father living in the group, and four others had a male that was a possible, but unlikely, father based on rank and behavior. Twelve other 2 year olds were conceived before, but born after the introduction of the current immigrant males.

The unit of analysis for this report is the dyad. Each dyad was characterized according to the sex of the 2-year-old focal subject and the age, sex, rank, and matrilineal kin relationship of the interacting subject. For adult males, whether he was natal or immigrant and his tenure in the group were also included. A male was considered to be "new" in the year he was introduced, and "resident" in subsequent years. Crossing each of the 56 2-year-old juveniles with each other group member produced 1362 dyads.

Effects of the independent variables on the frequency per hour of approach, play, and allomaternal caretaking were tested by analysis of variance, and post hoc t tests were used to differentiate which subgroups differed from one another. Technically, dyads were not independent because subjects and partners appeared in more than one dyad, but the patterns reported here all appeared to be robust and consistent across individuals and across groups.

ESTABLISHING RELATIONSHIPS WITH ADULTS

Do Juvenile Females Attempt to Associate with Adults Based on Their Expected Value as Alliance Partners?

If the value of an adult as an alliance partner is influenced by expectation of availability in the future, then 2-year-old females should be more attracted to adult females than to adult males. In the vervet colony, 2-year-old females approached adult females twice as often as they approached adult males (adult females: 2.32/hour; adult males: 1.14/hour; $t = 10.51$, $p < 0.001$).

Value as an alliance partner would be expected to be proportional to both rank and kinship. As illustrated in Figure 15.2, the attractiveness of adult females was strongly influenced by both of these variables (rank: $F = 41.68$, $p < 0.001$; kin: $F = 21.09$, $p < 0.001$). Juvenile females approached their mothers more often than their adult sisters, their adult sisters more often than more distantly related kin, and more distantly related kin more often than nonkin adult females (p's < 0.01). They also approached high-ranking adult females more often than middle-ranking adult females, and middle-ranking more often than low-ranking adult females (p's < 0.01).

The importance of rank and kinship was approximately equal and additive in contributing to the attractiveness of adult females. The highest approach rates by juvenile females were toward high-ranking mothers and sisters, females that combined the attributes of both high rank and close kin relationships. Juvenile females from middle- and low-ranking matrilines approached the highest ranking adult females as often as they approached their own mothers and adult sisters. The least attractive adult females were middle- and low-ranking nonkin that lacked both the power of high rank and the tendency to be helpful associated with kinship (p's < 0.01).

Juvenile females were less attracted to adult males overall, but within the adult male class, they preferentially associated with the highest

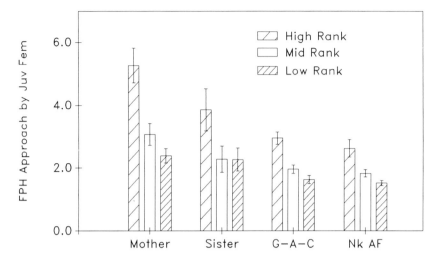

Fig. 15.2. Mean frequency per hour that 2-year-old females approached adult females, by matrilineal kinship and adult female rank. G-A-C, grandmother, aunt, and cousin; Nk AF, nonkin adult female.

ranking male (Fig. 15.3). Juvenile females initiated few friendly approaches to new males and were not attracted to the low-ranking resident males (tenure: $F = 7.25$, $p < 0.01$; rank: $F = 2.04$, ns; tenure $\times$ rank: $F = 4.75$, $p = 0.03$). The rate of approaches to high-ranking resident males exceeded the rate of approaches to all other categories of natal or immigrant adult males (p's < 0.01). Juvenile females did not discriminate their brothers or cousins from non-kin natal males in their rate of approach (kin: $F = 0.21$, ns).

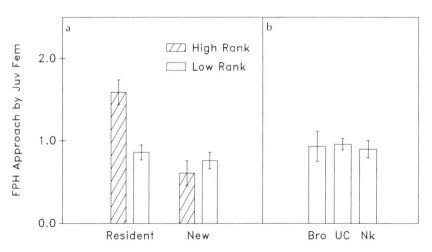

Fig. 15.3. Mean frequency per hour that 2-year-old females approached adult males (a) by male rank and tenure for immigrant adult males, and (b) by matrilineal kinship for natal adult males. Bro, brother; U-C, uncle, cousin; Nk, nonkin.

Table 15.1. Relative Contribution to Proximity between Juveniles and Adults

Proximity at age 2	Approach by juvenile (r)	Approach by adult (r)	Test for difference between correlations (p)
Juvenile females			
With all adults	0.85**	0.56**	<0.01
With adult females	0.82**	0.44**	<0.01
With adult males	0.76**	0.64**	0.05
Juvenile males			
With all adults	0.61**	0.40**	<0.01
With adult females	0.58**	0.31**	<0.01
With adult males	0.65**	0.56**	ns

Note: Correlations of frequency per hour of approaches by 2-year-old juveniles and by adults with the percentage time the dyad spent in proximity during the same year.
**$p < 0.01$.

The relative contribution of the juvenile females in initiating and maintaining relationships with their adult partners was evaluated by comparing the correlation of the rate of approaches initiated by the juvenile with the rate of approaches initiated by the adult in predicting the amount of time the dyad spent in proximity (Hinde 1974). This method used individual differences within age classes to predict proximity, and therefore was not biased by the differences between classes in the average rate of approaches.

Relationships of juvenile females with adults were determined more by the behavior of the juveniles than by the behavior of the adults. Adults in the colony were generally tolerant of approaches by juvenile females, and, as a result, the rate of approach by the juvenile female was a very good predictor of the amount of time that a dyad spent together (Table 15.1). Approaches by the adults also contributed significantly to the amount of time the dyads spent together, but the rate of approaches by the juvenile females was a significantly better predictor of dyadic proximity than the rate of approaches by the adults. This was particularly true for relationships between juvenile females and adult females. The relative contribution of juveniles and adults was more similar in juvenile female–adult male relationships.

Do Juvenile Males Associate with Adults Based on Their Expected Value as Alliance Partners?

If juvenile males are more attracted to adults that are more likely to be available to them in the future, they should approach adult males more often than adult females. In the colony, 2-year-old males approached adult males significantly more often than they approached adult females (adult males: 1.34/hour; adult females: 1.16/hour; $t = 2.33$, $p = 0.02$).

Among adult female partners, juvenile males preferred close kin that would be expected to be more likely to tolerate and assist them (kinship: $F = 21.14$, $p < 0.001$) (Fig. 15.4). Males approached their mothers more often than any other adult females, and their adult sisters more often than more distantly related adult females (p's < 0.01). Unlike juvenile females, 2-year-old males did not discriminate among adult females based on their dominance rank ($F = 0.62$, ns) (Fig. 15.4). Juvenile males did not preferentially approach high- or low-ranking adult females.

Relationships of juveniles with adult males were influenced by kinship, rank, and the likelihood of future availability (Fig. 15.5). Analysis of approaches in the colony indicated that 2-year-old males approached natal adult males significantly more often than they approached immigrant adult males ($t = 3.18$, $p < 0.01$). Among natal males, there was a strong preference for an older brother over more distantly related natal males (kinship: $F = 9.19$, $p < 0.01$). Among the immigrant males, rank was an important variable influencing attraction. Juveniles were almost twice as likely to approach high-ranking immigrant adult males as they were to approach low-ranking immigrants ($t = 3.18$, $p < 0.01$).

Juvenile males, like their female counterparts, played a large part in determining their own social associates. The amount of time that juvenile males spent near adult partners was closely related to the rate of approaches by the juveniles (Table 15.1). As with the juvenile females, the relationships that juvenile males formed with adult males were more reciprocal than their relationships with adult females. The contribution of both juveniles and adults to maintaining juvenile male–adult male spatial relationships was approximately equal.

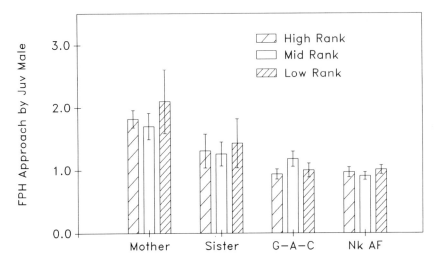

Fig. 15.4. Mean frequency per hour that 2-year-old males approached adult females, by matrilineal kinship and adult female rank. Abbreviations as in Figure 15.2.

Do the Social Relationships Established with Adults Persist over Time?

To determine whether the relationships that juveniles formed with adults as 2 year olds were maintained as they developed into adulthood, the juvenile focal subjects were followed for the next 2 years. The correlations were then computed between the rate that a juvenile approached a given adult at 2 years of age, and how often they were observed in close proximity to that same adult at 3 and 4 years of age. Juveniles born in 1986 were not old enough at the end of data collection in January 1990 to be included in the analysis at 4 years of age. The cohorts of juveniles born in 1982 and 1983 were also ex-

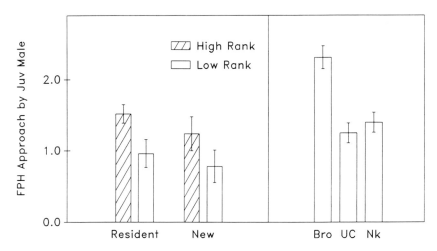

Fig. 15.5. Mean frequency per hour that 2-year-old males approached adult males, by residence, rank, and matrilineal kinship. Abbreviations as in Figure 15.3.

Table 15.2. Stability of Juvenile–Adult Spatial Relationships

Approach at age 2	Proximity	
	Age 3	Age 4
Juvenile female		
To adult female	0.68**	0.51**
Kin	0.68**	0.49**
Nonkin	0.55**	0.54**
To adult male	0.50**	0.86**
Juvenile male		
To adult female	0.54**	0.37**
Kin	0.54**	0.49**
Nonkin	0.28*	0.09
To adult male	0.31	0.34

Note: Correlations of frequency per hour of approaches by 2-year-old juveniles with the percentage time the dyad spent in proximity at 3 and 4 years of age.
**$p < 0.01$.
*$p < 0.05$.

cluded because their groups were fissioned in 1986, thus drastically changing the group composition and availability of adult partners. The remaining juveniles lived in relatively stable groups, with the only changes in adult composition occurring as a result of the maturation of immatures and occasional natural deaths.

During the 2-year period from 2 to 4 years of age, females maintained consistent relationships with older females. The rate at which a juvenile female approached a particular adult female at 2 years of age predicted the amount of time she spent in close proximity to the same adult female at 3 and 4 years of age (Table 15.2). This consistency could not simply be explained by the ongoing tendency for attraction to be based on kinship and rank. When the sample was divided into adult female kin and nonkin, the consistency of the relationships held equally for both groups (Table 15.2). When the effects of partner rank were partialled out, the rate of approaches at 2 years of age still predicted proximity at 3 and 4 years of age (age 3: $r = 0.62$, $p < 0.01$; age 4: $r = 0.51$, $p < 0.01$).

The relationships that juvenile males formed with adult females would necessarily be broken when they left the natal group, but until that time, they would have the opportunity to maintain consistent associations with preferred adult female partners. The amount of time that a 4-year-old natal male spent near individual adult females from his group could be predicted from his approaches 2 years earlier (Table 15.2), but a

breakdown of the dyads by kinship revealed that most of the consistency was due to the maintenance of attraction to close adult female kin, primarily the mother. There was no consistency in the males' relationships with nonkin adult females.

In the captive colony, as in the field, adult male tenure was relatively short, and most of the adult males that were present when the focal subjects were juveniles were no longer living in the group 2 years later. Within these constraints, the females showed a strong consistency in their preferences among the few immigrant adult males that were still available 2 years later (age 4: $n = 14$) (Table 15.2).

It is much more difficult to evaluate consistency in relationships formed by juvenile males with other males, because the real test of these relationships occurs after the males have left the natal group. All the older male kin (who were the preferred partners for the juvenile males), most of the immigrant males, and several of the 2-year-old male subjects themselves had been removed by the time the subjects were 4 years old, leaving only six juvenile male–adult male dyads still available 2 years later.

PRACTICING SKILLS FOR THE FUTURE

Characteristics of Preferred Play Partners

Play is a frequent activity for juveniles in the ecologically undemanding world of the captive social group, particularly for juvenile males. If one of the functions of play is to practice individual fighting skills, then males would be expected to play more often than females, and juveniles of both sexes should chose play partners that are closely matched in fighting ability.

Two-year-old males spent significantly more time in social play (13.3%) than did 2-year-old females (8.3%) ($t = 6.22$, $p < 0.01$). Choice of play partners was definitely not random. There were strong effects of age and sex on play-partner preferences that are consistent with the hypothesis that juveniles are selecting individuals that have high value as play partners.

Two-year-old males selected play partners that were comparable in size, strength, and motivation to play (Fig. 15.6). They played with other juvenile males more often than with juvenile females (sex: $F = 573.67$, $p < 0.01$) and with age peers more often than with older or

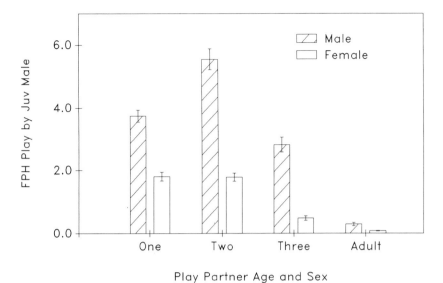

Fig. 15.6. Mean frequency per hour of play by 2-year-old males, by partner age and sex.

younger males (age: $F = 404.49$, $p < 0.01$). The choice of play partners was not influenced by kinship or matrilineal rank.

Juvenile females also selected play partners that were comparable in size and strength and were motivated to play (Fig. 15.7). Two-year-

old females played most often with age peers and with younger juveniles of both sexes (age: $F = 318.81$, $p < 0.01$). There was no tendency for females to selectively play with other females: in fact, females played with males more often than they played with other females (sex: $F =$

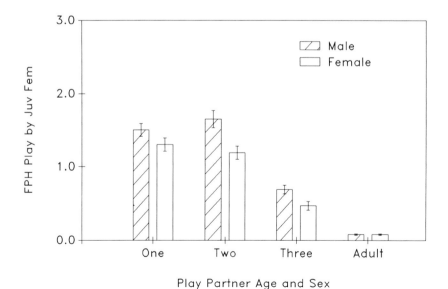

Fig. 15.7. Mean frequency per hour of play by 2-year-old females, by partner age and sex.

Table 15.3. Stability of Juvenile Play Relationships

	Proximity		
Play at age 2	Age 2	Age 3	Age 4
Juvenile female			
With juvenile female	0.27**	0.19*	0.18
With juvenile male	0.43**	0.12	0.12
Juvenile male			
With juvenile female	0.45**	0.22*	−0.05
With juvenile male	0.40**	0.31*	0.13

Note: Correlations of frequency per hour play by 2-year-old juveniles with the percentage time the dyad spent in proximity at 2, 3, and 4 years of age.
**$p < 0.01$.
*$p < 0.05$.

25.34, $p < 0.01$). The choice of play partners for juvenile females was not influenced by rank or matrilineal kinship.

Do Associations with Frequent Play Partners Persist into Adulthood?

The long-term consistency of associations with preferred play partners was tested in the same manner used to determine stability of relationships between juveniles and adults. The correlations between the rate of play with individual partners at age 2 and the amount of time that dyads spent together at 2, 3, and 4 years of age are shown in Table 15.3. At 2 years of age, there was a significant association between the rate of play and the amount of time a dyad spent in close proximity for both males and females. These preferential associations tended to decline over time, however. The individuals that were preferred play partners of the juveniles at 2 years of age were not the same individuals that sat together as young adults. Within sex classes, adult partner preferences could not be predicted from juvenile play-partner preferences.

Selecting Infants for Allomaternal Care

If allomaternal infant care allows a juvenile to practice mothering skills, then juvenile females would be expected to initiate caretaking activities more often than juvenile males. Both juvenile males and juvenile females showed interest in infants, but juvenile females initiated caretaking behaviors at a much higher rate (females: 1.48/hour/dyad; males: 0.33/hour/ dyad; $t = 14.3$, $p < 0.001$).

The characteristics that influenced attractiveness of infants for allomaternal care were also consistent with the hypothesis that juvenile females were practicing mothering skills. Selection of infants for caretaking was strongly influenced by the age of the infant (Fig. 15.8; $F = 56.6$, $p < 0.001$). Juvenile females were attracted to infants that were relatively young and helpless: the rate of caretaking was highest for 1- and 2-month-old infants, and dropped off dra-

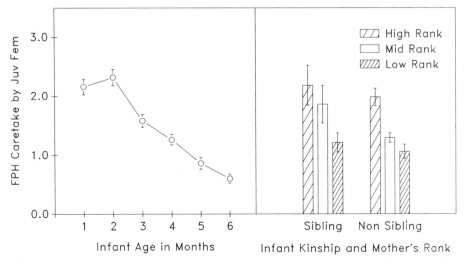

Fig. 15.8. Mean frequency per hour that 2-year-old females caretake infants, by infant's age, kinship, and mother's rank.

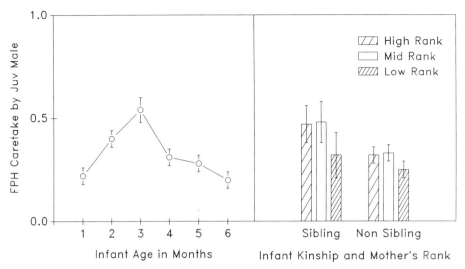

Fig. 15.9. Mean frequency per hour that 2-year-old males caretake infants, by infant's age, kinship, and mother's rank.

matically as the infants grew older and more independent. Selection of infants was also influenced by kinship (sibling vs. other; $t = 2.06$, $p < 0.05$) and by the rank of the infant's mother ($F = 18.01$, $p < 0.001$) (Fig. 15.8). Both of these factors were associated with accessibility. Mothers were more willing to release their young infants to their juvenile daughters, and high-ranking mothers gave up their infants to caretakers more often than middle- and low-ranking mothers. Selection of infants for caretaking was not influenced by the sex of the infant (male infants: 1.45/hour/dyad; female infants: 1.51/hour/dyad; $t = 0.44$, ns).

Infant caretaking by juvenile males was much less frequent, and selection of infants was governed by different factors. Juvenile males initiated more caretaking to 3-month-old infants than to younger or older infants (Fig. 15.9; infant age: $F = 9.14$, $p < 0.001$). At 3 months, infants were spending approximately 40% of the time more than 1 m away from their mother, and the intense interest of the juvenile females had begun to decline. Caretaking by males usually involved inspection rather than grooming or carrying and appeared to express curiosity about a new group member more than an interest in practicing parental care. Juveniles males were somewhat more likely to caretake their infant siblings than nonsibling infants ($t = 1.98$, $p =$

.05) (Fig. 15.9), but caretaking by juvenile males was not influenced by the rank of the infant's mother (Fig. 15.7; $F = 1.51$, ns) or by the sex of the infant (male infants: 0.33/hour/dyad; female infants: 0.32/hour/dyad; $t = 0.19$, ns).

Do Associations Between Infants and Caretakers Persist over Time?

The long-term consistency of associations between juvenile caretakers and the infants they selected for allomaternal care was tested in the same manner described above for relationships between juveniles and adults, and between play partners. The correlations between the rate of caretaking of individual infants at age 2 and the amount of time that dyads spent together when the juvenile subjects were 2, 3, and 4 years of age are shown in Table 15.4. The rate of caretaking for juvenile–infant dyads was positively correlated with the time spent in proximity during the same time period, for both female and male juveniles, but this relationship declined over time.

For the juvenile females, there was a significant association between the rate of caretaking and the amount of time that the dyad spent in proximity 1 year later, but when the effects of rank and kinship were controlled, the remaining partial correlation was not significant ($r =$

Table 15.4. Stability of Caretaker–Infant Relationships

Caretake at age 2	Proximity		
	Age 2	Age 3	Age 4
Juvenile female			
With all infants	0.63**	0.39**	0.19
With female infant	0.61**	0.32*	0.29
With male infant	0.65**	0.46**	0.12
Juvenile male			
With all infants	0.23**	−0.05	0.02
With female infant	0.20	−0.05	−0.08
With male infant	0.25*	−0.06	−0.06

Note: Correlations of frequency per hour of caretaking infants by 2-year-old juveniles with the percentage time the dyad spent in proximity at 2, 3, and 4 years of age.
**$p < 0.01$.
*$p < 0.05$

0.19). A year later, when the subject females were 4 years old, proximity relationships could not be predicted by prior caretaking.

Juvenile males showed no tendency to maintain preferential associations with particular infants, based on the rate of caretaking (Table 15.4). There were no significant correlations between the rate of caretaking by juvenile males and proximity 1 and 2 years later.

DISCUSSION

The impression that one gets when observing a vervet group is that the juvenile females are very busy establishing and reinforcing social relationships with older group members. They are highly discriminating in their partner preferences, and their choices are consistent with the hypothesis that juveniles are trying to establish relationships with adults that would be expected to be good alliance partners. Juvenile females sought access to the females that would be expected to be most able, based on rank, and most likely, based on kinship, to provide benefits. They also formed long-term relationships with the highest ranking adult males.

Juvenile males were also attracted to adults that were more powerful, more likely to tolerate and support them, and more likely to be available in the future. For males, these decision rules resulted in partner preferences that differed from those of juvenile females in ways that are consistent with sex differences in adult life history. Juvenile males selectively formed relationships with their older brothers and other natal males, who they would be likely to meet again as adults. Among group members that they would be leaving behind, juvenile males preferred close female kin and the highest ranking adult male as proximity partners.

The relationships that juveniles formed with these high-value partners persisted into the future and, for females, in particular, predicted proximity relationships in adulthood. These results comparing approaches by juveniles to adult group members with their proximity relationships with those same adults, 1 and 2 years later, supported the idea that juvenile females are forming relationships with high-value partners that are likely to benefit them as adults. When demographic circumstances allowed it, young adult males also continued to maintain the relationships that they had established as juveniles.

The characteristics that were used to select play partners were quite different from those influencing attraction to adults, and were consistent with the hypothesis that juveniles use play for exercise and to develop coordination and fighting skills. Preferred play partners were individuals that were similar in size, strength, and inclination to play. Preferential associations with play partners declined over the years as the motivation to play declined and did not persist into adulthood. There was no evidence that play was used by juveniles as a means of establishing long-term social relationships with particular individuals that would be useful in adult life. As was noted above, the best play partner for a 2 year old would not necessarily be expected to be the best alliance partner later in life. Play partners appear to be selected for their attributes as play partners and not for their expected value as adult affiliative and alliance partners.

Selection of infants for allomaternal care was consistent with the hypothesis that juvenile vervets use caretaking to practice infant-handling skills. Juvenile females selected the youngest infants for caretaking, and quickly lost interest as the infants matured. They also selected infants based on kinship and on the rank of the infant's mother, both attributes that influenced the willingness of the mother to release the infant. Observation of caretaker–infant dyads 1 and 2 years later indicated that allomaternal care was not being used to establish long-term proximity relationships between caregivers and recipients.

Comparison with Other
Cercopithecine Species

Adult Proximity Partners. To what extent
would we expect to find the same patterns of
attraction and affiliation in other cercopithecine
species? The available evidence suggests that
juvenile macaques and baboons are also attracted to adults that have high expected value as
alliance partners.

The preference for matrilineal kin, and association between kinship and alliance support,
has been found in virtually every cercopithecine
population that has been studied long enough for
kinship to be known (for review, see Walters
1987b). Kinship appears to be even more important in structuring social relationships for immatures than it is for adults. All studies that have
considered the topic report a continuing preferential association between juveniles and their
mothers (Kurland 1977; Cheney 1978b; Fairbanks & McGuire 1985; Rowell & Chism 1986;
Holman & Goy 1988; Pereira 1988a; Nakamichi
1989). The importance of matrilineal kin, other
than the mother, in affiliative associations of juveniles with other group members has also been
documented (Kurland 1977; Ehardt-Seward &
Bramblett 1980; Weigel 1980; Glick et al.
1986).

The preference for higher ranking adults has
also been demonstrated for other cercopithecines. Juvenile females are preferentially attracted to higher ranking adult females and approach and groom them more often than lower
ranking adult females in bonnet macaques and
baboons (Cheney 1978b; Silk et al 1981; Pereira
1988a). Access to the highest ranking adults
does appear to be more constrained than it is for
juveniles in the vervet colony, however. Competition to form relationships with high-ranking
animals, coupled with the threat of direct aggression, often inhibits low-ranking juveniles
from approaching the highest ranking adult
males and females in baboon and macaque
groups. Juvenile baboons and macaques are particularly subject to aggression from other group
members (Dittus 1977; Silk et al. 1981; Bernstein & Ehardt 1985b; Cords 1988b; Pereira
1988b; Cords & Aureli, Chapter 19, this volume), and, as a result, low-ranking juveniles are
likely to be afraid to approach and sit near adults
from the highest ranking matrilines. Colvin
(1985) noted that juvenile male rhesus monkeys

were attracted to higher ranking females, but the
greater likelihood of aggression from higher
ranking females prevented them from establishing tolerant relationships with their preferred
partners. De Waal (1986c) described a social-
class barrier between the lower and upper half of
the dominance hierarchy in a group of captive
rhesus monkeys: individuals from the lower half
of the hierarchy avoided situations that could
place them in conflict with members of the upper half of the hierarchy. These observations
suggest that although juveniles may be attracted
to form relationships with the highest ranking
adults in macaque and baboon societies, they are
likely to be constrained from doing so by the
nature of the dominance hierarchy.

The dominance style in the vervet colony is
less rigidly hierarchical than that usually described for baboons and macaques. Vervets, like
other guenons, do not use the formalized gestures of fear grimacing and mounting in social
interactions between dominants and subordinates (Kaplan 1987; Rowell 1988; de Waal,
Chapter 18, this volume). The highest ranking
females serve as the social focus of the group
and are relatively accessible to females from
other matrilines (Fairbanks, 1980; Seyfarth,
1980). Adult females and juveniles from different matrilines frequently cooperate in intergroup encounters and in coalitions against
adult males (Cheney 1981, 1983). Juveniles in
the captive colony do not receive more aggression than other group members, and the type of
aggression directed against juveniles is less severe than that directed against adults (Fairbanks,
unpublished data). Within this relatively un-
restrictive social environment, juvenile vervets
can and do form social relationships with high-
ranking adult females outside their own matrilines.

Understanding relationships between juveniles and adult males has been one of the most
neglected areas of primate research (Taub
1984a). In their study of free-ranging vervets on
Barbados, Horrocks and Hunte (Chapter 16,
this volume) found that juveniles of both sexes
had more affiliative and supportive relationships
with long-term resident adult males, who were
likely to be their fathers, than with challenging
immigrant males. A similar pattern was observed in the captive vervet colony: the highest
ranking resident male was approached more
often than recent immigrants. Male and female

juveniles approached the highest ranking resident male with equal frequency, both in the field and in captivity. The tendency for challenging adult males to be approached more often by juvenile females than by juvenile males found in Barbados (Horrocks & Hunte, Chapter 16, this volume) was not replicated in the captive colony, perhaps because the field study included older females that may have been approaching reproductive maturity, while the juvenile females in the data set reported here were only 2 years old. In both studies, there was evidence that juveniles recognize the attributes of adult males and that they initiate and maintain relationships with adult males that best serve their own self-interest.

Preference for adult partners found in the vervet colony was consistent with the hypothesis that partner preference should be influenced by the expectation of availability in the future. Juvenile vervets in the colony were more attracted to other individuals that they would be likely to be living with as adults. This pattern has also been reported for other cercopithecine species. Juvenile females in matrifocal societies devote more time and energy to establishing relationships with adult females (Kurland 1977; Cheney 1978b; Silk et al. 1981; Glick et al. 1986; Rowell & Chism 1986; Holman & Goy 1988; Pereira 1988a; Nakamichi 1989; Cords & Aureli, Chapter 19, this volume). Juvenile males are more attracted to older natal males that they would be likely to encounter again as adults after they left the natal troop (Dittus 1977; Meikle & Vessey 1981; Colvin 1983a; Vessey & Meikle 1984; Pereira 1988a; Nakamichi 1989).

Of course, there are numerous other hypotheses, besides future availability, that can be used to explain the fact that juveniles are attracted to like-sex adults (e.g., Pereira 1988a; Edwards, Chapter 22, this volume). In vervet monkeys it would not be possible to separate expectation of future availability from the many other differences between adult males and females that might influence juvenile partner choice. With more comparative studies of species with different patterns of emigration and dispersal, we can begin to separate out some of these potential influences (see Nash, Chapter 9, Strier, Chapter 10, and Watts & Pusey, Chapter 11, this volume).

In the vervet colony, juveniles established relationships of tolerance and familiarity with potential allies during the years when the need for support in intragroup conflict was relatively low. They then maintained these relationships into the time in their lives when alliance partners would be most valuable. In the 2-year period from 2 to 4 years of age, vervet monkey females grow to adult size and must cope with the problems of establishing themselves in the adult female hierarchy and with bearing and caring for their first infant. Stable predictable relationships with other adult females appear to facilitate these processes, whereas changes in adult female rank and the loss of key allies appear to hinder them (Horrocks & Hunte 1983a; Fairbanks & McGuire 1986; Cheney & Seyfarth 1987).

Although longitudinal data on the maintenance of social relationships formed by juveniles are not available for other cercopithecine species (but, for long-term stability of female social relationships among capuchins, see O'Brien & Robinson, Chapter 14, this volume), there is every reason to believe that the maintenance of affiliative relationships observed here would also be widespread among matrifocal primates. Preferential associations among female kin have been noted at all ages. Female baboons and macaques that have lost the support of close female relatives during the juvenile–adult transition have been shown to have greater difficulty in achieving their adult rank and successfully beginning their reproductive careers (Hasegawa & Hiraiwa 1980; Johnson 1987; Chapais 1988a; see also Crockett & Pope, Chapter 8, this volume, for howler monkeys). At this point, there is not enough information available to determine whether affiliative relationships that are established with nonkin adult females would also be maintained longitudinally in other cercopithecine species.

Play. The pattern of play partner selection observed in the vervet colony is similar to that reported in other studies of macaques, baboons, and vervets. Juvenile males usually spend more time engaged in rough-and-tumble play than juvenile females (Owens 1975; Bramblett 1978; Symons 1978; Raleigh et al. 1979; van Noordwijk et al., Chapter 6, this volume; Fagen, Chapter 13, this volume; see also Nash, Chapter 9, this volume, and Watts & Pusey, Chapter 11, this volume, for noncercopithecine species). When given the opportunity, juvenile males prefer male peers as play partners, while juvenile females are more likely to spread their play among age peers of both sexes and younger juveniles (Fedigan 1972; Owens 1975; Cheney 1978a; Pereira & Altmann 1985; Janus 1989). This pattern is consistent with the hypothesis that juveniles prefer play partners that are comparable in size and strength to themselves (Owens 1975; Fagen 1981; Pereira & Altmann 1985).

Engaging in activities that promote physical fitness could have immediate benefits for both males and females by promoting juvenile survival (Rubenstein, Chapter 4, this volume; Janson & van Schaik, Chapter 5, this volume), and both males and females could profit from developing communication skills and practice in social cooperation (Fagen, Chapter 13, this volume). But none of these functions explains why male cercopithecines generally play more often than females. The sex difference in the rate of play is predicted by the hypothesis that play functions to practice fighting skills if lifetime reproductive success of an adult male cercopithecine is relatively more dependent on individual fighting ability than is that of an adult female (Pereira & Altmann 1985). In addition, juvenile female cercopithecines can engage in a variety of different activities that are likely to influence their adult fitness, including infant caretaking and establishing long-term relationships with adult females. If these activities actually do promote female reproductive success, then juvenile males and juvenile females would be expected to budget their time differently and juvenile females should spend more time socializing and caring for infants, whereas juvenile males should spend more time playing.

The ideal play partner has very different characteristics from the ideal alliance partner. When juveniles mature and the motivation to play declines, there is no compelling reason to expect them to maintain close social ties with individuals that were chosen for their characteristics as play partners. As a result, play partnerships would not necessarily be expected to turn into long-term proximity partnerships. In the vervet colony, the frequency per hour of play at age 2 did not predict proximity 2 years later.

Allomaternal Care. The pattern and probable functions of allomaternal care appear to differ across species, and the general pattern observed

for most Cercopithecines is somewhat different from that reported for Colobines and Callitrichids. In most species of Cercopithecines, access to infants appears to be more restricted than it is for vervet monkeys (Hrdy 1976), but the pattern of sex differences and the factors that influence partner choice are similar. Younger infants appear to be universally more attractive than older infants, but juvenile allomothers may not always be able to gain access to the youngest infants (Struhsaker 1971; Chism 1986). In almost every case that has been reported, with the possible exception of the Barbary macaque, juvenile females show greater interest in allomaternal care than juvenile males (see also Edwards, Chapter 22, this volume, for sex differences in human children). Kinship, where it has been measured, is also usually found to be a factor in choice of partners for allomaternal care (Rowell et al. 1964; Hrdy 1976; Small & Smith 1981; Lee 1989; Fairbanks 1990). Studies differ in the importance of the mother's rank in determining infant attractiveness, but the pattern of results is consistent with the hypothesis that allomaternal care is influenced by accessibility: when high-ranking mothers provide greater access to their infants, then high-ranking infants receive more caretaking (Cheney 1978b; Lee 1989; Fairbanks 1990); when infants are taken against the will of the mother, then caretaking of low-ranking infants is more common (Silk 1980).

This pattern of results is consistent with the hypothesis that juvenile females are selecting infants to practice infant-handling skills. The practice hypothesis was further supported by data from the vervet colony demonstrating that females that had more experience in caretaking and carrying infants as juveniles were significantly more likely to rear their first live-born infant successfully (Fairbanks 1990).

Does allomaternal care also function to establish long-term social relationships between infants and caretakers? The effect of kinship on allomaternal care is consistent with the hypothesis that juvenile females use infant caretaking to reinforce bonds with siblings. But in the vervet colony, juvenile females that did not have infant siblings spent just as much time caretaking as juvenile females that did have infant siblings, suggesting that juvenile females are motivated to hold and carry infants for reasons other than establishing relationships with kin (Fairbanks 1990). If infants are being selected for their qualities as social partners, then juveniles should differentiate infants by sex, with juvenile females preferring female infants, and juvenile males preferring male infants. This was not the case in this study, and preferences for male or female infants for allomaternal care have generally not been reported in other studies. Individual case studies of persistent relationships between infants and caregivers have usually involved adult caregivers, not juveniles, and even these tend to decline with time (Gouzoules 1984; Chism 1986). Adoption of infants that have lost their mothers is better predicted by kinship than by previous allomaternal care (Berman 1982; Dolhinow & Krusko 1984; Chism 1986). In general, it seems unlikely that allomaternal care specifically functions to promote the development of persistent social bonds between infant and caregiver.

CONCLUSION

Adapting to the physical environment and learning how to avoid predators and to forage efficiently are obviously important factors in development (Janson & van Schaik, Chapter 5, this volume; van Noordwijk et al., Chapter 6, this volume), but they are only part of the "job" of juvenile development. As Crook (1970) pointed out over 20 years ago, "social selection is a major source of biological modification" for socially living animals. Juveniles not only must learn to cope effectively with features of the external world, but also must develop means of adjusting to and competing with other members of their social group, and must take advantage of opportunities to promote their own future welfare.

The results of this analysis on the behavior of juvenile vervet monkeys, and data from other species, suggest that juveniles do recognize the attributes of conspecifics and are highly selective in partner preferences for different behaviors. Partner preferences for play and allomaternal care are consistent with the hypothesis that these behaviors function to practice skills for later life, whereas preferences for adult proximity partners are consistent with the hypothesis that juveniles are attempting to form relationships with high-value alliance partners that are expected to be beneficial to themselves later in life.

ACKNOWLEDGMENTS

I would like to thank Karin Blau, Jill Kusnitz, Margaret Wortz, Diane Crumley, Michaela Heeb, and Dan Diekmann for assistance in data collection and animal care, and Michael McGuire for his continuing support for longitudinal research in the colony. Dorothy Cheney, Joan Silk, and Michael Pereira provided valuable comments on an earlier draft. This research was funded in part by National Science Foundation Grants BNS 84-02292 and 87-09765 to the author, and by a Veterans Administration Merit Review Grant to Michael McGuire. Computing support was provided by the Office of Academic Computing, UCLA.

16

Interactions Between Juveniles and Adult Males in Vervets: Implications for Adult Male Turnover

JULIA A. HORROCKS and WAYNE HUNTE

Studies of interactions between adults and juveniles in Cercopithecines have focused primarily on adult females and their daughters, perhaps because female philopatry and a matrilineal dominance hierarchy are conspicuous characteristics of this subfamily. Interactions between adult males and juveniles are less common and have received less attention. The objective of this chapter is to describe interactions between juveniles and adult males in vervet monkeys (*Cercopithecus aethiops sabaeus*) in Barbados, West Indies. Unlike *C. aethiops* in Africa, but characteristic of the social organization of other members of the genus (Rowell 1988), vervet troops in Barbados are typically unimale. Periodically, the adult male in a troop is challenged and replaced by an immigrant male. Consequently, there are at least three social contexts in which juveniles and adult males interact. These are as offspring interacting with their father during periods of no challenge by an immigrant male (stable periods), as offspring interacting with their father during periods of challenge by an immigrant male (challenge periods), and as juveniles interacting with an immigrant male that is challenging their father for troop tenure. This chapter compares patterns of aggression, support, and approaches between juveniles and adult males in these three social contexts. It comments on the implications of the interactions for the outcome of challenges for troop

tenure, and hence on the role of juveniles in influencing troop structure.

Functional interpretations of interactions between adult males and juveniles have been made primarily from the perspective of the male (e.g., Estrada 1984; Taub 1984a; for review, see Whitten 1987). They fall into two broad categories, paternal care and use of the infant to manipulate the behavior of adults—both adult females and other adult males.

Paternal care is most evident in monogamous species where certainty of paternity is high (Dunbar 1988). Monogamy is characteristic of the New World Callitrichidae, and paternal care takes the form of intensive caretaking, including frequent carrying and grooming, food sharing, and protection from predators and conspecifics (Fragaszy et al. 1982; Cebul & Epple 1984; Robinson et al. 1987). In most Cercopithecines, males are polygynous and females polyandrous. The consequence is that certainty of paternity is low, particularly in multimale troops. Consistent with this, but with the notable exception of Barbary macaques (Redican 1976; Taub 1984b), cercopithecine males do not show intensive caretaking. However, they often form strong associations with specific juveniles. Proximity is a characteristic feature of these associations, but the males also groom the juveniles (Kawai 1958; Mori 1979; Hill 1986) and protect them from conspecifics (Dunbar 1984; Gouzoules

1984; Whitten 1987). Partly because of this protection, and partly as a consequence of their own foraging activities, associated adult males may increase the juvenile's access to food (Hamilton 1984; Stein 1984a). Juveniles may therefore benefit from these associations, which is consistent with the observation that it is usually the juvenile that maintains proximity to the male (e.g., Gouzoules 1984; Vessey & Meikle 1984; Hill 1986; Keddy-Hector et al. 1989).

Paternity may not be the only, or even principal, cause of associative behavior, for associations often do not closely match paternal relationships (Snowdon & Suomi 1982; Smuts 1985). Males may associate with juveniles to manipulate the behavior of other males. Infant carrying in agonistic interactions between adult males, observed in some Cercopithecines (Hrdy 1976), may serve either to diffuse an opponent's aggression (Deag & Crook 1971) or to signal a male's willingness to escalate the conflict (Busse & Hamilton 1981). It may also protect offspring from potentially infanticidal immigrant males (Busse & Hamilton 1981), and the fact that infants cling willingly to the male (Packer 1980; Busse & Gordon 1983; Busse 1984; Stein 1984a) supports this.

Males may associate with juveniles to increase the likelihood of subsequently mating with their mothers (Seyfarth 1978; Stein 1984b; Smuts 1985), and males associating with juvenile females may increase their probability of mating with them when the females mature (Hrdy 1976). Finally, by associating with a juvenile, a male may increase the probability of support from its mother during disputes. This may facilitate the integration of males into multimale troops (e.g., Japanese macaques: Itani 1959; vervets: Struhsaker 1967b; Cheney 1981), and may be important in the acquisition and maintenance of rank by males (e.g., pigtail macaques: Massey 1977; bonnet macaques: Samuels et al. 1984; Japanese macaques: Packer & Pusey 1979; olive baboons: Strum 1983; vervets: Raleigh & McGuire 1989). Keddy-Hector et al. (1989) have shown that in captive vervets, an adult female is more aggressive to adult males that have been aggressive to her offspring, and least aggressive to adult males that have been most affiliative with her offspring.

Within the genus *Cercopithecus,* unimale troops, and hence male monopoly of matings, are common (Cords 1988a). Paternal care is believed to consist primarily of thwarting predatory attacks, and adult males are often spatially and socially peripheral (Cords 1987). Horrocks and Hunte (1986) suggested that sentinel behavior by adult males in unimale vervet troops is best interpreted as paternal care and mate protection. However, interactions between paternal males and offspring in unimale *Cercopithecus* troops have rarely been investigated.

The behavior of offspring toward paternal males may be particularly important during periods of challenge by an immigrant male for troop tenure, and there may be differences between juveniles in terms of the appropriate time for troop takeover. For example, maturing daughters of the resident male may be less resistant to troop takeover than sons or younger daughters, since father–daughter matings may lead to inbreeding depression (Wilson 1975). Moreover, juvenile males may be more resistant to troop takeover if, as apparently occurs in langurs (Sugiyama 1967; Boggess 1980; Rajpurohit & Sommer, Chapter 7, this volume) and black and white colobus monkeys (Oates 1977), they are forced to leave natal troops prematurely by successful immigrant males. Interactions between juveniles and challenging males have not been investigated, but may be particularly important since the probability of support of challenging males from adult females and juveniles may be influenced by the behavior of challenging males toward juveniles (Keddy-Hector et al. 1989). In general, the outcome of an immigrant male's attempt to join a troop may be more critical for juveniles in unimale than in multimale troops (e.g., Marsh 1979), and attempts by juveniles to influence the outcome may therefore be more marked.

Interactions between juveniles and paternal males in a unimale vervet troop during stable and challenge periods, and interactions between juveniles and challenging males, are the focus of this chapter.

STUDY SITE AND METHODS

Study Animals

Cercopithecus aethiops, the vervet monkey, is the most abundant and widespread of all African monkeys (Struhsaker 1967a). They are semiterrestrial, ranging from Senegal in the west to Somalia in the east, and from sub-Sahara in the

Fig. 16.1. Members of the study troop of *Cercopithecus aethiops sabaeus* in Barbados, West Indies.

north to South Africa. Within the past 350 years, the West African form *C. aethiops sabaeus* (Fig. 16.1) was introduced to the Cape Verde Islands off the west coast of Africa, and to Barbados and St. Kitts–Nevis in the Caribbean.

The vervet population in Barbados is now estimated at between 5000 and 8000, and is distributed throughout the island in both residential and agricultural areas. As indicated by observations of feral troops, by radio-tracking of a few troops, and by the composition of troops caught by the Barbados Primate Research Center in shooting nets, troop size in Barbados ranges from 10 to 25, with an estimated island mean of 15. Troop composition is typically one breeding male and four or five breeding females and their offspring. Females remain in their natal troops, producing their first infants when about 3.5 to 4 years old. Males emigrate at sexual maturity when about 5 years old. All males challenging breeding males for troop tenure therefore originate from troops elsewhere. The typical duration of troop tenure is about 3 years.

The duration of challenges for troop tenure in Barbados ranges from about 3 to 6 months. Con-

flicts between the two adult males are frequent and escalated, and continue until one of the males leaves the troop. The long period over which a single male is the only breeding male ensures that most or all of the juveniles in a troop are frequently the offspring of a single male.

Subjects

The study troop is located in a semiresidential area on the west coast of Barbados. It is a nonprovisioned troop that was habituated 12 years ago and has been under continuous observation for over 10 years. The genealogy of all troop members except that of adult males and females born prior to 1979 is known. The troop has been composed of four matrilines since 1979. As typical of vervets, offspring have acquired their mothers' ranks, and mothers have reversed the ranks of their daughters such that younger daughters outrank older (Horrocks & Hunte 1983a, 1983b). Over the 10-year period, there have been four resident paternal males, indicating three successful challenges for troop tenure.

Data were collected for 3 months in each of

three stable periods and three challenge periods over the 10 years of observation. Each stable and challenge period was chosen to ensure that data were being collected on a different adult male; that is, three different paternal males and three different challenging males were the adult male subjects of the study. Each paternal male had been the single male in the troop for the previous three breeding seasons prior to the start of data collection. This maximized the numbers of the male's offspring present in the troop at the time of data collection. Twenty-seven juveniles born to five adult females between 1979 and 1987 were the juvenile subjects of the study.

Data Collection and Analysis

Observation conditions are excellent, and observers can approach to within a few meters of most troop members. Data on aggressions, supports, and approaches between juveniles and adult males were collected on a behavior-dependent basis during observation sessions that lasted 2–3 hours. Aggressions recorded were threats, lunges, chases, hits, grabs, and bites. Aggressions were frequently over access to food, but often had no obvious proximate cause. Supports were defined as participation in an aggressive interaction on behalf of one animal against another. Approaches were defined as one animal moving to within 1 m of another and sitting down. At 5-minute intervals during each observation session, scan samples were conducted to record all individuals present in the observation area, allowing calculation of the number of behavioral acts within each adult male–juvenile dyad per amount of time both members of the dyad were simultaneously present (an average of 85 hours/dyad). Rates of aggressions and approaches within dyads were expressed as number/hour. Supports were expressed as the number by one individual on behalf of the other, divided by the number of aggressive interactions in which the recipient was involved, to give the percentage support for each member of each dyad. For this study, only those supports by adult males given during juvenile–juvenile interactions, and only those supports by juveniles given in adult male–adult male interactions were considered.

All paternal male–offspring dyads for the three stable periods were combined, as were all paternal male–offspring dyads and all challeng-

ing male–juvenile dyads from the three challenge periods. This provided a total of 31 paternal male–offspring dyads during the stable periods, 36 paternal male–offspring dyads during the challenge periods, and 36 challenging male–juvenile dyads during the challenge periods.

Effects of age, sex, and rank of juveniles on aggressions, supports, and approaches by juveniles toward adult males and by adult males toward juveniles were investigated by analysis of variance. Juveniles were classified as male or female, young or old, and high or low ranking. Young juveniles were those 18 months or less (youngest 5 months), and old juveniles were those older than 18 months (oldest female 36 months, oldest male 48 months). High-ranking juveniles were those belonging to the two highest ranking matrilines, and low-ranking juveniles were those belonging to the two lowest ranking matrilines. The number of juveniles in each age, sex, and rank category in stable and challenge periods is shown in Table 16.1. Each factor—age, sex, and rank—therefore, had two levels in the analysis of variance. Percentage support data were arcsine transformed to normalize their distribution prior to analysis; aggression and approach data were square-root transformed to improve the equality of variances. The values shown in the figures are means calculated from data prior to their transformation.

Table 16.1. The Number of Juveniles in Each Age, Sex, and Rank Category During Stable Periods and During Challenge Periods

| Offspring sex | Mother's rank | Offspring age | | Total |
		<18 months	>18 months	
		Stable Periods		
Male	High	3	4	7
	Low	4	5	9
Female	High	5	3	8
	Low	4	3	7
Total		16	15	31
		Challenge Periods		
Male	High	3	5	8
	Low	4	5	9
Female	High	4	6	10
	Low	3	6	9
Total		14	22	36

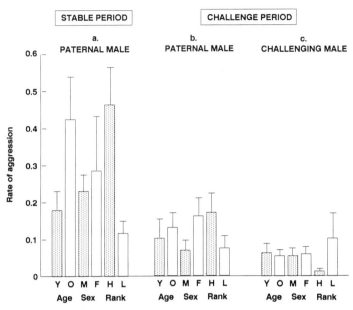

Fig. 16.2. Mean rates of aggression to juveniles of different age, sex, and rank by adult males, presented separately for paternal males in stable and challenge periods and for challenging males. Bars indicate +1 SE.

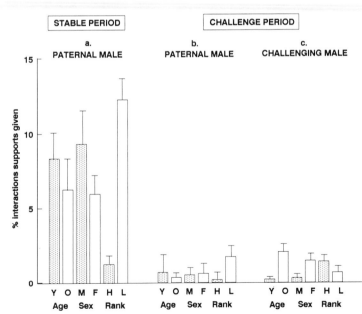

Fig. 16.3. Mean percentage support of juveniles of different age, sex, and rank by adult males, presented separately for paternal males during stable and challenge periods and for challenging males. Bars indicate +1 SE.

Table 16.2. Results of Analyses of Variance of the Effects of Age, Sex, and Rank of Juveniles on Rates of Aggression to Juveniles by Adult Males, and Percentage Support of Juveniles by Adult Males

| | Stable | | Challenge | | | |
| | Paternal male | | Paternal male | | Challenging male | |
	F	p	F	p	F	p
Aggression						
Age	4.45	0.041	1.94	0.179	0.03	0.873
Sex	0.32	0.519	4.95	0.038	0.18	0.588
Rank	7.90	0.012	4.32	0.047	12.60	0.002
Supports						
Age	0.33	0.515	0.03	0.844	5.35	0.026
Sex	2.15	0.161	0.13	0.556	4.05	0.051
Rank	11.10	0.005	4.38	0.049	3.09	0.070

Note: The results are presented separately for paternal males during stable periods and challenge periods, and for challenging males. *F, F* values from ANOVA; *p,* level of probability.

INTERACTIONS BETWEEN JUVENILES AND ADULT MALES

How Do Paternal Males Behave Toward Their Offspring During Stable Periods?

Paternal males were more aggressive to higher ranking than lower ranking offspring, and to older than younger offspring, but there was no effect of sex on paternal aggression to offspring (Fig. 16.2a; Table 16.2). Mean percentage sup-

port by paternal males was greater for lower ranking than higher ranking offspring (Fig. 16.3a; Table 16.2). Neither age nor sex of offspring significantly affected the support received from paternal males, although the values were higher for younger than older offspring and for male than female offspring (Fig. 16.3a). Approaches to offspring by paternal males were too rare to be analyzed for effects of age, sex, and rank of offspring (Fig. 16.4a). However, paternal males appear to approach younger offspring

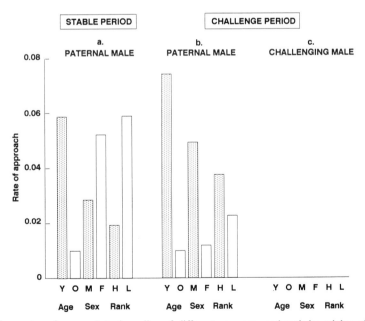

Fig. 16.4. Mean rates of approach to juveniles of different age, sex, and rank by adult males, presented separately for paternal males in stable and challenge periods.

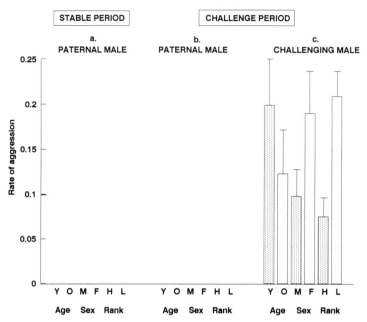

Fig. 16.5. Mean rates of aggression to challenging males by juveniles of different age, sex, and rank. Bars indicate +1 SE.

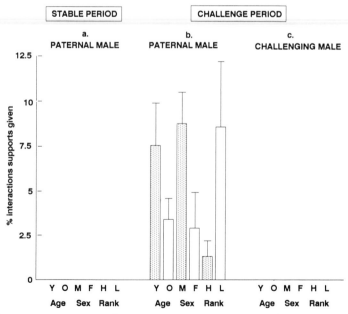

Fig. 16.6. Mean percentage support of paternal males by juveniles of different age, sex, and rank during challenge periods. Bars indicate +1 SE.

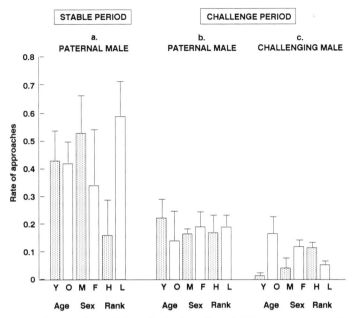

Fig. 16.7. Mean rates of approach to adult males by juveniles of different age, sex, and rank, presented separately for paternal males in stable and challenge periods, and for challenging males. Bars indicate +1 SE.

more often than older and lower ranking offspring more often than higher (Fig. 16.4a). The data therefore suggest that paternal males interact differently with their offspring depending on offspring rank and age.

How Do Offspring Behave Toward Paternal Males During Stable Periods?

Neither aggression to paternal males nor support of paternal males (the latter because no challenging males were present during these periods) by offspring was observed during stable periods (Figs. 16.5a and 16.6a). Lower ranking offspring approached paternal males more often than higher ranking offspring, but neither offspring age nor sex significantly affected approach rates (Fig. 16.7a; Table 16.3). This is consistent with the observation that paternal males were less aggressive to and more supportive of lower than higher ranking offspring, and may suggest that low-ranking offspring in particular benefit from proximity to the paternal male.

Does the Behavior of Paternal Males Toward Their Offspring Change During Challenge Periods?

As in stable periods, paternal males were more aggressive to higher than lower ranking offspring in challenge periods (Fig. 16.2b; Table 16.2). In contrast to stable periods, paternal males were more aggressive to their female than male offspring during challenge periods, and there was no effect of offspring age on aggression received from paternal males (Fig. 16.2b; Table 16.2). The greater aggression to female offspring may suggest a greater conflict of interest between paternal males and their female than their male offspring during challenge periods.

Overall rates of aggression by paternal males to offspring were appreciably lower during challenge periods (0.12/hour) than during stable periods (0.26/hour; Fig. 16.2).

As for stable periods, lower ranking offspring received more support than higher ranking offspring from paternal males during challenge pe-

Table 16.3. Results of Analyses of Variance of the Effects of Age, Sex, and Rank of Juveniles on Rates of Approach and Aggression to Adult Males by Juveniles, and Percentage Support of Adult Males by Juveniles

| | Stable | | Challenge | | | |
| | Paternal male | | Paternal male | | Challenging male | |
	F	p	F	p	F	p
Aggression						
Age					0.15	0.613
Sex					4.55	0.040
Rank					5.20	0.029
Supports						
Age			5.21	0.032		
Sex			6.34	0.023		
Rank			4.32	0.047		
Approaches						
Age	0.08	0.870	0.37	0.508	39.70	0.000
Sex	0.22	0.560	0.15	0.617	12.48	0.002
Rank	5.30	0.035	0.18	0.585	4.20	0.050

Note. The results are presented separately for paternal males during stable periods and challenging periods, and for challenging males. *F*, *F* values from ANOVA; *p*, level of probability.

riods, and there was no effect of either offspring age or sex on support received (Fig. 16.3b; Table 16.2). The frequency of support of offspring by paternal males during challenge periods (0.77%) was much lower than during stable periods (8.52%).

As in stable periods, approaches by paternal males to offspring were too rare to be analyzed during challenge periods. However, paternal males again appear to approach younger offspring more often than older (Fig. 16.4b), but they approach female offspring less often than male. This is a reversal of the pattern observed during stable periods (Fig. 16.4a), but is consistent with their greater aggression to female offspring during challenge periods (Fig. 16.2b).

Does the Behavior of Offspring Toward Paternal Males Change During Challenge Periods?

As in stable periods, offspring were not aggressive toward paternal males during challenge periods (Fig. 16.5b), but frequently supported them against challenging males (Fig. 16.6b). Paternal males received more support from lower ranking than higher ranking offspring, more support from younger than older offspring, and less support from their female than male offspring (Fig. 16.6b; Table 16.3). The greater support from lower ranking offspring reciprocates the behavior of the paternal male toward their offspring; the weaker support from

females again suggests conflict of interest between them and paternal males during challenge periods.

No effect of offspring age, sex, or rank on their rate of approach to paternal males during challenge periods was detected (Fig. 16.7b; Table 16.3).

How Do Challenging Males Behave Toward Juveniles?

Challenging males were more aggressive to lower ranking than higher ranking juveniles, but there was no effect of either juvenile age or sex on aggression received from challenging males (Fig. 16.2c; Table 16.2). The overall rate of aggression by challenging males to juveniles was lower (0.057/hour) than by paternal males to offspring during either stable or challenge periods (Fig. 16.2).

Support by challenging males was greater for higher ranking than lower ranking juveniles, for older than younger juveniles, and for female than male juveniles (Fig. 16.3c; Table 16.2; but note $p = 0.07$ for effects of rank). This is the reverse of the behavior of the paternal males toward offspring. In addition, there was a significant interaction effect of age and sex, indicating that older female juveniles received most support from challenging males ($F = 5.47$, $p < 0.05$). The frequency of support of juveniles by challenging males (1.4%) was substantially lower than by paternal males during stable periods (8.52%), but higher than by paternal males during challenge periods (0.77%; Fig. 16.3).

Challenging males were not observed to approach juveniles.

How Do Juveniles Behave Toward Challenging Males?

Higher ranking juveniles were less aggressive to challenging males than were lower ranking juveniles (Fig. 16.5c; Table 16.3). This may be the consequence of challenging males being less aggressive to and more supportive of higher ranking lineages. No effect of offspring age on aggression to challenging males was detected, but female juveniles were more aggressive to challenging males than male juveniles (Fig. 16.5c; Table 16.3). In addition, there was a significant interaction effect between age and sex ($F = 6.22$, $p < 0.05$) and between rank and sex ($F = 4.20$, $p < 0.05$), indicating that it was younger females and lower ranking females that were most aggressive to challenging males. Juveniles never supported challenging males against paternal males (Fig. 16.6c).

Higher ranking juveniles approached challenging males more often than lower, older juveniles approached them more often than younger, and female juveniles approached them more often than male juveniles (Fig. 16.7c; Table 16.3). This perfectly reflects the pattern of support by challenging males to juveniles, the males supporting higher ranking more than lower, older more than younger, and females more than males (Fig. 16.3c). There was also a significant interaction effect of age and sex, indicating that older juvenile females approached challenging males most often ($F = 12.73$, $p < 0.05$).

DISCUSSION

In unimale troops of the genus *Cercopithecus,* resident males monopolize matings (Cords 1988a). Consequently, most juveniles in a troop are offspring of the resident male. Resident males appear socially and spatially peripheral, and more concerned with troop protection than intratroop interactions (Cords 1987). However, since most juveniles are offspring, troop protection is itself an effective form of paternal care (Horrocks & Hunte 1986). Moreover, rarity of

interactions between paternal males and offspring need not imply either that they are unimportant or that they are not influenced by paternity. The data presented here suggest that paternal males interact differently with their offspring depending on offspring rank, age, and sex, and that the interactions change with social context, specifically with whether or not the male is being challenged for troop tenure by an immigrant male.

During stable periods, paternal males were more aggressive to their higher ranking than lower ranking offspring, and supported and approached the latter more often. They were more aggressive to older than younger offspring, and approached the latter more often. Offspring sex did not affect aggression, support, or approaches by paternal males.

Vervets are characterized by a strong matrilineal dominance hierarchy, lower ranking matrilines receiving frequent aggression from higher ranking matrilines (Horrocks & Hunte 1983a, 1983b) and being seldom supported by members of their own matriline (Hunte & Horrocks 1987). One consequence is that low-ranking juveniles have poorer access to troop resources, and this is reflected in lower growth rates and older ages at sexual maturity (Horrocks 1986). The cost of this differs for paternal males and high-ranking matrilines, since low-ranking juveniles are the offspring of the male, but are less related or unrelated to high-ranking matrilines. The paternal male's lower aggression and greater support for low-ranking juveniles may reduce the asymmetry in juvenile access to resources created by the matrilineal dominance hierarchy. We have frequently observed that the feeding rate of low-ranking juveniles accelerates in the presence of the paternal male, and that the juveniles typically position themselves such that the paternal male is between them and members of high-ranking matrilines. Moreover, when resources are clumped, low-ranking juveniles often do not feed until after the arrival of the paternal male. The fact that low-ranking offspring approach paternal males more often than high-ranking offspring suggests that the former in particular benefit from proximity to the paternal male. Gouzoules (1984), Vessey and Meikle (1984), Hill (1986), Keddy-Hector et al. (1989), and Fairbanks (Chapter 15, this volume) have all observed that proximity between juveniles and adult males is typically maintained by the juveniles, and Hamilton (1984) and Stein (1984a) have suggested that proximity to an adult male increases a juvenile's access to food. Finally, since size also influences a juvenile's ability to competitively monopolize a food item (J. A. Horrocks, personal observation), lower aggression and higher approach rates by paternal males to younger offspring may also reduce differences between offspring in access to resources.

Interactions between paternal males and offspring may change when the male is being challenged for troop tenure. Adult female support, particularly from high-ranking females, can facilitate immigration of males and influence their acquisition of rank in multimale troops (Itani 1959; Struhsaker 1967b; Massey 1977; Packer & Pusey 1979; Cheney 1981; Samuels et al. 1984; Raleigh & McGuire 1989). Since the behavior of an adult female toward a male may be influenced by the male's behavior toward her offspring (Keddy-Hector et al. 1989), paternal males might increase their support from high-ranking mothers by switching their own support from low-ranking to high-ranking offspring when being challenged for troop tenure. The behavior of offspring toward paternal males may also change during challenge periods. In particular, it may be in the interest of older female offspring to facilitate troop takeover by an immigrant male, since inbreeding depression may result from father–daughter matings. In interspecies comparisons, Clutton-Brock (1989) observed that the average length of tenure of an adult male in a troop is typically less than the average age of first conception.

In challenge periods, paternal males continued to be more aggressive to their higher ranking than lower ranking offspring, and to support and approach the latter more often. If favoring low-ranking offspring is a consequence of paternity, and favoring high-ranking juveniles is a strategy for retaining troop tenure, the data may suggest that paternity is the stronger influence on the male's behavior in challenge periods, and may make the male more vulnerable to troop takeover. The tendency of paternal males to continue favoring lower ranking offspring during challenge periods may be reinforced by the possible cost of a sudden reversal of established support patterns; that is, it may be an inappropriate time for the males to alienate offspring with whom they have established relationships.

In contrast to stable periods, paternal males were more aggressive to their female offspring than their male offspring in challenge periods, and approached the former less often. This suggests a greater conflict of interest between paternal males and their female offspring during challenge periods, and is consistent with the perspective that female offspring may be least resistant to troop takeover.

The overall rate of aggression by paternal males toward offspring during challenge periods was half that of stable periods. This may be a strategy by paternal males to minimize alienation of troop members during challenge periods. Alternatively, it may simply result from paternal males diverting more time and energy to contests with challenging males. The sharp decrease in frequency of support of offspring by paternal males during challenge periods is also open to both explanations, since decreasing the frequency of support for low-ranking over high-ranking offspring may reduce alienation of high-ranking matrilines.

Offspring were not aggressive toward paternal males during challenge periods, and frequently supported them against challenging males. Paternal males received more support from lower ranking than higher ranking offspring and from younger than older offspring. Low ranking and young offspring are favored by paternal males during stable periods, and may therefore have most to lose from troop takeover by an immigrant male. Although the behavior of paternal males toward male and female offspring did not differ during stable periods, paternal males received less support from their female than their male offspring during encounters with immigrant males. The reduced support may be both cause and consequence of the increased aggression by paternal males to female offspring during challenge periods; but the cycle toward greater conflict between paternal males and their female offspring is probably triggered by these offspring being the least resistant to troop takeover. We have occasionally observed older female juveniles grooming challenging males on the periphery of the troop.

Challenging males were more aggressive toward lower ranking than higher ranking juveniles and supported the latter more frequently. This behavior is precisely the opposite of that displayed by paternal males. If females are least aggressive to males that are most affiliative toward their offspring (Keddy-Hector et al. 1989), the behavior of challenging males may reduce the resistance they encounter from high-ranking matrilines. Reduced resistance is indicated by the fact that high-ranking juveniles approached challenging males more often, and were less aggressive to them, than were low-ranking juveniles. The tendency of challenging males to favor high-ranking over low-ranking juveniles may be both cause and consequence of the fact that the former are less aggressive to challenging males than the latter. However, we suggest that there may be both an a priori tendency of challenging males to favor high-ranking juveniles (to obtain the support of high-ranking matrilines) and an a priori tendency of low-ranking juveniles to be more aggressive to challenging males (they have more to lose from troop takeover, as they are favored by the paternal male).

Challenging males were more supportive of older than younger juveniles and of female than male juveniles. They were particularly supportive of older female juveniles, perhaps because the latter are future mates. Older female juveniles approached challenging males more often than other juveniles. The greater support of female juveniles by challenging males, their frequent approaches to challenging males, the lower support of paternal males by female offspring, and the increased aggression by paternal males toward female offspring during challenge periods all indicate conflicts of interest between female offspring and paternal males during challenge periods. The conflict of interest between paternal males and their female offspring is a second way in which paternity may increase the vulnerability of males to troop takeover; that is, they may be increasingly vulnerable as their female offspring approach maturity. Interestingly, juveniles never overtly supported challenging males over paternal males, and were never aggressive to paternal males.

The rate of aggression by challenging males toward juveniles was even lower than by paternal males toward offspring during challenge periods. This is consistent with the perspective that males attempt to minimize alienation of troop members during contests for troop tenure. Minimizing alienation of higher ranking matrilines may be particularly important, and rates of support by challenging and paternal males are of interest in this context. During challenge periods, immigrant males, who support higher over lower ranking juveniles, display higher overall support rates than do paternal males, who support lower ranking over higher ranking offspring.

ACKNOWLEDGMENTS

The comments of Dr. L. Fairbanks, Dr. M. Pereira, and two anonymous reviewers substantially improved the chapter. Financial support was provided by a University of the West Indies Study and Travel Grant to J. A. H., and an NSERC Operating grant to W. H.

Part IV

MANAGING SOCIAL CONFLICT AND THE DEVELOPMENT OF DOMINANCE RELATIONS

The following chapters explore the ways in which immature primates discover and develop their dominance relationships and otherwise manage social conflict. Group living per se exacerbates aspects of resource competition and can be expected to generate some conflict (Alexander 1974; van Schaik 1989). Aggression, submission, dominance relation, avoidance, appeasement, and postconflict reconciliation are among the principal mechanisms used by group-living animals, especially primates, to mediate conflict and thus potential access to resources.

Research on primates has contributed substantially to general knowledge on agonistic relations in animals. In many species, patterns of aggression and submission readily allow dominance relations to be quantified: individuals consistently express their subordination to particular partners, often in the absence of aggression. Fortuitously, the most studied monkeys—macaques, baboons, and vervet monkeys—exhibit an extraordinary hierarchical system of dominance based on matrilineal membership. Juveniles, females most reliably, acquire adult dominance rank adjacent to their mothers through receipt of systematic aggression and agonistic support: adult females regularly intervene during juvenile–adult conflicts on behalf of the member of the higher ranking family (Chapais 1992; Pereira 1992).

The first three chapters of this section strive toward yet a fuller understanding of dynamics within the cercopithecine system of dominance relation. Then, the cercopithecine system is contrasted to that of a prosimian primate that shares basic similarities in natural history with the monkeys. Together, the four contributions address three important themes in current research: first, primates contrast their own attributes with those of groupmates and some recognize attributes of others' relationships; second, primates integrate agonistic and affinitive behavior in their social relations; and third, presumptive cause–effect dynamics inferred from correlated observations should be tested, whenever possible, by proper experimental manipulations.

Chapais and Gauthier (Chapter 17) discuss cognitive aspects of perceiving and learning the relative attributes of oneself and others in the social group as they detail the information available to infant macaques before the actual period of rank acquisition. By determining the direct and observational experiences of young macaques, in relation to their own and others' kinship and potential dominance ranks, the authors begin to identify the information likely used by developing individuals in deciding whom to try to dominate and when. The importance of perceiving others' attributes for responding to aggression is also a theme for Cords and Aureli (Chapter 19) in their treatment of reconciliation in juvenile long-tailed macaques. Pereira (Chapter 20) suggests that neuroethological constraints explain the different process by which ringtailed lemurs acquire dominance. Immature ringtails rely heavily on their own agonistic capacities, seemingly because adults are unable to recognize the identities or dominance relations of fight participants instantaneously. This would preclude the rank-related patterns of support seen in cercopithecine societies, and such patterns in fact did not occur in either of two ringtail study groups.

In the past, aggression was often viewed as the polar opposite of affiliation, whereas current studies emphasize the interdependencies of these two social mechanisms. While individuals may be motivated to achieve the highest possible dominance rank, for example, it remains important for them to maintain tolerant, predictable relations with groupmates. The safety needs of juveniles, in particular, make it imperative that they not jeopardize their relations with older group members. Cords and Aureli show that juvenile long-tailed macaques are more likely to reconcile following aggression with the individuals with whom they maintain their friendliest relations, partners likely to be available and perhaps important to them in the future. But also,

juveniles did not reconcile preferentially either with kin or with recent agonistic supporters, suggesting that basic support patterns in cercopithecine societies make certain social relationships inherently more secure than others (see also Chapais 1992; Pereira 1992).

With his reconciled hierarchy model, de Waal (Chapter 18) also suggests interdependence between affiliation and dominance. He views dominance rank formalization as an essential aspect of social integration for most primates, arguing that the formal recognition of dominance asymmetry allows individuals to maintain predictable, tolerant, even cooperative relations. De Waal also shows that development of affinity between juveniles and unrelated adult females may facilitate the immatures' acquisition of matrilineal rank. Again, the ringtailed lemurs present a contrast. Although maturing female lemurs continue to associate with and receive some support from adult females, these positive aspects of social relations do not reliably influence females' acquisition of dominance or maintenance of membership in their natal group.

Finally, these chapters well represent the growing trend to integrate observational, comparative, and experimental research to answer difficult questions about primate social behavior. Detailed description of behavior in natural or seminatural social groups has long been a hallmark of primate sociobiology. But provocative post hoc imputation of behavioral function has too often represented the end of an investigator's work. Only rarely have cause–effect dynamics inferred from behavioral correlations been tested using experimental manipulations and procedures of control.

The study of aggression and dominance

is probably the primatological domain in which observational and experimental methods have been integrated most effectively. In this section, de Waal illustrates how a challenge situation can provide valuable information about resource access and social tolerance that cannot be inferred from the dominance hierarchy alone. Cords and Aureli verify their observational results with experimentally provoked aggression conducted under controlled conditions. Chapais and Gauthier describe preliminary experiments designed to test whether observation of the agonistic interactions of others can actually induce a change in an individual's rank-related behavior.

17
Early Agonistic Experience and the Onset of Matrilineal Rank Acquisition in Japanese Macaques

BERNARD CHAPAIS and CAROLE GAUTHIER

In many species of cercopithecine monkeys (macaques [*Macaca* spp.], baboons [*Papio* spp.], vervet monkeys [*Cercopithecus aethiops*]), adult females rank just below their mother in the dominance order, forming matrilineal hierarchies in which female kin occupy adjacent ranks. Such a degree of emancipation from the strictly individual determinants of rank (age, size, aggressiveness, etc.) reflects the fact that females are mutually interdependent (i.e., form alliances) for maintaining their rank above lower ranking females. These alliances are manifest in aggressive interventions, when females enter ongoing conflicts involving their kin or nonkin females and side with one opponent against the other. Such support is given most often to the higher ranking opponent against the lower ranking one, hence generating and reinforcing the matrilineal rank order (observational evidence: Cheney 1977, 1983; Walters 1980; Chapais 1983; Datta 1983a, 1983b, 1983c; Horrocks & Hunte 1983a; Netto & van Hooff 1986; Hunte & Horrocks 1987; Pereira 1989; Chapais et al. 1991; reviewed in Chapais 1992; experimental evidence: Chapais 1985, 1988a, 1988b, 1991, 1992; Chapais & Larose 1988; Chapais et al. 1991).

In these species, matrilineal rank inheritance appears to represent one of the main forces governing the social lives of females throughout their life cycle. Females strive for dominance per se (i.e., in the absence of any immediate resources) and show much opportunism when doing so (for a review, see Chapais 1992). This is first manifest during the process of rank acquisition, when they aggressively challenge females born to lower ranking mothers (hereafter *low-born* females) and join third parties against them. And it is still manifest afterward, during the period of rank maintenance, when they are observed to seize (infrequent) opportunities to outrank higher born females by forming either bridging alliances (with individuals ranking above their victim) or revolutionary alliances (with individuals ranking below the victim).

The importance of dominance in matrilineally organized societies is also apparent in that much of the distribution of affinitive interactions among females (e.g., social grooming) is correlated with the dynamics of dominance relations. The integration of affiliation and dominance is expressed, for example, in the formation of matrilineal rank relations (de Waal, Chapter 18, this volume), in the phenomenon of postconflict affiliation (or reconciliation; see, e.g., de Waal 1986a, 1989a; Cords & Aureli, Chapter 19, this volume) and in the dynamics of long-term alliances (see, e.g., Walters & Seyfarth 1987). Finally, at an ultimate level, dominance plays an important role in the differential access of females to resources (Wrangham

1981; Whitten 1983; van Noordwijk & van Schaik 1987) and their differential reproductive output (Harcourt 1987; Silk 1987).

Little is known about the very early phase in the development of matrilineal rank relations. Females are known to acquire their mother's rank, first in relation to their peers. This occurs soon after the female begins to interact agonistically with other individuals, usually before she reaches 7 to 10 months of age—that is, during or shortly after weaning (Berman 1980; Datta 1983a; Horrocks & Hunte 1983a). Females assume their mother's rank above older lower born females at a later age, at which time the difference in size between the younger and older females is reduced. At the end of this process, a young adult female is dominant to all females lower born to herself (Koyama 1967; Sade 1967; de Waal 1977; Lee & Oliver 1979; Walters 1980; Hausfater et al. 1982; Datta 1983a; Lee 1983b; de Waal & Luttrell 1985; Johnson 1987; Paul & Kuester 1987; Pereira 1988b). It is noteworthy that females differentiate among higher born and lower born females well before they are able to outrank the latter. This is evidenced by the fact that they selectively challenge lower born females (who typically resist) well before they outrank them (Walters 1980; Datta 1983a; Horrocks & Hunte 1983a; Pereira 1989).

What prompts infants to establish dominance relationships with their peers and challenge older lower born females? Three components of the early agonistic experience of infants have been suggested by previous authors. First, infants may learn about their maternal rank when third parties enter their skirmishes (which occur, for example, in play contexts) and side aggressively with them or their opponent. For example, Datta (1983a) reported that aggressive interventions were observed prior to the onset of the period of challenge in 66% of 45 dyads in rhesus monkeys (*Macaca mulatta*), and that these interventions were both more frequent and more effective for the higher born juvenile. Second, it has been observed that adult female rhesus and vervet monkeys directed aggression to lower born infants before the infants could be assigned ranks among their peers (Berman 1980; Horrocks & Hunte 1983a). On this basis, Berman (1980, p. 165) suggested that "infants may learn their rank through differences in amounts and sources of agonistic interactions directed toward them," and Horrocks and Hunte

(1983a) proposed that such aggression was a distinct mechanism of rank acquisition, which led to the formation of dominance relations among yearlings. Third, it has been suggested that infants may learn their position in the hierarchy by observing their mother interact with higher born and lower born females (Kawai 1958; Altmann 1980; Walters 1980).

In this chapter, we present an extended conceptual framework for studying the role of early agonistic interactions in rank acquisition and we use that framework to analyze the early development of rank relations in a captive group of Japanese macaques (*Macaca fuscata*). The analysis focuses on the agonistic experience that infants have *before* they initiate aggressive interactions with lower born individuals—that is, during the *preinitiative* phase of rank acquisition. In the case of rank relations between infants, that period of time was defined as preceding the onset of a stable dominance relationship (i.e., the onset of unidirectional aggression by the dominant and unidirectional submission by the subordinate). In the case of rank relations with older lower born females, the preinitiative phase of rank acquisition was defined as preceding the onset of aggression by an infant to an older female (Fig. 17.1).

Such early experience includes, in theory, any agonistic interaction initiated and controlled by other individuals, in the course of which infants may acquire information about their relative status, either through their own experience or by observing the participants. We identified six major categories of preinitiative agonistic interaction that could set in motion the process of maternal rank acquisition among infants, and four major categories of agonistic interaction that could contribute to the onset of aggression to older low-born females. These two categories of dyads were analyzed separately. The aim of this analysis was to quantify each category of interaction and assess its relative importance in rank acquisition.

CONCEPTUAL FRAMEWORK

Preinitiative Agonistic Experience and Rank Relations Among Infants

The six categories of preinitiative agonistic interaction that might initiate the process of matrilineal rank acquisition among infants are (1)

Fig. 17.1. Three members of the study group. The matriarch of the B matriline (aged 16 years) suckles her 5-month-old daughter (B7) and is being groomed by an older daughter (B5, aged 2 years). (Photo, C. Gauthier)

aggressive interventions by third parties in favor of high-born infants, (2) instances of aggression to low-born infants by the mothers of high-born infants, (3) displacements of low-born infants by the mothers of high-born infants, (4) retrievals of low-born infants by their mothers, (5) displacements of the mothers of low-born infants by high-born infants, and (6) agonistic interactions between high-born and low-born mothers. Table 17.1 summarizes for each category the information that the infants might potentially acquire through their own experience and through observing others.

Aggressive interventions comprise situations in which an individual enters an interaction between two infants, by siding aggressively with one of them (the beneficiary) against the other (the target). Two types of intervention were defined, depending on the context. Conflict interventions were subsequent to the beneficiary or the target performing a threat or a submissive behavior (e.g., a distress call). In nonconflict interventions, no noticible act of submission or aggression had preceded the intervention; the two infants were affiliating. In theory, ag-

gressive interventions provide both the beneficiary and the target with opportunities for recognizing the beneficiary's allies and for learning about the contexts in which allies intervene. The target infant also learns individually (and the beneficiary through observation) about the physical power of the beneficiary's allies.

In the second category of preinitiative agonistic interaction (high-born mothers directing aggression to low-born infants), the offspring of an aggressive mother might learn through observation about its mother's physical power relative to a low-born peer. On the other hand, a low-born victim might learn not only about the physical power of a high-born peer's mother, but about the connection between that powerful aggressor and the infant she carries. Whether such associative learning alone might lead to the low-born infant deferring to the high-born infant (as seems to be implied by Horrocks & Hunte 1983a) is not clear. However, such dyadic aggression from a high-born mother to a low-born infant can at least be expected to act on the victim as a powerful confirmation of the alliance between the mother and her offspring, *if* the vic-

Table 17.1. Information Potentially Acquired by Infants During the Preinitiative Phase of Rank Acquisition Between Peers

Categories of agonistic interactions	High-born infant learning		Low-born infant learning	
	By own experience	By observation	By own experience	By observation
Interventions against low-born infant	Social contexts of support. Identity of own allies	Allies' power relative to low-born peer	Social contexts of interventions. Identities and relative power of peer's allies	
High-born mother aggresses low-born infant		Mother's power relative to low-born peer	Relative power of peer's mother. If aggressor carries offspring, connection between them	
High-born mother displaces low-born infant		Low-born peer's submissiveness toward mother		
Low-born infant retrieved by its mother		Nonassertive attitude of peer's mother	Social contexts prompting retrievals	Nonassertive attitude of own mother
High-born infant displaces low-born peer's mother	Submissiveness of low-born peer's mother			Mother's submissiveness toward peer
High-born mother aggresses or displaces low-born mother		Mother's power relative to low-born peer's mother		Mother's submissiveness toward high-born peer's mother

tim has already been the target of an intervention by the same mother supporting her offspring. In other words, aggression from adult females to infants could be most efficient if subsequent to, or intermingled with, aggressive interventions.

In the third category of preinitiative agonistic interactions (high-born mothers displacing low-born infants), the offspring of high-born females might learn through observation about the capacity of their mother to induce submissiveness in low-born peers.

In the fourth category of preinitiative agonistic interaction, a low-born mother physically removes her offspring from proximity to a high-born infant. Two types of maternal retrievals were recognized. Conflict retrievals were subsequent to a skirmish between the infants, whereas nonconflict retrievals occurred when the recovered infant was being approached or affiliating with a high-born infant or its mother, and no act of submission or aggression had been noticed. The rationale for including retrievals in the preinitiative agonistic experience of infants is that in all cases the mother was acting nonassertively, and, in many cases, fearfully. In theory, therefore, retrievals might allow the re-

covered low-born infant to perceive that its mother avoided or feared the high-born infant or its mother, and the high-born infant to perceive the nonassertive attitude of the low-born mother. Note, however, that from the standpoint of the recovered infant, maternal retrievals were associated with maternal security, this factor interacting (conflictually?) with information about its mother's nonassertiveness.

In the fifth category of preinitiative agonistic interaction, low-born mothers are passively displaced by high-born infants. As a result, high-born infants might learn directly about the submissiveness of the mothers of low-born peers, whereas the offspring of the displaced mothers might acquire information about their mothers' status relative to high-born peers.

Finally, in the context of agonistic interaction between mothers, the infants of aggressive or supplanting mothers might acquire, through observation, information about their mothers' physical power relative to that of low-born mothers. Reciprocally, low-born infants might learn about their mother's submissive attitude toward the mothers of high-born peers.

Table 17.2. Information Potentially Acquired by Infants and Older Females
During Preinitiative Agonistic Interactions

Categories of agonistic interactions	High-born infant learning		Older low-born female learning
	By own experience	By observation	By own experience
Interventions against older low-born female	Social contexts of support. Identities of own allies	Allies' power relative to older female	Social contexts of interventions. Identities and relative power of infant's allies
High-born infant displaces older low-born female	Submissiveness of low-born female		
High-born mother aggresses older low-born female		Mother's power relative to low-born female	Relative power of infant's mother. If aggressor carries offspring, connection between them
High-born mother displaces older low-born female		Submissiveness of low-born female toward own mother	

Preinitiative Agonistic Experience and Rank Relations with Older Females

Four categories of preinitiative agonistic interaction that might lead high-born infants to direct aggression selectively to older lower born females were identified: (1) aggressive interventions in favor of high-born infants against older low-born females, (2) displacements of older low-born females by high-born infants, (3) instances of aggression to older low-born females by high-born mothers, and (4) displacements of older lower-born females by high-born mothers. Table 17.2 summarizes the information potentially acquired by high-born infants and low-born females in the course of each type of preinitiative agonistic interaction.

When a high-born infant receives support against an older low-born female, both the infant and the female might learn about the identity of the infant's allies, the contexts in which they intervene (conflict and nonconflict interventions; same definitions as above), and the mother's physical power relative to the other female. When a high-born infant passively displaces an older low-born female, it has a direct opportunity for learning its status relative to the latter. Finally, when a high-born mother carrying her infant directs aggression or displaces an older low-born female, the infant might acquire information about its mother's physical power and influence on the older female. On the other hand, the low-born female is reminded about the physical power of the high-born mother and the association between the latter and her offspring.

SUBJECTS AND RESEARCH METHODS

Subjects and Husbandry

This research was conducted on a group of Japanese macaques culled from the Arashiyama-West troop (Clark & Mano 1975) in the summer of 1984. The original group was composed of an adult male and three unrelated matrilines with similar age–sex compositions (Fig. 17.2). The subjects of the study were the seven infants born between March and June 1987. These infants were observed over their first year of life until May 1988. No experimental manipulations were carried out during that period. The data on the two male infants were lumped with those on the five female infants. Although the principles governing the rank relations of males and females in species forming matrilineal hierarchies may differ (Lee & Oliver 1979; Johnson 1987; Kuester & Paul 1988; Pereira 1988b, 1989, 1992), in the present case all seven infants acquired their mothers' rank. This pattern is common in Japanese macaques, where males rank at first according to their mothers' position and later (between 2 and 5 years) according to their relative age–size (Koyama 1967). The fact that the two male infants did acquire their mothers' rank implies that their preinitiative agonistic

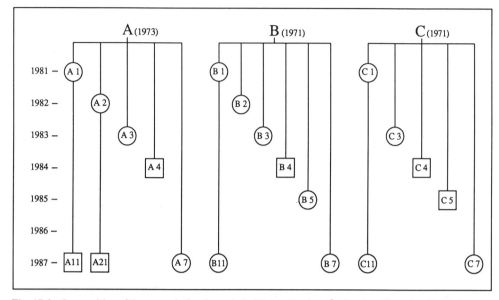

Fig. 17.2. Composition of the group during the period of data collection. ○, females; □, males. The alpha male is not represented.

experience must have been comparable to that of the five female infants; hence their inclusion in the present analysis.

The monkeys occupied three indoor rooms (77 m² × 3 m) and an outdoor pen (30 m² × 5 m) throughout the year. All rooms and the outdoor pen were equipped with cables and perches. The animals were fed daily with monkey chow and various types of seeds distributed in the deep woodchip litter of the indoor rooms. Fruits and vegetables were distributed daily or every 2 days, and water was available at will.

Agonistic Behavior

Aggressive behaviors included open mouth threats, staring, lunging, hitting at, grunting, chasing, pushing, grabbing, wrestling, and biting. Submissive behaviors included displacements, fear grimaces, flights, and screaming. A stable dominance relationship was defined as one in which submission was unidirectional from the subordinate and aggression was unidirectional from the dominant. The members of the A matriline have been consistently dominant to the members of the B and C matrilines, and the members of the B matriline have been con-

sistently dominant to the members of the C matriline since 1984. Most exceptions to this pattern at any one time (i.e., bidirectional submission or aggression) reflected a transitory stage in the dominance relations of immature highborn individuals (Chapais 1988b; Chapais et al. 1991).

For an aggressive intervention to be recorded, the third party had to interfere in an ongoing conflict (i.e., third-party aggressive acts that occurred subsequently to the conflict were excluded). When two or more individuals sided with the same beneficiary simultaneously, every triad (supporter–beneficiary–target) was counted as a separate intervention. When two or more individuals interfered sequentially for the same beneficiary, only the first to enter the fight was retained for the analysis. Subsequent supporters were excluded because it was difficult in that case to identify the individual they were supporting. Effective interventions are defined as those inducing the target to submit to the intervener and to stop threatening the beneficiary.

The intensity of aggression between two individuals was assessed by calculating the proportion of attacks with physical contact out of all aggressive interactions between these two individuals.

Data Collection

The monkeys were observed between 0900 and 1700 hours from a central observation booth. Data were coded on paper in a temporal sequence marked by 1-minute intervals. Each infant was observed daily (weekends included) over its first year using focal sampling (15-minute periods), and the concurrent agonistic interactions of all other infants were recorded *ad libitum*. The seven infants were observed over a total of 994 hours (763 hours of focal sampling, 109 per infant).

Data Analysis

Infant Dyads. Preinitiative agonistic interactions whose direction accorded with the matrilineal rank order (i.e., the interaction was in favor of the high-born infant) are referred to as *pro* interactions, and those whose direction was contrary to the rank order are called *contra* interactions. Under the hypothesis that preinitiative agonistic interactions are responsible for setting up the directionality of dominance among peers, we expect pro interactions to be more frequent than contra interactions (e.g., the high-born infant should receive more support than the low-born infant). This hypothesis was tested by means of two-tailed Wilcoxon signed ranks tests performed separately on each category and subcategory of preinitiative agonistic interaction.

Infant–Older Female Dyads. Dyads composed of a high-born infant and an older low-born female are called pro dyads, while dyads composed of a low-born infant and an older high-born female are called contra dyads. Under the hypothesis that preinitiative agonistic interactions incite infants to direct aggression selectively to lower born females, we expect these interactions to be more frequent for pro dyads. This hypothesis was tested using two-tailed Mann–Whitney tests (Student's *t* tests could not be utilized since the variances differed significantly in a number of cases; *F* tests).

Recall that frequencies of preinitiative agonistic interactions for every pro dyad were calculated for the time period preceding the onset of aggression by infants to low-born females. In the case of contra dyads, low-born infants did not direct consistent aggression to older high-born females. Therefore, frequencies of preinitiative interactions for contra dyads were calculated for the time period preceding the average age at which high-born infants began to direct aggression to lower born females (6 months, see below).

PREINITIATIVE AGONISTIC EXPERIENCE

Preinitiative Agonistic Experience and Rank Relations Among Peers

The seven infants formed 16 dyads of unrelated peers, of which 13 exhibited a stable dominance relationship during the study period (Fig. 17.3). The subordinate individual was observed to perform fear grimaces (the most unidirectional indicator of dominance; see, e.g., de Waal & Luttrell 1985) in 9 of the 13 dyads. Stability was reached at a mean age of 6.0 months (SE = 0.34). The three exceptions were confined to the same infant from the second-ranking matriline (B7), who did not submit to the three infants of the higher ranking matriline. This female received consistent and effective support from the adult male against at least five immature or adult members of the A matriline. This factor was clearly responsible for the three exceptions, since B7 was rapidly outranked by all members of the A matriline when the adult male was later removed temporarily.

Table 17.3 presents the results pertaining to the 13 dyads that reached stable dominance.

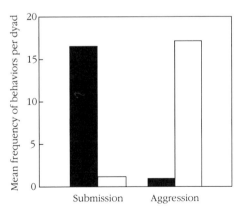

Fig. 17.3. Mean frequencies of submissive and aggressive behaviors per dyad of infants after the onset of stable dominance. ■, performed by low-born infant; □, performed by high-born infant.

Table 17.3. Frequencies of Preinitiative Agonistic Interactions that Might Affect the Onset of Stable Dominance Between Infants

Categories of agonistic interactions	n^a	$\bar{X}$ pro/dyad[b]	$\bar{X}$ contra/dyad[c]	% pro[d]	Wilcoxon tests n pairs	T
Aggressive interventions						
Conflict interventions	11	0.54	0.31	63.6		
Effective only	5	0.31	0.08	80.0		
Nonconflict interventions	37	2.08	0.77	73.0	9	34
Effective only	25	1.77	0.15	92.0	6	21*
Total	48	2.62	1.08	70.8	9	34
Effective only	30	2.08	0.23	90.0	7	28*
Infant's mother aggresses other infant						
Aggressive mother carries offspring	39	2.23	0.77	74.4	9	34
Aggressive mother is alone	108	6.62	1.69	79.6	10	54**
Total	147	8.85	2.46	78.2	12	70**
Infant's mother displaces other infant						
Displacing mother carries offspring	3	0.23	0.0	100.0		
Displacing mother is alone	16	1.23	0.0	100.0	8	36**
Total	19	1.46	0.0	100.0	9	45**
Maternal retrievals of infants						
Conflict retrievals	18	1.08	0.31	77.8	9	29
Nonconflict retrievals	54	3.0	1.15	72.2	12	63
Total	72	4.08	1.46	73.6	10	44
Infant displaces other infant's mother						
Displaced mother carries offspring	62	4.08	0.69	85.5	12	73**
Displaced mother is alone	25	1.38	0.54	72.0	9	35
Total	87	5.46	1.23	81.6	12	70**
Agonistic interactions between mothers[e]						
Aggressor carries offspring	26	2.0	0.0	100.0	9	45**
Victim carries offspring	34	2.62	0.0	100.0	10	55**
Both mothers carry offspring	16	1.23	0.0	100.0	8	36**
Total	76	5.85	0.0	100.0	11	66**

[a]Total frequency of recorded behaviors for that category.
[b]Mean frequency of pro interactions per dyad ($n = 13$ dyads).
[c]Mean frequency of contra interactions per dyad ($n = 13$ dyads).
[d]Percentage of pro interactions on the total of pro and contra interactions.
[e]Includes aggression and displacements.
*Two-tailed $p < 0.05$.
**Two-tailed $p < 0.01$.

This table gives for each category and subcategory of preinitiative agonistic interactions the total frequency of behaviors recorded (pro and contra), the mean frequency of pro interactions per dyad, the mean frequency of contra interactions per dyad, the degree of asymmetry between pro and contra interactions (as measured by the percentage of pro interactions), and the results of the Wilcoxon tests based on the null hypothesis of equal frequencies of pro and contra interactions.

Of the six major categories of preinitiative agonistic interactions, only maternal retrievals did not show a significant asymmetry for pro interactions. However, the difference for the most frequent type of maternal retrievals (nonconflict) was close to significance (two-tailed, $p = 0.06$).

Conflict interventions occurred too infrequently for a test to be carried out. Nonconflict interventions, however, and especially those that were effective, were distributed asymmetrically in favor of high-born infants (asymmetry of 92%). Overall, all effective interventions (conflict and nonconflict) showed a significant asymmetry (90%) in favor of high-born infants. The intensity of pro and contra interventions (i.e., the proportion of interventions with contact-aggression) did not differ significantly [$\chi^2 (1) = 0.10$, $n = 48$, $p = 0.75$; a Wilcoxon test could not be performed]. The mother accounted for 66.7% of all interventions ($n = 48$), the infant's other kin for 16.7%, and nonkin for 16.7%.

High-born mothers directed aggression to low-born infants when carrying their offspring or not, more often than low-born mothers did to high-born infants (asymmetry of 78.2%). The intensity of aggression by high-born and low-born mothers did not differ significantly (Wilcoxon test, $n = 9$, $T = 32$, $p = 0.3$).

The three other categories of preinitiative agonistic interaction showed an asymmetric distribution in favor of the high-born infant. The asymmetry was of 100% for mothers displacing infants, 81.6% for infants displacing mothers, and 100% for agonistic interaction between mothers.

Preinitiative Agonistic Experience and Rank Relations with Older Females

The seven infants formed 30 dyads with lower born females and 26 dyads with higher born females. Dominance relationships between infants and higher born females were stable in all cases: aggression was almost unidirectional from the higher born female (99.8%, $n = 824$ aggressive episodes) and submission was mostly performed by the infant (95.9%, $n = 892$) (Fig. 17.4a). Furthermore, the sub-

missiveness of high-born females toward lower born infants was circumscribed in time: females stopped submitting spontaneously to lower born infants when the latter reached 7 months on average (Fig. 17.5). In contrast, dominance relationships between infants and lower born females were unstable: Both aggression ($n = 649$) and submission ($n = 577$) were bidirectional (Fig. 17.4b).

Although infants were not dominant to older lower born females, they directed aggression selectively to these females. They did so in 24 of the 30 dyads (80%), whereas they directed aggression to higher born females in only 7.7% of the 26 dyads. Furthermore, infants directed aggression to lower born females on average 62 times more often than to higher born females (mean frequencies per dyad: 4.9 vs. 0.08, Fig. 17.4a and 17.4b). Infants began to direct aggression to older lowborn females at 6.2 months ($SE = 0.43$). Thus they recognized lower born females at about the same age they became dominant to lower born peers (6.0 months).

Table 17.4 presents the results on the distribution of preinitiative agonistic interactions affecting infants and older females. The structure of Table 17.4 is similar to that of Table 17.3, but recall that pro and contra refer here to different sets of dyads. The data for the pro dyads are restricted to the 24 dyads in which the infant began to direct aggression to the older female. Conflict interventions were infrequent and occurred as often pro as contra the matrilineal order (asymmetry of 51.6%); this was true for the subset of effective interventions as well (asymmetry of 55.8%). The rates of pro and contra interventions per conflict did not differ either (Mann–Whitney test, $Z = 0.20$, $p = 0.84$). In contrast, nonconflict interventions were more frequent and showed a marked asymmetry in favor of pro dyads (97.3%) (effective interventions: 98.4%). Overall, interventions (conflict and nonconflict) were more frequent for pro dyads. The intensity of interventions (proportion with contact aggression) did not appear to differ significantly between pro dyads (3.8%, $n = 78$) and contra dyads (10%, $n = 10$). The mother accounted for 76.1% of all aggressive interventions ($n = 88$), the infant's other kin for 9.1%, and nonkin for 14.8%.

Infants displaced older low-born females more often than they displaced older high-born females (asymmetry of 83.0%). Finally, ago-

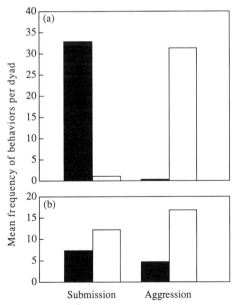

Fig. 17.4. Mean frequencies of submissive and aggressive behaviors (a) between infants and older high-born females and (b) between infants and older low-born females. ■, performed by infant; □, performed by older female.

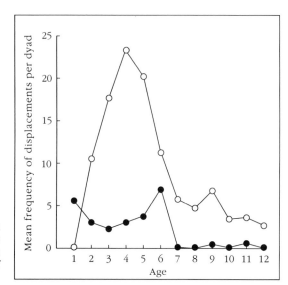

Fig. 17.5. Mean frequencies at which the seven infants passively displaced lower born females (○) and higher born females (●). Frequencies are expressed as rates per dyad per 1000 hours of observation. Age in months.

nistic interactions between mothers and older females (aggression and displacements) were almost totally concentrated within pro dyads.

DETERMINANTS OF THE ONSET OF STABLE DOMINANCE BETWEEN INFANTS

In this study, infants established stable dominance relationships among themselves at a mean age of 6 months (13 of 16 dyads; 9 dyads with fear grimace). Other authors have reported that by the age of 6 to 12 months, most infants

ranked among themselves according to their maternal rank (Berman 1980; Datta 1983a; Horrocks & Hunte 1983a). However, some degree of intergroup and interspecific variation in the age of rank acquisition among peers is to be expected. For example, de Waal and Luttrell (1985) reported that only about 18% of peer dyads had established stable dominance by the age of 1.5 years (criterion: fear grimace) in a captive group of rhesus monkeys. The formation of stable dominance relations among peers probably requires minimal rates of the relevant agonistic interactions (Tables 17.3 and 17.4)

Table 17.4. Frequencies of Preinitiative Agonistic Interactions that Might Affect the Onset of Aggression by Infants to Older Females

Categories of agonistic interactions	n^a	$\bar{X}$/pro dyad[b]	$\bar{X}$/contra dyad[c]	% pro[d]	Z^e
Aggressive interventions					
Conflict interventions	16	0.33	0.31	51.6	0.25
Effective only	13	0.29	0.23	55.8	0.24
Nonconflict interventions	72	2.92	0.08	97.3	4.99***
Effective only	59	2.42	0.04	98.4	4.85***
Total	88	3.25	0.38	89.5	3.97***
Effective only	72	2.71	0.27	90.9	3.96***
Infant displaces older female	132	4.5	0.92	83.0	4.22***
Mother carrying offspring aggresses older female	49	2.0	0.04	98.0	4.42***
Mother carrying offspring displaces older female	51	2.13	0.0	100.0	4.89***

[a]Total frequency of recorded behaviors for that category.
[b]Mean frequency of interactions per pro dyad ($n = 24$ dyads).
[c]Mean frequency of interactions per contra dyad ($n = 26$ dyads).
[d]Percentage of interactions for the pro dyads on all interactions for the pro and contra dyads (adjusted for the difference in their relative numbers).
[e]Z scores corrected for ties, obtained from Mann–Whitney tests.
***$p < 0.001$.

and, possibly, of affiliative interactions (de Waal, Chapter 18, this volume). Therefore, we expect the determinants of such rates (group and hierarchy size, structure of network of kin and nonkin alliances, population density, ranging patterns, proximity structure, etc.) to affect the pace at which rank relations are established. In the present case, group size was small and population density high, these factors increasing the rates of interactions and probably accelerating the process of rank stabilization and alliance formation.

What happened during the first 6 months in the lives of infants that led them to establish rank relations correlating with their mothers' rank? Table 17.3 indicates that preinitiative agonistic interactions were distributed asymmetrically in favor of the high-born infant for most categories. There were a few exceptions, however. Conflict interventions occurred infrequently, in part because the present sample was limited to the period preceding the onset of dominance. Maternal retrievals taking place in nonconflictual situations showed a marked tendency in the pro direction; that is, low-born mothers acted more often than high-born mothers as if anticipating a risky situation. Conflict retrievals, in contrast, were not distributed asymmetrically; that is, both high-born and low-born mothers retrieved their infants when they perceived that their infants might be running a risk.

To assess the relative importance of the various elements composing the early agonistic experience of infants, we shall examine separately the contribution of interactions between (1) high-born mothers and low-born infants, (2) high-born infants and low-born mothers, and (3) the mothers.

Role of Interactions Between High-Born Mothers and Low-Born Infants

Table 17.3 indicates that high-born mothers directed aggression to low-born infants in dyadic contexts (while carrying their offspring or not) and when supporting their offspring against a low-born infant. In theory, aggressive interventions provide a low-born infant with the most direct opportunities for recognizing the allies of a high-born infant (especially its mother, the most frequent ally). But interventions are not the only context fostering the recognition of alliances. When a low-born infant receives ag-

gression from a high-born mother that is carrying her offspring, this physical association may further confirm the alliance between the aggressor and its offspring, and any instance of aggression by the same high-born mother not carrying her offspring may reinforce that alliance, but indirectly. Therefore, these three categories of interactions might combine their effects and lead a low-born infant to defer to a high-born infant. This hypothesis finds support in the results of experiments in which juveniles were placed in situations where they were given more alliance power than a dominant peer. Soon after the allies of the juvenile (its mother or older sister) intervened on its behalf, the dominant peer began submitting to the beneficiary (i.e., before the latter was assertive) (Chapais 1988a). On the basis of this combined evidence, it seems likely that aggression and interventions by high-born mothers to low-born infants affect the directionality of dominance by inducing the low-born infant to submit to the high-born infant, this in turn prompting the assertiveness of the high-born infant.

It should be noted, however, that aggressive episodes between high-born mothers and low-born infants simultaneously allow the high-born infants to observe, and perhaps acquire, their mother's assertive attitude toward low-born infants. Such observational learning might well accelerate the formation of rank relations between infants, but whether it is sufficient for setting the directionality of dominance remains to be ascertained experimentally.

Relevant preliminary experiments have been conducted by Chapais (unpublished data). Experimental subgroups were formed such that a high-born juvenile was outranked by an adult low-born mother (following protocols described in Chapais 1988a). The juvenile daughter of the low-born female was then allowed to observe through a one-way screen her mother interact assertively with the high-born peer. If assertiveness can be learned through observation, then when the two juveniles are put together (without the mother), the observer would be expected to behave more assertively with, and perhaps outrank, the high-born juvenile. To make sure that any change in the rank relation of the juveniles would not be initiated by the high-born peer (who might be more vulnerable due to its having been outranked), the following test was first carried out. The two juveniles were put together after the high-born juvenile had been out-ranked by the low-born mother but *before* the low-born juvenile was allowed to observe them. The experiment was pursued only if the rank relation appeared stable. Three pairs of juvenile peers were tested in both directions; that is, if the juveniles reversed rank following the period of observation, the reciprocal experiment was carried out: the outranked peer was allowed to recover its rank by observing its own mother behave assertively with the other juvenile.

In the first dyad, the juveniles conserved their rank during the control test but reversed rank following the period of observation, and they did so in both directions. In the second dyad, the high-born juvenile submitted to the low-born peer during the control test, this pointing to some vulnerability and/or kin association effects (the experiment was interrupted). In the third dyad, the low-born peer outranked the high-born peer following the period of observation (as expected), but the reciprocal rank reversal did not take place. Overall, these results suggest that observing its mother behave assertively may play a role in one's rank acquisition, but many more experiments and controls are needed before any firm conclusion can be drawn about the origin of the changes induced and the nature of the information transferred.

We conclude that aggression and interventions by high-born mothers to low-born infants are likely to constitute a major component of the early agonistic experience of infants, primarily by inducing the low-born infant to behave submissively and, secondarily, by allowing the low-born infant to behave assertively.

Role of Interactions Between High-Born Infants and Low-Born Mothers

Low-born mothers avoided high-born infants in two contexts: when they were displaced by them and, to a lesser extent, when they retrieved their infants from proximity to a high-born infant (Table 17.3). Thus the offspring of low-born mothers had opportunities to observe their mother avoid high-born peers. On this basis a low-born infant might learn to avoid a high-born infant as, for example, rhesus monkeys watching their mother's reaction to snakes learn to fear snakes (see, e.g., Mineka & Cook 1988). Of course, these two types of situations differ in a number of respects. First, the nature of the be-

havioral responses, physiological mechanisms, and mental processes underlying social fears and the fear of animals may differ (see, e.g., Ohman et al. 1985). Second, an observer can acquire the fear of snakes just as effectively by watching its mother or unrelated models (Mineka & Cook 1988). In contrast, learning to avoid high-born females through observation would require by definition that the model be the mother (or other close kin).

It may nevertheless be reasoned that if a mother consistently avoided a given high-born female, her offspring might develop an avoidance of that female, regardless of the exact nature of the responses and mechanisms involved. But low-born infants received conflictual information from their mother about her dominance relationship with high-born infants: their mother not only was inconsistent in avoiding the high-born infant, but also could direct aggression to high-born infants (females typically resisted the aggressive challenges of high-born females; Fig. 17.4b: 77% of all aggression between high-born infants and low-born females was directed by the female). It would appear, then, that although the submissiveness of low-born mothers to high-born infants might foster some degree of vicarious learning by low-born infants, the submissive attitude of the low-born mother is probably not intense and unidirectional enough to account for rank acquisition among infants.

Role of Interactions Between Mothers

Agonistic interactions between mothers allowed the high-born infants to witness the dominance of their mother toward low-born mothers, and the low-born infants, the submissiveness of their mother toward high-born mothers. Given that these interactions were strongly unidirectional (Table 17.3), they can be expected to affect the dominance relationships of infants with the older females themselves; that is, a low-born infant might learn to defer to a high-born female only by observing its mother's response to that female (according to the above reasoning). But whether agonistic interactions involving mothers may allow infants to learn something about their rank relation with each other remains to be ascertained experimentally. The nature of the information transferred and of the mental processes involved (e.g., some forms of observa-

tional conditioning or cognitive social in-
ference) constitute interesting topics for future
research (see, e.g., Cheney & Seyfarth 1990).

DETERMINANTS OF THE ONSET
OF AGGRESSION TO OLDER
LOWER BORN FEMALES

The present evidence indicates that infants were
able to differentiate between high-born and low-
born females by 6 months of age, directing aggression selectively to the latter. Other authors
similarly reported that infants younger than 1
year were aggressive toward older low-born
females (Datta 1983a).

What kind of experience did infants acquire
during the preinitiative phase of rank acquisition
that led them to challenge older and stronger
low-born females? Low-born females submitted
spontaneously to high-born infants (Table 17.4).
Thus it could be that infants used that single
criterion and challenged the subset of females
that submitted to them. However, infants also
received spontaneous submission from high-
born females (although less frequently, Fig.
17.5), which, in contrast, they did not aggress.
Second and most important, although low-born
females submitted to high-born infants, they
also resisted their challenge (returning aggression), as mentioned above (Fig. 17.4b). Thus
high-born infants would soon learn to stop challenging older and stronger low-born females if
the risks involved were not in some ways reduced.

The data of Table 17.4 indicate that low-born
females received aggression from high-born
mothers carrying their offspring and from the
allies of high-born infants in the context of aggressive interventions. Conflict interventions
were infrequent (in part because the sample was
limited to the period preceding the onset of
aggression) and were not distributed asymmetrically. In contrast, nonconflict interventions were frequent and distributed asymmetrically in favor of the high-born infants.
Thus low-born mothers interfered for their offspring mainly when they appeared to need help,
whereas high-born mothers interfered both
when their infant might be at risk and when this
was not apparently so. Thus low-born females
had plenty of opportunities to learn about the
risks associated with attacking or merely in-

teracting with high-born infants (not to mention
their past experience with the other infants of the
same high-born females). It can be inferred,
therefore, that high-born infants learned to direct aggression selectively to those females they
could challenge at no appreciable costs.

CONCLUSION

The present results suggest that prior to the time
infants can be assigned ranks among themselves, and begin to behave assertively with
larger individuals, they have already acquired a
rich and diversified experience about the organization of power relations in their group. Among
other things, they may know the identities of
their own protectors and those of their group
mates, the social contexts in which these protectors are likely or unlikely to support or retrieve
them, the identities of the individuals they can
challenge at little cost and of those they should
avoid, the power of their mother relative to that
of other group members, and the identities of the
individuals to whom their mothers submit and of
those that submit to their mothers.

It is particularly striking that such knowledge
is imposed on the infants by their mother (and
also, but somewhat later, by their close kin and
some nonkin) through the highly active role
these individuals play in aggressing low-born
individuals in various situations (conflict interventions, nonconflict interventions, aggression
in various dyadic contexts). Such an assertive
(versus reactive) attitude by the allies of high-
born infants is particularly well exemplified by
the observation that most interventions occurred
when the infant did not seem to need help. Thus
high-born females appeared to control and initiate the matrilineal transmission of rank, rather
than act simply as models for the infants. Much
research (especially experimental) remains to be
done before it becomes possible to disentangle
the relative contribution of various individual
and observational forms of learning and information processing involved in rank acquisition.
The present evidence suggests, however, that
the primary source of information available to
infants may well reside in the agonistic interactions between high-born mothers and low-born
infants (interventions and dyadic aggression as
one complex).

On the basis of the present evidence, it appears that much of the fundamentals of matrilineal rank acquisition was already in place around weaning time in this group. In this perspective, the juvenile period appears basically as the continuation of this early phase of rank acquisition (see also Pereira, Chapter 20, this volume): as they grow older, juvenile females become progressively more effective in challenging older lower born females, until they outrank all of them (Walters 1980; Datta 1983b; Pereira 1988b). Nevertheless, an important new element in a female's strategy of integration in the matrilineal hierarchy will emerge during the juvenile period. Immature females will take a progressively more active role in the formation and maintenance of their alliances, by intervening themselves in the fights of others. They will do so by joining opportunistically their kin and nonkin most often against low-born individuals, thus building cooperative partnerships that will reinforce their own rank and the whole matrilineal rank order (for a review, see Chapais 1992). It is also mainly during the juvenile period that the life course of males and females will progressively diverge with regard to the patterning of both their dominance relations and the support they give and receive (reviewed in Pereira 1992).

ACKNOWLEDGMENTS

We thank Michelle Girard, Lucien Goupil, Alain Houle, Hugues Jean, Christiane Mignault, and Shona Teijeiro for their technical assistance, and Lynn Fairbanks, Michael Pereira, Jean Prud'Homme, Pascale Sicotte, and Jeffrey Walters for their helpful comments on the manuscript. This research was funded by the Natural Sciences and Engineering Research Council of Canada, the Fonds FCAR of the province of Québec, and the Université de Montréal.

18
Codevelopment of Dominance Relations and Affiliative Bonds in Rhesus Monkeys

FRANS B. M. DE WAAL

Because adult primates have social relationships that go back many years, even entire lives, the rules of interaction tend to become quite subtle and fine-tuned. This prevents the human observer from determining the causal connections between various components of the relationship, such as the relation between affiliative behavior and support in agonistic encounters (e.g., Seyfarth 1980; de Waal & Luttrell 1986). Do affiliative relationships cause partners to support one another, does support cause an increase in affiliation, or are there other explanations for the observed correlation? The only nonexperimental method to establish cause and effect in this regard is by documenting the sequence of stages through which social relationships change and develop. If characteristic X of a relationship usually develops before Y, and if X increases the likelihood of Y, there is good reason to believe that X facilitates Y. Hence the study of development opens a window on the causal mechanisms underlying primate social organization.

Given the multiplicity of variables, this view will remain somewhat foggy; firm conclusions require experimental evidence. Observational studies are crucial for theoretical development in this area, however, and may thus assist the design of experiments to decipher the causal infrastructure of primate social organization. If, for this purpose, we focus on the *process* of social development rather than on the developing individual, it is good to keep in mind that the two cannot be separated; the process is brought about by interacting individuals and affects their individual futures. The end product is a social position for each young primate, a position with certain characteristics in terms of social rank and affiliative ties with group mates. Inasmuch as these characteristics affect survival and reproduction, the process of social development is a game with high stakes. Rather than with its outcome, however, this study is concerned with the game's rules, particularly those relating to the establishment of dominance over others and the formation of social bonds. The main question is: Are these two processes integrated, or are they relatively independent? To address this question, we need finer distinctions than usually made between various aspects of social dominance.

First, we need to distinguish the communication of relative status by means of ritualized submission from other agonistic and competitive measures. The chimpanzee's (*Pan troglodytes*) bowing and pant grunting and the rhesus macaque's (*Macaca mulatta*) silent teeth baring differ from nonritualized expressions of fear by their extreme directional consistency (Angst 1975; de Waal 1977; Noë et al. 1980; de Waal & Luttrell 1985). That is, if individual A regularly directs one of these signals to B, B will virtually never do the same to A (whereas such reversals occur with a higher frequency in the direction of aggression, avoidance, or the outcome of competition). This has been taken to mean that the

direction of ritualized signals is relatively immune to transient contextual variation, as one
would expect if these signals serve to communicate "agreements" about who is dominant and
who is subordinate. In this view, then, social
dominance is not a mere human construct (Altmann 1981), but is recognized by the animals
themselves. That is, ritualized status signals reflect the animals' evaluation of their position
relative to that of other group members (de Waal
1986a, 1989b).

Most of the time, particularly in cercopithecine monkeys, the overlap in direction between
ritualized and nonritualized agonistic behavior
is so pronounced that the importance of the
above distinction may not be immediately apparent. Let me give one exception, therefore, to
illustrate the point. All members of a stumptail
macaque (*Macaca arctoides*) group, including
adult males, will scatter when a newborn infant
strays from its mother, yet submissive signals
during such events are shown exclusively by individuals ranking below the mother, and are
generally aimed at the mother, not the infant
(Gouzoules 1975; F. de Waal, personal observation). Thus unless one is prepared to assign top
ranks to helpless infants, ritualized submission
needs to be given greater weight than mere withdrawal in the determination of rank. Infants may
be feared (possibly relating to the risk of maternal intervention; Chapais & Gauthier, Chapter
17, this volume), but they are by no means recognized as dominant.

Second, it is important to consider inter- and
intraspecific variation in the nature of dominance relationships as reflected in the intensity
of aggression and the degree of social tolerance.
Maslow (1940) already compared the easy-
going, plastic dominance relations of the chimpanzee with the strictly enforced rank orders of
macaques and baboons. If *dominance style* is
the way that dominants treat subordinates in potentially competitive contexts, styles range from
"egalitarian" to "despotic," and differ between
even closely related species (de Waal & Luttrell
1989).

At the proximate level, dominance style is
expected to correlate with two variables: (1) the
security of the dominant's position, and (2) the
degree of mutual dependence between dominant
and subordinate. The primate literature contains
ample indications that a clear-cut acknowledgment of the dominant's rank by the subordinate

is a prerequisite for a relaxed relationship (reviewed by de Waal 1986a). If a potential rival
(usually an individual of the same age class and
sex) fails to communicate submission, the dominant appears to perceive its position as threatened resulting in aggressive attempts to settle
the issue. If, on the other hand, relative ranks
are clearly communicated, there is less need for
the dominant to enforce its position in an aggressive manner. Note, again, the cognitive interpretation; that is, social processes are explained on the basis of the animals' evaluations
of the state of their relationships. In this model,
named the *reconciled hierarchy* (de Waal
1986a), formalization of dominance relations
(i.e., their expression in ritualized communication) makes room for tolerance and reconciliation among competitors.

It is further to be expected that an important
determinant of dominance style is the degree to
which parties depend on one another for cooperation. In species with long-term social relationships, such as the primates, competition is
probably constrained by the partner's social value. Insofar as the dominant's rank depends on
agonistic support from a particular subordinate,
for instance, intolerant treatment of this ally by
the dominant may result in a discontinuation of
support, hence an undermining of the dominant's own position (e.g., de Waal 1986b). The
potential damage that competitiveness may
inflict on relationships partly explains why priority of access to resources is not always enforced among primates, and why correspondence between the formal rank order and the
distribution of resources is at best imperfect and
occasionally reversed (de Waal 1989b). In the
Arnhem chimpanzee colony, for example,
females often claim resources from adult males
that physically and formally dominate them. As
female support in this colony may influence the
rank order among males, female precedence has
been explained as the result of male concessions
to females because of the benefits, such as agonistic support, that females can provide to males
(Noë et al. 1980; de Waal 1982; see also Smuts
1987).

Proximate explanations of variation in dominance style across relationships, are paralleled
by socioecological explanations of the evolution
of species-typical dominance styles as a balance
between cooperation with and exploitation of
subordinate group members. This balance is as-

sumed to depend on the adaptive significance of group life, hence the cost of leaving the group (e.g., predation risks). The higher this cost, the more freedom dominants have in appropriating resources. Conversely, if dominants are genetically related to subordinates or derive important benefits from their presence, competitiveness will be mitigated (Vehrencamp 1983; van Schaik 1989).

In summary, primates engage in a variety of noncompetitive interactions and face ecological challenges that modify the nature of competitive relationships. Social dominance—the primary regulatory mechanism of competition—is therefore best understood while taking into account both competition and the need for affiliation and cooperation. The traditional biological interest in the relation between social dominance, on the one hand, and access to resources and reproductive success, on the other, covers only one aspect (e.g., Popp & DeVore 1979; Fedigan 1982; Shively 1985); it needs to be supplemented with research on socioecology, variation in dominance style, and the role of dominance in conflict resolution. My purpose here is to apply this wider perspective to the study of social development in rhesus macaques, particularly the process of rank acquisition and its relation to affiliative relationships.

This area is of particular interest because of the sharp distinction by some developmental psychologists between prosocial and aggressive behavior (e.g., Mussen & Eisenberg-Berg 1977; Rheingold & Hay 1978; but see Strayer & Noël 1986). By implication, aggressive and dominance-related behavior are considered antisocial. Whatever the heuristic relevance of such a dichotomy, it is of limited value in the real world as conflict is an inevitable and integrated part of human and animal sociality. Aggression is by no means incompatible with social bonding; on the contrary, the two variables are positively correlated in macaques (Kurland 1977; Bernstein & Ehardt 1986a; de Waal & Luttrell 1986, 1989). This correlation may seem paradoxical, yet we know that aggressive incidents are relatively often followed by reconciliation among kin as well as among nonkin with close relationships (de Waal & Yoshihara 1983; de Waal & Ren 1988; York & Rowell 1988; Aureli et al. 1989; Cords & Aureli, Chapter 19, this volume). The existence of these social mechanisms strongly argues for an ap-

proach in which aggression, dominance, affiliation, and cooperation are considered in conjunction.

ACCESS TO RESOURCES

Because it is considered the main reason why animals invest energy into achieving high ranks, access to resources is a central variable in the analysis of social dominance. This variable was investigated along with social variables, treated in subsequent sections, in order to trace the development of dominance relations of young rhesus monkeys.

A large breeding group of monkeys, housed in an indoor–outdoor enclosure, was subjected to drinking tests after 3 hours of water deprivation. During each test, water was accessible in a basin large enough for four to eight simultaneously drinking monkeys. These tests allowed determination of both the drinking order and tolerance relations (de Waal 1986c). The following results concern the performance of 42 young males and 55 young females, born from 1979 through 1988, in 66 drinking tests videotaped from 1983 through 1988.

Participants in a test were defined as individuals that either drank from the basin or showed other interest in the water (e.g., by licking spilled water off the rocks around the basin). All participants received a rank number dependent on their first sip from the basin, with participants that had not actually gained access to the basin sharing the bottom position. After averaging the drinking ranks for each individual per 1-year period, average ranks of mother and offspring were compared. A positive difference means that the offspring generally drank before the mother; a negative difference means that the offspring generally drank after her. The magnitude of the difference indicates the number of individuals ranking between the offspring and its mother. Figure 18.1 shows the results for daughters and sons separately. The average ranks of mother and offspring were compared pairwise by means of Wilcoxon tests.

As can be seen, both male and female offspring significantly outranked their mothers in the group's drinking order during the first 2 years of life. Because the drinking order and the formal hierarchy are correlated among adults (de Waal 1986c), the magnitude of the dif-

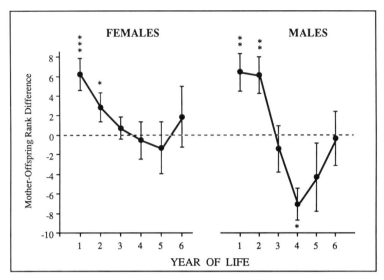

Fig. 18.1. Mean difference ($\pm$ SEM) between the positions of mother and offspring in the group's drinking order (i.e., order of first sip from the water basin). A positive difference means that the offspring drank on average before its mother. Mother and offspring mean drinking ranks have been compared pairwise with a Wilcoxon test (***$p < 0.001$; **$p < 0.01$; *$p < 0.05$, two-tailed).

ference indicates that infants and yearlings tended to outrank individuals that dominated their mothers. Following this period of great access to the basin, the drinking ranks of daughters became indistinguishable from their mothers', whereas those of sons dipped through a low point during their fourth year, significantly below their mothers'. Only a minority of males could be kept in the group beyond this age (the median age of male emigration in rhesus monkeys; Colvin 1983b); males that did remain in the group achieved drinking ranks similar to their mother's in their sixth year.

These results indicate that the relation between social dominance and access to resources is not simple. The high drinking ranks of infants and yearlings obviously reflect tolerance by dominant group members rather than dominance by the youngsters. Infants and yearlings never supplanted members of higher ranking matrilines; they just drank with or before some of the individuals that were avoided by their mothers. The youngsters themselves began to systematically avoid these individuals only in their third year of life. Apparently, rhesus monkeys are not fully integrated into the established hierarchy until this age, and priority of access is determined by more than one factor: social dominance is one, tolerance is another,

and resource value is no doubt a third factor.

As a result of the interplay among these factors there is room, even within the strict hierarchy of this species, for relationships that upset the general pattern of dominance-based access to resources. One of our females, for example, achieved drinking priority, even as an adult, over monkeys ranking well above her own matriline because of a special friendship she had developed at an early age with a particular high-ranking peer (de Waal 1989b). One might argue that this sort of exception, including the observed tolerance toward infants and yearlings, will disappear with increased motivation. Under sufficiently extreme food or water deprivation, the correspondence between social dominance and access to resources is indeed very strong (e.g., Richards 1974; Weisbard & Goy 1976; O'Keefe et al. 1983), and there can be no doubt that this condition does occur under natural conditions and has played a role in the evolution of social dominance (e.g., Dittus 1979; Wrangham 1981; van Noordwijk & van Schaik 1987). Yet this should not distract students of proximate social mechanisms from the fact that the outcome of competition is determined by the interplay of several factors that cannot be collapsed to social dominance.

To determine how infants and juveniles were

SECOND YEAR OF LIFE

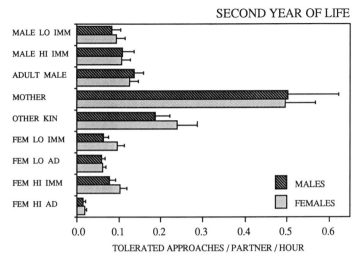

Fig. 18.2. Approaches by male and female yearling rhesus monkeys leading to tolerant interaction during drinking tests. The approach rates have been corrected for the number of potential partners per category and the total test duration; that is, rates are means (+ SEM) per subject per partner per hour. Nine partner categories are distinguished. Except for mother and other matrilineal kin, all categories concern unrelated individuals. FEM, female; LO, lower ranking matriline; HI, higher ranking matriline; IMM, immature; AD, adult.

treated by various partner categories, a comparison was made of interactions around the water basin. Predictions from Hamilton's (1964) kin selection model were tested—that juveniles will receive most tolerance from known kin, some tolerance from possible kin, and least tolerance from known nonkin. These three categories correspond with, respectively, matrilineal relatives, adult males, and unrelated adult females. Tolerance was expressed as the hourly rate of approaches leading to a tolerant interaction (i.e., sharing of the water basin) corrected for the number of individuals per partner category. Figure 18.2 confirms the predicted pattern for the second year of life. Juveniles of both sexes most frequently approached and drank with their mother and other matrilineal kin. The nonkin category most frequently approached were adult males, and the category least frequently approached were unrelated adult females, particularly females ranking above the juveniles' mother. This result also applied to the first and third year.

Interactions with unrelated dominant adults of both sexes were compared to explore these differences further (Fig. 18.3). The two measures considered were (1) the rate of tolerated approaches by juveniles, and (2) the rate of aggressive and nonaggressive exclusions preventing the juvenile's access to the resource (usually a mere approach by the adult caused retreat by the juvenile). The upper half of the graph shows that juveniles discriminated sharply between the two partner categories, approaching adult males far more frequently than high-ranking adult females. Tested for male and female juveniles separately, the difference was significant in both sexes during the first three years of life (Wilcoxon, two-tailed, $p < 0.001$ for the first and second years, and $p < 0.05$ for the third year).

The lower half of Figure 18.3 shows that exclusions by unrelated dominant adults tend to increase with the juvenile's age, except for a sharp drop during the fourth year for exclusions received by juvenile males. Juvenile females are excluded significantly more often by adult females than by adult males during the first 3 years of life (Wilcoxon, two-tailed, $p < 0.01$ during the first two years, and $p < 0.05$ during the third year). Male juveniles show a nonsignificant trend in the same direction.

The conclusion is that juveniles, in their relations with nonkin, meet more tolerance from adult male partners than from high-ranking adult female partners. Their relation with both categories changes over time with perhaps the most dramatic shift from the third to the fourth year. The extremely low interaction rates in

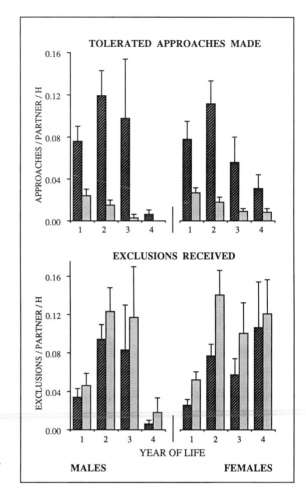

Fig. 18.3. Comparison of interactions around the water basin by young rhesus monkeys with unrelated adult males and unrelated adult females of higher ranking matrilines. The top graph provides for male (left) and female (right) subjects the tolerated approach rate toward adult male (hatched bars) and female (gray bars) partners. The bottom graph provides the rate of received exclusions from adult male and female partners. As in Figure 18.2, rates have been corrected for total test duration and number of potential partners of each type.

males, in terms of both tolerant and intolerant interactions, suggest that males occupy rather peripheral social positions during their fourth year. Whereas females do not show a similar withdrawal from the scene of competition at this age, their rate of tolerant encounters with adult males drops well below that of intolerant encounters, so that this relation comes to resemble the generally intolerant relation young females have since the second year with dominant adult females.

RANK ACQUISITION

Age Difference
and Formal Rank Establishment

Dominance ranks of female cercopithecine monkeys are determined by their mothers'

ranks. As adults, females usually outrank (1) their older sisters and (2) unrelated females dominated by their mother. These so-called matrilineal hierarchies are well documented in the genera *Macaca*, *Papio*, and *Cercopithecus*. Being remarkably stable over time (relative to male hierarchies), these hierarchies provide a highly predictable element to an otherwise variable social organization (e.g., Kawai 1958; Kawamura 1958; Sade 1967; Missakian 1972; Angst 1975; Hausfater et al. 1982; Chapais 1992). Dominance ranks in these species are in fact so predictable that the dependent variable investigated here is not whether but when young monkeys reach their expected positions.

The timing of this event depends very much on the employed criterion of dominance. If we accept threat behavior and approach–retreat interactions as indicators of dominance, infants of

Fig. 18.4. Two 14-month-old female rhesus monkeys. The one on the left shows bared teeth, a sign that her peer has reached formal dominance.

less than 1 year show already signs of dominance over adult members of lower ranking matrilines (Berman 1980; Horrocks & Hunte 1983a; Chapais & Gauthier, Chapter 17, this volume). Yet dominance over adults is obviously incomplete at this early age; finer grained analyses reveal a progressive domination that cannot be called complete until the youngster has demonstrated the capacity to elicit particular submissive gestures. Because, as discussed before, ritualized status signals are almost completely unidirectional, receipt of these signals marks a sharp, irreversible change in the relationship at which rank becomes relatively context free. Ritualized submission by members of lower ranking matrilines acknowledges a juvenile's formal dominance. In the rhesus

monkey (as in a number of other macaque species), the silent bared-teeth display is the criterion variable for this moment in time (Fig. 18.4). Except among age peers, the stage of "first submission" is usually not reached before the young monkey is several years old (de Waal 1977; Walters 1980; de Waal & Luttrell 1985; Pereira 1988b).

Data on rank acquisition were collected over a period of 7 years by means of *ad libitum* observation of bared-teeth displays during thousands of hours of observation. Results in the present and following sections concern 29 female subjects from the second through the fifth year of life. For many subjects, only part of this 4-year period was covered so that sample sizes vary with the ontogenetic time window considered.

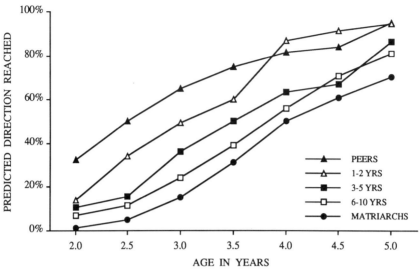

Fig. 18.5. Establishment of formal dominance by females born into higher ranking matrilines over same-age or older females of lower ranking matrilines. The graph provides the percentage of dyads, at a particular age of the subject, in which the subject's higher rank was formally acknowledged by the partner's silent teeth baring. Five partner categories are distinguished dependent on the age difference with the subject: peers (females born in the same year as the subject); 1–2, 3–5, and 6–10 yrs (number of years that the partner is older than the subject), and matriarchs (feral-born founding females of the group).

No data were collected on infants of less than 1 year, and males were not included because of the relatively small number of males kept in the group beyond the age of 2.

We would predict that rank acquisition depends on the balance of forces between the young female and the individuals to be dominated, such as their relative fighting ability and aggressive energy, agonistic alliances with both kin and nonkin, and social skills. One of the most important variables is probably the difference in age, and thus size and strength, between the youngster and her opponent. In addition, older individuals have greater social experience, and may have developed more effective and more stable alliances than younger individuals. For these reasons, we expect a negative correlation between the disparity in age and the speed with which young females establish formal dominance over older females. The first indications for such a relation were provided by de Waal (1977), Datta (1983a), and de Waal and Luttrell (1985). Data from the third study have been incorporated in this analysis.

Same-aged or older females of lower ranking matrilines were divided into five categories de-

pendent on their age when the subject was born. Figure 18.5 shows for each of these partner categories the proportion of dyads in which formal dominance was established by the young female, ranging in age between 2 and 5 years. The predicted correlation is clearly illustrated, with young females first outranking their age peers, and last the group's feral-born matriarchs (with estimated ages of 20 years or older). For 15 subjects with five or more female partners in lower ranking matrilines, Spearman correlations were calculated between the age at which formal dominance was first establishment over a partner (in months) and the disparity in age with this partner (in years). The mean ($\pm$ SEM) correlation coefficient was 0.45 ± 0.08, and a majority of individual coefficients exceeded zero (Wilcoxon, $z = 3.04$, $p < 0.01$).

Effect of Agonistic and Affiliative Relationships

The next question was how rank acquisition relates to other aspects of social relationships. For this more comprehensive analysis, three dyadic measures were considered in addition to the age disparity treated above: (1) matrilineal rank distance (i.e., the number of matrilines ranking between the subject's and the partner's matriline), (2) agonistic encounters per hour, and (3) proximity per 100 scan samples (i.e., contact modes of association, such as contact sitting, grooming, and huddling). The two behavioral measures were collected during standard observations (de Waal & Luttrell 1986). No distinction was made between agonistic incidents in which the high-born subject or her partner was the first to show aggression (i.e., 76.5 and 23.5% of 978 incidents, respectively), as the first act of overt aggression does not necessarily reflect confrontational initiative.

If rank acquisition depends on aggressive tactics, we predict a positive correlation between the rate of aggressive encounters and the speed of rank establishment by the member of a higher ranking matriline. We would further predict that partners at the bottom of the rank order can be outranked with less effort than partners ranking closely below the subject. This is based on the assumption that a matriline's position in the rank order is a function of its combined fighting power and the amount of support received from other matrilines, which is greater the higher the rank of the matriline (Chapais 1988b, 1992; Pereira 1992). Besides being dependent on aggressive strategies, however, we expect the establishment of dominance to depend on the nature of the affiliative relation between the two individuals concerned. According to the reconciled hierarchy model, bonding between two individuals and formalization of their dominance relationship are interdependent (de Waal 1986a).

Previous analyses, not reported here, demonstrated that both the rate of aggression and the proximity relation are significantly affected by age disparity and rank distance. A multiple regression design is necessary, therefore, to tease apart the role of each variable in the process of rank acquisition. For this purpose, a large number of dyadic relationships were analyzed while ignoring individual contributions to the variation. Information on 241 dyads covered a time window starting at or before the subject's age of 24 months and ending at or after she had reached the age of 42 months. The age at which a subject first elicited teeth baring in partners of lower ranking matrilines—the criterion variable in this analysis—was arbitrarily put at 43

Table 18.1. Multiple Regression Analysis on 241 Dyads Between Young Females and Unrelated Same-Age or Older Females of Lower Ranking Matrilines

	r	pr	t
Agonistic behavior	−0.286	−0.132	2.05
Proximity	−0.346	−0.284	4.55
Rank disparity	−0.067	−0.159	2.48
Age disparity	0.401	0.329	5.36

Note: The analysis compares four dyadic variables (i.e., hourly rate of agonistic behavior, percentage of scan samples spent in proximity, disparity in matrilineal rank, and disparity in age) with one criterion variable: the age at which the young female achieves formal dominance. The table provides Pearson correlations (r) and partial correlations (pr), as well as t values. The multiple correlation coefficient was $R = 0.518$.

months for subjects not achieving this stage during the selected time window.

Table 18.1 shows Pearson correlations (r) and partial correlations (pr) between the four independent variables and the age of dominance establishment. Negative correlations mean that a formal dominance relationship was established at an early age in dyads with a high score on the independent variable; that is, these correlations suggest a facilitating effect of the variable on rank acquisition. In the same way, a positive correlation suggests an inhibitory effect. The multiple correlation coefficient was 0.518. In order of decreasing importance (based on partial correlations), early dominance establishment over unrelated females of lower-ranking matrilines was associated with (1) similarity in age, (2) frequent affiliation, (3) disparity in matrilineal rank, and (4) frequent aggressive encounters. By far the most important independent contributions, however, were made by the first two variables, and the correlation with rank disparity was ambiguous.*

---

*Because the number of dyads exceeds the number of individual subjects, data points are not independent. The degrees of freedom lay between the number of individual subjects minus one and the number of dyads minus one; the t values in Table 18.1 are significant even by the second, more conservative estimate (df = 28, $p < 0.05$, two-tailed). In a second analysis of the same data, Spearman correlations were calculated separately for each subject with five or more partners. A significant bias was found toward positive correlations with age disparity and rank disparity, and toward negative correlations with agonistic behavior and proximity (Wilcoxon tests on individual

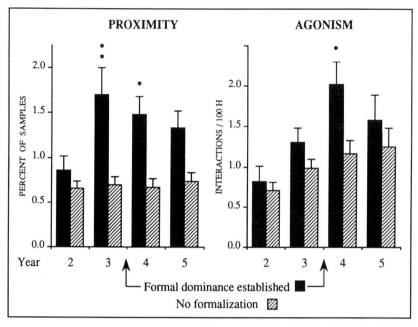

Fig. 18.6. Comparison of formalized dyads (i.e., formal dominance is established around the age of 3 years; black bars) and unformalized dyads (hatched bars). The graph on the left provides the mean (+ SEM) dyadic proximity rate,—that is, the percentage of scan samples spent in affiliative contact or grooming. The graph on the right provides the mean (+ SEM) dyadic rate of aggressive incidents per 100 hours. Means for the two dyadic categories have been compared across individual subjects with Wilcoxon tests (**$p < 0.01$; *$p < 0.05$, one-tailed).

Causal Directionality

The observed correlation between affiliative behavior and the establishment of formal dominance raises questions about causal direction. Does an affiliative bond facilitate the establishment of a clear-cut dominance relation, or does the dominance relation develop first, resulting in increased affiliation? Whereas in the previous analysis behavioral data were taken together over the entire study period to arrive at measures of aggression and proximity, the question of causality requires a sequential analysis from year to year. Two subsets of dyadic relationships were compared: (1) 60 dyads in which the young

correlations; $p < 0.05$). Note that this result (1) concerns direct correlations, not partial correlations, and (2) confirms three of the four correlations in Table 18.1 only. The negative overall correlation with rank disparity (Table 18.1) was contradicted by a majority of positive individual correlations.

female achieved formal dominance around the age of 3 years (i.e., within an age range from 30 to 42 months, with a mean of 36.4 months), and (2) 110 dyads in which formal dominance was not achieved before the time window closed at 42 months. These two categories are referred to here as formalized and unformalized dyads, respectively.

Comparison of these categories allows the determination of changes in affiliative and agonistic behavior before and after the establishment of formal dominance, and a comparison with a "control" group of dyads in which dominance relations failed to reach this stage. The statistical comparison is a one-tailed Wilcoxon test on mean dyadic rates per individual subject. In view of previous results, both proximity and aggression rates can be expected to be higher in formalized dyads; the question here is when this increase occurred.

Figure 18.6 presents the mean rates (+ SEM) for all dyads combined. The graph shows that during the second year of life the proximity rate in formalized and unformalized dyads was approximately the same. This changed in the year preceding the establishment of formal dominance: for a significant majority of subjects, the proximity rate of formalized dyads was higher than that of the control group. This situation continued into subsequent years, although the difference became less pronounced. A similar analysis of the rate of aggressive encounters shows a gradual increase of this rate in both dyadic categories, and a significantly higher rate of aggression in formalized dyads during the year immediately following dominance establishment.

It appears, then, that development of a close affiliative tie with a member of a lower ranking matriline facilitates the establishment of formal dominance over this individual, and that this event in turn facilitates aggressive interaction.

DISCUSSION

The affiliative component of dyadic relationships may be a better predictor of the speed of rank acquisition than the aggressive component. Such a link between affiliation and dominance relationships is expected within the reconciled hierarchy model, which regards the formal acknowledgment of rank differences as a prerequisite for cooperative relations among potential competitors (de Waal 1986a).

There exists a contradiction, however, in that the present data suggest that the affiliative relation develops before rather than after the establishment of formal dominance. This sequence needs further attention, as there are several possible explanations. A simple explanation would be that frequent proximity increases opportunities for agonistic interaction and hence of the establishment of a clear-cut dominance relation. It is also possible that relationships develop from social attraction to cooperation (both reflected in high proximity rates), and that this transformation requires the establishment of formal dominance by one of the two parties. Such a sequence would fit the reconciled hierarchy model, which views formalized dominance as a convention that serves to regulate conflict. The need for such a convention would obviously be greatest between individuals that combine conflict potential with a need for cooperation (de Waal 1986a).

When Kummer et al. (1978) demonstrated that male hamadryas baboons (*Papio hamadryas*) most readily initiate fights with individuals with whom they enjoy the closest relationships, the authors speculated that aggression might be less inhibited in intimate relationships, as these relationships are too firmly established to be vulnerable to its disruptive effects. A similar argument might apply to the developing relationships of young rhesus females in which aggressive behavior appears to reach its peak following the establishment of a close affiliative tie and a formalized dominance relation. In these relationships, the outcome of aggression may be more predictable, and the resilience to disruption greater, both of which may reduce the risks of open conflict. Probably, this holds for relationships among unrelated adult rhesus monkeys as well, as these relationships show a positive correlation between proximity and aggression rates (de Waal & Luttrell 1986). This leads to the hypothesis that aggression is expressed more freely between partners that have come to "trust" one another—that is, who know one another intimately and agree on the roles of dominant and subordinate within the relationship.

In the same way as the paradox of high aggression levels among closely associated individuals may be resolved by taking into account patterns of reconciliation, opportunities for conflict, the intensity of aggression, and so on, a finer grained analysis may resolve the apparent contradiction between the present findings and Walters's (1980) *targeting* hypothesis. Accordingly, young female baboons selectively aim aggression at particular (targeted) opponents in an effort to outrank them. The data on this phenomenon are ambiguous, and the investigator recognized the alternative possibility that "adolescents became more aggressive toward adults when the adults began exhibiting submissive gestures toward them" (Walters 1980, p. 71). This explanation also may apply to the present finding of high aggression rates following rather than prior to formal dominance establishment.

Yet instead of rejecting the targeting hypothesis on these grounds, it is entirely possible that targeting is expressed in a special type rather than in the overall frequency of aggressive encounters. Dependent on whether it is initiated by established dominants or challengers, aggressive behavior may differ in intensity, duration, and form. De Waal (1977) demonstrated that long-tailed macaques (*Macaca fascicularis*) show different aggressive patterns against subordinate opponents than against opponents that they do not (yet) dominate. The second type of aggression is characterized by vocal threats and appeals to third parties similar to the head-flagging observed in juvenile baboons, particularly females, against their targets (Walters 1980; Pereira 1989).

It is evident from the present study, however, that the process of rank acquisition cannot be understood on the basis of aggressive behavior alone. Affiliative and dominance relations in rhesus monkeys appear to *codevelop*; that is, these two components of the dyadic relationship develop in continuous interaction, each component influencing the development of the other. The process culminates in an integration of the young monkey into the adult hierarchy, meaning that the young monkey's dominance relations are firmly acknowledged, and intimate relationships are formed with both higher and lower ranking individuals. Despite reports of very early signs of a young monkey's dominance rank, full hierarchical integration is not reached until a female rhesus monkey is close to adult age. Before this time, the majority of her dominance relations are still informal, and access to resources is often greater than expected on the basis of matrilineal descent.

The latter phenomenon appears to be largely due to tolerance received from dominant individuals rather than to dominance by the youngsters themselves. The period of lenient treatment lasts approximately 2 years. We found that, similar to Japanese macaques (*Macaca fuscata*), tolerance is highest among kin (Yamada 1963). And similar to baboons (*Papio cynocephalus*) and capuchin monkeys (*Cebus*

sp.), adult males are the most tolerant nonkin (Janson 1986; Pereira 1988a). We failed to observe, however, the "favoritism" of dominant males toward male youngsters reported for toque macaques (*M. sinica*) (Dittus 1979); in our study, juveniles of both sexes had tolerant relations with adult males until the age of 4.

By the time tolerance toward them diminishes, the youngsters themselves begin to establish formal dominance over the members of lower ranking matrilines. As integration into the natal group is more important for females than for males, because of the natural tendency for male emigration in cercopithecine monkeys, we may expect the rules for formal rank establishment and its co-development with affiliative relationships to be different for each sex. Although this study hardly addresses sex differences, it does provide a first indication: a male's access to a contested resource drops significantly below that of his matriline when he reaches the age of 4, whereas a female's access does not. Whatever the male's fighting capacity at this age (which likely exceeds that of a same-age female), this suggests that he has not reached the degree of hierarchical integration that allows females to claim the access to resources commensurate with their matrilineal rank.

ACKNOWLEDGMENTS

I am grateful to Lesleigh Luttrell for many hours of data collection, and to Kurt Sladky for assistance with the drinking tests. The editors, Bernard Chapais, and two anonymous referees provided very helpful comments. Research was supported by grants from the National Science Foundation to the author (BNS-8311959 and BNS-8616853) and the National Institutes of Health to the Wisconsin Regional Primate Research Center (RR-00167).

19

Patterns of Reconciliation Among Juvenile Long-Tailed Macaques

MARINA CORDS and FILIPPO AURELI

This chapter focuses on one aspect of juvenile social life in macaques—the way in which juveniles cope with aggression. Aggression is a potential problem for all animals living in stable groups, for group integrity must persist despite the disruptive effects of aggression. For juvenile cercopithecine monkeys, coping with aggression may be particularly important for two reasons. First, juveniles are often the victims of aggression (Silk et al. 1981; Pereira 1988b), particularly of contact forms of aggression (Bernstein & Ehardt 1985b). Second, their size-related vulnerability, to both predators and aggressive conspecifics, makes them depend on relationships with their groupmates for protection and support (Pereira 1988a).

Before considering in greater detail the social and ecological demands facing juvenile macaques, we briefly review coping strategies available to animals generally when they are victims of aggression. Through appeasement and redirection onto other individuals, conflicts can be mitigated or shortened (Cords 1988b; Aureli et al. 1989; Aureli & van Schaik 1992a). We focus here, however, on "reconciliations" that occur after the exchange of aggressive behavior has ceased. The term "reconciliation" refers to a restorative, homeostatic process that sets a social relationship back on course after it has been disturbed by aggressive conflict (de Waal 1989a). Reconciliation has been recognized operationally in the friendly reunions between selectively attracted former opponents

that occur shortly after fights. Friendly reunions between former adversaries have been demonstrated explicitly in several species of nonhuman primates (de Waal & Yoshihara 1983; Judge 1983; Cords 1988b; de Waal & Ren 1988; York & Rowell 1988; Aureli et al. 1989; Cheney & Seyfarth 1989; de Waal 1989a; Judge 1991; Ren et al. 1991; Aureli 1992), and there is anecdotal support for their ubiquity among relatively long-lived social mammals (Kruuk 1972; Schaller 1972; Rasa 1977; de Waal & van Roosmalen 1979; Thierry 1986).

The reconciliatory function of friendly post-conflict reunions has long been assumed, but only recently demonstrated. Experiments on long-tailed macaques showed that operationally defined "reconciliation" did indeed restore social relationships after aggression (Cords 1992). A dyadic tolerance test, in which two long-tailed macaques could drink juice from closely spaced bottles, revealed that preceding aggression increased the time it took for the subordinate to co-drink with the dominant, and decreased the amount of time the subordinate spent co-drinking relative to baseline levels. If friendly reunions followed aggression, baseline latencies to and durations of co-drinking were reestablished. Appropriate controls ensured that it was the friendly reunion per se, rather than simple elapsed time, that increased compatibility after reconciliation. Reconciliation both increased tolerance by the dominant individual and decreased fear in the subordinate.

Friendly postconflict reunions also affect individual victims of aggression by reducing their tension (Aureli et al. 1989; Aureli & van Schaik 1992b). Behavior that indicates tension (references in Aureli & van Schaik 1992b), such as scratching, body shaking, and self-grooming, occurs at elevated levels following aggressive conflict. If postconflict reunion with the aggressor occurs, however, rates of these behaviors return to low levels. Changes in tension parallel changes in the likelihood of receiving further aggression from other group members (not necessarily original opponents). Relative to controls, victims are attacked at higher rates after the original conflict if no reunion took place, but not after reunions occurred.

The two studies summarized above were carried out on subjects of various age classes, but the same results are obtained when analysis is limited to juveniles. The results suggest that reconciliation is a behavioral mechanism relevant both to the well-being of individuals and to the well-being of their social relationships.

Our goal in this chapter is to investigate whether the occurrence of reconciliation after conflicts reflects the life history and the social interests of juvenile long-tailed macaques (*Macaca fascicularis*). To address these issues, we must consider what it is like to be a juvenile member of this species. There have been few reports on the life of juveniles in the wild. Based on spacing patterns in a Sumatran population, van Schaik and van Noordwijk (1986) suggested that juveniles' first priority is keeping safe, which means remaining in the main part of the group, and close to other monkeys (see also Pereira 1988a). Adults and subadults, by contrast, are more often found alone or in smaller splinter parties, apparently as the result of intragroup feeding competition, different dietary requirements, and reduced vulnerability. Like other macaques, females typically remain in their natal groups for life, whereas males emigrate when 5–6 years old (van Noordwijk & van Schaik 1985).

Juvenile social tactics are known primarily from captive groups (but see van Noordwijk et al., Chapter 6, this volume). Even here the data are limited, however, since only one study of long-tailed macaques has concentrated on the juvenile period (Netto & van Hooff 1986). The social trajectories of juveniles correspond to their lifetime residence patterns: juvenile females progressively concentrate their social activity on other females in their group. In their third year, they overtake their older sisters in dominance rank, assuming their "adult" positions in the hierarchy; coalition formation probably facilitates this process (de Waal 1977; see also Datta 1983a and Chapais 1988a for other macaque species). High rank in adult females is associated with more efficient foraging, greater safety from predators, and higher reproductive rates (van Noordwijk & van Schaik 1987). Juvenile males spend progressively less time with adult females, and associate relatively often with other young males (personal observation; also Handen & Rodman 1980 for *M. radiata*). The importance of relationships developed in the natal group to males' future lives is not known in this species, although there is some evidence from rhesus macaques that those relationships can be important later (Boelkins & Wilson 1972; Meikle & Vessey 1981; Berard 1990). *M. fascicularis* males tend to leave their natal groups in the company of familiar peers and join groups containing males they already know (van Noordwijk & van Schaik 1985; van Noordwijk et al., Chapter 6, this volume).

In this chapter, we present both observational and experimental data on aggression and subsequent reconciliation for juvenile macaque victims. Observational data come from spontaneous conflicts in intact social groups. Experimental data come from provoked conflicts in subgroups that were temporarily convened. Together these data provide complementary views of reconciliation behavior in our subjects.

OBSERVATIONAL DATA

Methods

Subjects. Data were collected from two captive groups of *M. fascicularis*. Both groups were established several years prior to study (Cords 1988b; Aureli et al. 1989). The Utrecht (U) group was observed from March to August 1987, when it consisted of 26 individuals (one adult male, five subadult males lacking full adult musculature, nine adult females, three juvenile females less than 3.5 years old, four juvenile males less than 4 years old, and four infants less than a year old). The Zurich (Z) group was studied from April to October in 1988 and 1989, when it consisted of 39 individuals (four adult males, six subadult males, 17 adult females, three juvenile males less than 4 years old, eight juvenile females less than 4 years old, and one infant less than a year old). Two females and one male were counted as juveniles in 1988, but as adult or subadult in 1989, when their ages exceeded 4 years.

In captive *M. fascicularis* females, menarche has been reported to occur at 2–3 years of age (Dang 1983; Honjo et al. 1984). In our captive groups, females typically gave birth for the first time at 4–4.5 years. We distinguished juvenile and subadult males according to morphological changes that occurred at about 4–5 years, when their canines began to grow larger than those of females, and testicular size began to increase above juvenile levels (van Schaik et al., in preparation).

Data Collection and Analysis. Observations were made when the monkeys occupied the outdoor areas of their quarters (75 m² for U group; 850 m² for Z group). We present two different types of observational data. First, baseline data on affiliative behavior (sitting within 1 m, contact sitting, or grooming), and on the frequency of aggression and interventions in aggression, were collected together with our co-workers M. Moonen and K. Vocking (U group) and C. Stamm (1990, Z group), using scan samples (for affiliation) and all occurrences records (for aggression and coalition formation). These data were collected in the months just preceding (U group) and concurrent with (both groups) the periods when postconflict data were taken. The affiliation data did not include reconciliations (see below).

Second, for both groups we followed the observation procedure of de Waal and Yoshihara (1983) to monitor the postconflict behavior of contestants in spontaneous aggressive conflicts. Briefly, one of the opponents was observed after an aggressive conflict, and its behavior was compared with that observed during a control period with no preceding aggression. There were some differences in the details of this protocol as applied to the two study groups. In U group, all observations were made on the victims of aggression—the animal that showed submissive behavior (or, where no submission was shown, the one initially attacked). Observation periods lasted 10 minutes. Control observation periods were made exactly 24 hours after the postconflict period, unless the subject was

involved in agonistic interactions ≤ 3 minutes before the scheduled start. In such cases, the start of the control period was postponed for 10 minutes. In Z group, about half the observations were made on victims of aggression, and half were made on aggressors. Observation periods lasted 15 minutes. Control periods occurred at least a day after postconflict periods, at roughly the same time of day (i.e., morning), when opponents had not been involved in aggression for at least 10 minutes, and when the two opponents had come to within 2 m of each other, were aware of each other's presence (as indicated by visual orientation), and were not engaged in social interaction with third parties. These conditions were adopted so that contestants would be as "available" to each other at the start of the control period as they would be just after a conflict. We acknowledge differences in the holding conditions of our animals and in our methods; where results from the two groups were similar, however, such differences only emphasize their robustness.

All social interactions between former opponents, and between the focal animal and other group members, were noted with the exact time of their occurrence since the beginning of the observation period. Only interactions in which there was affiliative physical contact (e.g., grooming, mounting, brief touching, sitting in contact) or in which friendly gestures were directed by the aggressor to the victim (e.g., lip smacking, genital presenting, or eyebrow raising) were considered as possible reconciliations (Fig. 19.1). (In this species, eyebrow raising signals readiness for friendly contact, and differs from raised-brow threats in terms of associated piloerection, ear position, and posture; see "mimen" in Angst 1974.) We designated as "reconciled" those conflicts in which the opponents had such a friendly reunion sooner after a conflict than in the matched control. (The relative merits of this and other operational definitions of reconciliation have been discussed by Aureli et al. [1989] and Cords [1993], who show that different definitions lead to remarkably similar conclusions.)

The length of observation periods limited a priori those reunions labeled as reconciliations: conflicts we classified as "unreconciled" might have been reconciled had we observed for longer periods. It seems to be a general pattern among macaques, however, that former oppo-

Fig. 19.1. Reconciliation between two juvenile males. The dominant aggressor (left) had approached his former opponent with raised eyebrows; here the opponent returns direct eye-to-eye contact while grabbing for the dominant's genitals and lipsmacking. (Photo, M. Maag)

nents increase their rate of affiliative interaction for only the first few minutes after a conflict (de Waal & Yoshihara 1983; Cords 1988b; de Waal & Ren 1988; Aureli et al. 1989; Aureli 1992). Some observations of Z group were extended to 45 minutes, but the percentage of reconciled conflicts as assessed by these data differed little from that calculated from 15-minute observation periods. We believe, therefore, that our observations were long enough to recognize reconciliation accurately.

In the analysis, we consider only cases in which juveniles were victims of aggression. We have analyzed data from the two groups separately. Data from U group were collected from 81 different dyads. One to eight paired samples were collected per dyad, so that there were 197 paired observation periods in the complete data set. Data from Z group were collected from 77 different dyads. Only one postconflict and one control sample were made for each dyad of animals. Most analyses were carried out at the level of individual victims. Where samples were small for individuals, we have used Fisher exact tests or G tests on pooled data. Where G tests gave significant results, we used William's correction to calculate G_{adj} (Sokal & Rohlf 1981). We regard these pooled results as preliminary, and consider them most meaningful when the patterns observed in both groups coincide. In such cases, we calculated a combined probability over both tests using Fisher's technique (Sokal & Rohlf 1981). Statistical tests were one-tailed unless otherwise noted.

Table 19.1. Observed and Expected Rates of Aggression (per Hour) Between Members of Different Age–Sex Classes

Recipients	Aggressor			
	Juvenile females	Juvenile males	Adult females	Adult and subadult males
Juvenile females	**0.704** ▲△	**0.112**	**2.296** ▲△	**1.716** ▲▽
	(0.372)	(0.124)	(1.120)	(0.684)
	[0.528]	[0.156]	[1.680]	[2.464]
Juvenile males	**0.112**	**0.032**	**0.480**	**1.040** ▲△
	(0.124)	(0.040)	(0.372)	(0.228)
	[0.180]	[0.056]	[0.580]	[0.848]
Adult females	**0.664** ▼	**0.232** ▼	**1.764** ▼	**2.876** ▲
	(1.120)	(0.372)	(3.364)	(2.056)
	[0.604]	[0.180]	[1.928]	[2.824]
Adult and subadult males	**0.072** ▼▽	**0.088** ▼	**0.408** ▼▽	**1.620** ▲△
	(0.684)	(0.228)	(2.056)	(1.256)
	[0.240]	[0.072]	[0.760]	[1.116]

Note: Data from Stamm (1990). Only interactions including pursuits, lunges, biting, and fleeing were counted. Threats usually occurred as well. Interactions consisting of threats only were not scored, because they would be too easily missed. Expected values based on the representation of different age–sex classes in the group are given in parentheses, while those based on marginal totals (see text) are given in brackets. Arrowheads indicate observed values that were significantly higher (pointing up) or significantly lower (pointing down) than expected values (see text for test of significance). Closed arrowheads relate to the expected values in parentheses, while open arrowheads relate to the values in brackets. Tests were made on frequency data; observed and expected frequency data have been transformed to hourly rates here to facilitate comparison with other studies.

Results

Distribution of Aggression. To establish the social milieu in which juveniles live, we first examined how often members of different age–sex classes gave and received aggression in Z group (250 observation hours, May–September 1989; Table 19.1; Stamm 1990). Expected values were based on the representation of different age–sex classes in the group. A second set of expected values was based on the probabilities of the actor age–sex class giving and the recipient age–sex class receiving aggression—that is, based on the marginal totals of the table. The significance of differences between observed and expected frequencies was calculated as described by Bernstein and Ehardt (1986b), with two-tailed p set at less than 0.003 for each comparison so that the overall probability of erroneously rejecting the null hypothesis in any test was less than 0.05.

Based on their proportional representation, juveniles of both sexes received more aggression than expected from adult and subadult males, and juvenile females received more than expected from adult females. That adult aggression is sex specific becomes clear when the overall tendency of different age–sex classes to par-

ticipate in aggression is taken into account: juvenile females received disproportionately more aggression from adult females, whereas juvenile males received disproportionately more aggression from mature males. Adults generally received less aggression than expected, except from mature males. Examination of scores for individual aggressors and recipients showed that these patterns of aggression were general.

Table 19.2 shows the rate of aggression received per time spent in proximity. The highest values are for juveniles receiving aggression from mature males and from unrelated adult females (seven of eight juveniles showed this pattern). These values are on average at least double those for adult recipients, or for juveniles receiving aggression from their peers. Also, juveniles received aggression at much lower rates from their mothers and adult sisters than from other nonrelated (or more distantly related) adult females (Wilcoxon matched pairs signed ranks test, $n = 8$, $T = 0$, two-tailed $p <$ 0.01).

Reconciliation by Juveniles After Spontaneous Conflicts. One step toward demonstrating rec-

Table 19.2. Rate of Aggression Received per Amount of Time Spent in Proximity by Different Age–Sex Classes

Recipients	Aggressors				
	Juvenile females	Juvenile males	Adult females		Adult and subadult males
			Kin	Nonkin	
Juvenile females	2.06	1.32	0.90	7.23	8.52
(n = 6)	(0.46)	(0.40)	(0.69)	(2.01)	(1.85)
Juvenile males	1.32	0.48	3.31	5.58	7.16
(n = 2)	(0.86)	(0.01)	(2.05)	(0.14)	(3.79)
Adult females	1.10	2.54	0.44		1.95
(n = 18)	(0.27)	(0.56)	(0.10)		(0.28)
Adult and subadult males	0.24	0.45	0.35		3.92
(n = 11)	(0.11)	(0.24)	(0.19)		(1.39)

Note: Values are averages of individual recipients, with standard errors given in parentheses below. For each individual, the number of aggressions received was divided by the percentage of that individual's samples in which a member of the aggressor age–sex class was within 1 m. For juveniles, the rate of aggression received from adult females that were related (mothers or sisters) or unrelated (all others) was differentiated.

onciliation is to show that former opponents are "attracted" after aggression (de Waal & Yoshihara 1983). "Attracted" pairs are those in which the former opponents had a friendly reunion only in the postconflict period or earlier in that period than in the matched control. "Dispersed" pairs are those in which reunion occurred only or earlier in the control period. In U group, there were 35 attracted and 5 dispersed pairs (binomial $p < 0.001$ for the combined data based on a 1:1 expectation). All seven juveniles were more often attracted to their former opponents than dispersed (sign test at individual level, $p = 0.008$). In Z group, there were 21 attracted and 11 dispersed pairs (binomial test, $p = 0.056$), and six of eight individuals were more often attracted to former opponents than dispersed (sign test, $p = 0.145$). If the sample size for Z group was increased by including those conflicts in which juveniles participated as aggressors ($n = 23$), then the tendency to be attracted to former opponents reached significance in this group as well (binomial $p < 0.009$, sign test $p = 0.02$).

In neither group did we find a difference in the frequency with which juvenile and adult victims reconciled (proportion of conflicts reconciled for juvenile vs. adult victims, 35 of 197 vs. 72 of 330 in U group and 21 of 77 vs. 38 of 167 in Z group, Mann–Whitney U tests, two-tailed $p > 0.05$). Nor did the form of reconciliation differ between juvenile and adult victims. In both cases, friendly reunions were most likely to occur within the first 3 minutes after a conflict, and they involved a variety of behavior types. Long-

tailed macaques do not use special behavior to reconcile (Cords 1988b; Aureli et al. 1989). Juveniles were more likely to play with their former opponent than adults, and adults were more likely to groom, but these differences would be expected based on the general behavioral repertoires of adults and immatures.

The question of who initiates reconciliation is difficult to answer with small data sets: comparison with control values is required, but former opponents interacted rarely in control periods. Considering victims of all age classes in U group, Aureli et al. (1989) found that victims initiated 75% of reconciliatory interactions with the former opponent, significantly more than the 42% of affiliative interactions they initiated during controls. In Z group, however, victims (of all ages combined) initiated similar proportions of reconciliations (26 of 51) and first affiliative interactions in controls (7 of 9; $G = 2.36$, two-tailed $p < 0.05$). In these comparisons, different dyads contributed to postconflict and control samples. In the nine Z-group dyads that interacted in both paired samples, the victim was equally likely to initiate reconciliatory and control encounters (McNemar test).

Reconciliation and Conflict Characteristics. In an attempt to explain variation in the occurrence of reconciliation, we first considered characteristics of the conflict itself that might correlate with subsequent reconciliation. Aggression in conflicts was divided into contact (e.g., hits, bites, and holds) and noncontact (e.g., threats and chases) forms. In U group,

25% of contact aggressions ($n = 71$) were reconciled, whereas only 14% of noncontact aggressions ($n = 126$) were reconciled. In Z group, 36% of contact aggressions ($n = 22$) and 23% of noncontact aggressions ($n = 57$) were reconciled. Although these results follow a similar pattern, the likelihood of reconciliation after contact and noncontact aggression is not significantly different in either group (Wilcoxon matched pairs signed ranks test, two-tailed, $n = 7$, $T = 5$ for U group and $n = 8$, $T = 15$ for Z group; combined $p > 0.05$).

Next we considered whether the decidedness of interactions was correlated with the likelihood of reconciliation. In decided interactions, only one opponent clearly signaled submission (e.g., teeth baring or fleeing). Aureli et al. (1989), considering victims of all ages in U group, had found that undecided interactions were especially likely to be reconciled. Among juvenile victims, however, the decidedness of the interactions was not related to the likelihood of reconciliation in either group. In U group, 29% of undecided and 17% of decided interactions ($n = 17$ and 180, respectively) were reconciled, while in Z group 29% of undecided and 26% of decided interactions ($n = 7$ and 69, respectively) were reconciled (Wilcoxon test, two-tailed, $n = 7$, $T = 5$ for U group and $n = 6$, $T = 9.5$ for Z group; combined $p > 0.05$).

Reconciliation and Relationship Characteristics.

The nature of the opponents' social relationship might influence the likelihood of reconciliation. We predicted that juveniles would seek reconciliation more often with opponents with whom they normally had their most valuable relationships—that is, those that most fostered their well-being (Kummer 1978).

First we compared the likelihood of reconciliation after conflicts in which juveniles had been the victims of related and unrelated aggressors (taking mothers or siblings as kin). In neither group did likelihood of reconciliation differ with aggressors of these two classes: 30% of conflicts with kin and 16% conflicts with non-kin ($n = 23$ and 146, respectively) were reconciled in U group (Wilcoxon test $n = 6$, $T = 10$), while 14% of conflicts with kin and 29% with nonkin ($n = 7$ and 70, respectively) were reconciled in Z group (Wilcoxon test $n = 4$, $T = 2$).

Although close kin are usually dependable social partners with whom juvenile macaques groom frequently and from whom they receive agonistic support, they are not the only animals with whom juveniles form such relationships: close relationships among peers and important support from unrelated adults have also been reported (e.g., Colvin & Tissier 1985; Netto & van Hooff 1986). We next looked at the effect of relationships more directly, by asking whether juvenile victims reconciled more with supporters than with nonsupporters. Baseline data indicated which groupmates had ever been observed to support a particular juvenile victim within a 6- to 7-month period that included the postconflict record. In neither group did juveniles reconcile more frequently with supporters than with nonsupporters. In U group, 19% of conflicts with supporters ($n = 106$) and 14% of conflicts with nonsupporters ($n = 88$) were reconciled ($G = 0.97$, $p > 0.05$), while in Z group 23% of conflicts with supporters ($n = 13$) and 28% of conflicts with nonsupporters ($n = 64$) were reconciled ($G = 0.143$, $p > 0.05$; combined $p > 0.05$).

Finally, we characterized relationships according to rates of affiliation and aggression. Affiliation scores reflected proximity relations measured during group scans: in U group, the score represented the time a juvenile spent grooming or sitting in contact with another group member, while in Z group, sitting within 1 m was also included. The rate at which a juvenile received aggression from a given opponent was calculated by dividing the number of aggressions received by the proximity score above. In "friendly" relationships, the opponent was in the top quartile of the juvenile's proximity scores, but in the bottom three quartiles of its aggressive rate scores. In "unfriendly" relationships, the opponent was in the top quartile of aggressive rate scores, and in the bottom three quartiles of proximity scores. Dyads whose affiliation and aggressive rate scores were both high or both low were dropped from the analysis. The top quartile consistently provided a natural break in the distributions of behavior scores for each juvenile.

In both groups, two to three times more conflicts were reconciled between partners with friendly relationships than between partners with unfriendly relationships (Fig. 19.2; Wilcoxon test, $n = 7$, $T = 1$, $p < 0.016$ for U group; $G_{adj} = 1.81$, $p < 0.10$ for Z group; combined $p < 0.02$).

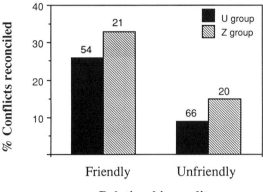

Fig. 19.2. The percentage of aggressive interactions followed by reconciliation in relation to the quality of the relationship between opponents. (See the text for criteria used for quality categories.) Numbers above the bars indicate the number of interactions.

Sex Differences. In view of the different life trajectories of males and females, we expected to find sex differences in reconciliation behavior. Overall, however, males and females reconciled equally often in both groups. In U group, 15% of conflicts of male victims ($n = 110$) and 21% of conflicts of female victims ($n = 87$) were reconciled ($G = 0.85$, two-tailed $p > 0.05$), while in Z group, 35% of conflicts of male victims ($n = 20$) and 25% of conflicts of female victims ($n = 57$) were reconciled ($G = 0.203$, two-tailed $p > 0.05$). Results were similar when analysis was limited to conflicts with nonrelatives; conflicts with relatives were too infrequent to analyze for sex differences. In nearly all cases (91% in U group, 93% in Z group), juvenile victims had lower genealogical ranks than their opponents; if only these cases are considered, there were still no sex differences in the likelihood of reconciling with (higher ranked) opponents.

We did find some sex differences when we considered the age–sex class of unrelated opponents. In U group, juvenile females reconciled more than juvenile males with adult females (9 of 37 vs. 1 of 39, $G_{adj} = 8.40$, two-tailed $p < 0.01$). In Z group there was a trend in the same direction (6 of 22 for female victims, 1 of 6 for male victims), but it was not significant ($G = 0.302$, two-tailed $p > 0.05$; combined $p < 0.05$). In neither group were sex differences evident in the rates of reconciliation with mature males.

Fights between juveniles were rare. Nonetheless, in both groups males reconciled more often than females when they were victims of peer attacks. In U group, 5 of 16 peer conflicts with male victims and 0 of 10 peer conflicts with female victims were reconciled ($G_{adj} = 5.02$, two-tailed $p < 0.05$). In Z group, 4 of 5 peer conflicts with male victims and 3 of 13 peer conflicts with female victims were reconciled (Fisher exact test, two-tailed $p < 0.05$). In both groups, male victims reconciled more than females regardless of the sex of the juvenile opponent.

EXPERIMENTAL DATA

Methods

In the experiments, aggressive conflicts were provoked by the experimenter. It was thus possible to make repeated measures in a uniform context on particular dyads, including some whose spontaneous aggressive interactions occurred infrequently (e.g., pairs of juveniles). Data were collected on dyads in subgroups of 7–10 animals (from Z group only) that were convened for 2–3 hours per day in an indoor compartment of the cage (24 m²). Aggression was provoked by offering the lower ranking member of a dyad a tidbit while its higher ranking partner was looking on. By carefully timing the provocation, it was possible to limit the interaction to a particular dyad (for details, see Cords

1988b). Typically, the sort of aggression shown included a lunge and open-mouthed threat, with occasional hits and grabs. Provoked aggression was never as violent as the extreme cases seen in spontaneous encounters.

The three subgroups studied consisted of (1) a mixture of adult females and juveniles, (2) all juvenile females (3–5 years old, all nulliparous), and (3) all juvenile males (3–5 years old, all with juvenile body musculature). The eight subject dyads in the first subgroup were mixed-age pairs in which a juvenile received aggression from a higher ranking adult female. Eight dyads of juvenile females and six dyads of juvenile males were sampled. Most dyads were sampled 10–12 times each. Within each subgroup, half of the sampled dyads were relatives (mother–offspring pairs in the mixed-age group, sisters among the females, and brothers or first cousin agemates among the males). In the all-juvenile subgroups, each of three males and four females were paired in one test dyad with a relative and in another with a nonrelative. To test the effect of kinship on reconciliation rate, Fisher exact tests were carried out for each of these seven individuals, and p values were combined, separately for males and for females, in the way described by Sokal and Rohlf (1981). The results on juvenile males have been detailed elsewhere (Cords 1988b).

Observation procedures were similar to those used for the entire Z group (above and Cords 1988b). Simple sitting in proximity was included as a possible reconciliatory behavior if the former opponents remained more than 2 seconds within 50 cm of each other. (Experiments on the function of reconciliation, briefly described in the introduction, indicated that simple sitting in proximity was as effective as more overt affiliation in repairing social relationships; Cords 1993) These interactions were included because others that were more overtly friendly (such as grooming or play) were rare, in any context, especially among female juveniles and among nonkin. Conflicts were designated as "reconciled" if the former opponents interacted earlier after a conflict than in the matched control.

Initiation of Reconciliation. Former opponents in the experimental situation interacted more often, in both postconflict and control periods, than opponents in spontaneous conflicts in the entire Z group (above). Nonaggressive interactions between former opponents (including sitting in proximity) occurred in 42% of postconflict periods in the whole group ($n = 77$) and in 81% of experimental postconflict periods ($n = 228$). Former opponents interacted in 34% of control periods in the whole group ($n = 77$) and in 73% of experimental control periods ($n = 231$). Because of greater limitations on physical and social opportunities, former opponents were probably less likely to ignore each other in the experimental situation.

Because interactions were relatively frequent in both observation periods, and because repeated measures were made on the same animals, the data from the experimental situation are especially suited for an examination of the initiative to reconcile relative to controls. For each of the 20 subject dyads in which reconciliation was observed, the proportion of reconciliations initiated by the juvenile victim was compared with the proportion of first friendly control interactions initiated by the same animal. Across dyads, victims initiated an average of 41% of reconciliatory interactions and 30% of first affiliative control interactions with their former opponents. In roughly half (12 of 20) of the dyads, the victim initiated a greater proportion of reconciliations than of first affiliative control interactions. Because several individuals appeared in more than one dyad, a summarizing statistic would not be valid; but no matter how one groups independent dyads, the likelihood that victims initiate first interactions with their opponents does not differ systematically after a conflict relative to controls. These results thus confirm those derived from observation of spontaneous conflicts in the entire Z group (above).

Reconciliation and Conflict Characteristics. Not all provoked conflicts were reconciled (0–100% across dyads). In 18 dyads showing both forms of aggressions, 47% of contact conflicts ($n = 75$) were reconciled, while 57% of noncontact conflicts ($n = 113$) were reconciled. Five dyads reconciled more often after contact fights, 11 reconciled less often after contact fights, and two showed no difference. The intensity of aggression was thus not closely related to the likelihood of reconciliation.

In only nine dyads were there "undecided" outcomes to conflicts. Overall, 30% of 23 undecided conflicts were reconciled, while 43% of

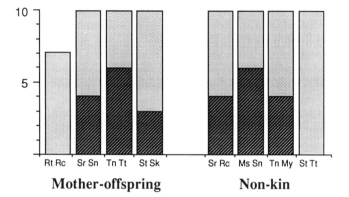

Fig. 19.3. Provoked conflicts that were reconciled by eight different adult–juvenile dyads in which juveniles were attacked. The dark hatched area represents reconciled conflicts, while the stippled area represents nonreconciled conflicts. The ordinate gives the number of conflicts.

68 decided interactions were reconciled. Six dyads reconciled less often after undecided conflicts, two reconciled more often, and one never reconciled. There was no close correlation between the decidedness of conflict and the likelihood of reconciliation.

Reconciliation and Relationship Characteristics. The main comparison in the experimental data is between kin and nonkin within each of the subgroups. In the mixed-age subgroup, reconciliation was equally likely when juveniles were victims of aggression by their mothers or by unrelated adult females (Fig. 19.3). In the juvenile subgroups, however, nonrelatives reconciled more often than did relatives (Fig. 19.4; Fisher exact test, two-tailed combined p's < 0.002 for females and for males), and males tended to reconcile with each other more often than did females (Mann–Whitney U test, two-tailed $p = 0.056$ for kin, two-tailed $p = 0.114$ for nonkin).

In the all-juvenile subgroups, fights between kin were as intense as those between nonkin (Kolmogorov tests, $p > 0.05$). Relatives, however, had more relaxed relationships than nonrelatives. Approach–retreat interactions between males were less correlated with formal dominance (as measured by bared teeth; de Waal & Luttrell 1985) among kin than among nonkin (Cords 1988b), and each of the four females that was paired with a sister and a nonrelative spent more time sitting in contact with her sister during control periods. Also, in choice tests conducted on the females, every subject spent more time sitting on the side of the cage nearest her

sister than the side nearest the unrelated partner (M. Cords, unpublished data).

DISCUSSION

Our results confirm other studies of cercopithecine monkeys (Silk et al. 1981; Bernstein & Ehardt 1985b; Pereira 1988b) that showed that juveniles receive much aggression, especially from unrelated adults. Because a top ecological priority for juveniles is avoiding predation, and this entails maintaining close proximity to other monkeys, one would expect juveniles to behave in ways that mitigate the effects of aggression received and that encourage tolerance by aggressors. We have shown that early in life, juvenile macaques do develop the ability to reconcile, and they reconcile at rates similar to those of adult victims. There is evidence that reconciliation allows juvenile victims to restore tolerance and individual tension levels quickly to baseline levels.

Before assessing the significance of the patterning of this behavior in juveniles' lives, it is useful to consider the function of reconciliation generally. So far, tests of the effects of reconciliation (see the introduction) have been limited to those of a very short-term nature (a few minutes postconflict). Although more difficult to demonstrate, it is possible and often assumed that reconciliation also has longer term effects. These may result from the cumulation of short-term effects, so that, for example, tolerance levels in the dominant partner increase over longer time spans, or chronic stress in the subordinate

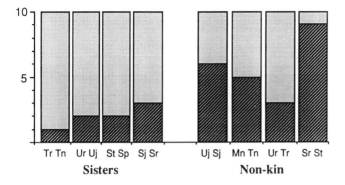

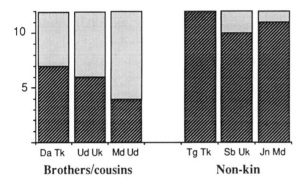

Fig. 19.4. Provoked conflicts that were reconciled by eight dyads of juvenile female peers (top) and six dyads of juvenile male peers (bottom). Legend for Figure 19.3 explains format.

is avoided. The occurrence of reconciliation may additionally communicate both to former opponents and to their groupmates a mutual interest in the persistence of their relationship (de Waal 1986a; for a similar idea, see Zahavi 1977). For juvenile macaques, both time scales are potentially relevant: reconciliation can have an immediate effect on their lives as juveniles, and may be part of an investment in future relationships. The relative importance of these two effects may differ depending on the opponent.

In principle, at least three characteristics of a relationship could influence the likelihood of reconciliation after conflicts. First, reconciliation may occur more often after conflicts between opponents that value their relationship. Value should be related to the ways in which the social partners benefit one another (Kummer 1978), over either the short or the long term. Insofar as benefits are asymmetrical, social partners may assess value differently. Second, even when relationships are valuable, reconciliation may be more frequent between individuals whose relationship is less secure (Cords 1988b). These social partners are less able to predict one another's dispositions (i.e., Kummer's [1978] "tendencies" and "availability"), and so their conflicts may require an explicit resolution that reconfirms mutual interest. Again, there may be asymmetries in how each partner assesses the security of the relationship. Finally, the accessibility of the partner may limit the occurrence of reconciliation: in particular, an individual may be more likely to approach its former opponent after a conflict if the dyad has a history of nonaggressive social interaction in other contexts; that is, if the partners are compatible. A less compatible opponent, who is more likely to react aggressively to an approach, is effectively less accessible. To explain the distribution of reconciliation according to a relationship's value and security is to explain it in ultimate terms; compatibility is a proximate constraint.

In practice, the value and security of relationships are difficult to quantify, whereas compatibility can be measured directly. Based on available reports of juvenile life, we made assumptions about the sorts of relationships that should be valuable to juvenile long-tailed macaques. Thus it is likely that social partners that provide agonistic support to juvenile victims are valuable social partners; however, reconciliation rates were not related to whether opponents supported juveniles in other fights. One explanation for this result is that we misjudged the importance of agonistic support. This seems unlikely, however, in view of observations that juveniles are frequent recipients of aggression, that safety requires them to stay in close proximity to groupmates, and that both juvenile rank and integration into the adult hierarchy in macaques depend on agonistic interventions (de Waal 1977; Datta 1983a; Chapais 1988b). A second explanation is that whereas juveniles value their supporters, supporters did not equally value their relationships with juveniles, reflecting the underlying asymmetry in support behavior. Reconciliation should occur most often when it is in the interest of both partners: a monkey can easily prevent reconciliation by ignoring the approaches of its former opponent, or by making itself unavailable through interaction with other individuals. Asymmetries in the value of relationships may thus explain why patterns of reconciliation did not parallel patterns of victim support to juveniles. Finally, the ability of juveniles to reconcile may have been limited by the compatibility of former opponents. Not all supporters are particularly friendly with juveniles: Netto and van Hooff (1986), for example, report considerable victim support to juveniles by the alpha male and by higher ranking unrelated adult females. Also animals with whom juveniles have friendly relations are not always supporters (Colvin & Tissier 1985). Overall, receipt of victim support was not related to the affiliative quality of relationships (i.e., friendly vs. unfriendly) for any of the juveniles in Z group.

Relationships with groupmates may provide juveniles with benefits other than agonistic support. Field reports have stressed the importance

that juveniles appear to place on maintaining spatial positions that provide safety from predators (Janson & van Schaik, Chapter 5, and van Noordwijk et al., Chapter 6, this volume): when group members disperse to feed from scattered, small food patches, juveniles remain in the main body of the group (van Schaik & van Noordwijk 1986). To do so, they must be tolerated by other individuals; by reconciling after conflicts they would promote continued tolerance. Our finding that reconciliation was more likely when a juvenile had a generally more affiliative relationship with its opponent is consistent with this idea. Moreover, at a proximate level, it may be easier for juveniles to reconcile when they have generally affiliative relationships with their former opponents: such animals should be relatively compatible after conflicts as well.

Compatibility may also be related to our findings that kinship did not affect reconciliation rate when juveniles were attacked by adults, but did have an effect when they were attacked by peers. These results also drew our attention to the security of relationships as a potentially important influence on the likelihood of reconciliation. A post hoc explanation of our results might be that kin relationships generally do not require reconciliation because they are inherently secure (Cords 1988b; de Waal & Ren 1988), and that compatibility of nonkin dyads determines rates of reconciliation (see also York & Rowell 1988): juveniles' well-documented compatibility with unrelated peers allows their preferential reconciliation, whereas low compatibility with unrelated adults precludes manifestation of the usual preference for nonkin.

Although kin relationships often appear to be more relaxed than those among nonkin (Colvin & Tissier 1985; Cords 1988b; de Waal & Ren 1988), the idea that reconciliation is generally more necessary for nonkin than for kin remains unconfirmed. Studies of monkeys have shown contrasting kinship effects on reconciliation: in some, kin reconciled more than nonkin (long-tailed macaques: Aureli et al. 1989; rhesus and stump-tailed macaques: de Waal & Ren 1988; patas: York & Rowell 1988), and in others the opposite was found (long-tailed macaques: Cords 1988b; vervets: Cheney & Seyfarth 1989). Because all these species are female bonded, the value of kin relationships should consistently exceed that of nonkin relationships

(Wrangham 1980), and is thus unlikely to explain the different results. In only three of these studies was the compatibility of dyads (measured variously in terms of physical contact, agonism, and agonistic support between dyad members) controlled: in long-tailed and stumptailed macaques, kin still reconciled more than nonkin (de Waal & Ren 1988; Aureli et al. 1989), whereas in rhesus macaques the dyads with more friendly relationships reconciled more often regardless of kinship (de Waal & Yoshihara 1983). It is possible that in different species (or groups) the balance between compatibility and security of kin relationships varies.

In comparing reconciliatory tendencies in dyads with different relationships, we have treated compatibility as characteristic of a dyad. Within dyads, however, there are probably differences in how each partner perceives the compatibility, and hence accessibility, of the other. These differences are likely to arise from asymmetries in relative rank, age, and power of partners, and they may help to explain why juvenile victims did not invariably initiate reconciliations. We would like to make two other comments with regard to initiation of reconciliation. First, successful reconciliation clearly depends on both partners, so ideally one should analyze attempts to reconcile; these are difficult to recognize operationally however. Even in the completed reconciliation attempts we did score, we cannot be sure that we did not miss some subtle cue on the part of the animal being approached. Second, data that fit the criteria for an ideal analysis (i.e., cases in which partners reconciled and interacted in the matched control) are sparse. For all these reasons, we believe that the lack of victim bias in initiation suggested by some subsets of our data does not cast doubt on the presumed benefits of reconciliation for victims.

Whereas our discussion has so far considered the value of juveniles' social relationships primarily over the short term, the subject of sex differences leads us to consider the long-term value of relationships as well. By the time they reach adulthood, males and females have very different relations with members of their natal groups. As juveniles, however, they showed no sex differences in the overall rate of reconciliation or in the likelihood of initiation. Sex differences did emerge, however, when the identity of the aggressor was considered. After

conflicts with unrelated adult females, juvenile females reconciled more often than juvenile males. Since juveniles of both sexes receive aggression from unrelated adult females at high rates, and yet depend on proximity to these (and other) members of their groups for safety, it seems unlikely that the sex difference in reconciliation rate reflects a sex difference in short-term value of relationships with unrelated adult females, or a sex difference in compatibility. The sex difference can be explained, however, by the higher mutual interest of at least some juvenile female–adult female dyads in restoring relationships with potential long-term benefits: only females' relationships with older, unrelated females endure. In addition, adult females may provide important agonistic support to unrelated juvenile females as they begin to integrate themselves into the adult rank hierarchy by challenging groupmates from lower ranking families (Netto & van Hooff 1986; Pereira 1992).

The second sex difference we found—that males reconcile more often than females with same-sex juvenile peers—is more challenging to understand in terms of modal life histories. Males eventually emigrate to other groups and become competitive with peers for mates. Familiar peers often emigrate together (van Noordwijk & van Schaik 1985), and it has been suggested that they may benefit from previously established relationships in the new group (Boelkins & Wilson 1972; Meikle & Vessey 1981; Berard 1990); not all familiar males establish bonds (van Noordwijk & van Schaik 1985), however, and actual evidence that male reproductive success depends on relationships with familiar peers is practically nonexistent. Female peers, by contrast, remain together for life. Although the females may compete for food and safety (van Noordwijk & van Schaik 1987), group living appears to be essential to their reproductive success, and group integrity depends on enduring, stable relationships. The sex difference in postconflict behavior that we found among juvenile peers seems to be quite general, having been found also in small isosexual groups of juvenile rhesus monkeys and among adults in larger groups of rhesus monkeys and chimpanzees (de Waal 1984). In a subsequent analysis of data from the large rhesus group, de Waal (1986d) concluded that the sex difference did not simply result from females reconciling less often, but from their reconciling more selectively with particular opponents. Since lifelong residence patterns of males and females are very different in macaques and chimpanzees, the sex difference in reconciliation is more likely related to some more fundamental sex difference in morphology or behavior. It could reflect common sex differences in risk aversion (males might more often attempt risky reconciliations), in potential for injury (longer male canines might make reconciliation more imperative), or in other benefits of resolving conflicts quickly (more dynamic networks of coalitions among males may require the ability to switch rapidly from aggressive to friendly motivations) (de Waal 1989a). Alternatively, it may reflect species-specific sex differences in the value and security of relationships between same-sex adults. Specifically, chimpanzee females are not gregarious and are less valuable to one another as social resources than are males. Among macaques, both males and females benefit from alliances and tolerance among same-sex individuals; among females, however, such benefits are usually based on kinship, which they will never lose, whereas adult male relationships are less closely tied to kinship and are thus less secure. These possibilities all concern adult characteristics and suggest that juvenile sex differences simply foreshadow them. A relative lack of affiliative interactions among juvenile females is also consistent with reports from other macaque species that these individuals withdraw socially from one another with age (Clarke 1978; Hayaki 1983). It would be informative to examine sex differences in reconciliation with sex peers in species in which females collaborate to control male behavior (e.g., *Erythrocebus, Miopithecus, Saimiri*).

There are no comparative data on juvenile reconciliation outside the genus *Macaca*. Insofar as juveniles of other species suffer as victims of aggression, however, we would expect them also to show conciliatory tendencies that could be investigated in relation to predictions based on normative life-history patterns. These life-history patterns allow one to estimate the value of relationships between juveniles and other group mates. Our results suggest, however, that patterns of reconciliation by juvenile victims reflect not only the value of particular relationships, but also their security and the ease with which reconciliation can occur. Furthermore, although some of our results are consistent with the idea that reconciliation may be used to repair relationships of likely long-term value, it has yet to be established that reconciliation has long-term effects.

To make specific predictions about patterns of reconciliation among juveniles of any species, then, we need detailed knowledge about juvenile life and juvenile life histories. Finally, reconciliation behavior must be put into context as one aspect of affiliative behavior, and as one mechanism among several that can be used to avoid or alleviate conflict.

ACKNOWLEDGMENTS

We thank the editors, reviewers, and R. Steenbeek and E. Sterck for comments on earlier versions of the manuscript. M. C. is grateful to the Harry Frank Guggenheim Foundation for financial support, and to the Swiss National Science Foundation for support of the Zurich colony. F. A. was supported by grants from the Dutch Ministry of Education, the NWO-CNR Dutch-Italian Agreement on Scientific Exchange, and the Ing. A. Gini Foundation. We are grateful to Hans Kummer, Jan van Hooff, and Carel van Schaik for giving us the opportunity to work in their labs, and to M. Moonen, H. Rodel, H. P. Schaub, C. Stamm, K. Vocking, and H. Westland for helping to collect baseline data. M. C. acknowledges E. Kruesi, M. Böhler, and N. Harley for assistance with experiments.

20
Agonistic Interaction, Dominance Relation, and Ontogenetic Trajectories in Ringtailed Lemurs

MICHAEL E. PEREIRA

In 30 years of research on the social behavior of immature and adult Cercopithecinae, important generalities in social organization have been revealed for these Old World monkeys. In the most studied species—savanna baboons, several macaques, and the vervet monkey—social groups typically contain many adult males and females (all subsequent reference to cercopithecines concerns these species; for overview of Cercopithecinae that form smaller groups, see Cords 1987). Females spend their lives in their natal groups, whereas most males disperse around puberty and attempt to transfer into other groups (Pusey & Packer 1987). At any given time in a group, virtually every pair of monkeys exhibits a stable dominance relationship; that is, one party voluntarily signals submissively to the other, often in the absence of aggression (Bernstein 1981). Also, both adult males and adult females tend to exhibit transitive hierarchies of dominance relations: if monkey A dominates monkey B, and B dominates C, then A also dominates C, and so forth.

The dominance hierarchies of cercopithecine adult females are much more stable than those of adult males, often persisting unchanged for many years (e.g., Sade 1972a; Bramblett et al. 1982; Hausfater et al. 1982). Two social forces are responsible for this stability: *kin support,* the female inclination to intervene during conflicts on behalf of closely related females, and "top-

down" *nonkin support,* the female inclination to intervene during conflicts between nonrelatives on behalf of the member of the higher ranking family (Chapais 1992; Chapais & Gauthier, Chapter 17, this volume; Pereira 1992). These forces also virtually ensure matrilineal "inheritance" of dominance status, whereby maturing females assume dominance ranks among adult females adjacent to those of close kin (e.g., Kawai 1958; Sade 1967; Walters 1980; Horrocks & Hunte 1983a).

Male life histories provide several basic contrasts. First, immature males' dominance relations in natal groups are less structured by adult female intervention in relation to matrilineal membership (Pereira 1989, 1992). After transferring to other groups, young adult males live among few, if any, kin and among unrelated peers whose identities typically change at least once each year, as males continue to transfer between groups. Males depend primarily on their abilities to win dyadic fights with peers to acquire and maintain their dominance rank in any given group, and their dominance relations often change several times a year (Hausfater 1975; Sugiyama 1976b; Dittus 1977; Packer 1979a, 1979b; Cheney 1983; van Noordwijk & van Schaik 1988; Noe 1989; Hamilton & Bulger 1990; but see Raleigh & McGuire 1989).

How general are these life-history patterns among group-living primates? Which parts of these behavioral systems are causally related? Do primates that maintain clearly defined dominance relations always exhibit transitive dominance hierarchies? When one sex is philopatric, is frequent kin support invariably seen in that sex? Does agonistic intervention generally function in primates to promote kin and to stabilize dominance relations? Primatologists have been unable to answer these questions because the vast majority of existing data on the development and adult expression of social behavior comes from baboons, macaques, and vervet monkeys, all of which share the entire suite of life-history traits described above. Comparative study of these species provides few tests of hypothetical cause–effect dynamics in these aspects of social organization (but see Pereira 1992).

Like others in this volume, this chapter contributes to the growing effort to amass behavioral data from primates showing distinctly dif-

ferent modes of social organization. It is the first report on behavioral development in a forest-living population of prosimian primates where all individuals' social and reproductive histories are known. My subjects were the juvenile and adolescent members of the two semicaptive groups of ringtailed lemurs (*Lemur catta*) maintained at the Duke University Primate Center (DUPC; see Fig. 20.1; Taylor & Sussman 1985; Pereira & Izard 1989; Pereira & Weiss 1991).

To provide an initial description of ontogenetic trajectories, I present (1) some basic data on physical and social development for eight separate birth cohorts, (2) several analyses of my first 15 months of data on juvenile and adolescent social behavior, and (3) the complete histories of female residence in the two DUPC study groups. Several behavioral analyses were designed specifically to determine whether the ontogenies of social conflict and of dominance relations in female ringtailed lemurs likely par-

Fig. 20.1. Lc2 Group members in DUPC Natural Habitat Enclosure 4: (left to right) adolescent males IK and HC, juvenile male AS, HC's mother B1, IK's mother D1, B1's adolescent sisters B2a and B2b, and the matriarch BB. Because the photograph was taken 1 year after the conclusion of the study, identified subjects were older here than when the present data were collected.

allel or contrast with patterns observed in cercopithecine societies. More detailed analyses of various aspects of ringtailed lemur social behavior are planned for later publication. But initial focus on female life histories is important at this early stage of lemur research because species-typical social structure depends strongly on basic patterns of female sociality (Wrangham 1980; van Schaik & van Hooff 1983; van Schaik 1989).

Ringtailed lemurs are an excellent species to compare with the Cercopithecinae because although they share with them several basic life-history features, they also show important differences. Like the monkeys, for example, ringtailed lemurs exhibit general female philopatry and male transfer between groups comprising several adult males, adult females, and immatures (Jones 1983; Sussman 1991). Also, all direct care for infants is provided by mothers, with extreme interest in infants shown by other females (Klopfer 1972; Klopfer & Boskoff 1979; Gould 1990). Unlike the monkeys, ringtails show brief periods of estrus (4–24 hours; Evans & Goy 1968), mating seasons as short as 1 to 2 weeks within groups (Jolly 1966; Pereira 1991; Sauther 1991), sexual size monomorphism (Kappeler 1990b), and invariable adult female agonistic dominance over males (Jolly 1966; Budnitz & Dainis 1975; Taylor & Sussman 1985; Kappeler 1990a). Structural analysis of social behavior has also revealed similarities and differences (Pereira & Kappeler, in preparation). Primates in both groups maintain stable, public dominance relations, using distinct visual and acoustic signals to communicate agonistic intent. Unlike among monkeys, however, dyadic dominance relations among ringtailed lemurs of either sex are not invariably transitive. When ringtail A dominates ringtail B, and B dominates C, ringtail A can be either dominant or subordinate to C (Fig. 20.2).

THE FOREST ENCLOSURES AND RESEARCH METHODS

Forest Enclosures and Study Groups

The DUPC enclosures total 14.2 ha and contain thousands of native North Carolina trees. Each lemur group's extensive naturalistic foraging is supplemented once daily with approximately

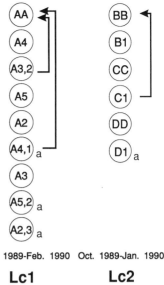

Oct. 1989-Feb. 1990   Oct. 1989-Jan. 1990

Lc1 Group   Lc2 Group

Fig. 20.2. Dominance relations among mature females in each group during the period of dominance stability in the middle of the study (end of premating season 1989 until beginning of birth season 1990). Each female shown dominated all those listed beneath her, except for the two "top-ranking" females, who were stably dominated by the females identified by their upward arrows. The pubescent (A4,1) and two adolescent females (A3,2; C1) each reversed their dominance relations with the fully adult matriarch of their group early in the premating season. Pubescent (young adolescent) females in each group are identified by lowercase *a*.

0.5 kg of Purina monkey chow (#5038) scattered widely at one of several provisioning sites in its core range. Further managerial intervention is limited to occasional exchange of males for outbreeding, removal of lemurs whose lives are jeopardized by injury or aggression (Vick & Pereira 1989), and detention of groups near heated shelters during brief periods of weather below −8°C. Four to 12 supplementary group captures, each 1 to 2 hours in duration, are conducted annually for morphometric measurements, pregnancy palpations, and health examinations.

The first long-term study group, Lc1 Group, was established at the DUPC in 1977 using six captive-born animals: one adult female, her infant and juvenile daughters, an unrelated adult male, and an unrelated juvenile of each sex. This group has lived in the Duke Forest since 1981 and has occupied its current 3.5-ha core area since 1983. Lc2 Group was formed in an adjacent 5.5-ha enclosure in 1987, using three unrelated captive-born females, their infant offspring, and two unrelated captive-born males (Pereira & Izard 1989). The members of each group are individually identifiable and fully habituated to accompaniment by one or two human observers. All matrilineal relationships are known, and, by the time of the present study, all but the oldest animals had lived in the forest since birth or prior to weaning. Three gates between the two groups' areas were opened permanently in March 1990, allowing direct daily contact between the groups and annual male transfer since then. Lc1 and Lc2 Groups behave territorially, allowing only limited range overlap, like the ringtailed lemurs at Berenty, Madagascar (Jolly et al. 1982; Mertl-Millhollen 1988). Further details of the groups' histories have been published elsewhere (Taylor & Sussman 1985; Taylor 1986; Pereira & Izard 1989; Vick & Pereira 1989; Pereira & Weiss 1991).

Subjects and Behavioral Sampling Methods

All lemurs born into the study groups are subjects for my focal sampling (Altmann 1974) between weaning (4–6 months of age) and young adulthood (~2.5 years of age). Thus the immatures of two sequential birth cohorts are always under study—one cohort of juveniles (weaned, prepubertal) and the other of adolescents (postpubertal, prereproductive). During the premating season each year, young adults are discontinued as focal subjects, the pubescent cohort is maintained, and the weanlings are added.

The behavioral data for this report were gathered in observation sessions conducted during daily periods of peak activity (0830–1130 hours and 1430–1830 hours) from March 1989 through May 1990, inclusive. Seventeen to 40 hours of focal sampling were accumulated for each of 25 subjects (Table 20.1), with comparable amounts of data gathered during each half-hour time slot in morning and afternoon sessions. I remained 3 to 10 m from all group members while conducting observations on foot using OS-3 Plus event recorders (GageTalker, Inc., Bellevue, Washington).

Every agonistic interaction involving focal subjects was recorded. Previous research had identified the aggressive and submissive behavior characteristic of ringtailed lemurs (Jolly 1966; Pereira & Kappeler, in preparation). Conflicts were scored whenever one lemur directed agonistic acts or signals toward another. For each interaction, all aggressive (A), submissive (S), and agonistically neutral behavior (O) exhibited by each participant were noted. A summary was recorded after both participants discontinued agonistic behavior for at least 3 seconds: each lemur was ascribed an A, an S, an AS, or an O to denote the types of behavior it had expressed. All complete interactions observed outside of focal samples also were recorded in this manner. During focal sampling, conflict intensities were scored on a five-point scale that considers both the kinds and durations of aggressive contact made and, for asymmetric interactions, the degree of effort made by the subordinate party to appease or evade its aggressor. The number of noninfant group members within 5 m of the conflict (0, 1, 2, > 2) also was recorded, and the two nearest were identified.

Interactions were considered *decided* when one lemur showed only submissive behavior while its opponent showed only aggression (A–S bout) or no agonistic behavior (O–S bout) (Hausfater 1975). Others were *undecided* (e.g., A–O, A–AS). Conflicts in the *ad libitum* record were tabulated monthly to help monitor dominance relations within each group (Fig. 20.2). Dyads exhibiting only unidirectional decided conflicts were considered to have exhibited stable dominance relations.

The criterion used previously to identify polydyadic conflicts (hereafter, "polyadic") among baboons (Pereira 1989, 1992) was also appropriate for this study: any lemur that directed aggressive behavior toward another already engaged in conflict with a third was considered to have intervened on behalf of the third lemur. The identities and roles of each participant were recorded for all polyadic conflicts observed.

Throughout focal samples, all approaches and departures involving subjects were scored (radius = 1.5 m) and all group members within 1.5 m of subjects were identified at 5-minute intervals. Also, each focal sample was rated low, moderate, or high for level of group activity

Table 20.1. Subjects and Hours of Observation

Age–sex class	Common name	Pedigree name	Group	Age[a] (months)	Seasons of study[b]	Hours of sampling
Adolescent males	EP		Lc1	26.5	BIR-LLAC	17.16
	CR		Lc1	26.5	BIR-LLAC	17.40
	NO		Lc1	26.5	BIR-LLAC	17.67
	AG		Lc1	24.0	BIR-LLAC	17.48
	PN		Lc2	26.0	BIR-LLAC	17.19
	JU		Lc2	26.0	BIR-LLAC	17.40
Adolescent females	KT	A3,2	Lc1	26.5	BIR-LLAC	17.38
	ML	C1	Lc2	26.5	BIR-LLAC	17.01
	CO	B1	Lc2	25.5	BIR-LLAC	17.31
Pubescent males	HP		Lc1	19.0	BIR-BIR+	39.78
	GL		Lc1	19.0	BIR-BIR+	39.22
Pubescent females	ER	A4,1	Lc1	18.5	BIR-BIR+	38.60
	AA	A5,2	Lc1	18.5	BIR-BIR+	39.60
	PH	A2,3	Lc1	14.0	BIR-BIR+	36.59
	CH	D1a	Lc2	18.0	BIR-BIR+	39.85
	SI	D1b	Lc2	12.0	BIR+	8.40
Juvenile males	BR		Lc1	10.0	PRE-BIR+	22.35
	AX		Lc1	10.0	PRE-BIR+	22.23
	IK		Lc2	10.0	PRE-BIR+	22.93
	HC		Lc2	7.5	PRE-BIR+	22.65
Juvenile females	NI	A5,3	Lc1	10.0	PRE-BIR+	22.17
	BU	A4,2	Lc1	9.0	PRE-BIR+	22.48
	SE	A2,4a	Lc1	9.0	PRE-BIR+	22.15
	BI	A2,4b	Lc1	9.0	PRE-BIR+	22.33
	DO	B2a	Lc2	10.0	PRE-BIR+	22.86
	AL	B2b	Lc2	10.0	PRE-BIR+	22.69

[a]Median age in months during subject's period of study.
[b]BIR, birth season (March and April); BIR+, birth season plus month of May; LLAC, late lactation (July and August); PRE, premating season (September and October).

(i.e., interindividual movement). A computer program used these ratings in tabulating neighbor data: it took data from only one point sample in sleepy focal samples, from the first and fourth point samples (15 min apart) in moderately paced samples, and from all four point samples in fully active samples. This accomplished three objectives. First, it increased statistical independence among point samples without unduly reducing the amount of data available for analysis. Second, it biased estimates of time spent together by any two social partners to represent primarily periods during which both agonistic and affinitive interactions were most likely to occur (M. Pereira, unpublished data). Finally, it minimized the effects of uncontrollable variation among subjects in proportions of data collected during periods of particular activity levels.

ONTOGENY OF SOCIAL CONFLICT AND DOMINANCE RELATIONS

Determination of Peer Dominance Relations Before Weaning

Invariably, dominance relations were established among birth peers by 4 to 5 months of age, 1 to 3 months before weaning. In most dyads, dominance appeared to be determined through experience in rough-and-tumble play, as this was virtually the sole mode of nonfilial social initiative exhibited by infants (see also Gould 1990). Intense grappling, however, was observed many times between 2.5- and 4.5-month-old infants (median: 99 days) that were particularly close in age (Table 20.2). Each time its onset was observed ($n = 4$), grappling erupted during rough-and-tumble play. Grap-

Table 20.2. Attributes of Infant Grapplers

Year	Pair[a]	Age[b]	Age difference[c]	Other age differences[d]	Rank difference at weaning
1982	BC–PL	82	0	2	?
1983	IR–KY	147	0	7,84,84	?
1984	CN–AN	150?	0	5,6,6,77	?
1986	PY–MI	103	0	35,62,76,76,79,79	1
	IS–MI	119	35	35,97,111,111,114,114	2
1989	NI–AX	111	5	7,33,33,36,36	1
		118			
		119			
	ZB–NI	74	33	0,3,3,38,40	NA[e]
	ZB–BR	81	40	0,3,3,38,43	NA
		99			
	SE–BI–BU[f]	79	0,3,3	3,36,41,43	1,2,1
1990	MG–NS	95	3	3	1
	CM–AS	105	22	NA	1

[a]Younger of two partners listed to left.
[b]In days, of younger animal.
[c]In days; 0 indicates twin siblings.
[d]In days, between younger animal and all other peers in cohort.
[e]ZB was wounded by an unidentified animal and died at 132 days.
[f]Simultaneously involved three infants; age and rank differences listed for each of three component dyads; SE and BI were twins.

plers fell to the ground, biting each other about the head and shoulders and raking each other with their feet. On every occasion ($n = 14$), some adult female or two—not always participants' mothers—responded by charging any noninfant within about 3 m. Only twice did a female attempt to intervene between grapplers (separate years, different females); each time, she was repelled instantly by another female.

Grappling established dominance relations (Table 20.2). In two pairs, two or three grapples were observed in a period of one to two weeks. Clear winners ($n = 4$) remained at conflict sites, taking over no resource, whereas their opponents ran to their mothers. Subsequently, winners readily intimidated former opponents during social play and routine conflicts over food or social partners. Eight of nine younger grapplers fought with their peer that was closest in age (see also below), and in six of eight pairs grapplers ultimately occupied adjacent dominance ranks in their cohorts. Eventual rank distance in the other two pairs was two.

Juvenile Dominance Relations

From weaning until puberty, juveniles maintained stable dominance relations with peers and with other group members (Fig. 20.3). In both groups, adolescents and adults of both sexes dominated all infants and juveniles, and juve-

niles dominated all infants. Also, no dominance relationship between juveniles reversed prior to puberty in any of the eight cohorts studied (Fig. 20.4).

Effects of age, sex, body weight, and maternal attributes on dominance among weanlings were investigated, taking into consideration effects of lemur reproductive biology. Estrous synchrony and females' capacity, barring conception, to ovulate up to four times at 39-day intervals across mating seasons (Jolly 1967; Pereira 1991) cause age differences among peers to be either small (< 3 weeks) or large (> 5, 10, or 15 weeks). Small age differences did not correlate reliably with weight differences at weaning, whereas large age differences invariably did (Fig. 20.4). All juveniles conceived during second or later sets of estrous cycles had the lowest weights in their cohorts at weaning, and eight of nine remained in the bottom half of cohorts by puberty, 1 year later (two-tailed sign test, $p = 0.04$).

Weanlings outranked their lighter peers (Fig. 20.4). Sex affected neither weanling weight nor dominance [males: $n = 16$, $X = 1054$ g, SD = 115; females: $n = 11$, $X = 998$ g, SD = 114; two-tailed t test, $t(25) = -1.27$, ns]. Also, the individual weights of twin weanlings matched those of singleton peers (Fig. 20.4), and litter size did not affect weanling dominance rank. Finally, no maternal correlates with weanling dominance ranks were evident. Maternal

	LOSER										
WINNER	AF1	AF2	AF3	AM1	AM2	JM1	JF1	JM2	JF2	JF3	JF4
AF1		34	64	30	35	11	7	21	11	14	13
AF2	1		24	8	5	20	30	14	6	22	10
AF3		1		15	5	22	30	18	11	24	14
AM1	22	28	42		79	20	17	13	10	13	9
AM2	1	30	39			23	14	14	10	21	8
JM1	1				1		43	57	34	28	29
JF1		2				2		38	31	40	21
JM2			1			1			29	42	37
JF2						1				37	28
JF3	1	1			1			3			13
JF4	1	2			2		2		1		

Fig. 20.3. Outcome of all decided dyadic conflicts observed among the juvenile and adolescent members of Lc1 Group during the study (March 1989–May 1990, inclusive). Five permanent reversals occurred between August and December 1989, when the three pubescent females overturned their male peers that had dominated them since weaning (see 1988 birth cohort, Fig. 20.4); that is, the ringtailed lemur phenomenon of female dominance developed.

weights at parturition did not predict the weight of singleton weanlings (mothers: $n = 8$, $X = 3037$ g, SD = 421; weanlings: $n = 8, X = 1110$ g, SD = 83; Pearson correlation coefficient = -0.046, ns) and maternal dominance relations did not predict offspring dominance relations: direction of dominance between mothers was the same as that between offspring in only 25 of 57 weanling dyads (Fig. 20.4).

Changes at Puberty

Marked in both sexes by genital growth and the onset of genital scent marking, puberty begins in the DUPC population at about 16 months of age, just after the summer solstice (Pereira 1991, unpublished data). Whereas no sex difference has yet been discovered in the growth or social behavior of prepubertal ringtailed lemurs (see below), except for exclusive male use of tail anointing and waving and ear flattening during play (M. Pereira, unpublished data), dramatic changes in social relations occurred after the onset of puberty that distinguished the sexes. Female dominance (over all males), for example, began at puberty (contra Gould 1990) and developed gradually across a period of about 7 months (Fig. 20.3). All females yet studied before and after puberty ($n = 10$) received submissive signals from adult males before they received them from formerly dominant male peers.

The other major changes were seen in agonistic relations with adult females. In four of the five cohorts that passed puberty during this study, one or more females was targeted for persistent attack by one to three adult females. No juvenile female or juvenile or adolescent male has ever been so targeted by adults (Vick & Pereira 1989; Pereira, unpublished data). All episodes against adolescent females began during the premating season or the following birth season, and the phenomenon contributed substantially to the general elevation of adolescent female rates of conflict with adult females (Fig. 20.5; two-tailed sign tests across the six annual seasons: females vs. male peers and females vs. juvenile females: both p's = 0.032). Adolescent male rates of conflict with older females were not significantly elevated over those of juvenile males.

SOCIAL ASSOCIATION AND AGONISTIC INTERVENTIONS

It is not uncommon for one ringtailed lemur to attack another already engaged in dyadic conflict. I accumulated data on 424 agonistic interventions during 721 hours spent with the lemurs: on average, one every 1.7 hours in a group. Comprehensive analysis of ringtailed lemur intervention behavior lies beyond our present scope, but a few preliminary tabulations provide

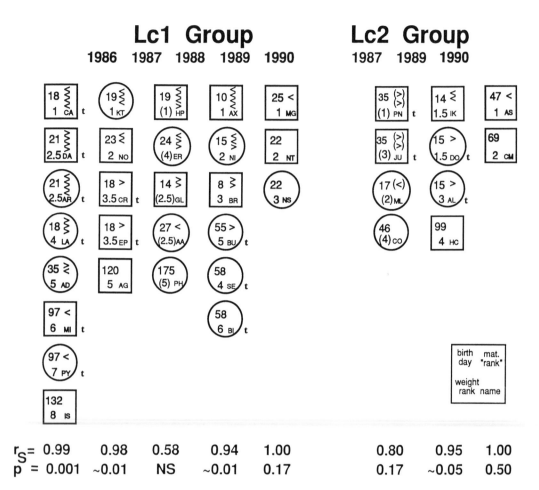

Fig. 20.4. Dominance hierarchies in eight cohorts of weanling ringtailed lemurs (birth year indicated). Subjects are listed top-to-bottom by decreasing dominance; males as squares, females as circles. Note key in lower righthand corner: March 1 is taken as the first day of birth season to calculate birthdays; each > sign for maternal "rank" indicates that the weanling's mother dominated the mother of a lower ranking peer; each < indicates that the mother was subordinate to the mother of a lower ranking peer. Spearman rank correlation coefficients indicate relationship between weight ranks and dominance ranks. Weight ranks and maternal "ranks" in parentheses were determined midway between weaning and puberty. *t*s indicate subjects that were twins.

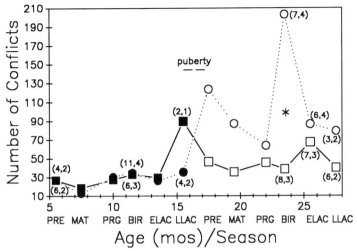

Fig. 20.5. Rates of conflict with adult females per hour spent in proximity to them (1.5 m). Closed circles, juvenile females; closed squares, juvenile males; open circles, adolescent females; open squares, adolescent males. Data were pooled from like-age subjects studied in separate groups and in consecutive years and median individuals' scores are plotted (6 hours of sampling per subject per season). Numbers of subjects and birth cohorts providing data are shown in parentheses each time they change, left to right. Seasons: PRE, premating (September–October), MAT, mating (November–December), PRG, pregnancy (January–February), BIR, birth (March–April), ELAC, early lactation (May–June), and LLAC, late lactation (July–August). Asterisk indicates significant two-tailed Mann–Whitney test of medians ($p < 0.05$).

context for the results to be presented (186 focal records all involved at least one juvenile or adolescent, whereas many of the 238 *ad libitum* records involved only fully mature adults). Third party aggressors were typically—but, not always—dominant to the lemurs they attacked (80%). But third parties did not systematically support the dominant parties of original conflicts (49%). Most interventions (80%) were against nonkin (Hamilton's $r < 0.25$), and most (73%) were on behalf of kin ($r \geq 0.25$). Fifty-eight percent of interventions were against nonkin on behalf of kin.*

Juvenile males and females received support

*For reasons presented in Pereira and Weiss (1991) and in the "Discussion," I generally classify ringtailed lemurs as "kin" or "nonkin" under three assumptions regarding paternity: natal group members born in different years were sired by different males, birth peers were sired by the same male, and every fully adult male present at the conception of a cohort is treated as though he sired the cohort.

primarily from their mothers, as did adolescent females (Table 20.3). Adolescent males received support primarily from other males, especially peers (including twin brothers). These patterns mirrored patterns of social association (Fig. 20.6). Whereas adolescent females continued to associate with their mothers at juvenile rates, adolescent males distanced themselves from females after the pubertal mating season, together with older males. This sex differentiation was accentuated when infants were born: females showed strong, reciprocal attraction with female kin at that time, whereas males began to depart their groups frequently, sallying together to and beyond range boundaries prior to actual group transfer (Fig. 20.6; Pereira & Weiss 1991).

Patterns of association, maternal intervention, and offspring response provided little evidence that female ringtailed lemurs actively influence the dominance relations of their offspring. First, maternal intervention was rare (mean: 0.6% of all conflicts). In approximately

Table 20.3. Top Supporters and Percentage Support Received from Mother

Age–sex class	Subject	Top supporter[a]		Second supporter		From mother (%)
Adolescent males	EP	Brother	(1)	Three	(1)	0
	CR	Brother	(4)	Two	(1)	17
	NO	Father	(2)	Four	(1)	0
	AG	Male peer	(2)	Three	(1)	0
	PN	Brother	(6)	Sister	(3)	0
	JU	Brother	(5)	Sister	(2)	0
Adolescent females	A3,2	Mother	(5)	Female cousin	(2)	50
	B1	Mother	(14)	Son	(3)	56
	C1	Adult female	(1)	None	(0)	0
Pubescent males	GL	Male peer	(4)	Female cousin	(1)	0
	HP	Mother	(1)	Two	(1)	33
Pubescent females	A4,1	Mother	(6)	Aunt	(3)	50
	A5,2	Mother	(13)	Aunt	(5)	62
	A2,3	Mother	(7)	Female peer	(2)	64
	D1a	Adult female	(4)	Two	(2)	0
	D1b	Mother	(1)	Two	(1)	33
Juvenile males	BR	Mother	(3)	Three	(1)	50
	AX	Sister	(3)	Mother	(2)	29
	IK	Sister	(2)	Brother	(2)	0
	HC	Mother	(7)	Adult female	(2)	64
Juvenile females	A5,3	Mother	(1)	Female cousin	(1)	50
	A4,2	Mother	(1)	Three	(1)	25
	A2,4a	Mother	(2)	Three	(1)	40
	A2,4b	Sister	(1)	None	(0)	0
	B2a	Mother	(7)	Sister	(4)	44
	B2b	Mother	(3)	Five	(1)	38

[a]Total observed supports from two top supporters shown in parentheses; number of group members listed when each of several provided same number of supports.

360 hours spent with each group, for example, only 8 of 26 immatures ever were observed to receive maternal support against peers, and 6 of the 8 received it only once (Table 20.4). Second, interventions were not strongly patterned by maternal dominance in relation to opponents. Mothers dominated the adult females against which they intervened in only 70% of cases ($n = 23$; daughters only, because males ultimately cannot dominate females), and they dominated peers' mothers in only 50% of interventions against peers (Table 20.4). Third, immatures intimidated by others never responded to maternal support by co-attacking their aggressor. Such co-attacking is a key component of rank acquisition for juvenile monkeys (e.g., de Waal 1977; Walters 1980). Fourth, maternal interventions overall were unpatterned by offspring sex or peer sex (Fig. 20.7; Table 20.4), and immature females evinced no greater anticipation of maternal support than did males. Despite receiving more adult aggression, adolescent females took no greater responsibility than did adolescent

males for the time they spent in association with their mothers (Fig. 20.6C), and only young juveniles—male and female alike—markedly increased rates of filial approach in direct response to female aggression (Fig. 20.8).

It thus appears that adolescent females received somewhat more maternal support than did adolescent males (Fig. 20.7) simply because mothers had far greater opportunity to intervene on their behalf: unlike adolescent sons, adolescent daughters were typically nearby and they frequently conflicted with other females. Proximity demonstrably influenced the probability of intervention. Fifty-five percent of all interventions during focal sampling ($n = 62$) came from one of opponents' two nearest neighbors at the moment of conflict. Nine of the 11 subjects that received maternal support were more likely to receive it when mother was one of their two nearest neighbors (sign test, $p = 0.03$), and this mode of proximity on average tripled the likelihood of maternal intervention.

Finally, interventions by nonkin females

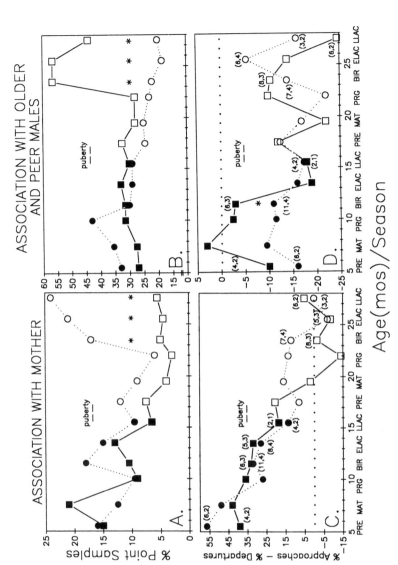

Fig. 20.6. Patterns of association with mother and with males: (A) percentage of time that mother was nearest neighbor; (B) percentage of time that any male peer or older male was nearest neighbor; (C) percentage of approaches between mother and offspring effected by offspring minus percentage of departures effected by offspring; (D) percentage of approaches between subjects and males effected by subjects minus percentage of departures effected by subjects. Points near zero line show percentage of departures effected by subjects minus percentage of departures effected by subjects. Points near zero line show animals equally responsible with partners for time spent in proximity. Consistent positive values indicate that subjects took more responsibility for time in proximity; consistent negative values indicate that partners took more responsibility. Other symbols and conventions as in Figure 20.5.

Table 20.4. Total Number of Maternal Supports Against Birth Peers Observed in 360 Hours Spent with Each Social Group

Age–sex class	Subject	Versus males	Versus females	Mother's status vs. peer's mother
Adolescent males	EP			
	CR		1	Dominant
	NO			
	AG			
	PN			
	JU			
Adolescent females	A3,2			
	B1	1		Dominant
	C1			
Pubescent males	GL			
	HP			
Pubescent females	A4,1	2		Dominant, subordinate
	A5,2	1		Subordinate
	A2,3	1		Subordinate
	D1a			
	D1b			
Juvenile males	BR	1		Dominant
	AX			
	IK			
	HC	1	1	Dominant, subordinate
Juvenile females	A5,3			
	A4,2			
	A2,4a	1		Subordinate
	A2,4b			
	B2a			
	B2b			

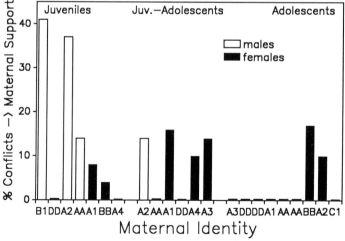

Fig. 20.7. Percentage of conflicts with adult females during which subjects received agonistic support from their mothers. To account partly for the effect of proximity on the probability of receiving support (see text), each subject's number of conflicts with adult females was multiplied by the proportion of time the immature spent in proximity to its mother.

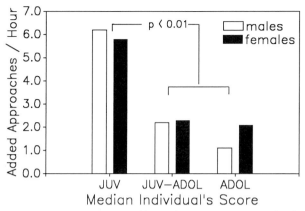

Fig. 20.8. Increase in rates of filial approach caused by moderate to severe aggression from other adult females. For each subject, the number of approaches to the mother that were made between each receipt of such aggression and the end of the focal sample was determined, along with the total amount of time represented by the terminal part of the sample. The total number of postconflict approaches and the total postconflict amount of sampling yielded the postconflict rate of filial approach. For each subject, the rate of filial approach across all sampling time then was subtracted from the postconflict rate. The only age or sex effect was the difference between juveniles and older subjects (two-tailed Mann–Whitney test of medians, $n,m = 11, 15, T = 2.98$).

could not have promoted the "inheritance" of maternal dominance status. Nonkin females' choice of supportee did not consistently parallel current dominance relations between subjects' mothers and present opponents (males: 33%, $n = 9$; females: 64%, $n = 28$). Adult female interventions also did not consistently favor female dominance over males (64%, $n = 17$).

ACQUISITION OF ADULT DOMINANCE AND RECRUITMENT OF NATAL FEMALES

In dense populations of ringtailed lemurs, females control the number of females that reside and reproduce in their groups using targeted aggression, (Vick & Pereira 1989; Gould 1990; Koyama 1990; M. Pereira, unpublished data). All episodes of targeting begin spontaneously during birth, lactation, or premating seasons, when one female begins to seek, follow, and attack one or two particular groupmates persistently (Vick & Pereira 1989; Pereira, unpublished data). Aggressors are often joined by one or two others. In the mildest episodes, a female initiates conflict with a dominant for one to several days, until she succeeds or fails to reverse their dominance relationship. In the

most intense episodes, dominants target subordinates for days, weeks, or months until the victims are expelled from the group. Most targets in such episodes incur severe canine puncture or slash wounds and/or sprained or broken limbs if they fail to leave their group. Impending or actual parturition triggers many birth-season episodes, and aggression sometimes wanes after victims' infants disappear or are killed by aggressors (Vick & Pereira 1989; Pereira, unpublished data).

The histories of our two study groups suggest that adolescent females in large or growing groups must become dominant to some adult females to be recruited to their natal group (Fig. 20.9). Lc1 Group was transferred to the DUPC enclosures in the premating season of 1981. During the first week, the young adult (A1a) and adolescent females (E1) targeted each other's mother (Taylor 1986) and were removed by technicians in favor of the older females. Sixteen other females were born that survived to mature in this group; but only five were recruited (successfully weaned their first infant). One recruit (E2) was expelled just after weaning her first infant (a son), and each of eight other females was expelled shortly after maturing (Fig. 20.9). Another female was removed due to

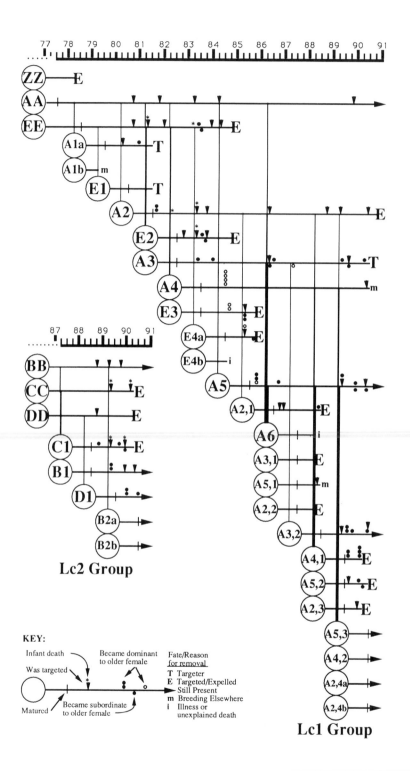

Lc2 Group

Lc1 Group

KEY:

Infant death
Was targeted

Matured
Became subordinate
to older female

Became dominant
to older female

Fate/Reason
for removal

T Targeter
E Targeted/Expelled
— Still Present
m Breeding Elsewhere
i Illness or
unexplained death

illness, and another was exported with males in 1988 to reduce levels of aggression and maintain species-typical group size. The remaining female (A3,2) lost her first infant during the 1990 birth season as she completed her rise to status of alpha female (Pereira & Kappeler, in preparation).

Each of Lc1 Group's four long-term recruits (A2, A3, A4, and A5) became dominant to two or more adult females just after maturing (Fig. 20.9). In contrast, none of the eight expelled females stably overturned any older female. These patterns were foreshadowed by the two aggressive young females removed in 1981 and have been replicated by females born recently in each study group (e.g., Lc2 Group, Fig. 20.9). All known dates of dominance reversals have been in seasons of targeted aggression.

A3,2 and A4,1 provided dramatic evidence of the importance of rising in dominance for adolescent females. In the birth season after she matured, A3,2 withstood a month of targeting by her grandmother AA and three aunts. Across the next year, she overturned all older females in dominance except her mother (no female has ever dominated her mother) (Pereira & Kappeler, in preparation). The following birth season, A4,1 also overturned all older females, including A3,2. Two months later, however, she was overturned and expelled by A3,2. First seen just after dawn with massive wounds to her head and limbs, A4,1 was attacked on sight by A3,2, who was the only other female that showed signs of having fought the night before (large, new hair pulls about the shoulders).

The group histories also provided much additional evidence that female ringtailed lemurs do not promote the dominance status of close kin (Pereira & Kappeler, in preparation). In Lc1 Group, for example, when A3,2 rose to alpha status her mother failed to rise in dominance with her (Fig. 20.2). In Lc2 Group, CC twice overturned BB to become alpha female, once for 4 months and again for 3 months; but neither time did her mature daughter C1 become domi-

nant to BB. Ironically, 1 month after CC fell in dominance again the second time, C1 *did* become dominant to BB, for 4 months, without affecting her mother's resumed subordination to BB (Fig. 20.2). After maturing in this group, D1 overturned CC and C1, although her mother (DD) had fallen to bottom rank more than a year earlier and was ultimately expelled. Two months later, D1 also reversed BB's second-ranking daughter (B1), but absolutely no reprisal could ever be detected from the alpha female. Similarly, in Lc1 Group, A3,2 targeted A4 for dominance reversal while she was subordinate to A4,1, but A4,1 never intervened on behalf of her otherwise powerful mother. As in several other cases now on record, she appeared not even to notice conflicts between A4 and A3,2, although many passed within 2 to 3 m of her.

DISCUSSION

Full interpretation of juvenile behavior requires familiarity with the array of adult behavioral strategies that exist in a primate population. For the ringtailed lemur, detailed behavioral data from individually recognized adults are just beginning to accumulate (Vick & Pereira 1989; Kappeler 1990a; Pereira 1991; Pereira & Weiss 1991; Sauther 1991; Sussman 1991); thus interpretation of data for any age class must yet be provisional. Pairing female reproductive and group membership histories with descriptions of prereproductive social development allows for such a prospective sketch of the ontogenetic trajectories that develop in female ringtailed lemurs.

Targeted Aggression and Female Ontogenetic Trajectories

I draw two major conclusions from the histories of the DUPC study groups:

1. Female ringtailed lemurs are not exclusively philopatric, as formerly thought (Jones 1983; Sussman 1991).

Fig. 20.9. Female residence histories of DUPC study groups. The founding members of Lc1 Group were mother (ZZ) and daughters (AA and EE). AA had been separated from her mother for as much as 1 year, however, beginning around the time of weaning. Lc2 Group's unrelated founding females (BB, CC, and DD) first met one another in the Duke Forest (see Pereira & Izard 1989), when C1 and B1 were 5.5- and 4.5-month-old infants, respectively. Closed circles above residence lines indicate that the dates when females rose in dominance over older females were known (see key); open circles indicate dates of dominance reversals that could only be estimated.

2. Females born into large or growing groups must overturn some adult females in dominance to begin their reproductive careers in their natal groups.

In large groups, females target particular peers for chronic attack and social eviction, and adolescent females are common targets (Fig. 20.9). Thus a bifurcation of ontogenetic trajectories is indicated at the start of female reproductive careers: recruitment to natal groups versus forced natal emigration. Females' assiduous resistance to eviction suggests that the costs of female emigration are high in the wild. At least 10 females in the Duke population and many in Berenty, for example, have lost infants before either departing from or persisting in their attackers' groups (Vick & Pereira 1989; Koyama 1990; Pereira, unpublished data).

Targeted aggression demands more attention in primate research. The phenomenon—an animal's focusing of persistent aggression upon a particular conspecific or set of conspecifics—lies at the heart of diverse social phenomena such as individual eviction, group fission, territoriality, infanticide, and even cercopithecine rank acquisition. From this perspective, targeted aggression takes its place alongside nepotistic affinity and support as one of the strong forces in primate social behavior. Study of its patterning in relation to variation in social and environmental circumstances and differences among individuals, age–sex classes, populations, and species would shed considerable light on the development and evolution of primate life histories.

Targeting between mature female lemurs entails a qualitative change in agonistic relations whereupon one female begins to seek and follow another to harass or attack her persistently. The phenomenon appears to show functional affinities with better known tactics in anthropoid life histories. Most often, dominants target subordinates, for example, with results that parallel those of targeted parental aggression toward like-sex offspring in the Hylobatids (Tilson 1981) and dominant females' targeting of subordinates in Callitricids (Abbott 1984): reproductive suppression, peripheralization, and potential eviction. When subordinates target dominants, however, the results often parallel those of targeting of low-born adults by maturing females in cercopithecine societies: dominance status reversal.

Further observations support the hypothesis of parallels between lemuroid and callitricid tactics of female reproductive competition. Attacks by dominant female lion tamarins (*Leontopithecus rosalia*) on subordinates tend to be severe; they disrupt the subordinates' reproductive efforts, but not their endocrine cycles (French & Inglett 1991). In contrast, senior *Saguinus* and *Callithrix* females less often attack their subordinates, who fail to ovulate in their presence (Abbott 1984; French & Inglett 1991). Similarly, female aggression controls reproduction and membership in groups of *Lemur catta* and *Eulemur fulvus* (Vick 1977; Taylor & Sussman 1985; Vick & Pereira 1989), while reproductive suppression appears to occur between female sifakas (*Propithecus verreauxi*) without severe aggression (Richard et al. 1992).

A recent analysis of all known episodes of targeted aggression in Lc1 Group showed that (1) mothers and daughters have never been opponents, (2) some sisters and grandmothers–granddaughters have targeted one another, (3) aunts and nieces and cousins of different ages have targeted one another frequently, whereas (4) same-age aunts–nieces or cousins (birth peers) have done so only rarely (Pereira, unpublished data). This pattern could be explained by typical degrees of relatedness between females. If reproduction in a group is usually restricted to one or two males (see Pereira & Weiss 1991) and reproductive male turnover typically occurs at shorter intervals than do births of surviving sisters, then different-age aunts–nieces and cousins would be virtual nonrelatives (Hamilton's $r = 0.13$ and 0.06, respectively) while most same-age aunts–nieces and cousins would be the most closely related females in a group ($r = 0.38$ and 0.31), except for mother–daughter pairs.

To control the number of females residing and reproducing in their groups, then, it appears that female ringtailed lemurs select for chronic attack those that are, on average, most distantly related. Adolescent females unable to overturn adults in dominance are evicted by their grandmothers and older aunts, nieces, cousins, and sometimes sisters.

Female Behavioral Tactics Prior to Reproduction

The development of agonistic relations in our study groups indicates that the only defense available to natal females against reproductive suppression and ultimate eviction following puberty is their own agonistic capacity. Because lemurs do not actively maintain dominance hierarchies (see below), this entails independent reversal of as many dyadic dominance relations as possible with older females. How might juvenile females best prepare for this effort? Maintenance of high dominance among birth peers is likely part of the answer.

Maximizing Growth. To my knowledge, knock-down, grappling fights between particular infants have been observed in no primate other than ringtailed lemurs. Grapples determine dominance relations between infants closest in size, whereas differences in body weights and presumably experience in social play determine all others. Dominance relations among healthy weanlings show remarkable stability thereafter. No changes were seen in any of eight birth cohorts across both juvenility and adolescence (first postpubertal year), except for the emergence of female dominance over high-ranking male peers by the end of the pubertal mating season.

Ringtailed lemurs experience the first harsh, dry season of their lives about 3 months after weaning (Jolly 1984; Sussman 1991), and the large majority of daily conflicts between juveniles determine access to feeding sites (Pereira, unpublished data). Under natural conditions, dominance over peers should help high-ranking weanlings avoid dry-season mortality (Pereira 1993) while maximizing growth and physical fitness, and thus reproductive and agonistic capacity at maturity. Those that are initially larger or otherwise agonistically superior become dominant, and appreciable size differences among peers are likely to become only more pronounced during prereproductive development in the wild.

Dominance may sometimes expand reproductive careers for wild female lemurs by advancing age at first reproduction. At the DUPC, where provisioning maximizes growth, females almost always bear their first infants at 2 years of age. Our rare female that fails to do so is invariably among the smallest in her cohort. Conversely, it is rare for females' first births to occur before 3 years of age in the wild (Sussman 1991), and the few females that do reproduce so early are likely to be large, and thus dominant among their peers.

Note, however, that the challenge of surviving and the importance of promoting one's own development should be equal for males and females in this monomorphic species (see also Pereira 1993). Thus the absence of female dominance among juvenile ringtailed lemurs, while fascinating, is not surprising.

Social Play. Particular functions for primate social play are difficult to determine (Fagen, Chapter 13, this volume), but many researchers agree that juvenile mammals can develop both physical fitness and fighting skills during competitive play (i.e., wrestling and chasing). If so, juvenile female ringtailed lemurs, whatever their dominance ranks in their cohort, may enhance their eventual chances of overturning older females in dominance by playing frequently with many partners.

Thus the practice fighting hypothesis predicts that no sex differences in rates or styles of competitive play should be found in ringtailed lemurs, where the reproductive success of both males and females appears to be promoted by superior agonistic capacity (Taylor 1986; Vick & Pereira 1989; Pereira & Weiss 1991; Sauther 1991). In contrast, in cercopithecine monkeys, where females but not males "inherit" their adult dominance status, males are expected to invest more time and energy in competitive play. In fact, a comparative analysis of social play has shown that none of eight predicted sex differences found in the competitive play of wild juvenile baboons (Pereira 1984; see also Nash, Chapter 9; Watts & Pusey, Chapter 11; Fairbanks, Chapter 15, this volume) could be detected among the eight juveniles born into Lc1 Group in 1986 (M. Pereira, unpublished data). Gould (1990) also reported a lack of sex differences in the social play of infant ringtailed lemurs in Madagascar.

Agonistic Intervention. Other than occasional shielding of close kin from aggression, no evidence of female support patterned toward any particular social end for recipients could be detected. Indeed, the most impressive—albeit unquantified—aspect of maternal support was the frequency with which mothers appeared to ignore offspring conflicts, even when they occurred within several meters. These results beg the question: Why do ringtailed lemurs intervene in others' fights? The cercopithecine monkey–ringtailed lemur comparison raises the more specific question: Why do female ringtails not intervene systematically in female conflicts, to promote the dominance status of close kin and to stabilize their dominance hierarchy? One imagines, for example, that by at least promoting kin a female could help her matriline dominate others, grow larger, and effect all subsequent evictions. Before considering potential roles for agonistic intervention in the life histories of female ringtailed lemurs, then, we must explore the basic nature of intervention in the species.

I believe that one factor helps answer both questions just posed: ringtailed lemurs are especially vulnerable to attack by peers while fighting. Subtle features of polyadic conflicts provide initial clues. Here are three examples:

8 July 1990, Lc1 Group: When adult female A5 rose to cuff a ruffed lemur over food, adult male DI charged from 3 m to co-attack it. When the group's other adult male, subordinate LE, stood up about 5 m away and began to trot toward the fray, DI instantly spun away from the ruffed lemur and attacked LE.

28 August 1990, Lc2 Group: When immigrant male CE stooped over a waterhole, subordinate immigrant HI stole behind and pounced on him, biting deeply into the back of his knee. As CE and HI grappled on the ground, lactating D1 sprinted over from about 10 m, leapt on the males, and bit hard into CE's shoulder. Both males fled explosively from the female.

31 August 1990, Lc1 Group: Natal male GL was the first to spot male SL approaching from Lc2 Group. An instant after he shot toward SL, so did male DI. While SL fled, GL turned, fell back against stiff grasses, and, fear grimacing, fought off the attack he appeared to anticipate from DI—who, in fact, continued past in pursuit of SL.

In none of these interactions was the third ringtail providing needed support to a mutualistic ally. In each case, the third party seemed to exploit the original conflict as an opportunity to co-attack a particular adversary. D1, for example, had already targeted CE for chronic attack; he was persistently trying to kill her infant (Pereira & Weiss 1991). Most important in all three and many other cases, there was compelling evidence that a combatant anticipated being attacked while fighting. The characteristic female attack of nearby noninfants in response to infant grappling also seems to anticipate the potential for opportunistic third-party aggression. For ringtailed lemurs, it seems, to fight is to risk sudden aggression from other group-

mates; agonistic intervention seems fundamentally opportunistic.

Why might this risk of fighting exist for lemurs but not for cercopithecine monkeys? First, consider in detail what the monkeys can do. Experiments have confirmed that the extraordinary stability of female cercopithecine dominance hierarchies is due to the female inclination to support even nonkin against others that they both outrank (Chapais 1992). Processes of rank acquisition suggest that the monkeys recognize the dominance hierarchy itself. High rankers, for example, even support the efforts of unrelated juvenile females to *become* dominant to the adult females outranked by the juveniles' mothers (Walters 1980; Netto & van Hooff 1986; Pereira 1989, 1992). Moreover, females so pattern their occasional interventions between unrelated play-wrestling infants—supporting the infant from the higher ranking family—before the infants themselves are likely aware of matrilineal dominance relations (Horrocks & Hunte 1983a; Chapais & Gauthier, Chapter 17, this volume).

The cercopithecine system of female agonistic intervention appears to be an evolutionary mutualism that benefits every pair of allies: the "top–down" collaborative aggression virtually ensures all females acquisition and maintainance of matrilineal dominance status (Chapais 1992; Pereira 1992). It also almost completely suppresses aggression from subordinates (Chapais & Shulman 1980; Walters 1980), precluding for them the opportunistic sort of dominance-reversing third-party aggression not uncommon among ringtailed lemurs.

Now, because high dominance promotes reproductive success in female lemurs (Vick 1977; Taylor 1986; Vick & Pereira 1989; Pereira, unpublished data), we might have expected female ringtails to show the same intervention mutualism. The accumulating data, however, indicate that mature females intervene in others' conflicts without regard to existing dominance relations (e.g., Table 20.4; Pereira, unpublished data), and consequently they do not invariably show transitive dominance hierarchies (Fig. 20.2; Taylor 1986; Pereira & Kappeler, in preparation).

I propose that female ringtailed lemurs do not exhibit top–down agonistic intervention because somehow they are unable to. To effect the intervention pattern, females must be able to (1) recognize group mates visually and (2) remember the directionality of dominance between them or their families. During their daily activities, ringtailed lemurs present abundant evidence that they recognize one another visually. Also, female targeting of other females (Vick & Pereira 1989) and of infanticidal males (Pereira & Weiss 1991) must involve some level of visual recognition. During several episodes of each mode of targeting, however, I have also seen aggressors err, sprinting to attack and suddenly stopping after approaching or even having initially grabbed the wrong animal (Pereira, unpublished data). The mistakes have invariably been—also to my eyes—among the group members most similar in appearance to known targets.

The Neuroanatomical Limitation hypothesis, then, proposes that ringtailed lemurs cannot reliably recognize groupmates that are fighting or moving rapidly, which would preclude systematic intervention on behalf of dominants. Lemurs do not show several of the anthropoid specializations of the visual system. Even ringtails, the most diurnal prosimians, for example, do not have fovea; the anatomy of their visual system provides only one-tenth to one-fifth the visual acuity enjoyed by macaques and humans (Neuringer et al. 1981, personal communication). Alternatively, the Sociocognitive Limitation hypothesis suggests that lemurs cannot remember the dominance relations between all pairs of groupmates, which would also preclude systematic top–down intervention. These two hypotheses are independent, but not mutually exclusive; each could be tested using response-task conditioning paradigms in relation to video-slide presentations (e.g., discrimination or match-to-sample tasks; Dasser 1987, 1988).

In any case, because top–down intervention does not occur among female ringtailed lemurs, their dominance hierarchies are inherently less stable than those of female cercopithecine monkeys. Subordinates are free to attack and intimidate particular dominants opportunistically, which leads to dominance reversals (see Chapais 1992). Reversals have been restricted to the seasons of targeted aggression. Several of the complete reversal processes we have witnessed began with spontaneous, intense interventive aggression, and all involved days or weeks of targeted aggression (Pereira, unpublished data).

This sketch of the organization of agonistic relations among female ringtailed lemurs has several implications for maturing females. First, I suggest, the ringtailed lemur inclination to exploit adversaries' conflicts as opportunities to co-attack them suppresses the rate and the effectiveness of kin support. Even dominant ringtailed lemurs must risk fighting only in relatively secure social contexts; thus many opportunities for kin support will be bypassed or exploited only with reservation. Indiscriminate intervention on behalf of maturing daughters would likely increase mothers' own chances of being attacked, targeted, overturned, and potentially evicted. Consequently, active female "inheritance" of maternal dominance status is not seen in this species.

By the same principle, however, pubertal females must somehow monitor agonistic dynamics among adult females to exploit opportunities to co-attack and overturn some of them. For adolescents, the risks of targeting adults must be high: adolescents are smaller than adults (Pereira 1993), and they are themselves prime targets for eviction. But the risks must be undertaken by adolescents to avoid reproductive suppression and eviction. Adult females in large groups face the complementary agonistic paradox: they must fight to avoid reversals and to expel adversaries, and they must avoid fighting to minimize their chances of being targeted themselves.

A spectacular example of this conundrum occurred in 1989, when a cohort of three females matured in Lc1 Group, which already included six adult females. When third-ranking adult A3 spontaneously attacked her pubescent niece A5,2, another pubescent niece, subordinate A4,1, immediately attacked her. A3 met and returned A4,1's aggression, but soon became intimidated when her own sisters (A2, A4, and A5) joined A4,1 as aggressors. A3's incipient targeting of A5,2 ultimately resulted in A3 herself falling to sixth dominance rank, becoming the second adult overturned by A4,1. A3 remained sixth ranking, and thus in greater danger of eviction, for 7 months (Fig. 20.2).

CONCLUSIONS AND FUTURE RESEARCH

With their system of agonistic relations, female ringtailed lemurs are, in a sense, caught midway between the systems of group-living nonprimates and of female cercopithecine monkeys. In nonprimates where dominance mediates success in feeding or reproductive competition, dominance relations are usually transitive and relatively stable because dominance rank correlates with a single phenotypic trait or constellation of traits (e.g., body size, antler size, and/or age) and because third-party influences—agonistic interventions—are nonexistent (e.g., Clutton-Brock et al. 1982; Byers 1986; Festa-Bianchet 1991). Female cercopithecine hierarchies are transitive and stable because of the deterministic modes of agonistic intervention outlined earlier.

Clearly, female ringtailed lemurs socially determine and maintain their dominance relationships, but equally clearly they do not *maintain* dominance hierarchies. Observers often can abstract hierarchies, especially after short studies or in groups with small numbers of females (cf. Jolly 1966; Sauther 1991). But also, ringtailed lemurs use opportunistic agonistic intervention to reverse relations with particular dominants, and their targets are not always adjacent on the "hierarchy" (e.g., Fig. 20.2). Thereafter, intransitive dominance relations often remain stable until the next season of targeted aggression (Pereira & Kappeler, in preparation).

Results from lemurs seem to confirm the old idea that dominance relations and hierarchies suppress the rates or severity of aggression. Targeted aggression in lemurs is most severe. It arises spontaneously and is infectious among females; even adversaries will collaborate against common adversaries, and victims are often badly injured (Vick & Pereira 1989). The potential for dominance reversal between almost any two females—that is, the absence of any mechanism to maintain a true hierarchy—likely contributes to the severity of the aggression. Also, females should resist eviction whenever possible because social or demographic circumstances may turn in their favor in a matter of weeks or by the next season for targeted aggression. In the same vein, when a subordinate female succeeds in acquiring dominance over a former superior and then ceases her aggression, she may be avoiding unnecessary risk of being targeted herself after having made progress toward continued residence in her group.

The present results and speculations underscore the need for long-term observations and detailed individual histories in the study of primates. If these can be obtained for many prosimian primates, as well as for the understudied New and Old World anthropoids—as is being accomplished for Cercopithecines, chimpanzees, mountain gorillas, and some langurs and *Cebus* monkeys—we certainly shall attain much fuller understanding of both the development and the evolution of primate social behavior.

In concluding, note that ringtailed lemurs traverse from one system of dominance to another in their lifetimes. Weanlings acquire dominance among peers according to their size and prowess. As juveniles, their transitive dominance hierarchies show impressive stability because, intriguingly, neither males nor females show the tactics of interventive aggression they will use to reverse and secure their dominance relations after maturing (Pereira, unpublished data).

Is this because small size renders juveniles too vulnerable to potential injury by third-party aggressors? Or because the metabolic challenges of surviving their first dry season (Pereira 1993) makes strategic social monitoring and interventive aggression too costly? Are ringtailed lemur cohorts born annually within periods of 2 to 3 weeks (Pereira 1991; Sauther 1991), rather than 3 to 6 months like other seasonally breeding primates, because this precludes an initial weight advantage for first-born infants (Fig. 20.4; see Pereira & Weiss 1991)? Do infant lemurs exhibit not only rapid linear growth but also "fatting" prior to weaning to engage the challenge of their first harsh season, as do some birds and seals (Perrings et al. 1973; Worthy & Lavigne 1983; Oftedal et al. 1987)? Is it rapid postnatal growth, along with the predictable harsh seasons, that has led female dominance and possibly other modes of female priority to evolve in lemurs (Jolly 1984; Richard 1987; Pereira et al. 1990; Pereira 1993; cf. Young et al. 1990)? Clearly, the Malagasy primates offer a fertile workshop for future research on roles of development in the evolution of primate life histories.

ACKNOWLEDGMENTS

I thank B. Chapais, A. Richard, J. Walters, and L. Fairbanks for critism of an earlier draft of this manuscript, and S. Nowicki, P. Klopfer, C. van Schaik, R. Wiley, and other members of the Research Triangle Animal Behavior Group for criticism of an oral presentation of the work. I thank

M. Neuringer and T. Norton for published and unpublished information on prosimian visual anatomy and acuity. Louise Martin contributed assistance with behavioral observations and management of the forest-living lemurs. All data on Lc1 Group during its earliest years in the Duke Forest (1981–1984), including those on targeted aggression, adolescent and adult dominance relations, reproductive success, and infant grappling, came from Linda Taylor's (1986) dissertation research and the DUPC individual history files. My research and writing are supported by the National Institute of Child Health and Human Development (R29-HD23243) and by the Duke University Primate Center (DUPC Publication No. 558). Finally, my deepest gratitude goes to T.-J. Pyer, whose generous support allows my research to proceed with few interruptions.

Part V
COMPARATIVE SOCIOECOLOGY OF CHILDHOOD

Part V demonstrates how some of the ideas of evolutionary biology have been applied in studies of juvenile humans—children. With our primary focus being the non-human primates, it was not possible to include a representative sampling of all the research being done today on child development and human socioecology. Instead, our hope is that by including these few chapters, we will help some biologists discover or better appreciate the commonalities that exist among cultural anthropology, child development, primatology, and biology. Likewise, we hope to introduce to investigators of other cultures and of child development the many parallels that exist between their work and that on animal behavior. Each of the chapters in this part shares themes and concepts with one or more chapters in earlier parts of the volume.

Blurton Jones (Chapter 21) uses the constructs of life-history theory (Part I) to make sense of differences between the parenting behavior observed in two hunter-gatherer societies and their consequences for child development. Among the !Kung people studied, parenting is described as permissive and responsive. Children re-

main in camp under adult supervision and are expected to work little to meet their own or others' needs. In contrast, the Hadza wean their children earlier, punish them more often, and leave them unsupervised for large parts of each day. Hadza parents demand considerable work from their children, including the gathering of a substantial portion of their own food and other tasks that are costly in energy and risk. Blurton Jones develops a model of trade-offs among three components of parental investment—production of offspring, promotion of their survival, and enhancement of their reproductive success—and argues that the parenting strategies shown by the !Kung and the Hadza can be understood in terms of trade-offs between production and offspring survival. He substantiates this assertion with corroborating evidence of the risks and payoffs of children's foraging, and of the relationship between maternal workload and interbirth interval in the two cultures.

In Chapter 22, Edwards tackles the question of how sex differences in social behavior develop, a topic of debate in human research and also one of considerable interest to primatologists, as reflected by at least half the chapters in the preceding sections. In primatology, sex differences in the behavior of juvenile primates are generally explained in terms of their ultimate functions. That is, they are primarily viewed as adaptations to the different selection pressures experienced by males and females. Although investigators often discuss the proximate mechanisms by which these sex differences might develop, they rarely test these ideas (see Fairbanks & Pereira, Chapter 24, this volume).

In a study of 15 different cultures, ranging from horticulturalist and pastoralist to industrialized modern societies, Edwards reports that the one sex difference that is most consistent across cultures is the tendency for girls to have more contact and social involvement with infants. To explain the development of this sex difference, she contrasts two hypotheses (1) that socialization pressure from the culture shapes appropriate sex-role behavior, and (2) that intrinsic attraction to same-sex adults and sex differences in responsiveness to nurturant modes of interaction lead boys and girls into different contexts of socialization. Edwards presents evidence that girls are more likely than boys to respond positively to their mothers' commands, and when placed with infants, girls are more likely to engage in nurturant and caretaking modes of interaction. In the author's view, society provides the role models and rewards for gender-appropriate behavior, but also girls and boys contribute to their own socialization with differential responsiveness to adults as socializing agents.

In the final chapter of the section, Worthman (Chapter 23) discusses a central theme of Part I—factors determining the timing of maturation. She characterizes the timing of puberty as the response of a developmental transducer influenced by a suite of environmental cues, and calls for an integration of research on growth and maturation with studies of psychosocial development. Worthman describes the hormonal processes that regulate human growth and maturation and discusses both how environmental factors augment or attenuate these processes and how sociocultural practices respond to maturational change. She then presents data for two late-maturing populations: the Kikuyu of Kenya and the Bundi of New Guinea. The basic physiology of puberty is shown to be similar across cultures, whereas its timing, underlying relationships among maturing subsystems, and sex differences are sensitive to environmental and social stimuli. This chapter effectively bridges gaps among academic domains concerned with ontogeny by illustrating the interdependencies of sociocultural processes, biological mechanisms, and ultimate functions.

21

The Lives of Hunter-Gatherer Children: Effects of Parental Behavior and Parental Reproductive Strategy

NICHOLAS BLURTON JONES

The !Kung and the Hadza are foragers of the sub-Saharan savanna. In both populations, men hunt Africa's well-known game animals, and women gather roots and fruits from a similar flora. But there are striking differences between the lives of !Kung and Hadza children. !Kung children are confined to camp in the constant presence of adults, yet they are allocated little that looks like "work." !Kung mothers are among the most indulgent and attentive ever described. They respond fast to crying and have never been observed to hit a child. Hadza children roam about the bush from an early age, spend whole days out of view of adults, and are often sent on errands and told to gather water and firewood or care for younger children. Hadza mothers are noisy, intimidating, and often not very responsive, and can be seen to smack children, sometimes threaten to whip them, and often ignore crying. How can we account for these differences in societies that are so similar in many features of behavior, social organization, and ecology?

I will argue that we can use the view that the behavior of parents serves the reproductive interests of the parents to understand the differences between !Kung and Hadza parental behavior.

TOWARD A MODEL OF PARENTAL BEHAVIOR AND REPRODUCTIVE STRATEGY

There is an important literature on parental strategies in evolution and animal behavior (Trivers 1972, 1974; Smith & Fretwell 1974; Lazarus & Inglis 1986; McGinley & Charnov 1988; Clutton-Brock 1991), and several papers illustrate the predictive value of an adaptationist perspective on human parental behavior (e.g., Daly & Wilson 1978; Draper & Harpending 1982; Turke 1987; McDonald 1988; Pennington & Harpending 1988; Draper 1989; Rogers 1990). Yet we still have little idea how to apply the perspective to many of the features of parental behavior that have most interested developmental and cross-cultural psychologists: parent–child interactions and "child-rearing practices." Psychologists emphasize the importance to the child's development of "contingent responsiveness," sensitivity to the child's moods and signals, face-to-face interaction, a stable caretaker, conversation with adults, a "stimulating" environment, and firm but kindly discipline. Anthropologists point out how often children are cared for by children, how little they receive direct instruction by adults, and how scarce is face-to-face interaction between mother and infant in many cultures. Yet the children nevertheless grow up healthy, sane, and successful.

I hope to show how we might begin to incorporate the kinds of parental behavior that interest psychologists in our adaptationist models. Ultimately this may enable us to understand the extensive variation that is observed in parental behavior and child-rearing practices.

Parental behavior evolved because it promoted the parent's reproductive success. Sometimes this will include responding to the offspring's self-interested requests; sometimes it may not. Natural selection favors individuals that leave greater numbers of descendants. Variation in number of descendants may be due to variation in several components of reproductive success (RS), such as length of reproductive career, fertility, offspring survival, and offspring RS (examples in Clutton-Brock 1988). In this chapter let us think about three such components: (1) producing a greater number of offspring (referred to as "production"); (2) promoting the survival of particular offspring (referred to as "survivorship" and "mortality"); and (3) enhancing the reproductive success of the offspring when they grow up (referred to as "ORS").

Reproductive success, when measured by number of grandchildren (RSG), results from the multiplication of each of these together, $RSG = N \times S \times R$, where N = number of offspring born, S = mean survivorship of offspring into their reproductive years, and R = mean number of offspring (grandchildren) born to each surviving offspring (ORS).

Parental behavior can promote these goals in a variety of ways. Clutton-Brock (1991) labels one important distinction as "depreciable" versus "nondepreciable" care. Depreciable care (like food) has its effect on only one offspring at a time; a second "dose" has to be added for each offspring. "Nondepreciable" care has the same effect on the fitness of each offspring, whether it is directed to many or to one. (I have tried to adhere to the welcome clarification of terms by Clutton-Brock, especially in using "expenditure" or "expend" instead of "effort." "Effort" has often been used not only as a cost to parent fitness but to refer to any parental behavior, and to the benefit of the behavior to offspring. I use "effort" to refer only to expenditure of re-

sources, not to hypothesized costs to fitness.) Another useful distinction is care that promotes only one component of reproductive success (protecting offspring from predators increases their survivorship but not their RS) versus care that promotes more than one component at a time, such as wealth (which may increase offspring survivorship and offspring reproductive success). Males and females may have very different parenting strategies. Most of my discussion in this chapter concerns females and depreciable care. I also attend only to behavior that promotes a single component of reproductive success, and thus I propose that a distinct set of parental behaviors promotes each component.

Behavior that may promote each goal could include the following:

1. Production: parents behave toward the child in a way that enhances the parent's ability to produce more children.
   a. Parents engage children in labor that will help support siblings and mothers; this saves the mother time and increases the resources available for her to allocate to pregnancy and lactation, even at the cost of loss of energy and time, and increased risk to children.
   b. Parents refuse to give resources requested by the child (perhaps leading to more conflicts and punishment, but see Lazarus & Inglis [1986] for a discussion of variation in parent–offspring conflict). Thus parents tend to be unresponsive and unprotective and to wean early.
   c. Parental beliefs might go along with behavior; thus, to be provocative, I suggest that the parental belief that indulgence "spoils" the child might accompany "production"-promoting behavior.
2. Survivorship: parents counter threats to the child's survival, for example, by preventing undernutrition, disease, predation, and accidents.
   a. Parents feed children; this affects their survivorship in various ways. Not only does it prevent starvation, but because of interactions between diet and disease, more food generally means less sickness and lower infant and child mortality.
   b. Parents respond to the child's requests, which we assume the offspring makes to

obtain resources such as food, warmth, and shelter that enhance its survival.

c. Parents protect the child, retrieving it and keeping it away from danger.

d. Parents give fewer energy-consuming or risky chores.

e. Parental behavior that enhances survivorship is perhaps less likely to produce disparity between the interests of parent and offspring, so we may expect fewer disputes and less parental hostility and punishment. Parents will be reported as nurturant, warm, taking precautions against disease if possible, saving children energy, and keeping children from predation and exposure.

3. Offspring reproductive success (ORS): this seems to be an aspect of parenting in which humans excel, and, of course, the most obvious example is the behavior that has most interested developmental psychologists.

a. Parents sometimes teach children, instruct them in subsistence skills, and improve the child's ability to earn a living, to compete for resources, and to deal with other individuals. Social interaction between parents and small children, such as the face-to-face interaction claimed to be so important for development, may belong in this category.

But anthropologists are familiar with a variety of additional ways in which parents may influence their children's RS:

b. Parents engage in grandparenting, taking direct care of and providing food for grandchildren.

c. Parents help select a mate, influence an offspring's choice of a mate, or increase a child's ability to compete for mates.

d. Parents accumulate wealth and endow children (Rogers 1990). Endowment is one way in which people increase their child's ability to compete for mates; however, it can also directly enhance a grandchild's survival and reproduction (for well-researched examples, see Hartung 1976 & Borgerhoff-Mulder 1987).

This list shows that it is easy to suggest adaptive consequences of the kinds of parental behavior that interest psychologists and anthropologists. Direct studies could be conducted to test the suggested effects. But there may be

more to be done with these ideas. For instance, we may wonder whether the kinds of parental behavior vary as functional groupings. Can we think of some populations as generally expending more time or energy on children's survivorship, others as expending more on bearing a large number of children, and still others as expending more on enhancing ORS by a variety of means? Can we associate patterns of parental behavior with demographic parameters?

Indeed, many of the world's cultures could be characterized as "production enhancers" (such as the many high-fertility agricultural societies in which children work a lot, are treated very strictly, and are thought of by some anthropologists as providing economic benefit for parents), "survivorship enhancers" (a few low-fertility hunter-gatherer populations, perhaps such as the !Kung and a few others with indulgent and attentive child rearing), and "ORS enhancers" (the most obvious fit being high-socioeconomic-status parents in industrial societies who LeVine [1977] describes as striving to develop their child's social and intellectual abilities above all else). There is even support for this by Hewlett (1988), who reported a negative relationship between fertility and "nurturance" in a sample of simple societies.

If we can characterize populations this way, and if there are links to reproductive and demographic variables, then whenever we know about the demographic parameters we can predict parental behavior and its variation. Conversely, since it often takes more time and effort to determine demographic and ecological parameters than to observe parental behavior, we may prefer to predict features of the ecology of reproduction from observing parental behavior.

It is important to construct a model to see what really follows from the basic principles. As a preliminary effort in this direction, I wrote a computer program that simulates just one limited issue: when it takes more of one kind of parental behavior to produce substantial gains in offspring survivorship, (1) Should we expect more expenditure on a second kind of behavior that also enhances offspring survival? This is a limited version of the question: Should we expect behavior that serves a common function (e.g., offspring survivorship) to vary together, as a group, across circumstances? (2) Will fitness be enhanced by a reduced number of births?

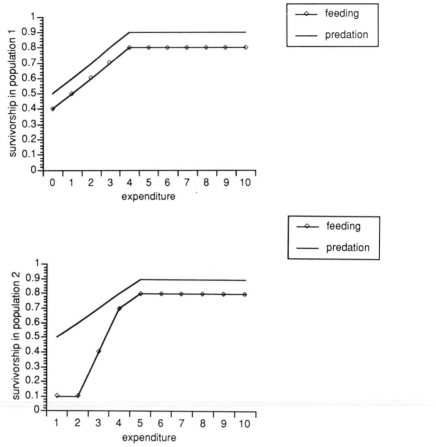

Fig. 21.1. Curves of offspring survivorship against parental expenditure used for the first simulation described in the text.

In the simulations, each mother enhances offspring survival by two kinds of behavior (Fig. 21.1). Each of these affects the probability that the offspring to which it is directed will survive the respective hazards—starvation or predation, for example. Each kind of behavior uses up arbitrary units of expenditure, which can be thought of as time or calories.

One kind of behavior (e1) simulates the expense of feeding children. To simulate one population (population 1), the effect of expenditure on survivorship falls little as expenditure declines; at low levels of expenditure, children do not starve. To simulate another population (population 2), the effect of this kind of behavior on survival is very strong (Fig. 21.1). Reduced care

has a massive effect on survivorship. The second kind of survival-enhancing behavior (e2) gives the same returns to expenditure in each population. This is intended to simulate behavior that avoids a rare but risky event such as predation. Survivorship is quite high, even at low levels of effort, and increases slightly and linearly with increased effort to a limit (returns to increased effort are then zero).

The computer program calculates the reproductive success that results from successively greater numbers of births (N), when a limited total number of arbitrary units of expenditure (E = 40 units) are allocated in all possible combinations between the two kinds of investment (e1 and e2) in each of the N offspring. The simulation concerns only "depreciable" care (in Clutton-Brock's [1991] terminology), such as food (two offspring require twice as much food as one offspring). The following equations are the core of the program:

$$E = e1 + e2$$

$$S = S1 \times S2$$
$$RS = N \times S$$
$$S1 = \text{some function of } (e1/N)$$
$$S2 = \text{some function of } (e2/N)$$

where $S1$ = probability of surviving hazards countered by parental behavior $e1$ (e.g., feeding); $S2$ = probability of surviving hazards countered by parental behavior $e2$ (e.g., predation); S = probability of average offspring surviving all hazards, to reproductive age; N = number of offspring born; RS = reproductive success measured in numbers of offspring surviving to reproductive age. The total available effort $E = e1 + e2$ (a version in which $E = e1 + e2 + N$, to simulate the allocation of effort to production, gave the same results).

In this model, I hold constant the allocation of expenditure to promoting ORS, and do not investigate different kinds of expenditure that promote production.

The results of the simulations are (1) higher RS is obtained from fewer births in the "starvation" simulation (population 2), in which feeding children has a strong effect on their survivorship; and more interestingly, (2) maximum RS is obtained by greater effort to avoid "predation" in this simulation (population 2) than in population 1. In population 1, RS was maximized by minimal effort invested in either kind of parental care. Further simulations using some 20 other pairs of curves of survivorship as a function of expenditure [given by the Basic statement $S1 = 1 - \exp[-(e1 - a1)]$, which gives a diminishing returns curve, with $a1$ representing various intersections with the x axis] gave the same result. Furthermore, in response to my questions Alan Rogers of the University of Utah began work on an analytical model of these issues. First indications are that it supports but greatly clarifies the results of the simulations. So we have some reason to believe that different kinds of behavior that promote survivorship can be expected to vary together.

More modeling might confirm that we can think of one strategy of parenting as investing generally in offspring survival and another as investing in production and investing less of each kind of effort in offspring survival.

Although computer simulations can be useful, they can leave us with relatively little understanding of why a result is obtained. The present result is probably heavily dependent on (1) the multiplication of probabilities of surviving different hazards, (2) the division of care among each of the N children, and (3) diminishing returns from each kind of investment. If this is so, then we must not generalize this finding to the different kinds of effort in production or ORS without further investigation. Although some kinds of behavior directed to ORS will be divided among N offspring, its effects are not obviously combined multiplicatively. Furthermore, wealth will be inherited by and divided among only $N \times S$ offspring. Much more modeling is needed before we can feel confident about the intuition-derived proposals stated above.

However, we now have more reason to pursue the idea that some patterns of parental behavior may be understood as a survivorship-enhancing strategy and others as a production-enhancing strategy. Can we understand the difference between !Kung and Hadza parenting in this way? If, as the characterization in the opening paragraph implies, !Kung mothers behave as though they are putting most effort into enhancing offspring survival, and Hadza behave as though they are putting most effort into enhancing production of large numbers of offspring, do the demographic and ecological data conform to the patterns that I have suggested? Before reporting field work on !Kung and Hadza women's reproductive strategies, I will introduce each population with a brief general description.

THE !KUNG

The 500 Dobe-area !Kung live in northwestern Botswana and northeastern Namibia in a flat and dry environment, as described by Yellen and Lee (1976). Intensive fieldwork has covered many aspects of !Kung life (e.g., Lee 1969, 1979; Lee & DeVore 1976; Howell 1979; Wilmsen 1989; Shostak 1981). Often noted is their short working week—women forage and men hunt about 2–3 days each week, acquiring around 2000 calories per camp member per day. Estimates of their foraging returns are 725 cal/hour (Blurton Jones et al. 1990), 670 cal/hour (Hawkes & O'Connell 1982), and, in the wet season, 1302 cal/hour (Sih & Milton 1982). One should note the toughness of those few days of work, which commonly include a 6-mile walk out from a dry-season camp near water and a 6-mile walk home carrying a substantial weight of food.

The population appeared to be almost constant (intrinsic rate of increase = 0.003) until the 1970s. Eight percent of the people were over 60, but life expectancy at birth is 30. Birth intervals were long, around 4 years between births, and average number of live births and total fertility rate (TFR) was 4.7 (Howell 1979).

Until recently, anthropologists believed that the !Kung exclusively occupied the Dobe area until the flight of the Herero through the area in 1905, and their return to settle in the 1950s. This has been challenged by Wilmsen (1989), and the degree and length of contact between the !Kung and Bantu and overseas populations are the subject of much debate (Solway & Lee 1990). It remains agreed that archaeological evidence shows that the area has been occupied for millennia; it is not a refuge into which people were driven in historic times.

Infancy

Konner (1972, 1977b; Bakeman et al. 1990) conducted intensive quantitative studies of !Kung infancy. Infants are seldom separated from their mothers but are often in the additional company of others, male and female, adult and child. When the mother goes gathering, the baby rides along and is taken on the mother's lengthy all-day gathering trips. Feeding is "on demand." Intervals between suckling bouts averaged 13 minutes (Konner & Worthman 1980). Children are weaned in the fourth year of life, and weaning can be an interlude of strife in an otherwise idyllic infancy (Shostak 1981). After weaning, children often protest being left home when the mother prepares to leave, and the mother often gives in and takes the child with her (Draper, personal communication), despite the obvious difficulty of carrying the child for

much of the great distance that !Kung women cover when gathering (Lee 1972).

Childhood

Draper (1976) and Draper and Cashdan (1988) reported their detailed quantitative observational studies of !Kung childhood (5–15 years). In bush camps, older children stay home, continuing a period of life in which they truly deserve to be called "dependents." They hang around camp and rest and play. They are almost constantly in sight or sound of adults, because there are usually several adults in camp.

Draper reports that !Kung children do almost no gathering and almost no work allocated by adults (0–3 minutes/hour). They are not even instructed to care for children (children caring for children is a prevalent feature of traditional agricultural societies) and actually do almost no childcare (0–3.8 acts/hour). !Kung say that in the wet season or whenever food is near, mixed-age groups of children may forage a little. Nonetheless, in Draper's year-round observations it was found that !Kung children foraged very little and were almost never out of camp (e.g., 8 of 76 "spot observations" for girls; 18 of 93 for boys). Thirteen of these 26 instances concerned children under 3 years who were carried by their mothers. Thus virtually all their food comes from adults.

Play includes all the various forms described in children from North America and Europe: play with objects, rough-and-tumble play, and imaginative and imitative games. The age of childhood idleness gives way in the mid-teens to very different careers for boys and for girls. Girls often marry around 15 (before menarche, reached on average at 17.5) (Howell 1979), and then begin to gather seriously. They accompany older women on major gathering excursions. Boys alternate between continued idleness as "owners of the shade" and beginning to learn about hunting, accompanying older relatives on hunts. Men do not marry until their late 20s.

Konner and Draper emphasize the !Kung parents' tolerance and availability to their infants and children. Other than during weaning, !Kung parents feed infants on demand, respond rapidly to a crying child by picking it up, allow children to climb onto them and rest in physical contact with them (apparently regardless of whatever else the mother is trying to do at the time), and ignore bad behavior. !Kung parents seldom allocate work to children, seldom utter commands, and seldom ask children to run errands. These features of !Kung child rearing are, of course, very close to those features proposed to promote survivorship.

THE HADZA

The 750 eastern Hadza, who live in the rock-strewn hills near Lake Eyasi in northern Tanzania, are best known from the work of Woodburn (1968a, 1968b) and the International Biological Program expeditions led by Barnicot (Bennett et al. 1975; Hiernaux & Hartono 1980).

Our demographic studies (Blurton Jones et al. 1992) support Dyson's (1977) conclusion that the Hadza population is increasing at about 1% per year, total live births and TFR is about 6.2, and mortality is not reliably different from that of the !Kung, with a life expectancy around 31 years.

The Hadza have been visited from time to time during this century, and useful accounts include Obst (1912) and Kohl-Larsen (1958). For most of this century, most Hadza hunted wild animals and gathered wild plant foods, traded with their neighbors by selling honey for cloth or metal, and owned virtually no livestock. At the southern end of their range, many Hadza have farmed for many years. Repeated attempts to persuade the Hadza to settle have led to several epidemics and to interruptions of the hunting and gathering way of life. Probably every adult has been involved in at least one such episode, although in many cases for quite short periods. But more than 95% of the food of Hadza that we studied was from wild animals and plants. The more remote history of the Hadza is completely unclear.

The fauna and flora on which the Hadza prey are not unlike those that the !Kung exploit. The Hadza environment, as Woodburn (1968a) remarked, is probably rather richer than the !Kung environment. The estimated wildlife production of this habitat was 232 kg/km²/year, or 3482 kcal/ha/year. Comparable figures are not available for !Kung country. Hadza men successfully

hunt buffalo, wildebeest, zebra (only a seasonal visitor to !Kung country), eland, kudu, kongoni, and impala. They seldom catch smaller game such as dik-dik, duikers, or spring hares, which !Kung take relatively frequently (Hawkes et al. 1991). Sections of Hadza country are densely covered in baobab trees; preliminary results of the 1989 aerial survey suggest 20–30 trees/km². Figures are not available for !Kung country, but there are probably fewer than one-tenth the number of trees per square kilometer.

Hadza women gather plant foods daily, mostly going out as a group on excursions lasting 4–7 hours, accompanied by one or more "guards," teenage boys or young men with poison arrows (Hawkes et al. 1989). Much of the year, the main quarry of the women are roots, tubers, and baobab pods. During several seasons (but not every year), berries are collected. Several other plant foods are known and used, including tamarind pods, also a favorite with vervets, baboons, and southern Californian food stores. The main harvest of the women in our study area is //ekwa (Vigna esculens). This grows deep in rocky soil, and its extraction is a lengthy subterranean jigsaw puzzle, sometimes involving the removal of large boulders and encounters with scorpions. Women meet to cook and eat roots while still in the bush, and then continue digging or return directly to camp with the remaining roots. Sometimes they cook roots just outside camp and may be discovered by more children than those who had accompanied them in the bush.

When foraging, Hadza women cover much smaller distances than !Kung women. Hadza women travel between 0.5 and 2 miles, based on a journey time of 4–60 minutes (average 25 minutes) each way (Hawkes et al. 1989). The climate in which Hadza women forage is less extreme. We have never encountered frost in Hadza country, although people complain about the cold at night. Daytime temperatures are lower in Hadza country than in !Kung country (Blurton Jones et al. 1989).

Hadza men hunt (like !Kung, with bow and poison arrows) in early morning and evening, or from a blind at night. Hawkes's preliminary analysis of foraging shows that in the wet season each Hadza receives at least 2000 cal/day, and, in the dry season, 4000 cal/day.

Infancy

Our observations are based on a large number of 1-hour focal follows of infants and children in and out of camp, balanced by time of day; the data have not yet been analyzed, so the following is an impression subject to correction when quantified. Quantitative information on children's gathering has been reported in Blurton Jones et al. (1989). The Hadza child's first year of life appears not to differ greatly from that of the !Kung infant. The mother is the principal caretaker. The baby spends most of its time riding on the mother's side or back. Suckling is frequent, and often, but by no means always, "on demand." One does, however, observe mothers break off suckling bouts, evoking protest from the infant. One also often hears crying and observes parents ignoring a crying infant. The baby is likely to be surrounded by relatives, old, adult, and young, and receives attention from them and sometimes is carried by them. It is quite common to see a baby carried during the gathering excursion by the mother's mother, young sister, or oldest child. Face-to-face interactions described in Western cultures (and in !Kung) can be seen between the Hadza mother and infant and other people and the infant. Hadza are weaned a good deal younger than !Kung, at around 2.5 years old. Soon after they are 2 years old, Hadza children begin to be left behind when the mother gathers, although they may be suckled before the mother leaves camp and as soon as she returns.

Childhood

Hadza children lack none of the charm and imagination of !Kung children. They have a robust humor and a pride in their life that we find attractive and impressive. Some show an unexpected and touching hospitality in their occasional concern that the observer should not get lost or too badly stuck in the thorns. Rough-and-tumble play, imaginative games (including portrayals of anthropometrists and epeme dancers, predators, and "foreigners"—Hadza from other regions or Datoga herders), and various play with objects are commonplace. Formalized games with rules also occur as well as target practice. Singing and dancing fills many evenings, particularly if there is a group of young teenage girls in the camp. Hadza children also play with dolls. These are made by the children themselves from rags and clay. Particular children seem to keep a collection of rags for this purpose.

Between the ages of 3 and 8, Hadza children seldom accompany their mothers on gathering excursions. Children over 8 may accompany their mothers, but do not always do so. They, like younger children, may be among the group that greets the women as they return to camp in the afternoon, with joyous shouts of "//ekwabe" and the dispatch of a delegate to summon the anthropologist to weigh the //ekwa roots quickly. However, in contrast to !Kung children, Hadza children are neither supervised nor confined to camp while the women are out gathering. Often the only adults in camp are men, who spend most of their day away in "the men's place." Some of the time there may be no men—for instance, if they are holding a men's feast, which must be held out of sight of women or children, or if there is a good report of a large animal being hit and they leave the camp to track it.

The children, usually in sizable mixed-age groups, may spend some hours out of camp. Sometimes they are at a favorite play site or at the water hole. More often they are gathering food, independently of the women (Figs. 21.2 and 21.3). We measured their productivity and observed their foraging during several field seasons. Dry-season returns were reported by Blurton Jones et al. (1989) and are quite substantial (Table 21.1). Subsequent observations suggest that children do just as well in the wet season.

In the foraging groups, even 3-year-olds try their hand at digging or picking up baobab pods and processing them (normally they process them back in camp, where there are rocks and the pounding stones left by the women). Instead of winnowing the shells, children dampen the pith with water, hammer it, suck the mush from a handful, and replace the residue for further hammering until only seed-shell fragments are left.

Unlike !Kung children, Hadza children appear to be given many errands and to perform useful tasks, bidden and unbidden. Such tasks cost the children time and energy, and sometimes expose them to the hazards of the bush. Children of either sex may be asked to hold a protesting toddler when the mother leaves camp

Fig. 21.2. Hadza children return from digging and eating Makalita roots in the bush not far from camp. The younger two are aged about 4 and 6.

to forage. Girls who accompany women on gathering excursions are likely to be asked to carry or entertain a toddler or baby. When they go with the women, their own foraging returns are much lower than when they forage near camp with the other children (Blurton Jones et al. 1989). Children commonly are sent to fetch water and sometimes firewood, and on a couple of occasions grass for a house (although if it is raining, this task seems to fall to the old women, major figures in many aspects of Hadza life; see Hawkes et al. 1989). Even toddlers are sent to

carry things from one house to another or one adult to another. I have twice seen children encouraged to drive a snake from the camp, once by throwing stones at it and once by shooting at it.

To the observer familiar with Konner's and Draper's accounts and with some experience of the !Kung parenting, the Hadza are strikingly different. We see Hadza parents use physical punishment, and we see and hear them shout prohibitions and commands at children. We see children given quite costly tasks and errands,

Fig. 21.3. Older Hadza children (aged 12–15) collecting //*ekwa* roots near camp. The two on the left are lifting a boulder to expose more roots. Some already excavated are piled on the rock at right. Adults obtain much larger examples.

including childcare and collecting water and firewood. We see infants refused the nipple long before weaning. We notice suckling sessions apparently broken off by the mother, not by the child. We see crying ignored or punished and children who are trying to get picked up refused. We see parents refuse, refuse, give in, and refuse again.

This bleak picture should not be exaggerated, and quantitative analysis may also redress the balance. Hadza children are active and cheerful most of the time and are welcomed in their home. Even among these people, who seldom publicly show affection or warmth, parents can

be heard to speak warmly of and to their child. But the overall picture is certainly not the developmental psychologist's dream presented by the !Kung.

Neither society shows much indication of behavior that enhances ORS, apart from grandmothering. We have observed a few instances of direct teaching of Hadza children. Fathers make bows for their small sons and lend axes to older sons, thus enduring some very slight cost to provide their children with learning experiences. Neither society has much access to wealth to endow. Grandmothering does seem quite important among the Hadza. Hawkes et al. (1989) show that older women forage for longer hours, and at least as effectively, as women of childbearing age. We expect tallies of our observations of who hands food to children to show that a grandmother or great-aunt is a significant source of food for Hadza children. The role of older !Kung women is not clear. They act as babysitters, but in interviews (Blurton Jones et al. 1989) they claimed to be more effective foragers than young women.

Table 21.1. Foraging Returns of Hadza Children

	Age group (years)	
	5–10	10–15
Baobab	629	1014
Makalita roots	163	559
Honey	0	339

Note: Expressed in calories per hour of gathering and processing.

!Kung and Hadza parenting correspond closely to two of the proposed functional groups of behavior. !Kung parents perform more of the behavior that is thought to enhance survivorship. Hadza parents perform more of the behavior that is thought to enhance production of greater numbers of offspring. Do data from ecology and demography conform to our expectations?

ECOLOGY AND DEMOGRAPHY OF !KUNG AND HADZA PARENTAL BEHAVIOR

Is there any evidence that !Kung parents obtain greater RS returns from their expenditure on children's survivorship than Hadza parents would from more nurturant and responsive behavior? Do Hadza parents gain greater returns from behavior directed toward production than !Kung parents would if they directed more resources to giving birth more often? What can we say about the ecological reasons for the differences in payoffs?

What Happens If a !Kung Woman Gives Birth at a Greater Rate?

Blurton Jones (1986), analyzing Howell's (1979) reproductive histories of !Kung women, showed that, as in many human populations, offspring mortality is higher at shorter interbirth intervals (IBI). Furthermore, at IBI below 4 years, mortality increased so much that it outweighed the gains from a faster rate of births. In a career of fixed length, fewer surviving offspring would be produced by intervals of less than 4 years between births than by intervals of around 4 years. The average IBI was close to the optimal IBI. This analysis of Howell's data showed that for the !Kung, giving birth at a greater frequency leads to lower RS. This indicates that a !Kung who expends more than average effort on production will end up with worse than average RS and thus might do better to direct that effort to offspring survivorship.

!Kung IBI are unusually long for a non-contraceptive-using population. Most populations have a shorter IBI. But most treat their children very differently from the !Kung. Accounts of child rearing in these shorter IBI, higher fertility populations suggest that observers see more of the behavior listed as enhancing

production of offspring, and less of that claimed here to enhance survival of offspring. Before we discuss whether the Hadza are such a case, let us examine the ecology that seems to lie behind the trade-off between IBI and mortality in the !Kung.

Why Do Shorter IBI Not Pay for the !Kung?

Blurton Jones and Sibly (1978) examined the consequences of shorter IBI for a mother's provisioning herself and the family. They used published data to extend Lee's (1972) model of the work entailed in raising a !Kung child. Lee also published data (1969, 1979) on the proportion of food provided by women, the calorie value per 100 g of various plant foods (including the staple mongongo nut), the proportion of time that children of various ages are carried or left home with babysitters, the age at which !Kung children are weaned, and the age at which teenagers become independent.

Blurton Jones and Sibly's (1978) computer model first calculated the weight of mongongo nuts needed to provide the mother's contribution to the calorie recommended dietary allowance (RDA) of one child as it grows from conception, past birth and weaning, to independence at 14 years old, including the calories to support the mother's pregnancy and lactation up to the normal age of weaning. The computer program accumulates the work required (weight of nuts and baby to be carried home from a gathering excursion if mother gathers daily, on alternate days, once every 3 days, etc.) to feed children under a variety of schedules of reproduction. The results show how work increases with shorter birth intervals, and how the increase accumulates and declines during the reproductive career (as children become independent).

The most striking result is that the work required to provide each member of the family with the RDA appropriate for his or her age changes very little with birth intervals longer than 5 years; but as IBI declines below 4 years, the work required increases dramatically. Blurton Jones and Sibly (1978) suggested that the work reached unsustainable levels (considering either back injury or heat stress, for which plenty of local data existed), and thus children would be undernourished and at great risk. They suggested that mortality might rise at short IBI so much that the observed average 4-year IBI of the !Kung might be the optimal IBI, the IBI that yielded the greatest product of survivorship and number of births in a fixed career (just as was later confirmed using Howell's data).

The curve of work (kilograms carried home from a gathering excursion) against IBI can be regarded as a plot of the work required to maintain constant offspring survivorship with increased rate of births (assuming the RDA endows constant survivorship). Since the mother's ability to work harder (carry more) must have limits, the curve can also be regarded as indicating an increased threat to survivorship as production of children increases. Seen this way, the curve shows sharply diminishing returns to investment in production of offspring. As IBI shorten (increased investment in production), the success in raising these offspring declines; that is, RS returns from each birth diminish.

To further attempt to cross the bridge between ecological and demographic data, we can compute from the model, and with an assumption about what !Kung women can carry for long distances through an extreme climate, the shortfall in food that would arise from shorter IBI. As IBI decline from 4 to 3 years, backload (weight that must be carried to feed the family for 2 days) increases by 5 kg. Five kilograms of mongongo nuts, the staple plant food of the !Kung, represents 4200 kcals. If this shortfall is shared among two adults and three children, it represents a shortfall of 840 kcals per person spread over 2 days. The calorie RDA of children aged 1–12 years (as used in Blurton Jones & Sibly 1978) ranges from 850 to 2400 kcals, of which the mother provides about half. Thus the shortfall represents between 17 and 49% of the RDA for a child. It is easy to believe that if intake is reduced this much, mortality increases substantially.

Thus for the !Kung, diverting resources to increased birth rate would entail such costs to offspring survivorship that RS would decline with increased investment in production. If allocating a larger part of the reproductive effort to production of offspring lowers RS, then allocating effort to survival or ORS must be more rewarding.

What Happens If Hadza Women Produce Children at Shorter IBI?

Hadza children appear to be very cheap to raise. The mother provides less than half the child's food (grandmother, men, and, above all, the child itself provide the remainder). The mother's foraging, although conducted daily, covers much smaller distances in an easier climate than !Kung women's foraging. The Hadza women acquire more calories per hour of gathering and processing (Hawkes & O'Connell 1982; Blurton Jones et al. 1989). The constraints that Blurton Jones and Sibly (1978) proposed for the !Kung cannot be anywhere near as stringent for the Hadza.

When we substitute the Hadza values reported in Blurton Jones et al. (1989) in the Blurton Jones and Sibly model, we get a much less striking effect of shorter IBI. For the Hadza, investing in a higher rate of births brings with it a much slower increase in threat to offspring survivorship. It seems likely that shortening IBI below !Kung levels would pay off in increased RS for the Hadza. I recently wrote a version of the Blurton Jones and Sibly model that calculates not weight carried, but time needed to gather and process the food to support families arising from different IBI. The results resemble those from the backload model.

So far we have been unable to conduct reliable interviews of reproductive history with Hadza women so we do not know how Hadza offspring mortality relates to IBI. But our demographic studies show that Hadza have greater completed family size (more births) than !Kung, although we cannot say whether this is caused by shorter IBI or longer reproductive careers. The greater number of births seems to be accomplished with no discernibly greater child mortality.

What Would Be the Consequences for !Kung and Hadza Children's Survivorship If the Mother Reduced Her Investment in Feeding Them?

The immediate consequence would be simple, obvious, and very different for !Kung and Hadza children. Hadza children would spend a little more time foraging each day. At the expense of a little play time, they would replace the missing food intake by their own efforts. !Kung children could try begging from a wider circle of kin and neighbors, but food in camp is already distributed generously, and often there is none. Draper's (1976) observations show that !Kung children do not forage. But if the mother fed them less, would they not forage? If a Hadza woman gains so much reproductive success (higher fertility with the same offspring mortality) because her children forage for themselves, *Why do !Kung women not match them by sending their children out to forage?* If we cannot answer this question, our understanding of these systems is shown to be very limited, and, indeed, our use of the adaptationist paradigm would be jeopardized.

We were, as is often the case in optimization studies, faced with asking "What if . . . ?" We know that !Kung children foraged very seldom. But our question was: What would happen if !Kung children did try to forage; if instead of actively discouraging them, their parents told them or allowed them to wander off in search of food? What costs might they face, and what benefits could they expect in return?

Blurton Jones et al. (1992) reported field work in Botswana on these questions. We proposed that in the hilly Hadza countryside, with its rocky outcrops and giant vistas of the rift valley escarpments, it would be easy for children to find their way about. Draper describes thick scrub coming right to the edge of !Kung camps, cutting out distant vistas; as the country is so flat, there is nothing much to look at if you do get the chance to see far. Draper and I had already seen the panic and instant mobilization of trackers that occur when a !Kung mother realizes that she does not know the exact location of one of her children. We expected that it was easy and dangerous for !Kung children to get lost. Thus Draper (with Hawkes or the author as note-takers and cross-examiners) interviewed adult !Kung about women's foraging and about children getting lost.

The interviews confirmed that !Kung parents show great concern about children getting lost. They said they tell children to stay in camp and try to impress on them the dangers of the bush. We were told of eight children who got lost, and two of them died. Eight lost is small compared with the numbers who die of sickness. But is the worry of the !Kung out of proportion to the risk? (The simulation suggested that we might expect !Kung mothers to do more to avoid risks than Hadza mothers.) When a !Kung child does get lost, it is hard to find it, and the risks once lost are high. The vigilance and quick response of !Kung adults may be just what ensures that few children get lost.

We also proposed that it was hard for !Kung children to find much food near camp. Because of local geology, there are few dry-season water holes in !Kung country, and camps, being near permanent water, are very far from the most productive food sources. Hadza country seems to lack this separation of food and water, even in the dry season. Everything is often close by. There are many water holes to choose from. We wondered whether !Kung children who tried to forage would have to go much farther than Hadza children to find as much food. Thus we set out to measure experimentally the returns from foraging in different habitats, particularly at different distances from permanent water.

We took groups of adults and children by truck to forage at various locations, weighing and identifying the food they obtained in a measured time. These "in patch" results show a striking increase in returns with distance from permanent water. The correlation between calories extracted from experimental quadrats and distance from permanent water was $0.3452, p < 0.006, n = 53$. More food can be obtained far from permanent water.

When we compare these results with the returns obtained by Hadza children (Table 21.1) and the distance they travel to forage, the difference is overwhelming. Hadza children forage very near camp and get returns greater than !Kung adults get anywhere. The benefits from foraging are much greater for Hadza children. To match them, !Kung children would have to travel many times farther, as far as 6 km across featureless country. The added distance itself would impose costs in energy consumption and heat stress, on top of which we can add the unmeasured risks from predators and the risks of getting lost.

Our data suggest that the balance of costs and benefits to !Kung and Hadza parents is different. The !Kung do not have the option of allowing their children to help raise themselves. The high risks of getting lost are balanced by only very low returns of foraging within conceivable child range of dry-season camps. !Kung parents who encourage their children to stay home and not to wander off in search of food must be expected to raise more children than those who permit this risky and unrewarding behavior.

Conclusion

These arguments suggest that Hadza women do better giving birth more often, and !Kung women do better limiting their rate of births and spending relatively more time or energy feeding children and keeping them safe. The results are in accord with the ideas about parental behavior that were expressed at the beginning of the chapter.

We also know quite a lot about the ecology that appears to determine this difference. The Blurton Jones and Sibly backload model (and its "time" version), the ecological and behavioral observations of the Hadza, and the foraging experiments with the !Kung enable us to see why optimal behavior differs between these populations. Feeding the offspring is an important part of the reason, and it presents different challenges in the two environments because Hadza children contribute so substantially to their own diet. They do this because food is abundant and close at hand, and it is not for the !Kung.

DISCUSSION

Summary

I have proposed that many kinds of parental behavior that interest psychologists and anthropologists can be viewed as having simple biological functions. Behavior that promotes the survival of offspring is likely to include responses to the offspring's own self-interested efforts to elicit resources and responses from the parent. Extracting useful work from children can be seen as enhancing the mother's ability to produce and raise more offspring. Effort of several kinds is put into enhancing the adult competitive ability and reproductive success of offspring.

I suggested that these three kinds of outcome might give rise to three dimensions of variation in child-rearing practices or to three functional groupings of parental behavior. It may be possible to deduce systematic relationships between these groupings and demographic parameters such as fertility. Thus high fertility might be expected to accompany low amounts of effort to promote survival and ORS.

I proposed that we might understand the differences between !Kung and Hadza parental behavior by linking the differences in behavior to the consequences of increased or lowered rates of births in their two ecologies. Several kinds of data and analyses fit these propositions. The proximity of food and water in Hadza country allows people to camp near both, and makes it

rewarding for children to forage for themselves. Children's foraging reduces the effect of decreased provisioning effort by the mother. Thus Hadza women gain greater RS by producing more births and showing less attentive care than !Kung women.

Problems

Much additional modeling is needed to develop these suggestions fully and to determine whether the intuitively derived predictions do follow from the basic propositions. It is particularly uncertain how much we should expect the behavior that promotes each goal (survival, production, ORS) to vary as three functional groups. For example, it is tempting, perhaps because it seems to explain the strange behavior of wealthy parents in industrial society, to think that effort directed to teaching children is to be expected to pay when parents are able to accumulate and endow wealth. But it is not at all clear that this really follows from my basic propositions. It may be that an equally good strategy in comparable conditions would be to invest effort only in wealth acquisition or only in teaching. Some high-SES parents in industrial societies may appear motivated both to accumulate wealth and to promote their child's education. But Wrigley (in preparation) shows that the landowning gentry of eighteenth- and nineteenth-century England, serious wealth endowers, showed little tendency to become personally involved in educating their children. In contrast, the newly emerging technocrats of the Birmingham "Lunar Group" (Priestley, Watt, Wedgewood, Erasmus Darwin), to whom the immediate returns of intellectual ability must have been evident, were both interested and involved in their children's intellectual development.

Comparing just two populations may fail to provide a stringent test of the propositions. For instance, a third hunter-gatherer population, the Ache of Paraguay, are described by H. Kaplan (personal communication) as showing both high fertility and high nurturance (it may be very significant in this connection that among the Ache, men provide most of the food, in contrast to other forager populations where women acquire more than men). It is also clear that we should not proceed without quantitative data on parental behavior to children of various ages and without models that accommodate age differences in the vulnerability of children.

Several specific issues that require further attention should be mentioned here.

1. I considered parental behavior that had only one function, it promoted survival, production, or ORS—not two or all three of them. There must be exceptions to this. In the Hadza, we have circumstantial evidence that the labor which Hadza children are persuaded to put into their mother's RS brings later benefits for their own RS (thus mothers' enrolling children in labor enhances both her production of offspring and her ORS). Hadza adults with more live siblings have more live children ($r = 0.42$, $n = 35$).
2. I considered only "depreciable" care, divided between the offspring so that, for example, two offspring need twice as much food as one. Some kinds of behavior are equally effective to each offspring regardless of number of offspring. For instance, collecting grass to roof a Hadza hut when the rain begins requires the same effort however many children huddle in the dry hut, and each of them will gain as much benefit from staying dry as would just one of them. Wealth has the interesting property that it can be used to help whichever offspring happen to survive until adulthood.
3. Local and specific factors may overwhelm general trends. For example, Kaplan and Dove (1987) emphasize the extraordinary efforts exerted by Ache mothers to keep infants and toddlers off the forest floor. This seems to be rewarded by protecting the children from numerous insect pests and carriers of infection. This is the main reason why Ache children are carried as much as they are, not anything more remote and subtly connected to reproductive strategies.
4. There are important omissions from my account. Expenditure has to be allocated to the parent's own survival. This is an issue in any animal because it influences the length of the reproductive career and thus the total number of offspring born. In humans and other primates, it is also important because of the parent's continuing effect on the survival and reproduction of all its existing offspring (Altmann 1980; Catanzaro 1991). An improved model would include effort allocated to maintaining and protecting the parent.

Alternatives

Alternative explanations of the differences between !Kung and Hadza child rearing are possible. Perhaps it is enough to offer an explanation based on proximate causes: Hadza mothers are simply busier than !Kung mothers and lack the time to be so nice to children! Of course, a proximate explanation need not exclude an explanation by ultimate causes. The ideas presented in this chapter offer an ultimate explanation of why Hadza women chose to be busier than !Kung women (if they really are). However, my argument does imply that a Hadza mother with one child will still behave differently from a !Kung woman with one child, and more like a Hadza woman with several children.

An important alternative explanation of the low fertility of the !Kung is offered by Harpending and Draper (1990, 1992), who suggested that the !Kung may suffer from the same supposedly disease-induced, low fertility as many other African populations. If this is the case, then there seems no need to employ complicated ecological, adaptationist explanations for their reproductive performance. The findings reported here and elsewhere about IBI, mortality, and ecology would either be erroneous or have no bearing on !Kung reproduction. Presumably, the greater fertility of the Hadza would be explained by postulating lower rates of disease among the Hadza, better access to medicines, or more time for the immune systems of the Hadza to learn to cope with these infections. The issue is unlikely to be settled by means better than Occam's razor until difficult medical and historical research is conducted.

Draper (1989) described an apparently general pattern of reproduction and child rearing in African societies. This pattern, high fertility and relatively brief parental care, closely resembles what I have called production-enhancing behavior. One can regard the !Kung as simply not conforming to this general African pattern. Draper associates the general African pattern with unpredictable resources (and unpredictable sources of mortality) and the perception of people as the most reliable and useful resource (for exchange, support, and cooperation). Draper's account seems to faithfully reflect many features of African rural society. The Hadza clearly resemble the general African pattern in fertility and child rearing. It seems likely (as many have suggested for many organisms) that unpredictable sources of mortality would render effort directed to survivorship unrewarding. Effort directed to production or ORS would then be more worthwhile. The issue is whether the difference between Hadza and !Kung is best understood by indicating their unexplained membership of different larger groupings (the African pattern or not the African pattern) or by postulating that the main ecological factor is not that proposed here (difficulty of obtaining enough food) but a difference in predictability of sources of mortality. This presents a real alternative to the interpretation given in this chapter and would be rewarding to pursue.

LeVine (1977) presented by far the most helpful framework for comprehending the variation in human child-rearing and parental behavior. He proposed that parents in all cultures have three aims, in order of precedence: (1) that their child should survive, (2) that as an independent adult it should be able to support itself and its family, and (3) that it should become a good member of its society. By reference to the striking difference in infant and child mortality, he accounted for the nurturance directed to infants by Third World parents and the spartan treatment of infants by wealthy parents in industrialized nations who expect rapid development of social and intellectual skills. High-SES North American or European parents can almost take it for granted that their infant will survive. They can put all their effort and attention into promoting its educational and social development and later competitive ability. LeVine's framework omits all reference to fertility and could not be applied to the differences between Hadza and !Kung.

An even more influential formulation in which child rearing is linked intimately with demography is Caldwell's wealth-flow theory of the demographic transition. Caldwell (1976) suggests that the "demographic transition," a decline in fertility held to accompany industrial and economic development, arises when parents change their ideology. They change from having children in order to increase the flow of wealth from child to parent (more children generate more wealth by their help around the farm), to wishing wealth to flow from parent to child,

which leads to greater emphasis on the care and development of a few "quality" children. Kaplan (in preparation) has recently shown that it is almost impossible for children to generate a positive flow of resources to their parents (see also Turke 1989). Rogers (1990) is a much more plausible alternative account of conditions in which it would be adaptive to reduce number of births and accumulate and endow wealth.

As is to be expected at such an early stage in the attempt to develop an adaptationist perspective on human child rearing, there are many ommisions and alternatives, no doubt more than those discussed above.

Implications for the Adaptationist Perspective

The adaptationist perspective on human child rearing and the constraints on the lives of juveniles by parents has several implications. There are implications for the way we set about comparison between human behavior and the behavior of our closest relatives. There are more fundamental implications for the ways in which we think about child rearing, its variation, and its impact on children's experiences and development.

The view proposed here differs from the evolutionary perspective promoted by Bowlby (1969) and others, including Blurton Jones (1972) and Konner (1972). Bowlby sought the "Environment of Evolutionary Adaptedness" of the human infant. This was the circumstance in which infants grew up during the bulk of our evolutionary history, and to which we supposed human infants were best adapted. These conditions included almost constant contact with one very responsive and attentive caretaker, normally the biological mother.

Tronick et al. (1987) criticized this view because of their observations of Efe Pygmy interactions with infants. Efe infants spent much time with people other than their mothers, and were even sometimes suckled by women who were not their mothers. Tronick et al. argued that observation of contemporary foragers does not support the Bowlby view and that rather than look for evolutionarily fixed patterns, we should look for flexible strategies that respond to local circumstances, although they seem to envisage a largely post hoc approach. My proposals do align with the view of Tronick et al. in that, even from an adaptationist perspective, we expect variation in important aspects of parental behavior. We expect many times and locations in our evolutionary history when infants and children had to cope with parents and experiences more like those of the Hadza than the !Kung. The "ideal" child rearing of our evolutionary history may not always have been so ideal. Presumably, infants and children were selected for the capacity to identify and cope with this whole range of variation (see also Belsky et al. 1991). This may not mean that they are unaffected by these different kinds of experience. There may indeed be a pattern of child rearing that is optimal from the point of view of the individual child. It is unlikely that any child has ever experienced it!

Teaching and learning are other arenas in which these proposals may help us reorganize our thoughts. Because children in simple traditional societies grow up successfully with little or no direct teaching, it was at one time concluded that this was the evolved "ideal" way to develop (e.g., Hertzog 1974). This helped support a fashion for less directive and interventive teaching practices, a philosophy of "every child its own Newton." This no longer seems to be the correct conclusion from cross-cultural or evolutionary arguments. The reason that so few parents (human and nonhuman) spend time in direct instruction of their children may be simply that it does not give such high fitness returns as spending the time getting more food to invest in more children, or in other forms of effort directed to production, survivorship, or ORS. Direct instruction may work very well but simply not enhance parents' RS under many circumstances. Even in the relative comfort of modern industrial society, there may often be more important things for parents to do than the behavior that psychologists might advise them to do.

Implications for Comparative Studies of Human Origins

Many primatologists are interested in the evolutionary origins of human behavior. Tooby and DeVore (1987) pointed out how much more complicated this is than we used to think. In addition to the known variation in behavior of monkeys and apes, great variation of contemporary hunters and gatherers is documented. The old view was that if we attended to the sub-Saharan savanna in which we evolved, we could discard many variants and attend to the !Kung model. This view has been challenged on almost every side (e.g., Schrire 1980; Foley 1982; Wilmsen 1989). Our research on the Hadza adds to the variation already known even from sub-Saharan savanna foragers.

The proposals presented here introduce yet more variation. Characteristic human parental behavior must for long have included the potential for varying the amount of expenditure on each component of reproductive success. Circumstances must often have favored more effort directed one way, and later more directed another way. The difficulty is further compounded by the likelihood that some modern ways of enhancing ORS are very new: What forms of wealth could an Australopithecine endow? Can we ever know how often it was worth its while to spend time teaching its offspring? However, nonhuman primates are known to promote ORS, most commonly by influencing the rank-order position of their offspring. Contemporary humans have added merely more methods for enhancing ORS. Some methods, such as grandparental provisioning, may have arisen because of the nature of human food sources (Hawkes et al. 1989). Humans seem to more often and more extensively exploit offspring to aid the mother's RS than do most other primates (although, again, primate counterparts may be discerned), which may also have to do with the large package size and extensive processing that Hawkes et al. indicate may characterize human food resources.

Role of Models and Experiments in the Optimization Approach to Studying Primate Adaptations

Any claim that behavior is optimal implies a claim that some other way of behaving would produce lesser contributions to fitness, or that the observed way of behaving might produce lesser returns if performed in different circumstances. Thus the approach implies that we can say something about outcomes of behavior that is *not* the most commonly observed in nature.

Experiment is the ideal way to pursue such claims. The work reported here includes two kinds of experiment. One is the computer simulation by Blurton Jones and Sibly (1978) in which, using the mass of information available about the !Kung and about human diets and physiology, we were able to make some confident suggestions about what would happen if !Kung women changed their reproductive schedules, suggestions apparently largely confirmed by Howell's (1979) direct data. This simulation can be used also to investigate the effects of other changes in women's behavior. The other kind is the series of foraging experiments conducted by Blurton Jones, Draper, and Hawkes in Botswana. !Kung no longer gain all their food by foraging; !Kung children never foraged much in the dry season, but (as long as some skilled individuals are available) that does not mean we cannot see what they would get if they did forage. Hawkes et al. (1991) conducted another experimental study with the Hadza, to examine returns from hunting and trapping species that are not normally pursued. Such experiments are relatively easy to conduct with people, given adequate thought and consideration for the subjects. But comparable experiments are likely to be possible, and would be at least as enlightening in studies of nonhuman primate adaptations.

ACKNOWLEDGMENT

The ideas expressed here owe much to discussion with J. S. Chisholm, H. Kaplan, P. Draper, H. Harpending, K. Hawkes, and A. R. Rogers.

22
Behavioral Sex Differences in Children of Diverse Cultures: The Case of Nurturance to Infants

CAROLYN POPE EDWARDS

A radical reconceptualization of sex-role development is taking place in the field of developmental psychology today (Maccoby 1988; Jacklin 1988). This restructuring involves both theoretical and empirical transformations. The theoretical aspect includes efforts to integrate social learning and cognitivist perspectives through a focus on self-socialization, which can be defined as the process whereby children influence the direction and outcomes of their development through selective attention, imitation, and participation in particular activities and modalities of interaction that function as key contexts of socialization. The empirical transformations involve a renewed focus on context. Whereas earlier studies of behavioral sex differences typically involved appraising individual behavioral dispositions across contexts, the new approach seeks ways to understand behavior within specific dyadic and activity settings.

The goals of earlier work were to understand how, why, and at what age girls and boys begin to vary behaviorally along such dimensions as "nurturance," "aggression," and "dependency," including determination of how sex-typical dispositions are influenced by cultural factors (exemplified by Maccoby & Jacklin 1974, and our early analyses of sex differences in children's behavior, in Whiting & Edwards 1973). In contrast, the new approach seeks to under-

stand (1) how different kinds of social behavior are elicited by different contexts of socialization (defined by the sex, age, status, and kinship of social interactants, ongoing activities, and other potent dimensions of setting); (2) how these contexts of socialization are distributed across cultures and associated with various adult subsistence strategies, family structures, household patterns, and forms of social networks; and (3) how boys and girls of each age in diverse cultures come to occupy different contexts of socialization.

Definite answers have yet to emerge on the causal mechanisms and processes in the ontogeny of sex differences. Instead, we remain at the preliminary stage of collecting a solid corpus of descriptive data. The data serve to show how sex, age, and cultural differences in children's interpersonal behavior are related to daily routines, including modes of subsistence, parental division of labor, household living arrangements, and social networks for work and leisure. These data closely parallel those data sought in current studies of nonhuman primates, where sex, age, and species differences in interpersonal behavior are studied in relation to sex-typical life-history strategies (as defined by patterns of dispersal and ranging, lifetime group membership, modes of sexual selection and parenting, and modes of achieving and maintaining group status).

In the remainder of this chapter, I attempt to illustrate the kinds of insights being achieved in cross-cultural studies of children's sex-typed behavior, and to show what kinds of comparisons can be drawn to findings on juvenile nonhuman primates. I focus on one dimension of interpersonal behavior—children's interaction with infants—as exemplary of the problem domain.

Only in recent years has alloparenting behavior received systematic study in either humans or nonhuman primates. Among the nonhuman primates, allomothering by nonparents, normally close kin, is common. The degree to which mothers permit it varies within and among species, however, as does the degree of interest in infants by nonparent animals (Hrdy 1976; McKenna 1987; Fairbanks 1990). Among the New World monkeys (Callithrichidae), where multiple births are common, both fathers and immature siblings play a prominent role in infant care. In red-bellied tamarins, for example, siblings and fathers compete with mothers to carry infants during the infants' first month of life (Pryce 1988). Among Old World monkeys, in contrast, the pattern of allomothering varies greatly across species. Extensive "aunting" by immature females has been described for a number of species, especially langurs and vervets, whereas free-living adult males may provide protection from predators or hostile monkeys but rarely carry or baby-sit infants (Redican & Taub 1981; Snowden & Suomi 1982).

THE *CHILDREN OF DIFFERENT WORLDS* STUDY

Method

The findings on child caretaking to be described come from our recent study, *Children of Different Worlds: The Formation of Social Behavior* (Whiting & Edwards 1988). The book is based on New Sample data collected from 1965 to 1975 by 10 collaborators and ourselves, as well as on data reanalyzed from the 1954–1956 Six Cultures Study (Whiting & Whiting 1975). The majority of the sample communities were traditionally part of tribal societies with subsistence based on horticulture and/or pastoralism ("middle-level" societies). The others were part of complex, stratified societies with economies based on intensive cultivation and/or industry. None was a hunter-gatherer group. Middle-level

societies normally make greatest use of children as infant caretakers, while the percentage of societies using child caretakers is smaller for hunters, gatherers, and fishers than for other subsistence types (Konner 1975).

The six New Sample communities, together with the Six Cultures, were located in the nations of Kenya (five communities), India (two communities), Liberia, Okinawa, the Philippines, Mexico, and the United States (one community each) (Table 22.1). In each of these communities, a sample of between 16 and 104 children aged 2 to 10 years old was selected for timed observation using Focal-Subject Sampling (see Altmann 1974). Behavior was recorded as written running records by trained members of the children's culture. In recording the focal child's social acts (event sampling), the observer followed the eyes of the focal child, identifying whenever possible not only the child's social interacts but also the event that invoked it and any response by a social partner. The records were taken in consecutive English sentences, for later coding. Behavior coding involved judgment of the apparent intention, which often could be made only when the entire sequence of events was known. Before an observation was started, the date, time of day, exact location, people present, and activities in progress were recorded. Time records were maintained along the left-hand margin of the paper, with notes as to when people entered or left the interactional space. With the exception of Bhubaneswar, India, observations were limited to the daylight hours and were distributed over four or five periods of the day. In the Six Culture Study, each record was 5 minutes in length; in the New Samples, they were 15 minutes to 1 hour in length, depending on the community. Methods of training observers and achieving interobserver reliability were roughly the same across communities.

In coding social events, we categorized each interact as a type of *mand,* defined as an attempt on the part of an individual to change the behavior of the social partner (Whiting 1980). We used six major categories of mands, each with subcategories: *ego dependent* (seeking comfort, physical contact, help, information, approval, food, other material goods, or permission); *ego dominant* (seeking to injure, annoy, insult, dominate, compete, or escape); *nurturant* (offering comfort, physical contact, help, information,

Table 22.1. Sample Communities in Which Behavior Records for Focal Children Were Collected

Location	Field researcher	Years of field work	Sample
Kien-taa, Liberia	Gerald Erchak	1970–1971	15 households; 20 children aged 1–6 (360 minutes of observation per child)
Kokwet, Kenya	Sara Harkness Charles Super	1972–1975	64 children aged 3–10 (120 minutes of observation per child)
Kisa and Kario-bangi, Kenya	Thomas Weisner	1970–1972	24 urban and rural families matched by age, education, and kinship ties; 68 children aged 2–8 (120 minutes of observation per child)
Ngeca, Kenya	Beatrice Whiting	1968–1970, 1973	42 homesteads; 104 children aged 2–10 (45–300 minutes of observation per child)
Bubaneswar, India (state of Orissa)	Susan Seymour	1965–1967	24 households (8 upper, 8 middle, 8 lower class); 103 children aged 0–10 (16 hours of observation per household)
Nyansongo, Kenya	Robert Levine Barbara LeVine Lloyd	1955–1956	18 homesteads; 16 children aged 3–10 (75 minutes of observation per child)
Juxtlahuaca, Mexico	A. K. Romney Romaine Romney	1954–1956	22 households; 22 children aged 3–10 (79 minutes of observation per child)
Tarong, Philippines	William Nydegger Corinne Nydegger	1954–1955	24 households; 24 children aged 3–10 (135 minutes of observation per child)
Taira, Okinawa	Thomas Maretzki Hatsumi Maretzki	1954–1955	24 households; 24 children aged 3–10 (74 minutes of observation per child)
Khalapur, India[a]	Leigh Minturn	1954–1955	24 households; 24 children aged 3–10 (95 minutes of observation per child)
Orchard Town, U.S.	John Fischer Ann Fischer	1954–1955	24 households; 24 children aged 3–10 (82 minutes of observation per child)

[a]Minturn returned to Khalapur in 1974–1975. The data used in this volume do not include the restudy.
Source: Whiting and Edwards (1988, p. 19).

approval, food, other material goods, or permission); *prosocial* (commanding an economic, a household, or a childcare chore, commanding hygiene or etiquette, reprimanding another's behavior); *sociable* (seeking or offering friendly response, including social play, laughing together, talking together, verbal or physical teasing, or horseplay); and *teaching* (offering general information, abstract knowledge, or information about skills necessary for a chore). After coding the interacts, the frequency totals were converted into proportion scores (proportion of all coded social acts by an actor or a category of actors).

A brief description of the sample communities is warranted. Nine of the 12 were rural peasant economies. A generation or two previously, the members of these communities had produced their own food through horticulture and animal husbandry, but by the time of study, the people had become involved in some cash cropping or wage work to buy products of the industrial world. The other three communities were urban; men worked as wage earners, entrepreneurs, or professionals. Women's work-loads varied greatly, with the heaviest in rural Kenya, where mothers desired many children and also performed heavy farm and household work. Settlement patterns of the communities varied from those composed of large farms, to hamlets, villages, large towns, and parts of cities. Average household size was smallest in the United States community of Orchard Town, New England, with three children per family, and largest in the Kenyan and North Indian samples, where 7–10 children were usual. Kin availability was greatest in the rural polygynous Kenyan households, which included as many as eight separate dwelling units. Kin availability was also high in many other samples, where relatives lived on contiguous or nearby land. Most isolated from kinfolk were the apartment dwellers of Kariobangi (a housing estate in Nairobi, Kenya) and the Americans in suburban Orchard Town.

In addition to these data, Whiting and Edwards (1988) report data collected under the direction of Ruth and Robert Munroe using a form of Instantaneous Sampling (Altmann 1974) called Spot Observation (Rogoff 1978; Munroe

Table 22.2. Sample Communities for Which Instantaneous Behavior Samples (Spot Observations) Were Collected

Location	Field researcher	Dates	Subject
Nyansongo (Gusii), Western Province, Kenya	Sara Nerlove	1967	10 girls, 12 boys
Vihiga (Logoli), Western Province, Kenya	Ruth Munroe	1967	8 girls, 8 boys
	Robert L. Munroe		
Ngeca (Kikuyu), Central Province, Kenya	Ruth Munroe	1970–1971	12 girls, 9 boys
	Robert L. Munroe		
Conacaste and Santo Domingo, Guatemala	Sara Nerlove	1971	28 girls, 25 boys
Santa Barbara (Canchitos), Peru	Charlene Bolton	1974	5 girls, 6 boys
	Ralph Bolton		
	Carol Michelson		
Claremont, California, U.S.	Amy Koel	1975	7 girls, 10 boys

Source: Whiting and Edwards (1988, p. 43).

et al. 1984). On designated days, and at set time periods during the day, the observer visited all the sample homesteads in turn and scored one set of records per subject child: proximity to home; predominant activity; sex, age, relatedness, proximity, and activity of all persons present in the child's interactional space; persons' social engagement with the subject; and whether the subject was being supervised by an authority figure. By this method, a total of 140 children aged 5–7 years old were studied during the years 1967 to 1975 in six sample communities located in Kenya, Guatemala, Peru, and the United States (Table 22.2).

Sex Differences in Children's Behavior

One of our main findings was that it is far easier to describe sex, age, and cultural differences in children's typical companions and activities than to find differences in their social behavior (relative proportions of nurturance, dominance, dependence, and sociability) after controlling for companions and activities. These findings were consistent across cultures.

1. Girls spend more of their day doing responsible or productive work, such as childcare, housework, and gardening; boys spend relatively more of their time in undirected activity or play, and these sex differences are seen from age 3 onward.

2. Sex segregation is the grand rule of social interaction during middle childhood (age 6–10): boys and girls segregate into same-sex peer groups whenever there are enough children available, and especially do so when they have already divided themselves into age-homogeneous groupings (for extensive discussion, see Maccoby 1988).

3. During middle childhood, boys reduce contact and interaction with their mothers and other adult females, and are observed at greater distances from home than are girls.

4. Girls have more contact and interaction with, and responsibility for, infants than do boys.

Girls' greater contact and interaction with infants was, perhaps, the most consistent behavioral sex difference we documented. In reanalyzing the Six Culture data for children 5–10 years old, and the Spot Observation data for children aged 5–7 (Table 22.3), we found that in 10 of the 12 samples girls were observed more

Table 22.3. Children's Involvement in Infant Care

Community	Girls (%)	Boys (%)	Difference
Six cultures[a]			
Nyansongo	32(5)	22(4)	+10
Juxtlahuaca	13(7)	6(8)	+7
Tarong	9(3)	9(3)	0
Taira	15(7)	6(3)	+9
Khalapur	9(7)	1(4)	+8
Orchard Town	10(1)	0(1)	+10
Spot observations[b]			
Nyansongo	14	7	+7
Vihiga	10	1	+9
Ngeca	7	3	+4
Conacoste/Santo Domingo	3	1	+2
Santa Barbara	6	0	+6
Claremont	0	0	0

[a]Percentage of observations in which children aged 5 and over (who have a sibling) are responsible for the infant sibling (the number of children with 1- to 18-month-old siblings are given in parentheses).
[b]Percentage of observations in which children aged 5–7 are holding an infant (1–18 months).
Source: Adapted from Whiting and Edwards (1988, p. 73).

often than boys taking care of infant siblings, and in the remaining two samples girls' and boys' scores were equal (sign test, $p < 0.001$, two-tailed).

Children's involvement in infant care was positively related to mothers' workload (see also Blurton Jones, Chapter 21, this volume). A rank ordering of the Six Cultures mothers' workload, based on ethnographic descriptions, is as follows from highest to lowest: Nyansongo, Tarong and Juxtlahuaca (approximately equal), Taira, Khalapur, and Orchard Town (Whiting & Whiting 1975, pp. 110–113). The rank ordering of children's involvement in infant care is Nyansongo, Taira, Juxtluahuaca, Tarong, Khalapur, and Orchard Town (Table 22.3). The correlation of these two rank orders approaches significance (Spearman $\rho = 0.81$). In the Spot Observation communities, similarly, both maternal workloads and child infant tending were highest in the Kenyan samples and lowest in suburban Claremont, California. Furthermore, Munroe et al. (1984) report Spot Observation data from four cultural communities showing a strong relationship of children's childcare and subsistence workloads to mothers', but not fathers', total workloads.

Whiting and Edwards (1988) also report the more detailed findings of our collaborators after the Six Culture study, and these findings strongly confirm the sex difference in children's involvement with infants. Seymour's (1988) discussion of her findings for Bhubaneswar, North India, is particularly interesting because she also demonstrated the interaction of maternal workload and child sex (mediated by family status) in determining the amount of responsibility assigned to children. She conducted timed observations in 24 households (all with infants) containing 43 children aged 6–10 years. In middle- and upper-status families, childcare responsibility was negligible for boys and low for girls. In lower-status families, however, where the mothers worked outside the household, girls cared for, assisted, and disciplined their infant siblings almost four times more frequently than did brothers and six times more frequently than did the middle- and upper-status girls. In Bhubaneswar, mothers of all status groups considered it undesirable to involve sons in childcare; therefore, in the lower-status homes, daughters bore the load of assisting their mothers.

Harkness (1975) conducted spot observations

in Kokwet, a rural Kipsigis farming community in western Kenya. The observations made of children's companions and activities were randomized over three periods of the day and indicate that girls aged 2–6 years tended babies in 9% of their observations, versus 1% for the boys. Similarly, girls aged 7–14 were observed tending babies in 8% of observations, versus 3% for boys.

Sieley (1975) made spot observations in Kiplelji, the community adjacent to Kokwet, where she collected home observations on each of 58 children aged 6–7 years and 47 children aged 10–11. In the younger age group, 52% of the girls versus 17% of the boys were observed at least once caring for, entertaining, or protecting a baby [$\chi^2(1) = 7.63\ p < 0.01$]. In the older age group, the differences were less, with 60% of the girls observed caring for babies versus 35% of the boys [$\chi^2(1) = 2.65$, nonsignificant].

Wenger (1983, 1989) conducted 1328 spot observations on 105 children aged 2–11 years in a rural Giriama community near Mombasa, Kenya. She found that work was the activity that increased most sharply with age and differentiated the sexes. After age 8, girls' time was heavily structured to serve the needs of the household as a productive and reproductive unit, whereas boys gained new freedom to roam away from the home compound. At age 8–11, girls were observed supervising or caring for an infant or a toddler in 8% of their observations, versus less than 1% for boys. Furthermore, in observing focal children, observers noted whether an infant (18 months or younger) was present in the focal child's interactional space. Infants were present in the interactional space of girls more than boys at all ages (2–3 year olds: girls 18% of observations, boys 11%; 4–5 year olds: girls 21%, boys 13%; 6–7 year olds: girls 22%, boys 19%; 8–11 year olds: girls 27%, boys 5%).

Clearly, girls perform more infant care and are more involved with infants than boys in many subsistence-based societies in which busy mothers recruit help from older children. Because children under age 5–6 are not usually considered mature enough to care for infants, the sex difference appears at about that age and increases over the middle-childhood years.

Children's involvement in infant care bears no consistent relationship to fathers'. Concerning the Six Cultures communities, for example, an approximate ranking of fathers' involvement with infants and toddlers puts Tarong first, Juxtlahuaca second, Orchard Town and Taira in the middle, and Khalapur and Nyansongo lowest (Whiting & Edwards 1988). This ranking is orthogonal to that of children (and also boys, considered separately; Table 22.3). In the Spot Observation samples, a similar approximate ranking places Claremont first, the three Kenyan communities last, and the Latin American communities in between—a ranking negatively associated with that of the children (Table 22.3).

In our sub-Saharan African communities, fathers traditionally had little contact with infants, whereas children (including boys) cared for or supervised infants and toddlers. Husbands and wives often ate, slept, and socialized separately. Young boys prior to initiation spent much more time with mothers than fathers and helped with all kinds of tasks defined as feminine, especially when families lacked daughters (Ember 1973). Even in more modern times, these patterns have endured. Indeed, in all our Kenyan rural communities, boys prior to the age of initiation were involved with infants: in Kien-taa, Kokwet, Kisa, Kariobangi, and Ngecha, 17% of boys aged 4–10 years old versus 34% of girls were observed interacting with an infant (Whiting & Edwards 1988, Appendix D).

It is interesting to compare these Kenyan findings with those from other societies. Katz and Konner (1981) reviewed the role of the father cross-culturally and found that fathers' involvement with infants depended on subsistence adaptation, family organization, and general cultural definitions of male and female. Using Barry and Paxson's (1971) scales on 186 historically and linguistically independent cultures of all subsistence types, Katz and Konner found that fathers were more likely to be rated as "close" versus "distant" to infants in societies whose primary mode of subsistence was gathering, fishing, shifting agriculture, or horticulture, as opposed to hunting, herding, or advanced agriculture [$\chi^2(6) = 17.9, p < 0.01$]. They also found that fathers were more likely to be rated as close in societies that did *not* combine polygyny, patrilocal residence, and the extended family type of organization [$\chi^2(1) = 7.67, p < 0.01$].

Konner (1976) reports ethnographic field work among a traditional Kalahari !Kung group in which foraging was still the mode of subsistence, where women gathered much of the family's food on long hikes into the desert and fathers were intimately involved in domestic life. Men held and played with their babies around the camp but left routine caregiving to the women. Children were little involved in responsible work of any kind, including childcare. Konner collected 15-minute observations (6 per age-point per infant) for infants aged 1–94 weeks. Counting any sort of participation by older children in the course of an infant observation, Konner found that girls played more with infants of all ages and both sexes than did boys. (In addition, girls were more involved with girl infants and boys with boy infants.) Thus although there was no formal use of children as infant nurses, !Kung girls interacted with infants more than did boys. Girls preferred to spend time in camp near adults and little children, whereas boys spent more time playing outside the village (Draper 1975).

Blurton Jones (Chapter 21, this volume) discusses child-rearing strategies in a contrasting group of foragers, the Hadza of Tanzania. He claims that children do little formal childcare, but that children of either sex may be asked to hold a protesting toddler as the mother leaves camp to go foraging and that girls who accompany women are likely to be asked to carry or entertain an infant or a toddler.

Draper and Cashdan (1988) studied hunter-gatherers in transition; they compared nomadic foraging !Kung with nonnomadic !Kung who had recently come to settle near permanent water resources to tend goats and donkeys and to raise crops. The move from a technologically simple to a more complex mode of production was associated with more hierarchical authority patterns, increased sexual division of labor, greater use of child labor, and greater differentiation among individuals. In the settled !Kung context, mothers engaged in intensive subsistence activity within the village, and their daughters (near at hand) became ready targets for heightened requests to do chores and run errands. As a consequence, the behavior of sedentary !Kung children was much more sex differentiated than the behavior of comparable bush children. Draper (1975) reports on 55 hours of elapsed-time observations on 38 focal children aged 2 to 14. Children aged 2 to 6 were observed to do virtually no childcaring acts whether their parents were nomadic or sedentary. However, this changed at age 7–14, especially for the sedentary girls whose mothers drew them into domestic work. The average number of childcaring acts performed per hour were bush boys, 2; bush girls, 1; sedentary boys, 0.4; and sedentary girls, 5.3.

PROBING FOR CAUSES AND CONSEQUENCES

What explains such consistent results? There are two major, mutually compatible hypotheses, with evidence for each. First, there is appreciable evidence of socialization pressure. Girls are often preferentially assigned to care for infants because mothers think girls make better caregivers or want to train girls for their future mothering roles. In the rural Kenyan communities, for example, mothers prefer to use daughters or nieces aged 6–8 as "child nurses," but will use sons of similar age if no girls are available (Ember 1973). In a classic cross-cultural study, Barry et al. (1957) rated published ethnographies for degree of socialization pressure received by boys and girls in different domains. Ratings of "pressure toward nurturance" were judged largely by ethnographic statements about the assignment of childcare, and the results indicated greater pressure toward nurturance for girls than boys in 82% of 33 societies.

There is also considerable evidence of self-socialization, a process whereby children's own choices of models for imitation or identification and of preferred social companions, settings, and activities influence their developing behavior. Thus one avenue of self-socialization regarding interaction with infants involves sex-role identification: as children gradually develop concepts of "masculinity" and "femininity," and understand to which group they belong, they attempt to match their behavior to their conceptions (Maccoby & Jacklin 1974; Maccoby 1990). They identify with one or more adults of their sex and selectively attend to and imitate same-sex models. Because in every society mothers are more involved in infant care than are fathers, we would hypothesize that sex-role identification would lead girls to seek infant-caretaking opportunities and boys to ignore or resist them.

Our evidence for this hypothesis is not as direct as would be desirable but does indicate that girls are more cooperative with mothers' prosocial commands (including commands involving child care) starting as young as age 2–3 years (Table 22.4). We examined age and sex differences in percentages of children's compliance to maternal commands or suggestions in 12 cultural samples. "Compliance" was defined as either immediate or shortly delayed obedience. In some cultures, the sex difference was significant, and across the 28 comparisons girls were more cooperative or compliant in 20, boys in 7, with one tie (sign test $z = 2.31$, $p < 0.05$, two-tailed). Girls' more rapidly developing language skills and empathy toward others (Hoffman 1977; Fischer & Lazerson 1984) could contribute, along with processes of sex-role identification, to girls' greater cooperation with mothers' commands; nevertheless, it is our sense of the data that boys were less involved with mothers' work goals than were girls and that around age 4–5 they began to avoid their mothers and resist maternal authority (see Whiting & Edwards 1988 for observations).

A second important avenue for self-socialization involves children's preferred partners or styles of interaction. Thus girls may be more attracted to infants or mother–infant dyads than are boys, and/or they may prefer nurturant modes of interaction. Only the first of these has been directly studied in either children or the nonhuman primates, although many studies of children's fantasy play show girls from an early age preferring to act out family dramas involving themes of nurturance, whereas boys prefer role play involving monsters, animals, or superheroes focused on themes of dominance and aggression (e.g., Spiro 1980; Pitcher & Schultz 1983). In most nonhuman primate species yet studied, juvenile females more often associate with adult females and exhibit greater interest in neonates and young infants than do males (Pereira & Altman 1985; Pereira 1988a; Fairbanks, Chapter 15, this volume).

Sex differences in children's interest in infants have been studied by a number of American investigators but with an emphasis on children's responses to unfamiliar infants, outside a family context (Fogel & Melson 1986). In a series of studies, Berman and her associates (Berman 1986) found that during both early and middle childhood boys and girls approach and respond positively to unfamiliar infants and toddlers. During the preschool years, sex dif-

Table 22.4. Percentage of Children's "Total Compliance" (Immediate or Delayed) to Mothers' Prosocial Commands and Reprimands (A) and to Mothers' Instigations (of Any Type) (B)

[A] Community (New Samples)[a]	Age 2–3 years			Age 4–5 years			Age 6–8 years		
	Girls (%)	Boys (%)	Difference	Girls (%)	Boys (%)	Difference	Girls (%)	Boys (%)	Difference
Kien-taa	70	[73]	−3	89	83	+6			
Kokwet	91	75	+16*	84	84	0	96	89	+7
Kisa	83	62	+21	91	83	+8	83	76	+7
Kariobangi	76	95	−19	87	76	+11	86	[92]	−6
Ngeca	81	70	+11*	86	84	+2	96	92	+4*
Bhubaneswar, lower class	62	67	−5		68		77	62	+15
Bhubaneswar, middle and upper classes	57	53	+4	63	50	+13	78	58	+20

[B] Community (Six Cultures)[b]	Age 3–6 years			Age 7–10 years		
	Girls (%)	Boys (%)	Difference	Girls (%)	Boys (%)	Difference
Nyansongo	71	53	+18	63	45	+18*
Juxtlahuaca	69	79	−10	78	[44]	+34**
Tarong	68	48	+20**	77	61	+16+
Khalapur	51	60	−9	67	50	+17
Orchard Town	48	53	−5	76		

Note: All tests of significance are based on φ tests, derived from χ^2. Tests are two-tailed: $+p < 0.10$, $*p < 0.05$, $**p < 0.01$, $***p < 0.001$.
[a]Kien-taa (girls and boys 6–8) and Bhubaneswar Lower Class (girls aged 4–5) have been omitted because they were based on one child–mother dyad only. Scores from Kien-taa (boys 2–3) and Kariobangi (boys 6–8) have been placed in brackets because they were each based on two child–mother dyads only.
[b]Taira (all subgroups) and Orchard (boys 7–10) have been omitted because there were too few acts. The score from Juxtlahuaca (boys aged 7–10) has been placed in brackets because it was based on fewer than 10 child–mother acts.
Source: Adapted from Whiting and Edwards (1988, p. 151).

ferences in attraction to infants are not clearly evident. However, strong sex differences appear around age 5, at which time children enact stereotyped "parental scripts" based on perceived adult masculine and feminine roles. Thus American boys become watchful and protective but generally passive toward babies unless the babies need direct instrumental help, whereas girls become highly interactive and nurturant.

In one study conducted in a day-care center, older (4.5–5.5 years) and younger (3.5–4.5) boys and girls were compared in spontaneous play with a young toddler and then when explicitly asked to look after the child (Berman & Goodman 1984). Sex and age differences emerged only after the caretaking request. Then, older girls interacted most (and with the most varied and nurturant styles), older boys interacted least, and younger girls and boys at an intermediate level. In another study (Reid et al. 1989), girls and boys aged 4–6 years were asked to pose for a photograph with an infant. Girls smiled, touched, and stood closer to the baby than did boys, but the differences were strongest when children were specifically directed to act

as the baby's Mommy or Daddy. Remarkably, the command to pose as parent caused boys to move farther away, whereas it caused girls to move closer. These responses were independent, because each child was individually tested.

Overall, American psychologists suggest differences in style of responsiveness rather than basic attraction toward infants. They argue that adult socialization and children's own identification with same-sex adults together foster the development of the sex difference in responsiveness during middle childhood. Do the cross-cultural data also show a sex difference in children's style of interacting with infants?

The data in Whiting and Edwards (1988) are useful in providing a close look at settings where large families and kin-centered life-styles offer both boys and girls opportunities to interact with infants and toddlers. We examined the question by comparing the proportions of boys' and girls' (aged 4–10 years) social acts to infants that were nurturant (defined earlier) in eight communities for which enough data were available. Six were located in Kenya. In the other sample communities, boys interacted with infants too seldom to compare their relative nurturance. (Girls aged 3–11 exceeded boys in percentage of aggregated social acts made to infants in each of the Six Cultures, sign text, $p < 0.05$; Whiting & Whiting 1975, Table 19.)

The girls' mean nurturance was higher than the boys' in six of the eight samples (Table 22.5; two-tailed Wilcoxon matched-pairs signed-ranks test, $W = 30, p < 0.04$), and perusal of the observations indicated that girls took greater interest and pleasure in caring for infants. Boys, although fully competent, were eager to leave their charges and play with peers, often in a rambunctious style (Whiting & Edwards 1988). However, meeting infants' needs in a skillful way clearly had positive consequences for all children: infants ceased crying, smiled, and reached out for those who responded appropriately to their signals. Because girls were preferentially assigned tasks of childcare, they had greater opportunities to learn and practice nurturant styles.

This issue brings us finally to the issue of the behavioral consequences of juvenile caretaking. What are the short- or long-term effects? A primary hypothesis in primatology is that juvenile alloparenting functions to ensure later parental competence (Hrdy 1976; Fairbanks, Chapter 15, this volume). Unfortunately, there are few data from free-living primates to demonstrate the effect (Pereira & Altmann 1985). Recent research on captive vervets, however, has shown a correlation between allomothering and eventual reproductive success (Fairbanks 1990). Also, data from wild tamarins suggest that females may indeed benefit from juvenile experience in handling infants (Baker 1990).

Our data did not allow long-term effects to be examined, but did permit investigation of potential short-term effects. We compared social relations with peers for groups of 6- to 10-year-old children that simultaneously had relatively great versus little interaction with infants. If experience with infants helps children to develop nurturant social styles, then children who interacted most with infants were expected to be more nurturant as well to same-sex peers, their preferred social partners during middle childhood.

The 12 subgroups of boys and girls aged 6–10 from the six communities in the Six Culture Study provided the data (Table 22.6). Peers were defined as children of the same 6–10 age grade as the focal children but not siblings. The predicted correlation was found (two-tailed Spearman rank correlation, $\rho = 0.61$, $p < 0.05$), supporting the hypothesis of transfer of nurturance from infant to peer contexts for both boys and girls. Although these correlational data do not demonstrate a cause–effect relationship and are not directly analogous to the nonhuman primate studies of allomothering and increased reproductive fitness, they are consistent with the hypothesis that performing infant care may influence children's nurturant behavioral capacities.

Table 22.5. Sex Differences in Mean Proportion of Nurturance to Lap Children by Girls and Boys 4–10 Years of Age

| Community | Number of dyads (number of all social acts to lap children in parentheses) | | Percentage of nurturance (as a proportion of all social acts to lap children) | | |
	Girls	Boys	Girls (%)	Boys (%)	Difference
New Samples[a]					
Kien-taa	5 (108)	3 (69)	38	25	+13
Kokwet	8 (89)	6 (48)	27	31	−4
Kisa	3 (16)	1 (5)	49	0	+49
Kariobangi	8 (131)	3 (27)	39	29	+10
Ngeca	10 (109)	4 (36)	47	29	+18
Six Culture samples[b]					
Nyansongo	6 (174)	3 (35)	45	51	−6
Juxtlahuaca	5 (72)	4 (33)	62	36	+26
Tarong	6 (173)	4 (186)	52	41	+11

[a]Nurturance in the New Samples includes only initiated nurturance. In the Six Culture samples it includes both initiated and responsive nurturance.
[b]Only the communities where the age of the lap child could be identified are included in this table.
Source: Whiting and Edwards (1988, p. 178).

Table 22.6. Interaction of 6- to 10-Year-Old Girls and Boys with Lap Children (0–24 months) and Proportion of Nurturance to Same-Sex Peers

Interaction with lap children		Proportion of nurturance to same-sex peers	
Group	Percentage	Group	Percentage
Nyansongo girls	37	Tarong girls	26
Juxtlahuaca girls	34	Nyansongo girls	20
Tarong boys	28	Juxtlahuaca girls	17
Juxtlahuaca boys	26	Tarong boys	13
Khalapur girls	20	Nyansongo boys	11
Tarong girls	18	Juxtlahuaca boys	10
Taira girls	7	Khalapur boys	10
Khalapur boys	6	Orchard Town girls	9
Nyansongo boys	6	Orchard Town boys	9
Orchard Town girls	5	Khalapur girls	8
Taira boys	4	Taira boys	8
Orchard Town boys	0	Taira girls	5
Mean	16		12

Source: Whiting and Edwards (1988, p. 264).

Weisner (1987) presents supportive data on one of our New Sample groups, the Abaluyia of western Kenya (communities Kisa and Kariobangi). He collected 168 timed observations on 63 children aged 2 to 6. Mothers made more requests and instructions regarding child-caretaking to girls than to boys. Girls, in turn, were more nurturant and prosocially commanding to other girls (age unspecified) than boys were to boys.

Finally, Munroe et al. (1984) compared boys' and girls' workloads in four societies with their modes of social behavior when not working. The children lived in four communities with heavy subsistence workloads: Logoli of Kenya, Garifuna of Belize, Newars of Nepal, and American Samoans. In each community, 48 children aged 3 to 9 were studied by means of spot observations (30 per child) and, on separate occasions, timed observations of social behavior. For both girls and boys, significant positive correlations were found between caretaking levels and nurturance scores when not caretaking, strongly supporting the hypothesis of transfer of nurturance from the infant-caregiving setting to behavior with other partners.

DISCUSSION AND RECOMMENDATIONS FOR FUTURE RESEARCH

In sum, involvement with infants is one of the most consistent sex-differentiated behavioral domains of middle childhood, yet neither the causes nor the consequences of the phenomenon are well understood. Only recently has the phenomenon attracted the attention of developmental psychologists, due perhaps to their earlier focus on parent–child and peer relationships and to North American ethical values (cultural biases) favoring universalistic over kin-centered norms of care and concern. Child–infant relationships are gaining increased attention and can be seen as central rather than peripheral to the understanding of sex-role and sex-identity development. Indeed, we propose that studying the child–infant dyad in the context of the family provides a natural window into the self-socialization processes that augment and extend behavioral sex differences.

In the prevailing psychological opinion, sex differences in children's involvement with infants are understood primarily as the result of sex-role identification and societal expectations, which jointly lead girls to attend selectively to and imitate adult females and to internalize general expectations for feminine behavior, including (in specific circumstances) nurturance of and preference for infants. In the absence of societal expectations, no differences in nurturance of or preference for infants are predicted, according to this perspective. For example, Fogel (1984, p. 178), a leading expert on infancy and development of nurturance, concludes his review:

It seems that differences in male versus female responsiveness to infants appear in middle childhood and late adolescence and again in middle age. Young [preschool] boys, men of child-rearing age, and grandfathers seem about equally responsive to babies as their female counterparts. These studies suggest that sex differences in interest in babies may be related to society's expectations of how males and females should behave rather than to some underlying biological [genetic] predisposition favoring females. Males' interest in babies seems to correspond to times in the life cycle when men are exposed to babies and expected to take an interest in them. Women are more likely to be expected to be interested in babies all through their lives.

We would like to modify this account of socialization by placing preferred social partners at the head rather than the tail of the system of causal theorizing. What we call the "company we keep hypothesis" states that age, sex, and cultural differences in children's typical social partners lead to individual differences in social behavior (because of the eliciting properties of various types of dyadic interaction) with long-term implications for social development. On the basis of accumulated cross-cultural observations, we claim that distinctive generic behavior can be identified in particular types of dyadic interaction. These elicited responses appear to be easily learned and to resist extinction, although their specific form is influenced by learning processes that modulate, channel, attenuate, or amplify them in the culturally appropriate direction. Thus in searching to explain the development of sex differences, rather than limiting our analysis to asking whether adults treat boys and girls differently in face-to-face interaction, we begin with the fact that from an early age, girls and boys are typically observed in different settings where they have differential opportunity to interact with various age–sex–kinship categories of companions. They receive differential opportunities to acquire and practice behaviors such as nurturance and aggression—habits that may become lasting and general.

To further differentiate our position from the prevailing account, we consider the possibility of genetically based predispositions. In raising the prior question of why boys and girls frequent different settings and dyadic interactions, we (Whiting & Edwards 1988) have stressed socialization factors such as task assignment. However, on the basis of the primate literature, we also recognize the possibility of complementary predispositions, especially intrinsic prefer-

ence for same-sex partners. The "company we keep hypothesis" fits well with Pereira's (1988a, p. 201) claim: "Animals are predisposed to learn particular types of behavior, and behavioral ontogeny functions, in part, to facilitate this selective learning and ensure that certain early social experiences are virtually inevitable" (see also Pulliam & Dunford 1980).

In sum, we propose that involvement with infants is an important source of sex-typed behavioral development in children. The system of self-socialization is founded on attractedness to like-sex community members, followed by identification. Girls are predisposed in their development to maintain proximity to adult females, where they receive maximal opportunity to attend selectively and maintain proximity to infants, with the result that they gain knowledge and practice in nurturing styles of interaction. As they gain knowledge and practice, they become more skillful caregivers, with the result that nurturant interaction becomes differentially rewarding to them. The intrinsic rewards of skillful caregiving, along with cognitive-developmental processes motivating girls to increase self-esteem by mastering sex-appropriate competencies, augment and extend the underlying sex differences in social preference and selective attention.

This argument clearly parallels Maccoby's (1988, 1990) and Jacklin's (1989) conclusion that preference for same-sex-peer interaction is a primary facilitator of sex-differentiated development in children. They provide much empirical evidence for the emergence of a preference for same-sex peers in children by age 3, several years before consolidation of sex-role identification. The same-sex preference involves selective attention, selective responsiveness to the reinforcements of same-sex others, and (by middle childhood) tendencies to avoid or exclude the opposite sex when in groups. Maccoby concludes that sex segregation in children's playgroups is a central mechanism in the development of sex-typed behavior and communication styles because of the distinctive "cultures" of boys' versus girls' groups. Our theory complements theirs by highlighting the importance of proximity to mothers and infants in girls' sex-typed development.

To test the proposed theory about the influence of interaction with infants on sex-role development, the following empirical questions need answers. First, at what age does girls' greater involvement with infants emerge? Does the evidence support the hypothesis that sex-differentiated social preferences and selective attention emerge in early childhood, before sex-role identification is consolidated? Second, if sex-differentiated involvement with infants does emerge early, what process (or combination of processes) causes it? We have suggested that girls are more involved than boys with infants, not as a result of greater basic attraction to infants, but as a secondary result of their attraction to female adults and/or nurturant and caretaking modes of interaction. Careful observation of children's behavior is required under a series of conditions, with improved measurement of how and when children initiate interaction or proximity-seeking and which social partners they selectively observe and imitate. Methods used in studying spacing behavior, proximity, and allomothering in nonhuman primates (e.g., Hinde & Atkinson 1970; Pereira 1988a; Fairbanks 1990) would enable observation of children in natural family situations while maintaining sufficiently refined data to separate the three factors.

Finally, what are the short- and long-term sequelae of children's involvement with infants? Data are needed to test potential hypotheses about caregiving experiences as source of (1) nurturant style of relating to peers, (2) later parenting skill, (3) generalized preference for needy social partners, and (4) the centrality of caring for others to emerging sex-role self-concepts, and other potential consequences.

Studies addressing these questions should clarify the role of interaction with infants in sex-differentiated behavioral development and provide the data we need to understand the fundamental processes of self-socialization.

The original version of this paper was presented as part of the Symposium on the Cultural Construction of Gender, at the annual conference of the American Psychological Association, Boston, August 1990.

23

Biocultural Interactions in Human Development

CAROL M. WORTHMAN

The study of human juveniles has come to be divided into two compartments, one the domain of psychologists who study psychosocial development, and the other that of pediatricians and human biologists who study physical processes of growth and maturation. This compartmentalization, although historically understandable, has impeded our understanding of development as a biosocial dynamic.

This chapter will explore the interface of biology and society in ontogeny, emphasizing the unexpected ways in which culture shapes physical development in juveniles, which, in turn, affects social process. Various lines of evidence have converged—from psychobiology, neurology, ethology, life-history theory, and comparative human biology—to suggest that human development is fundamentally a biosocial process. The biocultural approach is taken to promote the view of biological processes as, often, nondeterministic. Phenotype, rather than simply genotype, is the site at which natural selection acts; any organism, but particularly slowly maturing, long-lived social species such as humans, reflects as a phenotype the ongoing, continuous process of ontogeny, in which expression of genetic material is mediated by and through its milieu. The study of human development is thus the study of ontogeny, and should properly yield insight into the organization and operation of adaptation. These considerations have led many developmental biologists to emphasize the contingent, multivalent properties

of ontogeny, expressed in current attention to alternate live-history strategies, evolutionarily stable strategies, epigenesis, prepared learning, and the view of physical development as a process of selective stabilization (rather than stabilizing selection) (Changeux 1985; Chisolm 1988). Overall, the increasing emphasis in human biology is to *explain* patterns of human variation. This goal entails integration of data on various levels of analysis, from evolutionary theory to physiology.

Here, we will consider the evolutionary background of human developmental adaptedness. Then, the biology of human juvenility will be discussed, and its relation to life-history strategy will be detailed. Next, influences of the (culturally constructed) environment on ontogeny will be reviewed and their dialectic nature stressed. Finally, two studies of later-maturing populations will be described. This research underscores the need for more fine-grained research on development that incorporates biological and contextual–behavioral measures. It also demonstrates the necessity for more comparative study of ontogeny to better elucidate the regulation of human development.

HUMAN LIFE-HISTORY STRATEGY

Historically, comparisons of humans with other taxa sought to identify distinctive or unique human characteristics. More recent comparative research, especially with nonhuman primates, has revealed important similarities and continuities with other taxa. Humans, like their relatives the great apes, are relatively large bodied and long lived. Also like their relatives, they bear single offspring with relatively extended periods of juvenile dependency and wide interbirth intervals. Human infants rely on caregivers for thermoregulation, feeding, stimulation, mobility, and sanitation during at least the first year of life. Weaning occurs at around 2 years and marks the beginning of the juvenile period, which is here defined as lasting until full physical maturation is attained at ages 15–25 (ages 13–18 for sexual maturity). First birth averages at 18 years in females and 25 years in males.* Along with other large-bodied species, humans show marked maturational delay with concomitant expansion of the juvenile period. Uniquely, humans provision juveniles and do not require them to be self-sufficient at least until maturity. Although immatures can, and often do, assume important productive roles (Blurton Jones et al. 1989; Blurton Jones, Chapter 21, this volume), parents rarely expect them to be nutritionally independent. This feature greatly increases the amount of parental investment required for successful reproduction.

Delayed maturation may be the product of low adult and high relative (to adult) juvenile mortality rates, as well as increased body size (Promislow & Harvey 1990; Pagel & Harvey, Chapter 3, this volume). Logically, slow maturation and long dependency would appear to expand the risk to juveniles of prereproductive mortality, not only directly through increased exposure to mortality risk, but also indirectly through demographic, ecologic, and epidemiologic vicissitudes that may erode their support and future prospects. Yet the extended period of dependency in infancy and childhood may actually reduce juvenile mortality, so that a payoff for intensified parental investment is the enhanced probability of offspring survivorship despite the lack of juvenile self-sufficiency (Lancaster & Lancaster 1983; Janson & van Schaik, Chapter 5, this volume). Long-term studies of baboons, chimpanzees, and contemporary human foragers show the prereproductive loss to be 60–70% (Altmann 1980, pp. 34–35), 48%

(Goodall 1986, pp. 112–113), and 38–42% (Howell 1979; Hill & Kaplan 1988), respectively (see also Crockett & Pope, Chapter 8, and Rajpurohit & Sommer, Chapter 7, this volume).

Although juvenile survivorship is relatively high in humans, early mortality is brisk enough to show that the juvenile period is under strong selective pressure. For instance, it is estimated that 40,000 young children die in developing countries each day, and that 2 million children die annually of measles alone (UNICEF 1985, pp. 3, 70). Clearly, behavior or attributes that mitigate the probability of prereproductive mortality is highly selected in both parents and offspring. Parents, though, are not universally nurturant, but balance costs of intensifying investment in existing offspring against opportunities to increase total offspring number (Hill & Kaplan 1988; Hurtado & Hill 1992). Although behavioral ecologists have focused mainly on adult–parental behaviors, juvenile behavior and development should also be considered as highly selected, inasmuch as all who live to reproduce are also survivors of the juvenile period.

Avoidance of mortality may not be the only reason for delayed maturation—or extended childhood—in humans, for the content of childhood itself is thought to be crucial to adult viability. Human reproduction is decidedly a process of transmitting information through two channels, the genetic and the cultural (Boyd & Richerson 1985). For humans, that environment is dominated by culture, representing arrays of practical knowledge (skills, rules of behavior, social knowledge, and so forth) that must be acquired to become a competent member of the social group. Juveniles, then, are dependent on others not only for material support, but also for information. Positive selection pressure for expansion of the prereproductive phase thus is reinforced when postnatal learning and socialization contribute to future fitness (see Pagel & Harvey, Chapter 3, this volume). Relative freedom of juveniles from the food quest and provision of a protected learning environment in the form of a home base create enhanced opportunity for exploration, play, and learning to acquire subsistence and social skills essential for successful adult life. Such an arrangement may also repay parents by providing a setting in which children can "mind" one another.

Delaying maturation may incur a further cost,

---

*Culture definitively and variably determines age at first birth.

inasmuch as it increases generation time and reduces the length of the reproductive life span. Women are especially constrained in this regard, for the obligate time invested in each reproductive act (gestation, parturition, lactation) is long, being at least 2 years. Humans, like other large-bodied species with delayed maturation and low adult mortality, can outweigh costs of maturational delay with increased reproductive life span (see Pagel & Harvey, Chapter 3, this volume). But if age at last birth is fairly constant (Bongaarts & Potter 1983, pp. 42–43), then delayed maturation in girls may erode reproductive potential. Further, as the amount of additional parental investment required per offspring increased through expanded care and socialization of juveniles in human evolution, the importance of time over which that investment was made was likewise amplified. It has been suggested that delayed maturation is a reproductive strategy, in which slow maturation increases opportunity for sibling caretaking and reduces caloric demand of children per unit time (Bogin 1988, pp. 92–95). In partial support of this notion, parents have been observed during periods of food shortage to shift food distribution and workloads to children in whom the equivalent work is energetically less expensive than in adults (Leonard & Thomas 1989). In societies where sons marry later than daughters and become net providers during and after puberty, one would expect to see a "helper at the nest" effect. Among Ache foragers, sons become net calorie producers at 17 and marry only 3 years later. But Hill and Kaplan (1988) found that Ache women whose oldest surviving child was male had significantly lower reproductive success. In this case, childcare assistance from daughters possibly confers parity-specific advantages.

The selective pressures impinging on two life-history components—childhood and reproductive career—therefore are reflected in the timing of puberty, for puberty marks the end of childhood and the beginning of the reproductive career. Some environmental circumstances may favor delayed maturation and prolongation of the period of slow growth, learning, and socialization, whereas others may favor earlier initiation of reproduction. This concept is not novel: as Stearns and Koella (1986, p. 895) note, the "idea that age at maturation is determined by a balance between the advantages of short generation time and the advantages of

large size is widespread in the life history literature." In this literature, it is generally assumed that deferral of maturation is linked to an adaptive advantage—decreased juvenile mortality rates (see Janson & van Schaik, Chapter 5, this volume). However, for humans, maturation delay may be accentuated by adult reproductive success, because skills and information acquired in childhood increase adult viability, mating success, and/or parenting (see Pagel & Harvey, Chapter 3, this volume).

Puberty can therefore be characterized as a transducer for a suite of selection pressures. Timing of the onset and maturational events of puberty reflect a set of adaptive tensions generated by the convergence at puberty of important trends in the evolution of the human life-cycle strategy—developmental retardation and a K-selected reproductive strategy. The former is manifested in a lengthening of gestation time and the preadult phase; the latter is characterized by fewer, single offspring with high investment in each. Puberty, then, occurs at a nexus of selective forces, and has doubtless itself been selected to act as a transducer of ontogenic and population cues. In other words, the timing of puberty has been selected to be variable; physiologic processes responsible for the timing of pubertal changes have been subject to selective pressure to respond to certain salient environmental and experiential features. That may be why the factors determining timing of puberty, secular trends, and population differences in pubertal attainment have been so difficult to isolate: they are multifactorial and contingent in ontogeny. This concept will be developed in the subsequent section.

There are, furthermore, sex differences in the constraints shaping juvenile development. Although these differences are often interpreted as relatively minor in humans, consistent disparities in size and morphology, maturational schedules (Worthman 1987b), and mortality rates (Daly & Wilson 1983, p. 298; Wilson & Daly 1985) suggest basic divergences in life-history strategies. Development of females is thought to be more canalized (i.e., more refractory to environmental insult) than that of males, because females are constrained to achieve minimal size and mass to support pregnancy, parturition, and lactation (Stini 1975, 1982). Further, timely achievement of reproductive maturity and initiation of reproductive career may confer a fitness advantage, as time is a major limitation on the number of offspring females can bear. Sex-differentiated constraints of size on reproduction could be reflected in maturational schedules: girls achieve reproductive competence well after, whereas boys attain first emission before, peak height gains. This difference may also reflect size- or age-specific reproductive tactics that are available to males but not to females. Societies universally allow serial monogamy or/and polygyny, which enhance potential male lifetime reproductive success (e.g., Wood et al. 1985). Because of altered size and timing constraints, males are considered more sensitive to environmental quality and more likely to defer growth under suboptimal conditions.

Empirical evidence to support these concepts is rather thin and is complicated by the fact that so many societies systematically mandate gender-differentiated treatment of children (McKee 1984; Stinson 1985; Hrdy 1990). The latter observation alone suggests that the sexes are subjected to disparate selection pressures mediated by social factors. "Environmental quality" is, moreover, a global term covering disparate elements and heterogeneous microenvironments. Parents are both expected and observed to modify investment in any individual offspring according to immediate constraints that determine relative cost and potential fitness value of that offspring (Sieff 1990). Rather than expecting retardation in males or canalization in females, we should consider which environmental factors may differentially affect the development of either sex. One might more confidently predict that males and females will differ in the environmental factors to which they are susceptible, as well as in their degree of sensitivity to any given feature.

THE BIOLOGY OF HUMAN JUVENILITY

Human growth is characterized by developmental delay and the presence of a marked adolescent growth spurt (Watts 1986). The degree to which these features are shared with other primates has been debated (Watts & Gavan 1982; reviewed in Bogin 1988, pp. 57–68), in part because much of the primate data concern changes in weight, not length. A recent report establishes the presence of a distinct skeletal

growth spurt in puberty for rhesus and notes that the sequence of skeletal changes at puberty was the same for rhesus and human (Tanner et al. 1990).*

The relative growth of tissue types in pre- and postnatal development (Fig. 23.1) provides yet more insight into the biological substrates of delayed maturation. At birth, infant length is about 20% of adult height. Growth rates are high neonatally but fall sharply over the first 2 years and then decline slowly but steadily until a transient, pronounced increase occurs at puberty, so that adult height is attained by age 15–25 (Tanner & Whitehouse 1976). The lymphatic system, by contrast, reflects adaptation to the heavy selection pressure of disease in childhood: its constituents expand rapidly during infancy and childhood to reach nearly 200% of adult weight by puberty, regressing thereafter to adult size by the end of the second decade. Although overall growth is markedly delayed in humans, that of the brain is not: as in other mammals, it develops in advance of the body, so that children reach 90% of adult brain weight by age 6 and achieve adult size before puberty. There must be strong selective advantage to early development of the brain, for it is an energetically expensive organ to maintain, accounting for over 50% of total basal metabolic rate in the first year of life, and a third or more thereof throughout childhood (Holliday 1986). Anthropologists have generally assumed that rapid brain development supports the pattern of information acquisition in the juvenile period (Bogin 1988, p. 61). Last, growth of the reproductive system mirrors the lengthening of the prereproductive phase, by showing little change—except for a brief flurry of activity after birth—until puberty, when maturity is reached at around age 15. In addition, fat tissue has an unusual developmental pattern in humans, who are in general unusually fatty, only matched among mammals by some sea-living taxa. Fat is laid down in the final trimester of gestation, so that infants have 12% body fat at birth. After infancy, children often become leaner. Both sexes increase lipid storage during puberty, but males subsequently lose it, leaving a marked gender difference at the end of adoles-

*Of interest is the finding that the spurt in skeletal growth was more marked and overall maturation more rapid in indoor- than in outdoor-housed females (Wilson et al. 1988).

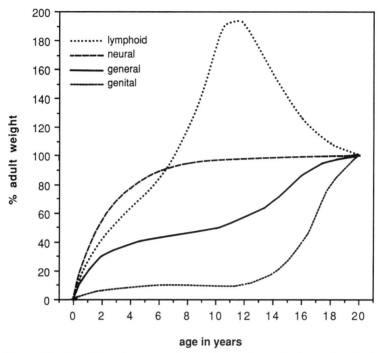

Fig. 23.1. The growth of four different systems—lymphoid, neural, reproductive, and total body—in terms of percent of adult weight. (Redrawn from Scammon 1930)

cence, at which time males have about 12% and females 25% body fat by weight (Holliday 1986).

Human immatures also evidence a distinctive endocrine profile that guides and supports development (Fig. 23.2). Shortly after birth, the hypothalamic–adrenal–gonadal axes exhibit a burst of activity that declines, by age 6 to 8 months, to the very low levels of gonadal steroids and the moderately low levels of adrenal androgens that characterize childhood (Forest et al. 1974). A marked increase in adrenal androgens, called adrenarche, occurs around ages 6 to 8 (Sizonenko & Paunier 1975). Thereafter, the onset of puberty is signaled by changes in the pattern and amount of gonadotropin (luteinizing hormone, LH; follicle-stimulating hormone, FSH) production, which trigger increases in levels of gonadal steroids (gonadarche) that, in turn, operate synergistically with other hormones to contribute to the adolescent growth spurt and stimulate the appearance of adult sex characteristics (Stanhope & Brooke 1988; Ducharme 1989). Adrenal androgens also continue to rise during the course of puberty.

Although the role of gonadotropins and gonadal steroids in puberty is reasonably well understood, the part of adrenal androgens remains unclear. It is thought that adrenal androgens promote skeletal maturation, for there is a linear relationship between them during puberty (Nottelmann et al. 1987). There is some suggestion that they exert a direct effect on growth, partially independent of gonadal steroid action (Weirman et al. 1986). Certain of them (notably dehydroepiandrosterone sulfate, DHEAS) are also linked to body-fat patterning that emerges during this period (Katz et al. 1985).

Comparison of these endocrine patterns with those in other mammals and primates indicates that adrenarche is shared by humans, chimpanzees, and gorillas (Collins et al. 1981; Smail et al. 1982). It appears that primates are distinctive from other mammals in showing high concentrations of adrenal androgens (Cutler et al. 1978). However, young humans and chimpanzees have low levels of these hormones and exhibit onset of andrenal androgen rises at the mid-juvenile stage.

Perhaps the greatest uncertainty regarding ju-

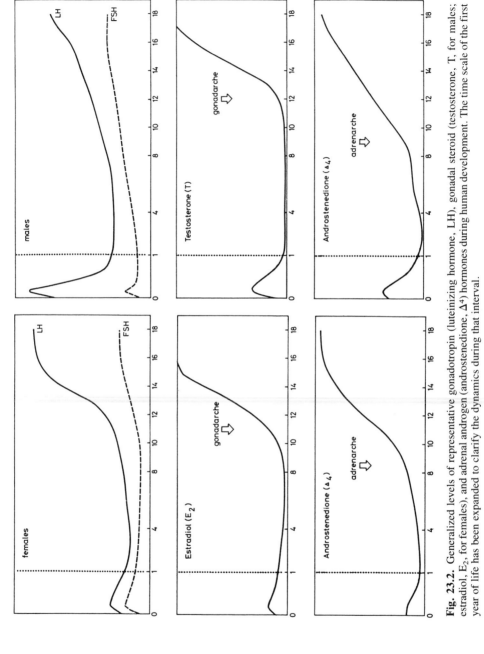

Fig. 23.2. Generalized levels of representative gonadotropin (luteinizing hormone, LH), gonadal steroid (testosterone, T, for males; estradiol, E_2, for females), and adrenal androgen (androstenedione, Δ^4) hormones during human development. The time scale of the first year of life has been expanded to clarify the dynamics during that interval.

venile development in humans concerns the nature of the trigger or cascade of events responsible for initiating pubertal maturation. Such an impetus is apparently required because of the very extent of developmental retardation: establishment of childhood as a prolonged developmental phase also apparently required a physiologic (probably neuroendocrine–endocrine) mechanism to exit that phase and initiate the next. What is that mechanism? Although endocrine precursors and concomitants of puberty are known, the nature of the pacemaker that regulates the timing of puberty by controlling its onset is obscure. Numerous hypotheses have been offered, but none has received definitive support (reviewed in Weirman & Crowley 1986). Since adrenarche precedes onset of puberty by about 2 years, it was at first thought that it may represent the trigger for pubertal onset (Collu & Ducharme 1975); however, evidence from clinical and normal populations has shown adrenarche and gonadarche to be dissociable (Weirman et al. 1986; Worthman 1987b). It has been suggested that the pineal gland plays a role in pubertal onset through changes in melatonin production that derepress the neuroendocrine regulatory system (Wurtman & Waldhauser 1986), but the relevant data are mixed. Another suggestion has been that percent body fat acts as a metabolic index that regulates onset of puberty (Frisch 1984). Then, an alteration in enzyme activity that increases the amount of bioactive gonadotropin-releasing hormone (GnRH) has also been proposed as a mechanism. Bogin (1980) has advanced a catastrophe model for the discontinuity in endocrine–neuroendocrine regulation that occurs with puberty. At present, it appears that puberty is initiated endogenously by the brain, perhaps the hypothalamus. Recently, Stanhope (1989) has concluded that there is no discrete trigger of puberty; rather, the hypothalamic–pituitary–gonadal axes gradually mature during childhood, culminating in progressive changes in GnRH secretion and the onset of gonadal maturation. The onset of puberty, then, is characterized by rising levels of gonadotropins and gonadal steroids, and the resulting first appearance of genital or breast growth.

So far, this chapter has sketched the developmental picture in terms of an overall human average, but it should be noted that considerable individual, population, and temporal variation

exists, along with gender differences, in the course and content of maturation. These differences are particularly pronounced and well documented at puberty. Population variation in timing of puberty is large: median age at menarche has been observed to range from 12.5 years for well-fed urban groups to 17.5–18.5 years in certain traditional Highland New Guinea settings (Eveleth & Tanner 1976, pp. 214–215). Maturational timing has also been observed to change within populations over time: secular trends to increased growth rates in childhood and reduced age at menarche were first noted in Western industrializing settings (Eveleth & Tanner 1976, pp. 218, 260–261; Roche 1979; Brundtland et al. 1980), and have also been evident in developing countries (Zemel et al. 1993). At any given time, within-population differences in timing and course of development—especially puberty—are consistently found, so that maturational heterogeneity among peers is common (Marshall & Tanner 1969, 1970).

CONTEXT SENSITIVITY OF PHYSICAL DEVELOPMENT

Although much individual and population variation is attributable to genetic sources, much of it also reflects sensitivity of physical development to the context in which it occurs. Indeed, sampling of environmental quality and adjustment of development to prevailing conditions commence *in utero*. Birth weight, for instance, is largely influenced by maternal factors (Tanner 1978). Variables known to influence human development include nutrition, disease, and psychosocial stressors; other variables—such as activity and work patterns, environmental lighting, thermal stress, and patterns of infant stimulation—are also thought to have an effect (Eveleth & Tanner 1976; Konner & Shostak 1986).

These factors can influence maturation by modulating both developmental rates and outcomes. The secular trends to increased size for age in children and reduced age at menarche, mentioned above, appear to be largely the result of changes in rates of development. That is, children under "improved" environmental circumstances grow more rapidly and achieve adult status earlier: in the case of industrializing populations, the trend was for age at adolescent growth spurt to decline by 4 months per decade over the last 100 years (Meredith 1976; Brundtland et al. 1980). Acceleration in skeletal development was paralleled by that in reproductive maturation: age at menarche declined at the same rate as age at growth spurt, and has appeared to plateau at menarcheal ages of 12–13 (Brundtland & Walloe 1973). Thus age at menarche in these populations has declined by 2–4 years over the last 150 years. Concurrently, some change in outcome has occurred, including a notable but more modest trend to increased adult size (about 4–6 cm, or up to 1 cm per decade) (Eveleth & Tanner 1976, p. 261; Meredith 1976; van Wieringen 1986). The increase in adult size is thought to be due in part to the relative absence of influences that retard growth (e.g., illness, parasites, malnutrition) (reviewed in Bogin 1988, pp. 127–130).

Secular changes in population averages for growth and maturation rates apparently parallel overall amelioration of conditions under which juveniles develop. Because of differences in these conditions within populations, high social status has been consistently associated with earlier maturation, with the rare exception of countries such as Sweden and Norway in which class-based essential resource inequities are minimal (Bielicki 1986). Relationships of developmental rates to social class, changing overall living conditions, and family size all suggest the sensitivity of growth to the environments children experience. Halts and reversals in secular trends in Western Europe observed in association with worsening conditions, most notably during World War II, are confirmatory (Brundtland et al. 1980).

Various circumstances make it difficult to generalize about sex differences in degree of environmental sensitivity. As noted above, many societies show consistent sex bias in treatment of children, for parental investment tends to favor the gender that brings greater productive or reproductive rewards (reviewed in Hrdy 1990; Sieff 1990). Thus the sexes are rarely faced with identical developmental conditions; rather, parental perception of need or value, or systems of gender socialization themselves, lead to differential experience in workload, care, housing, and nutrition (Worthman 1993). Furthermore, the data from which to make good cross-population comparisons are sparse. Variation in menarcheal age is widely documented; there is no equivalent milestone for males. Menarche is in some ways, however, the best-studied "nonevent" in human development, for it is not equivalent to reproductive maturity (Vihko & Apter 1980; Lemarchand-Béraud et al. 1982), and little is known about its relevance for gauging variation in reproductive maturation among males. Physiologically, menarche is a way station in the ongoing process of female reproductive development that indicates potential, rather than actual, reproductive capacity. Postmenarcheal subfecundity is well documented: as much as 5 years is required after menses commence for young women to attain mature rates of ovulatory frequency (Fig. 23.3). Close examination of endocrine change, physical growth, and morphologic change in both sexes can provide a far richer basis for unraveling population and gender differences in juvenile development.

THE ROLE OF SOCIAL CONTEXT IN PHYSICAL DEVELOPMENT

As the previous section shows, even gross measures of maturation rate show that physical development is facultatively modified by environmental conditions. Here, we will pursue more detailed evidence of how physical ontogeny is "designed" to unfold in a dynamic relationship with milieu. What might be the basis for developmental plasticity? As Chisolm (1988, p. 88) summarizes, "the adaptive significance of phenotypic plasticity comes . . . from the phenotype's ability to *track* its developmental environment, to be affected by it in ways that promote fitness." Plasticity is advantageous when the individual is faced with an environment that is "predictably unpredictable." Stearns and Koella have offered an explanation for the environmental sensitivity of growth rates that is based on life-history theory. They suggest that this sensitivity evolves because "a change in growth rate acts as a reliable signal to the organism that well-defined changes in the demographic environment can be expected. Growth rate thus acts as a cue for upcoming environmental events that cause changes in adult or juvenile mortality and have done so dependably through evolutionary time" (Stearns & Koella 1986, p. 895).

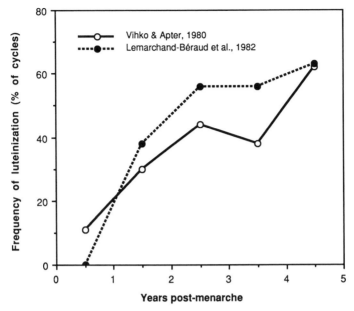

Fig. 23.3. Ovulatory frequency (proportion of cycles showing luteinization) by gynecologic age (number of years postmenarche) in Swiss and Finnish young women.

Human societies vary considerably in the content of juvenile experience (e.g., Blurton Jones, Chapter 21, this volume). The differences start early, with variation in birthing practices (Lozoff 1983); feeding, bedding, and transport of infants (Whiting 1981); size of social unit and opportunity for learning and play (Draper 1976); and myriad other areas. The impact of those experiential differences in behavioral, physical, or cognitive outcomes has been only partially explored (Edwards & Whiting 1980; Edwards, Chapter 22, this volume). However, the "culture of childcare"—that is, the social organization of attitudes toward and treatment of children—represents an important adaptive domain. The behavior of parents, other adults, siblings, and peers constitutes the field of action in which the juvenile obtains valuable resources, be it information, food, social support and status, shelter and clothing, or health care. Here we will detail how the juvenile both influences and developmentally responds to the field of social action. The contention is that not just parental strategies, but also juvenile ontogeny and behavior are highly selected to further survival of the individual.

Although the present focus is on juveniles, infancy is the period in which interactions among caretaking environment, developmental biology, and behavior have been most intensively studied. Models and concepts developed through this work will be considered here because they can inform our thinking about juvenile development.

Infancy

From birth, infants emit signals for eliciting care: their neotenous appearance acts as a releaser of nurturant behavior and a reducer of aggression, and their fussing and crying notify others of physical needs. But, in many additional ways, infant behavior and pattern of emerging competencies facilitate relationships with caregivers. Infant smiles have been studied intensively, demonstrating that the ontogeny of smiling is clearly geared to provide an early, positive social cue to caregivers, who interpret first smiles as affective communicative gestures (Bower 1982, pp. 260–268). Caretaker responses to the infant smile as a social cue then stimulate the infant and shape future smiles, so that the infant becomes not only neurologically capable of generating full social smiles, but also socially competent to produce them appropriately, in a relational context (Kaye 1982).

It is well recognized that infants and parents quickly habituate one another. Components of caretaker behavior and infant behavior and physiology can interact to influence the infant's relationships with caretakers. For instance, very fussy babies are likely to raise anxiety and frustration in their caregivers, who may then alter the care they provide. But that very fussiness may be a product of feeding or carrying patterns: Barr and colleagues (Hunziker & Barr 1986; Barr & Elias 1988) have found that both feeding frequency and amount of carrying correlate inversely with amount of infant crying and fussing.*

The contexts to which infants have been reliably exposed in evolutionary history may also entrain differentiation of physical development (Hofer 1987; Blass 1990). Infants are, for example, reliably presented with an array of visual material from the time of birth; covering an eye (e.g., congenital cataracts) for even a brief interval during the first months results in permanent visual impairment. Studies in the cat and monkey have shown that the impairment is due to dysfunctional organization of the visual cortex: apparently, the ontogeny of the visual cortex is designed to occur in response to patterns of incoming visual stimuli (Wiesel 1982; Changeux 1985, p. 235). If these stimuli are disrupted, then structural–functional differentiation is permanently affected.

A final example of long-term ontogenetic impact of context in infancy is the apparent effect of early physical stress on maturation rate and adult size. Following work in the rat that had shown effects of early moderate stressors (cold, handling) on time of vaginal opening and adult size (Levine et al. 1967), Whiting and colleagues performed a cross-cultural study that revealed an association of physically stressful infant treatment practices (e.g., circumcision, scarification, mother–infant separation)† with

reduced age at menarche and increased adult height (Landauer & Whiting 1981). The endocrine profiles presented in Figure 23.1 may provide an endocrine basis for this effect. As noted above, there is a burst of activity among the hypothalamic–gonadal–adrenal axes in the first 6 to 12 months of life, during which time it is thought that functional relationships among them are established. Might it be that this time represents a sensitive period, during which stressors that challenge these axes permanently affect their set points and alter their activity? This question could be addressed empirically to determine if the stress effect is indeed causal.

Thus infancy includes sensitive periods during which salient environmental variables may strongly influence subsequent development. At later ages, the same variables may either not affect development or affect it differently. The case of infant stress exemplifies this point: moderate or transient infant stress exerts a *stimulating* effect on subsequent development, whereas growth of older children responds *negatively* or not at all to physical or psychosocial stress. The effect in infancy appears to be organizational and thus long term, whereas that in childhood is more acute and may be mitigated by growth-stabilizing processes when the stressor is removed. Further, the stress models for infants have involved tactile or sensory stimuli, whereas stressors known to influence children include those that compromise energy balance, challenge immune systems, or create psychic distress.

Childhood

Environmental sensitivity of ontogeny continues in childhood. As was reviewed in the previous section, the pace of physical growth is acutely influenced by environmental quality. Episodes of illness, food shortage, and intense psychosocial stress are accompanied by slowing or cessation of growth (reviewed in Mascie-Taylor 1991). These developmental vicissitudes are buffered by stabilizing growth processes. Abatement of conditions causing restricted growth results in accelerated height gain, or

---

*The effect of feeding frequency appears to be mediated by gas formation from incomplete carbohydrate absorption, which is differentially influenced by frequency of milk ingestion, as well as by individual differences in metabolism.

†Infant stress effects are a matter of degree: mild–moderate or short-term stress is growth enhancing, whereas severe or prolonged stress is growth inhib-

iting. Prolonged maternal separation (or, more precisely, deprivation of tactile stimuli) is known to have a negative, lasting effect on infant growth (Schanberg & Field 1987).

catch-up growth, that pushes height toward the normal population range of size for age (Prader et al. 1963). Growth rates achieved during catch-up periods can be far higher than any observed during normal development (Tanner 1978, pp. 154–160). This phenomenon supports the notion that growth is a target-seeking process. Amelioration of negative conditions results in differential allocation of energy to growth when that becomes possible. Catch-up growth can compensate for growth deficits, but only if conditions improve within a moderate period; chronic malnutrition results in reduction of final attained height (reviewed in Martorell & Habicht 1986).*

The adaptive advantages of small size under limited energetic resources have been debated (reviewed in Bogin 1988, pp. 121–125; Dennett & Connell 1988). Rapidly growing children genetically predisposed for tallness may be more subject to malnutrition under food restriction (Bogin 1988, pp. 121–122); indeed, Ellison (1981), noting the persistent cooccurrence of reductions in infant mortality and accelerated maturation, has conjectured that the secular trends in maturation rate may be partly the result of the differential survival benefit accruing to rapid developers when environmental conditions improve. Then, some studies suggest that in populations subject to malnutrition and poor public health, smaller women have greater reproductive success than do taller ones. But other reports indicate that longer, heavier children may have less mortality under heavy parasitic and disease pressure; however, larger size under these conditions may indicate resistance or lowered exposure to parasites and pathogens.

Social stratification and demographic variables also influence child growth. Size and growth rate correlate with class where class is associated with differential access to essential resources (Bielicki 1986). Further, family size (Goldstein 1971) and birth order (Belmont et al. 1975) have been observed to correlate negatively with child height. This relationship has been observed in Western urbanized countries; one wonders whether the opposite may be true in horticultural and peasant societies, where the family constitutes the labor base for production and household size is related to wealth. On the

other hand, child workload increases with household size (consumer–worker ratio), whereas parental work load does not (Munroe et al. 1984). This, with the finding that children's work effort correlates negatively with maternal or paternal work level, indicates the pervasiveness of child-labor recruitment. Munroe et al. (1984) found, in a four-culture study, that children worked 23% of the time; even 3-year-olds did chores in 10% of observations. Workload can represent an energy drain that interacts with nutrition to determine net energy available to the child; net energy balance, rather than food intake alone, may be a better predictor of physical well being (Hurtado & Hill 1990). One notes in passing that the secular trends in industrialized countries occurred over a period when child-labor patterns changed dramatically, and children were increasingly sedentized in school. Research on human development should perhaps study child workload more systematically (see Blurton Jones, Chapter 21, this volume).

Surprisingly subtle ways in which social arrangements can influence child growth continue to be uncovered. For instance, a close association between number sharing a child's bed and reduced child height has been reported, a relationship that persists when effects of multiple covariates (e.g., social class, family size) are removed (Terrell & Mascie-Taylor 1991). These investigators suggest a causal pathway from crowding to sleep disturbance and disruptions of sleep-related growth hormone release.

An even more pervasive influence on environmental quality than social class or family size is gender. As noted above, gender differences in treatment of children are common, encompassing differential allocation of workload, nutrition, health care, and risk exposure. The classic comparative project on child socialization, the Six Cultures Study led by the Whitings in the 1950s and 1960s, documents virtually universal gender-specific role and task assignment (Whiting & Whiting 1975; Whiting & Edwards 1988; Edwards, Chapter 22, this volume). Boys and girls spend much time in different social settings; in part, they are in different contexts because they are given disparate tasks to do. Beliefs about needs and vulnerabilities of each gender may legitimate and reinforce selective treatment (Worthman 1993). Although effects of this divergence on behavioral, psychosocial, and life-history outcomes

---

*Catch-up growth has also been observed in non-human primates (Kerr et al. 1975).

have been explored, less consideration has been given to influences on physical ontogeny, although sex-differentiated morbidity and mortality have been shown to track differential treatment (Dickemann 1979; Boone 1988; Cronk 1989).

Systematic sex bias in treatment of children has been shown to create differences in growth as well as survivorship. In rural Bangladesh, strong male preference supports differential food allocation and provision of health care. Male food intake is greater at all ages (weight-adjusted intake was strongly biased only to age 5), and medical treatment visits for males were higher. Accordingly, female exceeded male mortality at all ages beyond 1 month postpartum, and male children's height for age, weight for age, and weight for height were markedly greater and malnutrition indices lower (Chen et al. 1981). Analysis of data from the same study area indicated that bias in food allocation increased during periods of shortage (Bairagi 1986). Although always poorer than that of boys, nutritional status in girls was more acutely affected during famine, as was degree of stunting. In line with evolutionary explanations, sex bias was most pronounced in nutritional status in families of high socioeconomic status. Study in a Mexican agrarian region with less extreme sex bias also showed that dietary quality of girls was poorer, and that their distributions of weight for height and weight for age, but not height, were lower (Dewey 1983).

Puberty and Adolescence

Puberty is a life phase in which interactions between physical development and social context are prominent. The distinction between puberty and adolescence characterizes this interplay: puberty is the sequence of physical changes that leads to biological maturity, whereas adolescence is a socially designated phase between childhood and adulthood. Puberty, therefore, is universally present, although its timing is variable. The presence, content, and timing of adolescence, on the other hand, vary greatly. As a culturally recognized developmental period, adolescence ranges from the virtually nonexistent (e.g., Amhara: Levine 1965), to the highly elaborated and prolonged (e.g., East African pastoralists such as Masai). It is more common for males to spend an extended period in a liminal

preadult status than it is for females, who are generally more rapidly inducted into adult productive and reproductive roles.

Throughout the juvenile period, physical and behavioral changes act as cues to the social milieu that entrain socialization practices. That is, members of the social group are attentive to developmental status in determining expectations, demands, and statuses accorded the juvenile. The physical changes of puberty often act as cues for timing of social interventions (Worthman 1987b, 1993). In many societies, menarche prompts rites of passage—marriage, or change of age grade or other formal status change (Burbank et al. 1986). Other physical changes (e.g., height gain, muscle bulking and facial hair in males, breast development in females) may also elicit cultural responses, in terms of changing roles, statuses, and experiences. Because of their salience to the behavior of others, patterns of visible development can also be considered as evolved signals to group members: the gender difference in timing of maximum growth with respect to reproductive maturation illustrates this point (Fig. 23.4).

As with growth in childhood, timing and course of pubertal development closely track markers of environmental quality such as social class, nutrition, family size, morbidity, and mortality. Ages at pubertal growth acceleration and menarche have been repeatedly observed to vary with these factors (Eveleth & Tanner 1976; Bielicki 1986; van Wieringen 1986). But context may exert more subtle influences on maturation. For instance, father absence (Surbey 1990) and psychosocial distance from parents (Steinberg 1988) may accelerate maturational timing of daughters—but not of sons. Across the primate order, stress and other psychosocial stimuli have been regularly observed to affect pubertal timing and progression to reproductive maturity (Worthman 1990). In addition, marriage customs may alter maturational timing: girls of polygynous societies that prescribe early, premenarcheal marriage to older men reach menarche earlier than in monogamous societies with delayed marriage (Bean 1983). The effect may be mediated pheromonally (reviewed in Worthman 1990). But such polygynous societies are also likely to have greater father–daughter distance and father absence from the maternal domestic unit.

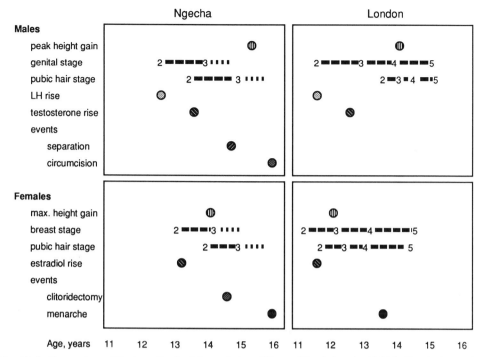

Fig. 23.4. Comparison of timing of pubertal events in a Kikuyu (Ngecha) and a British (London) sample. Auxologic and pubertal staging data are from Marshall and Tanner (1969, 1970). Because this study did not include endocrine measures, data on a comparable British sample are drawn from Preece (1986). The rating scales for genital, breast, and pubic hair development run from stage 1 (childhood condition) to stage 5 (adult condition). Numbers of pubertal stages are positioned at the median age at their first appearance: thus breast stage 2 and genital stage 2 mark the morphologic onset of puberty in either sex. For the three hormones charted, the age at first notable rise is marked. The only culturally recognized puberty "event" that the Kikuyu and British populations have in common is menarche; Kikuyu still clitoridectomize 40% of young women, circumcise all young men, and remove most young men to separate residences from the nuclear household (noted on the chart as "separation"). Circles denote the median age at occurrence of these culturally determined transitions; patterned fill of the circles indicate identical or analogous measures.

Differences in maturation rates tend to be stable across the juvenile period: children who grow quickly also tend to reach puberty earlier and mature sooner than those who grow more slowly (van Wieringen 1986). What consequences do these differences have for adult fitness? This question has been investigated almost solely among women, because number of offspring is more readily verified than among men. Attempts to link timing of menarche to future fertility have been complicated by intervening social variables, for timing of maturation influences timing of social induction into the reproductive career. Nevertheless, some studies

have found clear associations among age at menarche, time of first birth, and lifetime fertility (Udry & Cliquet 1982; Borgerhoff-Mulder 1989), whereas others have not confirmed these relationships (Sandler et al. 1984). Biological differences in fecundity that may underlie maturational timing-related differences in fertility are not yet clearly established. Apter and Vihko (1983) reported that duration of postmenarcheal subfecundity covaries with age at menarche in European women. Conversely, a large study in Bangladesh has found that length of subfecundity is negatively related to age at menarche (Foster et al. 1986).

Biosocial Dynamics
in Juvenile Development

This section has surveyed how the socially constructed circumstances under which the juvenile develops are systematically affected by developmental status, which, in turn, is affected by environmental conditions. Juvenile development is apparently affected by myriad dimensions of social organization and behavior, including social class, nutrition, illness, family size and composition, birth order, and psychosocial stress. Therefore, mating strategies that determine family composition, reproductive strategies that determine family size and sex ratio, parental investment patterns that influence child nutrition and labor, cultural bases for sex-biased treatment, and subsistence patterns that influence energetics in children can all be linked to variation in child growth and maturational timing. As was suggested at the outset of this section, the complex web of multiple influences on growth and maturational timing argues for the interpretation of development as a transducer of multiple cues to environment qualities that influence future fitness. The implications of developmental differences for adult fitness have, however, been less systematically studied.

STUDIES OF LATE-MATURING
POPULATIONS

Construction of a complete adaptive model of human ontogeny has been hindered by a lack of sufficient comparative data. Our understanding of the physiology of childhood and puberty is based almost entirely on information from industrialized, usually European populations. These populations are, however, considerably earlier maturing than is the case in many other settings, and, because of the secular trend, markedly earlier maturing on average than their own ancestors had been. By concentrating our study of adolescence in early-maturing populations, we have not obtained a representative profile of the range of human ontogenetic adaptedness. Questions that would be answered by adequate comparative data include the following: Are there circumstances under which the relative timing of maturation shifts between girls and boys? What are the fitness consequences of variation in maturational timing? Are

there physiologic differences in the course of juvenile development in later versus earlier maturing peoples? Furthermore, detailed information on the physiology of late maturation would improve our understanding of the causes and concomitants of the secular trend to accelerated maturation in particular and the regulation of developmental rates in general.

Puberty can be precisely tracked with anthropometry, morphometrics, and endocrine measures. A useful indirect measure of sexual maturation in males and females has been rating of sex-character development (genitals, breasts, and pubic hair), from which pubertal stage can be inferred. Formulated by J. M. Tanner in the late 1950s, the rating scales have been widely applied in Western research settings and show good validity with respect to underlying endocrine changes (e.g., Nottelman et al. 1987). They have, however, been less frequently and consistently applied in populations in less-developed countries. Hormones provide an excellent gauge of the physiologic processes that control reproductive and skeletal maturation in both sexes (e.g., Lee & Migeon 1975; Lee et al. 1976), and thus offer high potential for the comparative study of human development. But such measures have seldom been performed outside Western industrialized and urban clinical settings, probably for technical and logistical reasons.

The following sections will briefly describe studies of later maturing populations that represent steps toward expanding our comparative perspective on human development.

Kikuyu

Between 1979 and 1981, I conducted a study of late childhood and adolescence among the Kikuyu, Bantu-speaking agriculturalists of central highlands Kenya (described in Worthman 1987a, 1987b). The study area was Ngecha, a sublocation that had been a site for earlier studies of infant and child development, under Beatrice and John Whiting (Whiting & Whiting 1975; Landauer & Whiting 1981). For this reason, precise ages and background data were available for all participants. The population is reasonably well nourished and has relatively low morbidity and mortality rates. Cross-sectional data on endocrine, growth, and pubertal status were obtained from 54 females and 48 males between the ages of 9 and 16, 32 and 26 of whom, respectively, contributed repeat samples after an interval of 15 months, at ages 10 to 17. In addition, questionnaires ascertaining menarcheal status, circumcision status, intentions and status with respect to clitoridectomy, and residence in separate boys' houses were drawn from 867 individuals ages 10 to 20, equally divided by sex.

Based on published reports from other East African populations (reviewed in Eveleth & Tanner 1976, p. 214), it was anticipated that the rural Kikuyu population would be later maturing. This was indeed the case: median age at menarche was 15.9 years. To provide a comparison with a well-studied earlier-maturing population, pubertal development of Kikuyu and British youth has been charted in Figure 23.4. Age at onset of puberty is generally determined from endocrine changes or pubertal ratings, based on age at first significant elevation of gonadal steroids (testosterone, estradiol) or at onset of genital or breast development (transition from pubertal stage 1 to 2). Based on both these measures of pubertal onset, Kikuyu boys entered puberty 1 year later and girls nearly 2 years later than did British children. From an alternative means of estimating age at entry into puberty from pubertal rating,* the estimate of pubertal onset in Kikuyu girls may be downward revised to 12.5 years. Even so, it means that Kikuyu boys entered puberty at the same time as, or perhaps just a little ahead of, girls. The causes of this situation are unclear. Although one would most readily expect that differential treatment may be responsible, there is no documentation that such is the case and the question has not been addressed empirically. But the 1.5- to 2-year delay of entry into puberty among Kikuyu is not sufficient to explain the 2.5-year difference between the populations in

*Age at first appearance of stages could be calculated for adolescents who participated in the longitudinal part of the study: these contributed monthly measures over a period of 15 months. Rather than use average age at first appearance of breast stage 2, one can considerably increase the number of data points entered into a probit analysis by examining the median age at disappearance of breast stage 1. In this way, all breast stage ratings can be used, rather than only those that mark first appearance of the stage.

age at menarche; puberty appears to be more protracted for Kikuyu girls, so that they take 0.5 year (or 1.0 year, if one uses the lower estimate of age at entry into puberty) longer to reach menarche after entering puberty (Fig. 23.5).

While the timing of puberty in the two populations is different, the physiologic process of puberty is quite similar. The timing of initial significant rises of LH and testosterone or estradiol with respect to pubertal stage progression appears to be substantially the same. Furthermore, patterns of endocrine change (prolactin and gonadal steroids) by pubertal stage in Kikuyu adolescents resemble those reported for Western populations (Worthman 1986).

Although Kikuyu boys enter puberty either just before or simultaneously with girls, the local perception is that they are later maturing than girls. Why? The answer appears to be that the overt signs of physical maturation are evidenced in boys later than in girls. That is, as shown in Figure 23.4, peak height gains occur nearly 1.5 years later in Kikuyu boys than they do in girls. Analogously, peak height velocity occurs 2 years later in British boys than it does in girls. In general, boys are considerably more advanced in reproductive maturation at the time that they show the most visible secondary signs of puberty—height gain, beard growth, muscle bulking, and voice change. For instance, Kikuyu boys show maximal height gains nearly 3 years after entering puberty, whereas they occur after just over 1 year in girls.

Kikuyu parents aver that it is best to let their sons be circumcised after they have finished most of their growing. So, although parents can neither detect endocrine changes directly nor scrutinize genital development in their sons, they observe the overt signals of maturation to gauge developmental progress, and time a major rite of passage on that basis. The ideal timing for the culturally analogous operation for girls, clitoridectomy (Worthman & Whiting 1987), is just before menarche. Interestingly, although parents do not say they do so, this culturally significant operation occurs on average just after the average age at first appearance of breast stage 3 (which is when breasts become readily visible as developed, even under clothing). As it happens, breast stage 3 is probably the best proximal indicator of onset of menses (Marshall & Tanner 1969). Of considerable interest is the

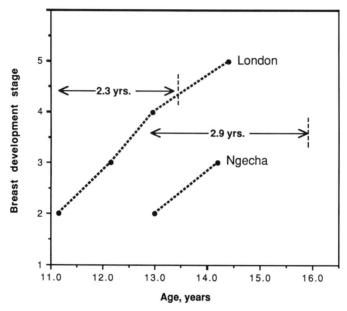

Fig. 23.5. Comparison of durations of puberty in the Kikuyu (Ngecha) and the British (London) samples. Duration of puberty is estimated from time elapsed between age at onset of puberty (based on median age at appearance of breast development [stage 2]) and age at menarche. Vertical dashed lines indicate median age at menarche; data points indicate median age at first appearance of breast stage.

observation that although Kikuyu parents do not overtly track reproductive maturation in their sons, the age at which sons are moved into a separate boys' house is best correlated with age at first emission. Thus the fact that (despite outward appearances) boys and girls mature reproductively at about the same time is covertly recognized here, for boys are moved into separate housing at the same age as girls are clitoridectomized. These relationships are visible in Figure 23.4.

These observations fit the expectation from life-history theory that maturational patterns are attended to by group members, and are thus to be understood as evolved under social conditions to act in part as signals that help time adjustments in social status, restrictions, and resources. Rather than being indifferent, social process appears to be potentially quite sensitive to physical ontogeny. Contrary to expectation, maturation of males was not later than that of females. The data hence indicate that the relative timing of puberty in males and females is not fixed, but can show population variation. Findings from this later-maturing population further show that the duration of puberty may

vary; in this case, it appeared to be prolonged, as girls took longer to reach menarche.

Factors that may promote maturation delay here include high population density, small plot size, low cash income, large family size (mean completed fertility 7.5), and periodic food shortages in dry years. Frank malnutrition is rare, and health care is available. Schooling is universal, for parents hope that children of either sex will enter the modern economy and contribute remittances. There is no overt sex bias, but the end of tribal warfare and extensive herding has decreased boys' workload, whereas domestic and farm work for girls has, if anything, intensified. Women and girls haul firewood and water over seasonally long distances. A sibcare system, with daughters the preferred caretakers, is used, and it frees the mother for work away from home. Sex differences in work and freedom increase during puberty, when boys move into separate quarters and girls are expected to be increasingly responsible and productive. Household composition has changed with the virtual disappearance of the separate men's house, attenuation of polygyny, and creation of single-family dwellings. Net offspring contributions to family wealth have shifted: girls no longer bring much bridewealth into the home, and bridewealth costs for sons have diminished (Worthman & Whiting 1987). Finally, traditional egalitarian treatment in puberty operation rites vanished with the illegalization of clitoridectomy in the 1930s. Boys are circumcised in a clinic under local anesthesia and sterile conditions, and are publicly feted; girls are clitoridectomized in relative secrecy and without anesthesia or antibiotic, by a socially peripheral female operator. These subtle shifts in relative offspring costs and status by gender may promote the delayed maturation of girls.

Bundi

The Bundi are Gende-speaking swidden horticulturalists located in highland Papua New Guinea, a region where the most extremely delayed of maturation profiles have been observed. Based on a survey taken in 1966–1977, Malcolm (1970) reported that the Bundi had a median age at menarche of 18.0 years, which is one of the latest recorded population menarcheal ages in the world. In 1983–1984, Zemel returned to the same group and resurveyed them, with the help of Carol Jenkins of the Papua New Guinea Institute of Medical Research (detailed in Zemel 1989). Anthropometric measures and menarcheal status were obtained from 345 rural boys and girls with known ages 10–24 years;* matching serum samples were drawn on 193 of those surveyed. Sera were assayed for gonadal steroids and adrenal androgens.

In the 17 years since Malcolm's survey, Bundi have undergone a decrease in menarcheal age. Median age at menarche for rural girls is 17.2 years; in urban girls, it is markedly earlier, at 15.8 years (Zemel et al. 1993). This discussion will focus on the later-maturing rural sample. Rises in gonadal steroids occurred about 3 years later than in published studies of Western adolescents. In girls, estradiol showed its first notable rise at age 14–15 years; in boys, testosterone also first increased significantly at age 14–15 years. Similarly, adrenal androgen (specifically, DHEAS) increases of adolescence were delayed by 2–3 years in males and at least 3 years in females. These increases occurred at the same

age in boys and girls, unlike the pattern of earlier elevation in girls reported for Western populations. DHEAS was associated with growth status: it correlated well with height and body widths in early adolescence (ages 10–15 years), whereas in late adolescence it correlated with body widths.

The data document a dramatic drop in age of menarche: 0.8 years in less than 2 decades. This pace is more rapid than the secular trend in Western Europe, where menarcheal age declined by about 4 months per decade (Tanner 1978). The trend may be partly caused by improved diet and health conditions, a notion supported by the markedly earlier (although still relatively late) age at menarche in urban Bundi, whose nutrition and health are better still. Heavy reliance on sweet potato is increasingly attenuated by imported foods, especially rice and tinned fish. Increases in protein, sugar, and fat intake are greater in urban Bundi. It is apparent that, like the rural Kikuyu, rural Bundi show little sex difference in the timing of puberty. It also appears in both populations that adrenal activity has a role in skeletal maturation of both sexes that is distinct from that of gonadal steroid hormones. Because puberty ratings were not performed in this study, we cannot estimate age at entry into puberty precisely. However, Zemel (1989) has modeled ages at peak height velocity for rural Bundi boys and girls, showing them to be 16.0 and 13.8, respectively. This is a similar gender difference in timing to that reported for a British sample (Marshall & Tanner 1969), but it appears that attainment of menarche in Bundi is remarkably delayed with respect to time of maximal growth, for there is a 3.4-year gap between the two events. The British sample that we have been using as a comparison representative of earlier-maturing groups is reported to have a gap of just 1.3 years separating these events. This suggests that in Bundi girls, gonadal maturation may be markedly delayed with respect to skeletal maturation. In contrast, Bundi males show close correlation of testosterone and height, and major increments in both are coincident. This implies no such partial dissociation of gonadal and skeletal maturation in boys. Further, it appears that, like Kikuyu, Bundi girls take longer to go through puberty than do early-maturing populations.

Dietary and social factors that could contribute to persistent delayed development in Bundi

*Only children with birth records were included in these analyses.

children include undernutrition, with low protein and fat intake (Malcolm 1970; Dennett & Connell 1988), and customary treatment of children. That traditional Bundi are a gender-segregated society with an ethos of female inferiority (Fitz-Patrick & Kimbuna 1983) suggests that differential treatment may cause greater delay in girls. But girls' nutritional indices are equivalent to boys' across the first two decades (Zemel 1989). Household composition could contribute to delay in rural girls. Traditional fathers are absent, living mainly in a men's house removed from the mother–child domestic-production units. The absent father is perpetuated in changing rural areas by male labor migration; nuclear families, on the other hand, characterize urban settlements. Bundi had elaborate, prolonged male initiation that stretched over late childhood to young adulthood, commencing with removal from the mother and moving into a male residence, proceeding through stressful indoctrination (including fear, beating, isolation, and tests of physical endurance), and culminating in nose piercing. Girls, by contrast, were inducted early into the heavy female domestic and gardening workloads, with a brief seclusion at menarche. Such practices are widespread in highland Papua New Guinea and may have been prompted by the extent of maturational delay, for most groups believed that initiation was *necessary* to make boys grow. Over the last 40 years, intertribal warfare has ceased; health care, cash cropping, wage labor, and schooling were introduced; and initiation cycles have been attenuated among Bundi.

Another factor may influence Bundi development, illustrating the rate-limiting effects that single essential nutrients can exert: Bundi children may be zinc deficient. Zinc deficiency is known to cause developmental delay and stunting (reviewed in Prasad 1991), and has recently been reported in children of another Papua New Guinea district (Gibson et al. 1991). If zinc privation plays a major role in small size and slow maturation of Bundi and other Papua New Guineans, then interpretation of these patterns as adaptive is no longer appropriate. Rather, this would be a case of phenotypic modulation induced by a critical shortage that overwhelms internal growth regulatory mechanisms (Smith-Gill 1983). In a broader sense, the degree to which small size and maturational delay in hu-

mans is a manifestation of evolved plasticity rather than phenotypic modulation remains open for investigation. Clarification of the components of human variation that represent evolved plasticity rather than passive modulation is of obvious importance for explaining that variation.

LESSONS FROM LATE-MATURING POPULATIONS

Here, study of later-maturing groups has been used to explore and seek explanations for the range of human developmental variation. Endocrine, anthropometric, morphometric, behavioral, and cultural data have been brought to bear; the utility of proximal (endocrine) measures in probing functional relationships and providing a basis for comparison is apparent. So far, the basic physiology of puberty appears robust across populations: relationships among hormones, morphologic change, and physical growth appear substantially the same as in existing reports. But the timing, relationships among maturing subsystems, and sex differences may shift.

The findings have indicated a need for more comparable data on males and females to elucidate the ontogeny of gender differences. These two studies have shown both that the two sexes can vary in secular change in timing of pubertal maturation and that their relative rates of maturation may differ across populations. In neither group did boys enter puberty later than girls, which is the usual pattern reported. In both Kikuyu and Bundi, the interval during puberty before menarche is prolonged, suggesting that the rate of gonadal maturation was slowed. Bundi girls also demonstrated that the relative rates of gonadal and skeletal maturation may differ.

There are numerous potential adaptive reasons why females may delay menarche under the same suboptimal conditions as those that foster slowed growth. The mechanism of this delay has been an object of conjecture. One developmental attainment essential to female reproductive success is a sufficiently large pelvic girdle to accommodate pregnancy and childbirth (reviewed in Rosenberg 1988). Ellison (1982) has suggested that gonadal maturation may track pelvic width and, based on analysis of Berkeley Guidance Study data, that 24 cm is a threshold width required for attainment of menarche.* A regression of biiliac diameter—the distance between the iliac crests—by age in Kikuyu girls crosses the age at menarche (15.9 years) at a pelvic width of 24.1 cm (95% confidence limits, 23.4–24.8 cm). This closely conforms with Ellison's prediction. If pelvic size is a limiting factor, then it may help explain the pronounced delay in menarche with respect to peak growth in Bundi. Like most children of Papua New Guinea, Bundi are quite small, falling at or below the National Center for Health Statistics 5 percentile for size at all ages. Therefore, because they start from smaller sizes, Bundi girls may need to delay menarche longer into skeletal maturation in order to allow the pelvis to reach a critical dimension. In fact, a linear regression of biiliac width grouped by age in rural Bundi girls indicated that a pelvic width of 24 cm would be reached at age 16.83 years (95% confidence limits, 22.5–25.6 cm; age 13.4–20.3 years). Linkage of pelvic dimensions with menarche is probably not causal, but mediated by a common endocrine pathway.

Moreover, what both studies have illustrated—although in quite different ways—is the dynamic relationship of physical development to socially constructed context. Timing of major rites of passage and daily living arrangements were markedly affected among Kikuyu by developmental status of the individual. Cultural models for differential gender socialization interacted with the sex differences in outward maturation to construct quite different experiences for girls and boys, and fed the perception of males as later maturing even when, if anything, it was the females who were more delayed. These perceptions may further fuel differential treatment by gender. Bundi, on the other hand, illustrate strikingly how context influences rates of development. Acculturation in rural areas (diet change, health care, schooling) was accompanied by a rapid drop in menarcheal age, and urban settings were associated with further accelerated physical growth and reduction in age at menarche. Previous sections have shown that this experience is common. Although parents may value the health and other benefits of quality of life in an urban setting, the consequences for maturation rates in their children are usually unintended and unanticipated.

*Only 7% of girls reached menarche before attaining 24-cm biiliac diameter.

OVERVIEW

This chapter has elaborated the view of human development as an inherently biosocial process, and presented ontogeny as designed to progress through a dialectic between biological and contextual processes. It has examined how juvenile developmental schedules manifest the operation of both biological and social constraints. It also demonstrates ways in which outward signs of maturational status can serve as cues to others that influence access to social and material resources. Evidence of manipulation of these signals includes partial detachment of ovarian maturation from menarche and of testicular maturation from physical growth. Areas that require exploration include:

1. The connection of maturational differences entraining social responses with immediate costs, benefits, and long-term outcomes for juveniles
2. The behavioral ecology of juvenile energetics with measures both of intake and expenditure and of their relationship to growth
3. The effects of change in family size and composition on child growth and maturation rates
4. The identification of gender-differentiated developmental sensitivities to specific environmental cues
5. The impact of differential treatment on maturation, using prospective data
6. The degree to which outward signs of maturity are "manipulated" with respect to actual reproduction and growth status

In general, these questions need to be linked to measures of parental and offspring mortality and fitness. Under what conditions and in what domains are adults more indifferent or attentive to juvenile maturation status, and what are the developmental outcomes? The effects may be demography and frequency dependent.

Growth can act as a summary process that transduces multiple inputs, sensitivities, and constraints through time into an ongoing developmental outcome. An outstanding question is the degree to which facultative delayed maturation and small body size in humans are evolved adaptive responses to the environment. Comparative research has isolated life-history factors associated with delayed maturation across taxa

(Pereira, Chapter 2, Pagel & Harvey, Chapter 3, and Rubenstein, Chapter 4, this volume), but these do not directly specify factors responsible for intraspecific ontogenetic plasticity and adaptive variation. Observed effects of nutrition or morbidity on human growth are generally assumed, but not unequivocally proved, to be adaptive. We may expect evolved ontogenetic tracking of an environmental feature when its variation serves as a reliable cue to determinants of future fitness. Features that would be predicted to act in this manner include quality, quantity, and regularity of food supply; indices of mortality probabilities; and energy drains (thermal stress, workload). These might influence not only the pace of growth and timing of reproductive maturity, but also the long-term organization of physiologies that regulate fecundity and stress response.

The goal—to explain, not simply describe, human variation—is widely shared by biologists. One dimension that has not been pursued here is how cultural processes interact with individual variation. Social responses to or interventions in biological process can act as magnifiers of small differences or compressors of large ones. One can readily think of cultural dynamics that normalize development in the face of individual variation. However, others may accentuate individual differences by increasing perception or actual degree of variation. A human biology for the "real' contemporary world has to address these issues in a thoroughgoing manner, to help us more completely understand what shapes our similarities and differences.

ACKNOWLEDGMENTS

I thank the Kikuyu children and their parents who participated in the study, as well as the Kenyan Government and Department of Paediatrics, University of Nairobi Medical School, under whose auspices the research was performed. Assistance in the field from Dr. Julius Meme and Samuel Karanja is gratefully acknowledged, as is that of Roberta Todd, who helped perform the hormone assays for both the Kenyan and New Guinean studies. The Kenya project was conducted with support from NIMH postdoctoral training Grant MH 4088-03. Written with support from a W. T. Grant Foundation Faculty Scholarship.

24
Juvenile Primates: Dimensions for Future Research

LYNN A. FAIRBANKS and MICHAEL E. PEREIRA

This volume covers a lot of ground and, as with many scientific endeavors, raises more questions than it answers. The theories of modern evolutionary biology have created a stimulating framework for functional interpretations of behavior, leading our contributors to suggest a diverse array of hypotheses concerning how specific behaviors and preferences by juveniles function to promote fitness. This final chapter highlights some of the themes that reappear throughout the book. It contrasts different interpretations that have been used to explain juvenile behavior, and suggests how some of the ideas put forth in these chapters might be tested. Our hope is to stimulate future research that will contribute to our understanding of what juveniles do and why they do it.

The prevailing view of primate development has changed over the past decade or two as the concepts of evolutionary biology have been integrated into our interpretation of developmental patterns and events. The field and laboratory studies of the 1960s and early 1970s provided qualitative and quantitative descriptions of juvenile behavior and social relationships (Pereira & Fairbanks, Chapter 1), but the premises underlying their interpretation were rooted largely in prevailing theories of the time. What follows is a generic description of "the juvenile primate" that emerged during this earlier period.

Juvenile Primates: The Old Narrative

The juvenile primate matures in the context of the social group. While the physical and physiological processes of growth and maturation are changing its body, socializing agents in the group are influencing its behavior. Species differences in social organization cause individual juveniles to have different kinds of experiences, which, in turn, channel them into appropriate adult social roles.

The mother–infant bond is the individuals' first and most important social attachment, and it serves as a prototype for all later relationships. The ideal mother is sensitive and responsive to the needs of her developing offspring. An indifferent or rejecting mother can produce an offspring that is insecure, overly aggressive, and less able to develop relaxed affinitive relationships with other group members.

As the juvenile develops independence from its mother, other socializing agents gain in importance. Through play, the developing individual has the opportunity to practice skills and form close social bonds with peers. The socialization process facilitates the integration of the individual into the social group and promotes social cohesion and group unity.

Because primates are long-lived animals living in complex social systems, immatures need time to learn the behavioral and communicational skills that will make them successful adults. The long period of time before sexual maturation gives them time to ob-

359

serve and imitate adult group members and to practice behavior that will be useful later in life.

These traditional descriptions of the juvenile primate emphasize the ways that the social group shapes the behavior of the individual. They view the juvenile as a relatively passive recipient of social influence, and they focus on proximate sociological and psychological causes of behavior. Although biological determinants of behavior are recognized, individual and even species differences in adult behavior are seen as partly the result of differences in early experience that pass on local traditions and shape the behavior of succeeding generations. Implicit assumptions are that the social group has been designed to socialize its young and that the maintenance of a cohesive and unified group is a central function of the socialization process.

The development of theories of selection and life-history evolution in biology has changed the way we interpret the behavior of juvenile primates. In the past 2 decades, the emphasis has shifted from proximate to ultimate causes, from group selection to individual and kin selection, and from the view of social development as a relatively passive to an active process designed to accomplish specific strategic goals.

Juvenile Primates: The New Narrative

The juvenile primate is an autonomous individual developing in the context of its social group. As such, it must cope and compete with, accommodate and exploit surrounding group members. Juveniles must learn what to expect from other individuals, what they are likely to do, who is likely to help and who to harm them. At the same time, juveniles must learn to forage independently and to avoid the environmental dangers of injury and predation. In most populations, more than half of all infants fail to reach adulthood. Those that do must emerge from the juvenile period as competent adults, able to compete successfully with conspecifics.

Juveniles grow up within a network of kin relationships. They derive benefits from associating with kin that are likely to provide protection, tolerance, and support. Not all kin groups are equal, however, and variation in availability and attributes of kin contribute to asymmetries in developmental opportunities.

The juvenile contributes actively to its own development. Juveniles seek the company of adults for safety, for access to food and shelter, and for information. When adult reproductive success depends on fighting ability, juveniles seek opportunities to develop fighting skills through competitive play. When reproductive success can be improved through practice in infant handling, juveniles seek opportunities to hold and carry infants. Juveniles actively pursue dominance status, beneficial affiliative relationships, and appeasement and reconciliation following aggression.

The tactics for accomplishing developmental goals depend partly on the local ecology and social structure, and partly on inherent species-typical perceptions, preferences, and abilities. While adult behavior can be influenced by early experiences, the individual reacts to environmental variation throughout development in ways designed to achieve species-typical objectives.

The length of the juvenile period is determined by the time it takes to develop a large brain and grow an adult body, and by the trade-off between the advantages of short generation time and those of increased juvenile survival and adult life span associated with waiting.

The old and new views of juvenile primates differ in subtle but important ways. The current view emphasizes the potentially high toll of juvenile mortality and the difficulties of surviving the juvenile period, and it sees behavior as fundamentally shaped by natural selection for ultimate functions. It recognizes that competition both among peers and between juveniles and adults can limit juvenile opportunity and success. Behavior is more canalized than in the earlier view, but it is also more intelligent; that is, behavior is goal-seeking rather than passively responsive to external environmental stimuli.

But, as with the earlier view, most of the ideas that structure current thinking about juveniles remain untested. In most areas of research on juvenile behavior, we are still in the descriptive stage. We describe behavior qualitatively and quantitatively, and then suggest functional interpretations of observed patterns in the discussion.

It has been difficult for research on juvenile primates to proceed beyond this first phase for several reasons. Longitudinal research on development is expensive and time-consuming for species that take up to a decade or more to mature and then typically have an adult reproductive career of another 10–20 years. Field experiments to test hypotheses derived from observation and theory have long been used by biologists working with other taxa, but researchers are generally more reluctant to interfere with primate populations. A combination of factors—including size, long life span, low fecundity, and close phylogenetic relationship to ourselves—contributes to the perception that primates' individual lives should be valued and not disturbed. In addition, primate field sites are usually used for several projects simultaneously, and there is concern that manipulation of social or environmental variables will alter the site for other research.

There has been a long history of experimental studies of primate development in the laboratory (Pereira & Fairbanks, Chapter 1), but the hypotheses of interest to laboratory researchers have generally been different from those considered by field biologists (Rosenblum et al. 1989). Logistic difficulties and differences in the background and training of field and laboratory researchers have hindered the testing of hypotheses derived from field research in captivity. A combination of some justifiable, and some unjustifiable, bias against research on captive animals has added to these problems. Certain questions, such as the effect of juvenile spacing behavior on vulnerability to predation, are difficult to transfer to the laboratory. But many others would be conducive to verification with carefully designed experiments in controlled captive settings (Kummer et al. 1990).

These hurdles are beginning to be overcome in some areas as primatologists develop creative designs for field experimentation and more effective integration of field and laboratory research. Recent examples of research programs that combine evolutionary theory and naturalistic observation with experimental verification include studies of primate cognition (Cheney & Seyfarth 1990; Kummer et al. 1990), reconciliation (Aureli et al. 1989; Cords 1992),

predator recognition (Macedonia et al. 1989), and the role of alliances in rank attainment (Chapais 1992). But still, the large majority of stimulating hypotheses concerned with the functions of juvenile behavior and development have yet to be systematically challenged.

FUNCTIONS OF JUVENILE BEHAVIOR

In the current view, the two basic concerns for a juvenile are to survive the juvenile period and to develop into a reproductively competent adult. Although the importance of these two goals is a truism in biology, there is considerable difference of opinion among researchers as to how the observed patterns in juvenile development and behavior function to satisfy them.

In this section, we discuss several themes that have come up repeatedly in these chapters, including why juveniles grow slowly, why they preferentially associate with particular group members, why they play, and what is learned from conspecifics. For each of these questions, we have tried to contrast how different authors explain the patterns they observe in terms of life-history trade-offs, the socioecological context, and the goals of juvenile development. We then suggest how some of the functional hypotheses proposed in this volume might be evaluated.

Why Grow Slowly?

The comparative analyses of mammalian and primate life histories in Part I each approach the problem of unraveling the mesh of interrelated life-history variables differently, but all reach the general conclusion that physical constraints on growth and maturation in the context of variability in the developmental environment influence the amount of time it takes to attain adult size and reproductive maturity (Pagel & Harvey, Chapter 3; Rubenstein, Chapter 4; Janson & van Schaik, Chapter 5). This same conclusion was also reflected in the cross-cultural comparisons of the timing of maturation in human societies (Worthman, Chapter 23).

The life-history models help to define the parameters that are likely to be important in understanding variation in the age of maturity. As such, they highlight areas for empirical research. For example, Janson and van Schaik (Chapter 5) argue that slow growth reduces the risk of juvenile mortality due to malnutrition, but they also acknowledge that small size reduces juveniles' ability to compete for food and increases their risk of mortality due to predation. What do we need to know in order to determine how these conflicting factors operate in animal and human populations?

In humans, slow juvenile growth followed by accelerated growth at puberty are well documented (Worthman, Chapter 23; Fig. 1.3). In contrast, growth data that are sufficiently fine grained to detect changes in growth velocity with age for free-living nonhuman primates are rare (Janson & van Schaik, Chapter 5; Crockett & Pope, Chapter 8). With more detailed information on species variability along this dimension, comparative analyses of relationships between growth rate and other features of development would be possible. For example, Pereira (Chapter 20) found that dominance among weanling ringtailed lemurs depends primarily on relative size. In Japanese and rhesus macaques (Chapais & Gauthier, Chapter 17; de Waal, Chapter 18), by contrast, juvenile dominance relations among peers depend primarily on alliances determined by matrilineal membership. If juvenile rank is related to fitness, there should be greater selection pressure for fast growth rates in the lemurs than in the macaques. Lemurs, in fact, do have fast postnatal growth rates (Pereira et al. 1987; Pereira 1993), whereas macaques show the suppressed growth characteristic of large anthropoids.

To elucidate the role of growth limitations in determining the length of the juvenile period, we need more information about variation in food supply, energy budgets, and the costs of growth for individuals of different ages and sizes, both within and between populations. Information of this type can be gained by monitoring naturally occurring variation in food quality and predation in relation to growth, health, and mortality. Several methods currently available could be applied toward these ends, such as precisely defined observational scales to estimate body fat (Berman & Swartz 1988), analysis of urine and fecal samples for parasites and metabolic indicators of health and physical condition (Eley et al. 1989; Altmann et al. 1993), and use of the doubly labeled water technique to measure rates of energy expenditure (Milton et al. 1979). Age-specific growth rate responses to changes in food supply can be examined in pop-

ulations of free-ranging primates that can be weighed periodically to discover whether certain sizes or stages of development are more vulnerable or responsive than others (Altmann & Alberts 1987; Moses et al. 1992; Altmann et al. 1993; Pereira 1993).

Comparative and correlational methods can give insight into likely relationships between growth rate, size, and vulnerability to mortality, but controlled experiments with manipulation of independent variables are ultimately necessary to verify cause and effect (for examples in other taxa, see Pereira, Chapter 2). Given the prominence of assumptions about these life-history variables in the current hypotheses on why primates take so long to mature, this would be an extremely fruitful area for further empirical research.

Selecting Social Associates

In every study that measured spatial relations, juveniles were found to associate nonrandomly with other group members. Young juveniles of both sexes spend a large portion of their time near their mothers, and when the social system allows it, they also associate with adult males that are likely to be their fathers. As they mature, time near the mother declines and older juveniles tend to be attracted to same-sex adults: juvenile and adolescent males are attracted to older males, whereas females seek the company of older females. These patterns were remarkably consistent across species, but the proposed explanations varied.

Many authors consider safety to be a primary motive for partner preferences, particularly for young juveniles (van Noordwijk et al., Chapter 6; Nash, Chapter 9; Watts & Pusey, Chapter 11; Horrocks & Hunte, Chapter 16). The need for safety is considered to be so important that juveniles will compromise their foraging efficiency to maintain spatial proximity to adults (Janson & van Schaik, Chapter 5), and will risk approaching adults following aggression to reestablish tolerant relationships (Cords & Aureli, Chapter 19). In human societies, differential mortality risks are considered to be the basis of differences in parenting behavior and the amount of work allocated to children (Blurton Jones, Chapter 21).

A second function frequently used to explain

partner preferences is social learning. Watts and Pusey (Chapter 11) see adolescent male chimpanzees' associations with adult males as creating opportunities to increase social skills through observation and interaction. Edwards (Chapter 22) proposes that the preference of children for like-sex adults leads children into situations and contexts where they can learn gender-appropriate behavior.

Other authors explain the preference for like-sex adults as investments in relationships that will persist into adulthood. Strier (Chapter 10) interprets the preference of older juvenile and subadult muriquis males to associate with other males as establishing long-term reciprocal relationships, and Fairbanks (Chapter 15) argues that the juvenile period is a time when female vervet monkeys are trying to establish relationships with high-value adult females that will be useful to them later in life.

How do we determine when and if these purported functions operate in primate life histories? Support for the safety hypothesis can be found in field studies of juveniles that are separated from particular group members through death or emigration. Juvenile female howler monkeys that emigrate have much higher risks of mortality than do those that stay in the natal group with their mothers and other close kin (Crockett & Pope, Chapter 8). Juvenile chimpanzees are unlikely to survive after the death of their mothers, whereas juvenile gorillas that are still living with their fathers do have a good chance of surviving maternal death (Watts & Pusey, Chapter 11). Similarly, juvenile Hanuman langurs that are forced out of the natal troop by new males often seek the company of their fathers, and those that reside with their fathers after emigration may be more likely to survive than those who do not (Rajpurohit & Sommer, Chapter 7).

These results suggest that parents play a part in reducing the risks of mortality for their juvenile offspring, but considerably more data are needed before we can understand the extent of this effect and the factors that influence its expression (Blurton Jones, Chapter 21). Accurate data on causes of mortality are extremely difficult to obtain, particularly in relation to patterns of association within groups (Cheney & Wrangham 1987). One research strategy that might provide more complete information on how predation shapes the spatial relations of ju-

venile primates would be to study their principal predators. Focusing on a predator's hunting tactics would be likely to provide insight into how proximity patterns of juvenile primates contribute to vulnerability.

Seeking the company of adults for safety, social learning, and to establish relationships are functions that can be served simultaneously. There are circumstances, however, that would lead to different predictions for different hypotheses. Emigration patterns determine whether a juvenile will have an opportunity to invest in long-term relationships that are likely to promote adult fitness. Males of most species, and females of some, will leave the natal troop to breed, and they will have little or no opportunity to establish relationships that will be useful in the future. Contrasts between pairs of animals that are likely to be available to each other in the future and those that are unlikely to be can help separate the short- and long-term benefits of particular kinds of relationships. In most studies of matrifocal Cercopithecines, female–female bonds have been found to be stronger than male–male bonds. In one of the few studies of a species where males are philopatric and females disperse, Strier (Chapter 10) found that male–male bonds were stronger than female–female bonds, thus supporting the hypothesis that juveniles are investing in future relationships.

Longitudinal studies of social relationships in capuchins (O'Brien & Robinson, Chapter 14), vervets (Fairbanks, Chapter 15), rhesus macaques (de Waal, Chapter 18), and ringtailed lemurs (Pereira, Chapter 20) also suggest that there is continuity in relationships formed during the juvenile years, and that advantages established early persist into adulthood. But individuals of all these species are also capable of adjusting to change by forming new relationships. The costs and benefits of maintaining relationships and establishing new ones are areas that would be conducive to study through tracking of demographic changes in undisturbed groups and through experimental manipulation of group membership in captivity.

Why Play?

One of the most noticeable features of juvenile behavior is play. Several chapters in this volume present new descriptive data, and others discuss its structure and functions (Rubenstein, Chapter 4; van Noordwijk et al., Chapter 6; Nash, Chapter 9; Watts & Pusey, Chapter 11; Fagen, Chapter 13; Fairbanks, Chapter 15). Functions that have been proposed for play over the years include developing muscles, cardiovascular capacity, locomotor skills, fighting ability, communication, dominance, and social bonds, as well as teaching the developing individual about its own attributes in relation to those of others (Fagen 1981, Chapter 13). Some of the hypothesized functions of play are thought to benefit the juvenile as a juvenile, and others are believed to increase adult capacities that are related to fitness.

Rubenstein (Chapter 4) argues that play is mostly about producing a competent juvenile, one that will survive to maturity. Other contributors assume that play provides practice for behavior that will be useful later in adult life (Nash, Chapter 9; Watts & Pusey, Chapter 11; Fairbanks, Chapter 15). Fagen's (Chapter 13) initial model of play allowed equal benefits at all ages, but the failure of this model to predict the generally observed pattern of decline in the rate of play with age led him to propose that the costs and benefits of play probably change with age.

One way to decide when play functions primarily to promote juvenile versus adult fitness is to look at its form and pattern. Sex, age, and context differences, for example, can support one hypothesis over another. If adult males can profit more than adult females from individual fighting skills, and if play as a juvenile helps to develop those skills, then we might predict that juvenile males should play more often than juvenile females. If play functions primarily to develop speed and coordination in predator avoidance, and if these benefits are greater for smaller, less mature individuals than for older, more developed individuals, then younger animals should play more than older animals, but no differences would be expected in the rate of play for males and females of the same size. Testing these hypotheses requires not only examination of the form of play, of course, but also verification of the accompanying assumptions—for example, that adult males benefit more than adult females from fighting skills.

The comparative method is illustrated by comparing results from four species presented in this volume. Partner preferences for play-

fighting were remarkably similar for gorillas, galagos, and vervet monkeys (Nash, Chapter 9; Watts & Pusey, Chapter 11; Fairbanks, Chapter 15). In each case, juvenile males played more often than juvenile females, and males preferred male partners of the same age or older, whereas females played proportionally more with younger individuals. These three species differ widely in phylogeny and social organization, but for all three it is considered likely that adult male reproductive success is more influenced by individual fighting ability than is adult female reproductive success. In ringtailed lemurs, by contrast, this asymmetry in the importance of male and female fighting ability does not hold, and play data indicate that juvenile females play as much and as hard as juvenile males (Fagen, Chapter 13; Pereira, Chapter 20). Given the above assumptions, these comparisons support the hypothesis that play-fighting functions to augment fighting skills.

The comparative method is only a first step, of course, in identifying causal mechanisms and evolutionary functions. Controlled experiments need to be performed that are specifically designed to determine the relationship between play and other aspects of juvenile or adult competence (e.g., Potegal & Einon 1989). For example, males raised in social groups with or without other juvenile males available as play partners, but whose experiences were otherwise similar, could be matched in dyadic encounters. If play with male peers helps juvenile males to hone their fighting skills, then the deprived males should be at a disadvantage in later contests with more experienced males.

Play would be a promising focus for research teams to study proximate effects on metabolism, muscle mass, and growth for males and females of different ages. Experiments could also be designed to discover what one animal learns about the attributes of others through the process of social play. Juveniles' willingness to play with almost anyone would permit play trials among animals that were otherwise unfamiliar with one another. Partner preference choice tests and discrimination test methods (e.g., Dasser 1988; Kummer et al. 1990) could then be implemented to discover how juveniles categorize other individuals on dimensions such as familiar–stranger, dominant–subordinate, or preferred–unpreferred.

What Do Juveniles Learn from Conspecifics?

Most authors in this volume believe that social learning during the juvenile years has a definite impact on juvenile survival and eventual adult attributes. Both Janson and van Schaik (Chapter 5) and Nash (Chapter 9) give examples of potential influences of observational learning on food location, selection, and handling. Edwards (Chapter 22) believes that children develop gender-appropriate behavior by being responsive to the rewards and expectations of same-sex adults, and Watts and Pusey (Chapter 11) suggest that social learning is so important for adolescent male chimpanzees that they sacrifice foraging efficiency and risk exposure to increased aggression to observe and interact with older males.

Studies that have specifically investigated social effects on foraging skills have often contradicted the conventional wisdom by concluding that many skills thought to be attributable to social learning are better explained as examples of individual trial-and-error and practice (Cambefort 1981; Galef 1988; Visalberghi & Fragaszy 1990b). Milton (Chapter 12) provides evidence that spider monkeys develop species-typical behavior, including basic preferences for food types, social partners, and physical environments, without the aid of adult role models.

Given the questions raised by these studies, and the prevalence of social-learning explanations for development, the time is ripe for research specifically designed to evaluate social-learning hypotheses. When a juvenile female approaches a mother and infant, she could be learning mothering skills through observation (Pereira 1988a). She could be seeking opportunities to practice infant handling (Fairbanks 1990, Chapter 15). She could be placing herself in a context where she is likely to be rewarded for performing nurturant behavior (Edwards, Chapter 22). Or she could be engaging in a nonfunctional activity that later matures into an adaptive adult behavior (Quiatt 1979; van Noordwijk et al., Chapter 6). Even if observation of adults is not essential for normal development (Milton, Chapter 12), does this mean that behavioral development is not influenced by observational learning when adults are present?

One of the ways to differentiate among functional hypotheses and evaluate the action of social learning is demonstrated by Chapais and Gauthier (Chapter 17). They begin by carefully documenting the specific kinds of experiences available to immature Japanese macaques to learn their birth rank before they begin to assert that rank. They follow with a preliminary report of an experiment designed to determine if information gained through observation actually influences how juveniles act toward peers. This process, from observation and correlation to experimental verification of hypothesized effects and functions, is needed in all areas of research on primate development.

Whereas Chapais and Gauthier see birth rank as being imposed on immatures at a young age, other contributors emphasize the active part played by juveniles in placing themselves in appropriate contexts for learning species- and sex-typical behavior (Nash, Chapter 9; Watts & Pusey, Chapter 11; Fairbanks, Chapter 15; Edwards, Chapter 22). Current views suggest that there are inherent species and sex differences in perceptions, selective attention, and the valence of different types of rewards. Can we verify such differences independent of the behavior they predict? This would be an exciting area of collaboration with neuroscientists looking for relationships between behavior and neurological structure or function (Bateson 1991). For example, measurement of differential activity in the brains of male and female primates while observing same-sex or opposite-sex partners would provide support for the hypothesis of selective attention to same-sex conspecifics. Assessment of the arousing or calming effect of particular stimuli could be used to evaluate predictions of differential reward value according to species or sex differences in natural history and socioecology. The techniques of the neuroscientist could be combined with knowledge of the natural behavior of the animals to identify proximate mechanisms for ultimate functions.

CONCLUSION

When we began this volume, our hope was not to summarize an area of knowledge, but to re-inspire interest in a topic that we felt had been unjustly neglected. Studies of infant development were stimulated first by attachment theory and more recently by the theories of parental investment and parent–offspring conflict. Research on adult behavior focused on ideas from sociobiology and behavioral ecology. By comparison, the juvenile period seemed to have no compelling theories to motivate research. With the integration of life-history theory, the study of juveniles has been linked to the rich and rigorous theoretical framework of evolutionary biology. Now, juvenile behavior is viewed as responsive to trade-offs between different components of the life history within the context of ecology and social organization. Using this conceptual framework, our contributors have gener-

ated many stimulating new hypotheses about how individuals meet the goals of surviving the juvenile period and developing into competitive adults.

At this point, the comparative database on juveniles is growing, but almost all the new hypotheses on development remain to be tested. We have outlined the type of effort that will be needed to evaluate the current ideas pertaining to juvenile primates. The success of this book will be measured by how effectively it stimulates researchers to reconsider the juvenile period and to direct their talents and energies toward understanding this fascinating stage of life histories.

Bibliography

Abbott, D. 1984. Behavioral and physiological suppression of fertility in subordinate marmoset monkeys. Am. J. Primatol. 6, 169–186.

Agoramoorthy, G. 1987. Reproductive behaviour in Hanuman langurs, *Presbytis entellus*. Ph.D. dissertation, University of Jodhpur.

Agoramoorthy, G., & Mohnot, S. M. 1988. Infanticide and juvenilicide in Hanuman langurs (*Presbytis entellus*) around Jodhpur, India. Hum. Evol. 3, 279–296.

Ahumada, J. A. 1989. Behavior and social structure of free-ranging spider monkeys (*Ateles belzebuth*) in La Macarena. Field Studies of New World Monkeys, La Macarena, Colombia 2, 7–31.

Alados, C. L., & Escos, J. 1991. Phenotypic and genetic characteristics affecting lifetime reproductive success in female Cuvier's, dama and dorcas gazelles (*Gazella cuvieri, G. dama* and *G. dorcas*). J. Zool., London 223, 307–321.

Albon, S. D., Clutton-Brock, T. H., & Guinness, F. E. 1987. Early development and population dynamics in red deer. II. Density-independent effects and cohort variation. J. Anim. Ecol. 56, 69–81.

Aldis, O. 1975. Play Fighting. New York: Academic Press.

Alexander, R. D. 1974. The evolution of social behavior. Annu. Rev. Ecol. Syst. 5, 325–383.

Alford, R. A., & Harris, R. N. 1988. Effects of larval growth history on anuran metamorphosis. Am. Nat. 131, 91–106.

Allen, S. H. 1984. Some aspects of reproductive performance in female red fox in North Dakota. J. Mammal. 65, 246–255.

Alm, G. 1959. Connection between maturity, size, and age in fishes. Rep. Inst. Freshwater Res., Drottingholm 40, 1–145.

Als, H., Tronick, E., & Brazelton, T. B. 1979. Analysis of face to face interaction in infant adult dyads. In: The Study of Social Interactions (Ed. by M. E. Lamb, S. J. Suomi, & G. Stephenson), pp. 33–75. Madison: University of Wisconsin Press.

Alterman, L. 1990. Isolation of toxins from brachial gland exudate of *Nycticebus coucang*. Am. J. Phys. Anthropol. 81, 187.

Alterman, L., & Hale, M. E. 1991. Comparison of toxins from brachial gland exudates of *Nycticebus coucang* and *N. pygmaeus*. Am. J. Phys. Anthropol. Suppl. 12, 43.

Altmann, J. 1974. Observational study of behaviour: Sampling methods. Behaviour 49, 227–267.

Altmann, J. 1978. Infant independence in yellow baboons. In: The Development of Behavior: Comparative and Evolutionary Aspects (Ed. by G. M. Burghardt & M. Bekoff), pp. 253–278. New York: Garland STPM Press.

Altmann, J. 1979. Age cohorts as paternal sibships. Behav. Ecol. Sociobiol. 6, 161–164.

Altmann, J. 1980. Baboon Mothers and Infants. Cambridge, Mass.: Harvard University Press.

Altmann, J., & Alberts, S. 1987. Body mass and growth rates in a wild primate population. Oecologia 72, 15–20.

Altmann, J., Hausfater, G., & Altmann, S. A. 1985. Demography of Amboseli baboons. Am. J. Primatol. 8, 113–125.

Altmann, J., Hausfater, G., & Altmann, S. A. 1988. Determinants of reproductive success in savannah baboons, *Papio cynocephalus*. In: Reproductive Success (Ed. by T. H. Clutton-Brock), pp. 403–418. Chicago: University of Chicago Press.

Altmann, J., & Samuels, A. 1992. Costs of maternal care: Infant-carrying in baboons. Behav. Ecol. Sociobiol. 29, 391–398.

Altmann, J., Schoeller, D., Altmann, S., Muruthi, P., & Sapolsky, R. 1993. Variability in body size and fatness in a free-living nonhuman primate population. Am. J. Primatol. In press.

Altmann, S. A. 1962. A field study of the sociobiology of the rhesus monkey, *Macaca mulatta*. Annu. N.Y. Acad. Sci. 102, 338–435.

Altmann, S. A. 1965. Sociobiology of rhesus monkeys. II. Stochastics of social communication. J. Theor. Biol. 8, 490–522.

Altmann, S. A. 1974. Baboons, space, time, and energy. Am. Zool. 14, 221–248.

Altmann, S. A. 1981. Dominance relationships: The Cheshire cat's grin? Behav. Brain Sci. 4, 430–431.

Altmann, S. A. 1988. Darwin, deceit, and metacommunication. Behav. Brain Sci. 11, 244–245.

Altmann, S. A. 1991. Diets of yearling female primates (*Papio cynocephalus*) predict lifetime fitness. Proc. Natl. Acad. Sci. U.S.A. 88, 420–423.

Altmann, S. A., & Altmann, J. 1970. Baboon Ecology. Chicago: University of Chicago Press.

Altmann, S. A., & Altmann, J. 1977. On the analysis of rates of behaviour. Anim. Behav. 25, 364–372.

Altmann, S. A., & Altmann, J. 1979. Demographic constraints on behavior and social organization. In: Primate Ecology and Human Origins (Ed. by I. S. Bernstein & E. O. Smith), pp. 47–64. New York: Garland STPM Press.

Altmann, S. A., & Wagner, S. S. 1978. A general model of optimal diet. Rec. Adv. Primatol. 4, 407–414.

Anderson, J. R. 1990. Use of objects as hammers to open nuts by capuchin monkeys (*Cebus apella*). Folia Primatol. 54, 138–145.

Andrew, R. J. 1963. The origin and evolution of the calls and facial expressions of the primates. Behaviour 20, 1–109.

Angst, W. 1974. Das Ausdrucksverhalten des Javaneraffen *Macaca fascicularis* Raffles. Fortschritte Verhaltensforsch. (Beiheft Z. Tierpsychol.) 15, 1–90.

Angst, W. 1975. Basic data and concepts in the social organization of *Macaca fascicularis*. In: Primate Behavior, Vol. 4 (Ed. by L. A. Rosenblum), pp. 325–388. New York: Academic Press.

Anon. 1991. Seeing is believing: American Ballet Theatre brings a new production of the magical Coppelia to life. Stagebill (John F. Kennedy Center for the Performing Arts), January 22–23.

Apter, D., & Vihko, R. 1983. Early menarche, a risk factor for breast cancer, indicated early onset of ovulatory cycles. J. Clin. Endocrinol. Metab. 57, 82–86.

Ashmole, N. P. 1963. The regulation of numbers of tropical oceanic birds. Ibis 103b, 458–473.

Atkinson, A. M. 1963. Accident avoidance behavior in preschool children: Free play vs. controlled situation. M.S. thesis, University of Wisconsin, Madison.

Aureli, F. 1992. Reconciliation, redirection and the regulation of social tension in macaques. Ph.D. dissertation, University of Utrecht.

Aureli, F., & van Schaik, C. P. 1992a. Post-conflict behaviour in long-tailed macaques. I. The social events. Ethology 89, 89–100.

Aureli, F., & van Schaik, C. P. 1992b. Post-conflict behaviour in long-tailed macaques. II. Coping with the uncertainty. Ethology 89, 101–114.

Aureli, F., van Schaik, C. P., & Van Hooff, J. A. R. A. M. 1989. Functional aspects of reconciliation among captive long-tailed macaques (*Macaca fascicularis*). Am. J. Primatol. 19, 39–51.

Axelrod, R., & Hamilton, W. D. 1981. The evolution of cooperation. Science 211, 1390–1396.

Badrian, A. J., & Badrian, N. L. 1984. Group composition and social structure of *Pan paniscus* in the Lomako Forest. In: The Pygmy Chimpanzee: Evolutionary Biology and Behavior (Ed. by R. L. Susman), pp. 325–346. New York: Plenum.

Badrian, N. L., & Malenky, R. R. 1984. Feeding ecology of *Pan paniscus* in the Lomako Forest, Zaire. In: The Pygmy Chimpanzee: Evolutionary Biology and Behavior (Ed. by R. L. Susman), pp. 275–299. New York: Plenum.

Baerg, W. J. 1938. Tarantula studies. J. N.Y. Entomol. Soc. 46, 31–43.

Baerg, W. J. 1963. Tarantula life history records. J. N.Y. Entomol. Soc. 71, 233–238.

Bairagi, R. 1986. Food crisis, nutrition, and female children in rural Bangladesh. Pop. Dev. Rev. 12, 307–315.

Bakeman, R., Adamson, L. B., Konner, M., & Barr, R. G. 1990. !Kung infancy: The social context of object exploration. Child Dev. 61, 794–809.

Baker, A. J. 1990. Adaptive significance of infant care behavior by nonreproductive members of wild golden lion tamarin (*Leontopithecus rosalia*) groups. Abstracts from XIIIth Congress of the Int. Primatol. Soc., p. 177.

Baker-Dittus, A. 1985. Infant and juvenile-directed care behaviors in adult toque macaques, *Macaca sinica*. Ph.D. dissertation, University of Maryland.

Baldwin, J. D. 1969. The ontogeny of social behavior of squirrel monkeys (*Saimiri sciureus*) in a seminatural enviornment. Folia Primatol. 11, 35–79.

Baldwin, J. D. 1971. The social organization of a semifree-ranging troop of squirrel monkeys (*Saimiri sciureus*). Folia Primatol. 14, 23–50.

Baldwin, J. D. 1986. Behavior in infancy: Exploration and play. In: Comparative Primate Biology, Vol. 2, Part A. Behavior, Conservation and Ecology (Ed. by G. Mitchell & J. Ervin), pp. 295–326. New York: Liss.

Baldwin, J. D., & Baldwin, J. I. 1976. Effects of food ecology on social play: A laboratory simulation. Z. Tierpsychol. 40, 1–14.

Baldwin, J. D., & Baldwin, J. I. 1977. The role of learning phenomena in the ontogeny of exploration and play. In: Primate Bio-Social Development (Ed. by S. Chevalier-Skolnikoff & F. E. Poirier), pp. 343–406. New York: Garland Press.

Baldwin, J. D., & Baldwin, J. I. 1979. The phylogenetic and ontogenetic variables that shape behavior and social organization. In: Primate Ecology and Human Origins: Ecological Influences on Social Organization (Ed. by I. S. Bernstein & E. O. Smith), pp. 89–116. New York: Garland STPM Press.

Barr, R. G., & Elias, M. F. 1988. Nursing interval and maternal responsivity: Effect on early infant crying. Pediatrics 81, 529–536.

Barry, H., III, Bacon, M. K., & Child, I. L. 1957. A cross-cultural survey of some sex differences in socialization. J. Abnormal Social Psychol. 55, 327–332.

Barry, H., III, & Paxson, L. M. 1971. Infancy and early childhood: Cross-cultural codes. Ethnology 10, 467–508.

Bateson, P. P. G. 1976. Specificity and the origins of behaviors. In: Advances in the Study of Behavior (Ed. by J. S. Rosenblatt, R. A. Hinde, E. Shaw, & C. Beer), pp. 1–20. New York: Academic Press.

Bateson, P. P. G. 1978. How does behavior develop? In: Perspectives in Ethology, Vol. 3. Social Behavior (Ed. by P.P.G. Bateson & P. H. Klopfer), pp. 55–66. New York: Plenum.

Bateson, P. P. G. 1981. Discontinuities in development and changes in the organization of play in cats. In: Behavioral Development (Ed. by K. Immelmann, G. W. Barlow, L. Petrinovich, & M. Main), pp. 281–295. New York: Cambridge University Press.

Bateson, P. P. G. 1988. The active role of behaviour in evolution. In: Evolutionary Processes and Metaphors (Ed. by M. W. Ho & S. W. Fox), pp. 191–207. New York: Wiley.

Bateson, P. P. G. 1991. The Development and Integration of Behaviour: Essays in Honour of Robert Hinde. New York: Cambridge University Press.

Bateson, P. P. G., & Klopfer, P. H. 1982. Perspectives in Ethology, Vol. 5. Ontogeny. New York: Plenum.

Bean, J. W. 1983. Cross-cultural variation in maturation rates in relation to marriage system. M.A. thesis, University of Chicago.

Bearder, S. K. 1987. Lorises, bushbabies, and tarsiers: Diverse societies in solitary foragers. In:

Primate Societies (Ed. by B. B. Smuts, D. L. Cheney, R. M. Seyfarth, R. W. Wrangham, & T. T. Struhsaker), pp. 11–24. Chicago: University of Chicago Press.

Bearder, S. K., & Doyle, G. A. 1974. Field and laboratory studies of social organization in bushbabies (Galago senegalensis). J. Hum. Evol. 3, 37–50.

Bearder, S. K., & Martin, R. D. 1980. The social organization of a nocturnal primate revealed by radio tracking. In: A Handbook on Biotelemetry and Radio Tracking (Ed. by C. J. Amlaner, Jr., & D. W. MacDonald), pp. 633–648. Oxford: Pergamon Press.

Beatty, W. W. 1984. Hormonal organization of sex differences in play fighting and spatial behavior. In: Sex Differences in the Brain (Progress in Brain Research, 61). (Ed. by G. J. DeVries, J. P. C. DeBruin, & H. B. M. Uylings), pp. 315–330. Amsterdam: Elsevier.

Becker, C. 1982. Bemerkungen zum Spielverhalten bei Orang-Utans (Pongo pygmaeus) in Kleingruppen. Z. Kölner Zoo 25, 35–43.

Becker, C. 1984. Orang-Utans und Bonobos im Spiel. Munich: Profil Verlag.

Beer, C. G. 1973. Species-typical behavior and ethology. In: Comparative Psychology: A Modern Survey (Ed. by D. A. Dewsbury & D. A. Rethlingshafer), pp. 21–77. New York: McGraw-Hill.

Bekoff, M. 1977. Mammalian dispersal and the ontogeny of distinctive phenotypes. Am. Nat. 111, 715–732.

Bekoff, M. 1984. Social play behavior. BioScience 34, 228–233.

Bekoff, M. 1988. Motor training and physical fitness: Possible short- and long-term influences on the development of individual differences. Dev. Psychobiol. 21, 601–612.

Bekoff, M., & Byers, J. A. 1985. The development of behavior from evolutionary and ecological perspectives in mammals and birds. Evol. Biol. 19, 215–286.

Bekoff, M., Diamond, J., & Mitton, J. 1987. Life-history patterns and sociality in canids: Body size, reproduction and behavior. Oecologia 50, 386–390.

Belmont, L., Stein, Z. A., & Susser, M. W. 1975. Comparisons of associations of birth order with intelligence test score and height. Nature (London) 255, 54–56.

Belsky, J., Steinberg, L., & Draper, P. 1991. Childhood experience, interpersonal development, and reproductive strategy: An evolutionary theory of socialization. Child Dev. 62, 647–670.

Bennet, A. F. 1978. Activity metabolism of the lower vertebrates. Annu. Rev. Physiol. 40, 447–469.

Bennett, F. J., Barnicott, N. A., Woodburn, J. C.,

Pereira, M. S., & Henderson, B. E. 1975. Studies on viral, bacterial, rickettsial, and treponemal diseases of the Hadza of Tanzania, and a note on injuries. Hum. Biol. 2, 61–68.

Bennett, P. M., & Harvey, P. H. 1988. How fecundity balances mortality in birds. Nature (London) 333, 216.

Berard, J. 1990. Juvenile dispersal: Maintaining relationships in a changing world. Am. J. Primatol. 10, 172–173.

Bergner, L., Mayer, S., & Harris, D. 1971. Falls from heights: A childhood epidemic in an urban area. Am. J. Public Health 61, 90–96.

Berman, C. M. 1978. Social relationships among free-ranging infant rhesus monkeys. Ph.D. dissertation, Cambridge University.

Berman, C. M. 1980. Early agonistic experience and rank acquisition among free-ranging infant rhesus monkeys. Int. J. Primatol. 1, 153–170.

Berman, C. M. 1982. The social development of an orphaned rhesus infant on Cayo Santiago: Male care, foster mother–orphan interaction and peer interaction. Am. J. Primatol. 3, 131–141.

Berman, C. M., & Swartz, S. 1988. A nonintrusive method for determining relative body fat in free-ranging monkeys. Am. J. Primatol. 14, 53–64.

Berman, P. W. 1986. Young children's responses to babies: Do they foreshadow differences between maternal and paternal styles? In: Origins of Nurturance (Ed. by A. Fogel & G. F. Melson), pp. 25–52. Hillsdale, N.J.: Erlbaum.

Berman, P. W., & Goodman, V. 1984. Age and sex differences in children's responses to babies: Effects of adults' caretaking requests and instructions. Child Dev. 55, 1071–1077.

Bernstein, I. S. 1981. Dominance: The baby and the bathwater. Behav. Brain Sci. 4, 419–467.

Bernstein, I. S., & Ehardt, C. L. 1985a. Agonistic aiding: Kinship, rank, age and sex influences. Am. J. Primatol. 8, 37–52.

Bernstein, I. S., & Ehardt, C. L. 1985b. Age–sex class differences in the expression of agonistic behavior in rhesus monkey (*Macaca mulatta*) groups. J. Comp. Psychol. 99, 114–132.

Bernstein, I. S., & Ehardt, C. 1986a. The influence of kinship and socialization on aggressive behaviour in rhesus monkeys (*Macaca mulatta*). Anim. Behav. 34, 739–747.

Bernstein, I. S., & Ehardt, C. L. 1986b. Modification of aggression through socialization and the special case of adult and adolescent male rhesus monkeys (*Macaca mulatta*). Am. J. Primatol. 10, 213–227.

Bernstein, I. S., & Smith, E. O. 1979. In summary. In: Primate Ecology and Human Origins: Eco-

logical Influences on Social Organization. (Ed. by I. S. Bernstein & E. O. Smith), pp. 341–349. New York: Garland Press.

Bertram, B. C. R. 1978. Living in groups: Predators and prey. In: Behavioral Ecology: An Evolutionary Approach (Ed. by J. R. Krebs & N. B. Davies), pp. 64–96. Sunderland, Mass.: Sinauer.

Biben, M. 1986. Individual- and sex-related strategies of wrestling play in captive squirrel monkeys. Ethology 71, 229–241.

Biben, M. 1989. Effects of social environment on play in squirrel monkeys: Resolving Harlequin's dilemma. Ethology 81, 72–82.

Biben, M., Symmes, D., & Bernhards, D. 1989. Vigilance during play in squirrel monkeys. Am. J. Primatol. 17, 41–49.

Bielicki, T. 1986. Physical growth as a measure of the economic well-being of populations: The twentieth century. In: Human Growth, 2nd ed., Vol. 3 (Ed. by F. Faulkner & J. M. Tanner), pp. 283–305. New York: Plenum.

Bingham, H. C. 1932. Gorillas in a native habitat. Publ. Carnegie Inst. 426, 65.

Bishop, J. M., & Bishop, N. H. 1978. An Ever-Changing Place. New York: Simon & Schuster.

Bishop, N. 1979. Himalayan langurs: Temperate colobines. J. Hum. Evol. 8, 251–281.

Bittner, S. O., & Delissovoy, V. 1964. Accident patterns in a nursery school sample of children. J. Nursery Ed. 19, 194–197.

Blackwell, K. F., & Menzies, J. I. 1968. Observations on the biology of the potto (*Perodicticus potto*, Miller). Mammalia 32, 447–451.

Blass, E. M. 1990. Suckling: Determinants, changes, mechanisms, and lasting impressions. Dev. Psychol. 26, 520–533.

Blurton Jones, N. G. 1972. Comparative aspects of mother child contact. In: Ethological Studies of Child Behavior (Ed. by N. G. Blurton Jones), pp. 305–328. London: Cambridge University Press.

Blurton Jones, N. G. 1986. Bushman birth spacing: A test for optimal inter-birth intervals. Ethol. Sociobiol. 7, 91–105.

Blurton Jones, N. G., Hawkes, K., & Draper, P. 1993. Differences between Hadza and !Kung children's work: Original affluence or practical reason? In: Key Issues in Hunter Gatherer Research (Ed. by E. S. Burch). Oxford: Berg. In press.

Blurton Jones, N. G., Hawkes, K., & O'Connell, J. F. 1989. Modelling and measuring costs of children in two foraging societies. In: Comparative Socioecology (Ed. by V. Standen & R. Foley), pp. 367–390. Oxford: Blackwell.

Blurton Jones, N. G., & Sibly, R. M. 1978. Testing

adaptiveness of culturally determined behaviour: Do bushman women maximise their reproductive success by spacing births widely and foraging seldom? In: Human Behaviour and Adaption (Ed. by N. Blurton Jones & V. Reynolds), pp. 135–158. Society for Study of Human Biology Symposium, No. 18. London: Taylor & Francis.

Blurton Jones, N. G., Smith, L. C., O'Connell, J. F., Hawkes, K., & Kamuzora, C. L. 1992. Demography of the Hadza, an increasing and high density population of savanna foragers. Am. J. Phys. Anthropol. 89, 159–181.

Boccia, M. L., Laudenslager, M., & Reite, M. 1988. Food distribution, dominance, and aggressive behaviors in bonnet macaques. Am. J. Primatol. 16, 123–130.

Boelkins, R. C., & Wilson, A. P. 1972. Intergroup social dynamics of the Cayo Santiago rhesus (*Macaca mulatta*) with special reference to changes in group membership by males. Primates 13, 125–140.

Boesch, C., & Boesch, H. 1990. Tool use and tool making in wild chimpanzees. Folia Primatol. 54, 86–99.

Boggess, J. 1980. Intermale relations and troop male membership changes in langurs *Presbytis entellus* in Nepal. Int. J. Primatol. 1, 233–274.

Boggess, J. 1982. Immature male and adult male interactions in bisexual langur (*Presbytis entellus*) troops. Folia Primatol. 38, 19–38.

Bogin, B. 1980. Catastrophe theory model for the regulation of human growth. Human Biol. 52, 215–227.

Bogin, B. 1988. Patterns of Human Growth. Cambridge: Cambridge University Press.

Boinski, S. 1988. Sex differences in the foraging behavior of squirrel monkeys in a seasonal habitat. Behav. Ecol. Sociobiol. 23, 177–186.

Boinski, S., & Fragaszy, D. M. 1989. The ontogeny of foraging in squirrel monkeys, *Saimiri oestedi*. Anim. Behav. 37, 415–428.

Bongaarts, J., & Potter, R. G. 1983. Fertility, Biology, and Behavior. New York: Academic Press.

Bonner, J. T. 1965. Size and Cycle: An Essay in the Structure of Biology. Princeton, N.J.: Princeton University Press.

Bonvicino, C. R. 1989. Ecologia e comportamento de *Alouatta belzebul* (Primates: Cebidae) na mata Atlântica. Rev. Nordestina Biol. 6, 149–179.

Boone, J. 1988. Parental investment, social subordination, and population processes among the 15th and 16th century Portuguese nobility. In: Human Reproductive Behavior: A Darwinian Perspective (Ed. by L. Betzig, M. Borgerhoff Mulder, & P. Turke), pp. 201–219. Cambridge: Cambridge University Press.

Borgerhoff-Mulder, M. 1987. On cultural and reproductive success: Kipsigis evidence. Am. Anthropol. 89, 617–634.

Borgerhoff-Mulder, M. 1988. Reproductive success in three Kipsigis cohorts. In: Reproductive Success (Ed. by T. H. Clutton-Brock), pp. 419–443. Chicago: University of Chicago Press.

Borgerhoff-Mulder, M. 1989. Menarche, menopause, and reproduction in the Kipsigis of Kenya. J. Biosoc. Sci. 21, 179–192.

Borries, C. 1988. Patterns of grandmaternal behaviour in free-ranging Hunuman langurs (*Presbytis entellus*). Hum. Evol. 3, 239–260.

Bower, T. G. R. 1982. Development in Infancy, 2nd ed. San Francisco: Freeman.

Bowlby, J. 1969. Attachment and Loss, Vol. 1. Attachment. London: Hogarth.

Bowlby, J. 1973. Attachment and Loss, Vol. 2. Separation. London: Hogarth.

Box, H. O. 1975. Quantitative studies of behaviour within captive groups of marmoset monkeys (*Callithrix jacchus*). Primates 16, 155–174.

Box, H. O. 1984. Primate Behaviour and Social Ecology. London: Chapman & Hall.

Boyce, M. S. 1979. Seasonality and patterns of natural selection for life histories. Am. Nat. 114, 569–583.

Boyce, M. S. 1981. Beaver life-history responses to exploitation. J. Appl. Ecol. 18, 749–753.

Boyce, M. S. 1988. Evolution of life histories: Theory and patterns from mammals. In: Evolution of Life Histories of Mammals (Ed. by M. S. Boyce), pp. 3–30. New Haven, Conn.: Yale University Press.

Boyd, R., & Richerson, P. J. 1985. Culture and the Evolutionary Process. Chicago: University of Chicago Press.

Bramblett, C. A. 1978. Sex differences in the acquisition of play in juvenile vervet monkeys. In: Social Play in Primates (Ed. by E. O. Smith), pp. 33–48. New York: Academic Press.

Bramblett, C., Bramblett, S. S., Bishop, D. A., & Coelho, A. M. 1982. Longitudinal stability in adult status hierarchies among vervet monkeys (*Cercopithecus aethiops*). Am. J. Primatol. 2, 43–51.

Bramblett, C. A., & Coelho, A. M. 1987. Development of social behavior in vervet monkeys, Sykes' monkeys, and baboons. In: Comparative Behavior of African Monkeys (Ed. by E. L. Zuker), pp. 67–79. New York: Liss.

Brett, J. R. 1979. Environmental factors and growth. In: Fish Physiology, Vol. 8 (Ed. by W. S. Hoar, D. J. Randall, & J. R. Brett), pp. 599–677. New York: Academic Press.

Brody, S. 1945. Bioenergetics and Growth. New York: Reinhold.

Bronson, F. H. 1989. Mammalian Reproductive Biology. Chicago: University of Chicago Press.

Bruce, H. M. 1959. An extroceptive block to pregnancy in the mouse. Nature (London) 184, 105.

Brundtland, G. H., Liestol, K., & Walloe, L. 1980. Height, weight, and menarcheal age of Oslo schoolchildren during the last 60 years. Ann. Hum. Biol. 7, 307–322.

Brundtland, G. H., & Walloe, L. 1973. Menarcheal age in Norway: Halt in the trend towards earlier maturation. Nature (London) 214, 478–479.

Budnitz, N., & Dainis, K. 1975. Lemur catta: Ecology and behavior. In: Lemur Biology (Ed. by I. Tattersall & R. W. Sussman), pp. 219–235. New York: Plenum.

Bulger, J. B., & Hamilton, W. J., III. 1988. Inbreeding and reproductive success in a natural chacma baboon, Papio cynocephalus ursinus, population. Anim. Behav. 36, 574–578.

Burbank, V., Ratner, M., & Whiting, J. W. M. 1986. The duration of maidenhood across cultures. In: School-Age Pregnancy and Parenthood: Biosocial Dimensions (Ed. by J. B. Lancaster & B. A. Hamburg), pp. 273–302. New York: Aldine.

Buss, L. 1987. The Evolution of Individuality. Princeton, N.J.: Princeton University Press.

Busse, C. D. 1984. Triadic interactions among male and infant chacma baboons. In: Primate Paternalism (Ed. by D. M. Taub), pp. 186–212. New York: Van Nostrand Reinhold.

Busse, C. D., & Gordon, T. P. 1983. Infant carrying by adult male mangabeys (Cercocebus atys). Am. J. Primatol. 6, 133–141.

Busse, C. D., & Hamilton, W. J., III. 1981. Infant carrying by male chacma baboons. Science 212, 1281–1292.

Byers, J. 1986. Natural variation in early experience of pronghorn fawns: Sources and consequences. In: The Individual and Society (Ed. by L. Passera & J. P. Lachaud), pp. 81–92. Toulouse: Privat I.E.C.

Bygott, J. D., Bertram, B.C.R., & Hanby, J. P. 1979. Male lions in large coalitions gain reproductive advantages. Nature (London) 282, 839–841.

Caine, N. G. 1986. Behavior during puberty and adolescence. In: Comparative Primate Biology, Vol. 2, Part A. Behavior, Conservation, and Ecology (Ed. by G. Mitchell & J. Erwin), pp. 327–362. New York: Liss.

Caine, N., & Mitchell, G. 1979. A review of play in the genus Macaca: Social correlates. Primates 20, 535–546.

Caine, N. G., & Mitchell, G. 1980. Species differences in the interest shown in infants by juvenile female macaques (Macaca radiata and M. mulatta). Int. J. Primatol. 1, 323–332.

Calder, W. A. 1984. Size, Function, and Life History. Cambridge, Mass.: Harvard University Press.

Caldwell, J. C. 1976. Toward a restatement of demographic transition theory. Pop. Dev. Rev. 2, 321–366.

Cambefort, J. P. 1981. A comparative study of culturally transmitted patterns of feeding habits in the chacma baboon Papio ursinus and the vervet monkey Cercopithecus aethiops. Folia Primatol. 36, 243–263.

Camperio-Ciani, A., & Chiarelli, B. 1988. Age and sex differences in the feeding strategies of a free-ranging population of Macaca mulatta Zimmerman, 1788 (Primates Cercopithecidae), in Simla (India). Monitore Zool. Ital. NS 22, 171–182.

Candland, D. K., French, J. A., & Johnson, C. N. 1978. Object-play: Test of a categorized model by the genesis of object-play in Macaca fuscata. In: Social Play in Primates (Ed. by E. O. Smith), pp. 259–296. New York: Academic Press.

Capitanio, J. P. 1986. Behavioral pathology. In: Behavior, Conservation, and Ecology (Ed. by G. Mitchell & J. Erwin), pp. 411–454. New York: Liss.

Caro, T. M. 1988. Adaptive significance of play: Are we getting closer? Trends Ecol. Evol. 3, 50–53.

Carpenter, C. R. 1934. A field study of the behavioral and social relations of howling monkeys (Alouatta palliata). Comp. Psychol. Monog. 10, 1–168.

Carpenter, C. R. 1935. Behavior of red spider monkeys in Panama. J. Mammal. 16, 171–180.

Carpenter, C. R. 1940. A field study in Siam of the behavior and social relations of the gibbon (Hylobates lar). Comp. Psychol. Monog. 116, 1–205.

Carpenter, C. R. 1964. Naturalistic Behavior of Nonhuman Primates. University Park: Pennsylvania State University Press.

Carpetis, R. C., & Nash, L. T. 1983. The communicative potential of urine-washing in Galago senegalensis braccatus. Am. J. Primatol. 4, 339.

Case, T. J. 1978. On the evolution and adaptive significance of postnatal growth rates in the terrestrial vertebrates. Quart. Rev. Biol. 53, 243–286.

Caswell, H. 1983. Phenotypic plasticity in life-history traits: Demographic effects and evolutionary consequences. Am. Zool. 23, 35–46.

Caswell, H., & Hastings, A. 1980. Fecundity, developmental time, and population growth rate: An analytical solution. Theor. Pop. Biol. 17, 71–79.

Catanzaro, D. de. 1991. Evolutionary limits to self-preservation. Ethol. Sociobiol. 12, 13–28.

Caughley, G. 1966. Mortality in mammals. Ecology 47, 906–917.

Cebul, M. S., & Epple, G. 1984. Father–offspring relationships in laboratory families of saddle-back tamarins. In: Primate Paternalism. (Ed. by D. M. Taub), pp. 1–19. New York: Van Nostrand Reinhold.

Chadwick-Jones, J. K. 1991. The social contingency model and olive baboons. Int. J. Primatol. 12, 145–161.

Chagnon, N. A., & Irons, W. (Eds.). 1979. Evolutionary Biology and Human Social Behavior: An Anthropological Perspective. North Scituate, Mass.: Duxbury Press.

Chalmers, N. R. 1980a. Developmental relationships among social, manipulation, postural, and locomotor behaviors in olive baboons (*Papio anubis*). Behaviour 74, 22–37.

Chalmers, N. R. 1980b. The ontogeny of play in feral olive baboons (*Papio anubis*). Anim. Behav. 28, 570–585.

Chalmers, N. R. 1984. Social play in monkeys: Theories and data. In: Play in Animals and Humans (Ed. by P. K. Smith), pp. 119–141. Oxford: Blackwell.

Chalmers, N. R. 1987. Developmental pathways in behaviour. Anim. Behav. 35, 659–674.

Changeux, J.-P. 1985. Neuronal Man. New York: Pantheon.

Chapais, B. 1983. Dominance, relatedness and the structure of female relationships in rhesus monkeys. In: Primate Social Relationships: An Integrated Approach (Ed. by R. A. Hinde), pp. 205–217. Oxford: Blackwell.

Chapais, B. 1985. An experimental analysis of a mother–daughter rank reversal in Japanese macaques (*Macaca fuscata*). Primates 26, 407–423.

Chapais, B. 1988a. Experimental matrilineal inheritance of rank in female Japanese macaques. Anim. Behav. 36, 1025–1037.

Chapais, B. 1988b. Rank maintenance in female Japanese macaques: Experimental evidence for social dependency. Behaviour 104, 41–59.

Chapais, B. 1991. Matrilineal dominance in Japanese macaques: The contribution of an experimental approach. In: The Japanese Macaques of Arashiyama. (Ed. by L. Fedigan & P. Asquith), pp. 251–273. Albany: State University of New York Press.

Chapais, B. 1992. The role of alliances in social inheritance of rank among female primates. In: Coalitions and Alliances in Humans and Other Animals (Ed. by A. Harcourt & F. de Waal), pp. 29–59. New York: Oxford University Press.

Chapais, B. Girard, M., & Primi, G. 1991. Non-kin alliances and the stability of matrilineal dominance relations in Japanese macaques. Anim. Behav. 41: 481–491.

Chapais, B., & Larose, F. 1988. Experimental rank reversals among peers in *Macaca fuscata*: Rank is maintained after the removal of kin support. Am. J. Primatol. 16, 31–42.

Chapais, B., & Schulman, S. 1980. An evolutionary model of female dominance relations in primates. J. Theor. Biol. 82, 47–89.

Chapman, C. 1987. Flexibility in diets of three species of Costa Rican primates. Folia Primatol. 49, 90–105.

Chapman, C. 1988a. Patterns of foraging and range use by three species of neotropical primate. Primates 29, 177–194.

Chapman, C. 1988b. Patch use and patch depletion by the spider and howling monkeys of Santa Rosa National Park, Costa Rica. Behaviour 105(1/2), 99–116.

Chapman, C. A. 1990. Ecological constraints on group size in three species of neotropical primates. Folia Primatol. 55, 1–9.

Chapman, C., & Chapman, L. J. 1986. Behavioural development of howling monkey twins (*Alouatta palliata*) in Santa Rosa National Park, Costa Rica. Primates 27, 377–381.

Charles-Dominique, P. 1974. Vie sociale de *Perodicticus potto* (Primates, Lorisidés). Étude de terrain en forêt équatoriale de l'ouest africain au Gabon. Mammalia 38, 355–379.

Charles-Dominique, P. 1977a. Ecology and Behaviour of Nocturnal Primates. London: Duckworth.

Charles-Dominique, P. 1977b. Urine marking and territoriality in *Galago alleni* (Waterhouse, 1837—Lorisoidea, Primates)—A Field study by radio-telemetry. Zeit. Tierpsychol. 43, 113–138.

Charles-Dominique, P. 1978. Solitary and gregarious prosimians: Evolution of social structures in primates. In: Recent Advances in Primatology, Vol. 3. Evolution (Ed. by D. J. Chivers & K. A. Joysey), pp. 139–149. London: Academic Press.

Charles-Dominique, P., & Bearder, S. K. 1979. Field studies of lorisid behavior: Methodological aspects. In: The Study of Prosimian Behavior (Ed. by G. A. Doyle & R. D. Martin), pp. 567–629. New York: Academic Press.

Charles-Dominique, P., & Petter, J.-J. 1980. Ecology

and social life of *Phaner furcifer*. In: Nocturnal Malagasy Primates: Ecology, Physiology, and Behaviour (Ed. by P. Charles-Dominique et al.), pp. 75–95. New York: Academic Press.

Charlesworth, B. 1980. Evolution in Age Structured Populations. New York: Cambridge University Press.

Charnov, E. L. 1982. The Theory of Sex Allocation. Princeton, N.J.: Princeton University Press.

Charnov, E. L. 1986. Life history evolution in a "recruitment population": Why are adult mortality rates constant? Oikos 47, 129–134.

Charnov, E. L. 1990. On evolution of maturity and the adult lifespan. J. Evol. Biol. 3, 139–144.

Charnov, E. L. 1991. Evolution of life history variation among female mammals. Proc. Natl. Acad. Sci. U.S.A. 88, 1134.

Charnov, E. L., & Schaffer, W. M. 1973. Life-history consequences of natural selection: Cole's result revisited. Am. Nat. 107, 791–793.

Chen, L. C., Huq, E., & D'Souza, S. 1981. Sex bias in the family association of food and health care in rural Bangladesh. Pop. Dev. Rev. 7, 55–70.

Cheney, D. L. 1977. The acquisition of rank and the development of reciprocal alliances among free-ranging immature baboons. Behav. Ecol. Sociobiol. 2, 303–318.

Cheney, D. L. 1978a. The play partners of immature baboons. Anim. Behav. 26, 1038–1050.

Cheney, D. L. 1978b. Interactions of immature male and female baboons with adult females. Anim. Behav. 26, 389–408.

Cheney, D. L. 1981. Inter-group encounters among free-ranging vervet monkeys. Folia Primatol. 35, 124–146.

Cheney, D. L. 1983. Extra-familial alliances among vervet monkeys. In: Primate Social Relationships. An Integrated Approach (Ed. by R. A. Hinde), pp. 278–285. Oxford: Blackwell.

Cheney, D. L., & Seyfarth, R. M. 1981. Selective forces affecting the predator alarm calls of vervet monkeys. Behaviour 76, 25–61.

Cheney, D. L., & Seyfarth, R. M. 1983. Nonrandom dispersal in free-ranging vervet monkeys: Social and genetic consequences. Am. Nat. 122, 392–412.

Cheney, D. L., & Seyfarth, R. M. 1987. The influence of intergroup competition on the survival and reproduction of female vervet monkeys. Behav. Ecol. Sociobiol. 21, 375–386.

Cheney, D. L., & Seyfarth, R. M. 1989. Redirected aggression and reconciliation among vervet monkeys, *Cercopithecus aethiops*. Behaviour 110, 258–275.

Cheney, D. L., & Seyfarth, R. L. 1990. How

Monkeys See the World. Chicago: University of Chicago Press.

Cheney, D. L., Seyfarth, R. M., Andelman, S. J., & Lee, P. C. 1988. Reproductive success in vervet monkeys. In: Reproductive Success (Ed. by T. H. Clutton-Brock), pp. 384–402. Chicago: University of Chicago Press.

Cheney, D. L., & Wrangham, R. W. 1987. Predation. In: Primate Societies (Ed. by B. B. Smuts, D. L. Cheney, R. M. Seyfarth, R. M. Wrangham, & T. T. Struhsaker), pp. 227–239. Chicago: University of Chicago Press.

Chevalier-Skolnikoff, S. 1982. A cognitive analysis of facial behavior in Old World monkeys, apes and human beings. In: Primate Communication (Ed. by C. T. Snowdon, C. H. Brown, & M. R. Peterson), pp. 303–368. Cambridge: Cambridge University Press.

Chism, J. 1986. Development and mother–infant relations among captive patas monkeys. Int. J. Primatol. 7, 49–82.

Chism, J. 1991. Ontogeny of behavior in humans and nonhuman primates: The search for common ground. In: Understanding Behavior: What Primate Studies Tell Us About Human Behavior (Ed. by J. D. Loy & C. B. Peters), pp. 90–120. New York: Oxford University Press.

Chisolm, J. S. 1988. Toward a developmental evolutionary ecology of humans. In: Sociobiological Perspectives on Human Development (Ed. by K. B. Mac Donald), pp. 78–102. New York: Springer.

Chiszar, D. 1985. Ontogeny of communication behaviors. In: The Comparative Development of Adaptive Skills: Evolutionary Implications (Ed. by E. S. Gollin), pp. 207–238. Hillsdale, N.J.: Erlbaum.

Chivers, D. J. 1974. The Siamang in Malaya: A Field Study of a Primate in Tropical Rain Forest. (Contributions to Primatology, 4). Basel: Karger.

Chivers, D. J., Raemaekers, J. J., & Aldrich-Blake, F. P. G. 1975. Long-term observations of siamang behaviour. Folia Primatol. 23, 1–49.

Chulee Chawalsilp. 1981. Effect of mother's age, infant's age and the time of day on the grooming of the infant and the infant play in the long-tailed monkey (*Macaca fascicularis*). M.Sc. thesis, Mahidol University, Bangkok.

Clark, A. B. 1978a. Olfactory communication, *Galago crassicaudatus*, and the social life of prosimians. In: Recent Advances in Primatology: Evolution, Vol. 3 (Ed. by D. J. Chivers & K. A. Joysey), pp. 109–116. New York: Academic Press.

Clark, A. B. 1978b. Sex ratio and local resource competition in a prosimian primate. Science 201, 163–165.

Clark, A. B. 1982a. Scent marks as social signals in *Galago crassicaudatus*. I. Sex and reproductive status as factors in signals and responses. J. Chem. Ecol. 8, 1133–1151.

Clark, A. B. 1982b. Scent marks as social signals in *Galago crassicaudatus*. II. Discrimination between individuals by scent. J. Chem. Ecol. 8, 1153–1165.

Clark, A. B. 1985. Sociality in a nocturnal "solitary" prosimian: *Galago crassicaudatus*. Int. J. Primatol. 6, 581–600.

Clark, A. B. 1991. Individual variation in responsiveness to environmental change. In: Primate Responses to Environmental Change (Ed. by H. O. Box), pp. 91–110. London: Chapman & Hall.

Clark, A. B., & Ehlinger, T. J. 1987. Pattern and adaptation in individual behavioural differences. In: Perspectives in Ethology, Vol. 7. Alternatives (Ed. by P. P. G. Bateson & P. H. Klopfer), pp. 1–47. New York: Plenum.

Clark Le Gros, W. E. 1971. The Antecedents of Man. Edinburgh: Edinburgh University Press.

Clark, M. M., Spencer, C. A., & Galef, B. G., Jr. 1986. Reproductive life history correlates of early and late sexual maturation in female Mongolian gerbils (*Meriones unguiculatus*). Anim. Behav. 34, 551–560.

Clark, T. W., & Mano, T. 1975. Transplantation and adaptation of Arashiyama A troop of Japanese macaques to a Texas bushland habitat. In: Contemporary Primatology (Ed. by S. Kondo, M. Kawai, & E. Ehara), pp. 358–361. Basel: Karger.

Clarke, M. R. 1978. Social interactions of juvenile female bonnet monkeys, *Macaca radiata*. Primates 19: 517–524.

Clarke, M. R. 1990. Behavioral development and socialization of infants in a free-ranging group of howling monkeys (*Alouatta palliata*). Folia Primatol. 54, 1–15.

Clarke, M. R., & Glander, K. E. 1981. Adoption of infant howling monkeys (*Alouatta palliata*). Am. J. Primatol. 1, 469–472.

Clarke, M. R., & Glander, K. E. 1984. Female reproductive success in a group of free-ranging howling monkeys (*Alouatta palliata*) in Costa Rica. In: Female Primates: Studies by Women Primatologists (Ed. by M. F. Small), pp. 111–126. New York: Liss.

Clarke, M. R., & Zucker, E. L. 1989. Social correlates of timing of sexual maturity in free-ranging howling monkeys (*Alouatta palliata*). Am. J. Primatol. 18, 140.

Cleveland, J., & Snowdon, C. T. 1984. Social development during the first twenty weeks in the cotton-top tamarin (*Saguinus o. oedipus*). Anim. Behav. 32, 432–444.

Clutton-Brock, T. H. (Ed.). 1977a. Primate Ecology: Studies of Feeding and Ranging Behaviour in Lemurs, Monkeys, and Apes. London: Academic Press.

Clutton-Brock, T. H. 1977b. Some aspects of intraspecific variation in feeding and ranging behaviour in primates. In: Primate Ecology: Studies of Feeding and Ranging Behaviour in Lemurs, Monkeys, and Apes (Ed. by T. H. Clutton-Brock), pp. 539–556. London: Academic Press.

Clutton-Brock, T. H. 1985. Size, sexual dimorphism, and polygyny in primates. In: Size and Scaling in Primate Biology (Ed. by W. L. Jungers), pp. 51–60. New York: Plenum.

Clutton-Brock, T. H. (Ed.). 1988. Reproductive Success. Chicago: University of Chicago Press.

Clutton-Brock, T. H. 1989. Female transfer and inbreeding avoidance in social mammals. Nature (London) 337, 70–72.

Clutton-Brock, T. H. 1991. The Evolution of Parental Care. Princeton, N.J.: Princeton University Press.

Clutton-Brock, T. H., Albon, S. D., & Guinness, F. E. 1986. Great expectations: Dominance, breeding success and offspring sex ratios in red deer. Anim. Behav. 34, 460–471.

Clutton-Brock, T. H., Guinness, F. E., & Albon, S. D. 1982. The Red Deer: Behaviour and Ecology of Two Sexes. Chicago: University of Chicago Press.

Clutton-Brock, T. H., Guinness, F. E., & Albon, S. D. 1983. The costs of reproduction to red deer hinds. J. Anim. Ecol. 52, 367–383.

Clutton-Brock, T. H., & Harvey, P. H. 1976. Evolutionary rules and primate societies. In: Growing Points in Ethology (Ed. by P. P. G. Bateson & R. A. Hinde), pp. 192–237. Cambridge: Cambridge University Press.

Clutton-Brock, T. H., & Harvey, P. H. 1977. Primate ecology and social organization. J. Zool. (London), 183, 1–39.

Clutton-Brock, T. H., & Harvey, P. H. 1978. Mammals, resources, and reproductive strategies. Nature (London) 273, 191–195.

Clutton-Brock, T. H., & Harvey, P. H. 1980. Primates, brains and ecology. J. Zool. (London), 190, 309–323.

Coelho, A. M., Jr. 1985. Baboon dimorphism: Growth in weight, length and adiposity from birth to 8 years of age. In: Nonhuman Primate Models for Human Growth and Development (Ed. by E. S. Watts), pp. 125–159. New York: Liss.

Cole, L. C. 1954. The population consequences of life history phenomena. Quart. Rev. Biol. 29, 103–137.

Collins, D. C., Nadler, R. D., & Preedy, J. R. K.

1981. Adrenarche in the great apes (abstract). Am. J. Primatol. 1, 344.

Collu, R., & Ducharme, J. R. 1975. Role of adrenal steroids in the regulation of gonadotropin secretion at puberty. J. Steroid Biochem. 6, 869–872.

Colvin, J. D. 1983a. Description of sibling and peer relationships among immature male rhesus monkeys. In: Primate Social Relationships (Ed. by R. A. Hinde), pp. 57–64. Sunderland, Mass.: Sinauer.

Colvin, J. D. 1983b. Influences of the social situation on male migration. In: Primate Social Relationships (Ed. by R. A. Hinde), pp. 160–171. Sunderland, Mass.: Sinauer.

Colvin, J. D. 1985. Breeding-season relationships of immature male rhesus monkeys with females. I. Individual differences and constraints on partner choice. Int. J. Primatol. 6, 261–287.

Colvin, J., & Tissier, G. 1985. Affiliation and reciprocity in sibling and peer relationships among free-ranging immature male rhesus monkeys. Anim. Behav. 33, 959–977.

Conlan, K. E. 1989. Delayed reproduction and adult dimorphism in males of the amphipod genus Jassa (Corophioidea: Ischyroceridae): An explanation for systematic confusion. J. Crust. Biol. 9, 601–625.

Conover, W. J. 1980. Practical Nonparametric Statistics, 2nd ed. New York: Wiley.

Cords, M. 1986. Interspecific and intraspecific variation in diet of two forest guenons, Cercopithecus ascanius and C. mitis. J. Anim. Ecol. 55, 811–827.

Cords, M. 1987. Forest guenons and patas monkeys: Male–male competition in one-male groups. In: Primate Societies (Ed. by B. B. Smuts, D. L. Cheney, R. M. Seyfarth, R. W. Wrangham, & T. T. Struhsaker), pp. 98–111. Chicago: University of Chicago Press.

Cords, M. 1988a. Mating systems of forest guenons: A preliminary review. In: A Primate Radiation (Ed. by A. Gautier-Hion, F. Bourliere, J-P. Gautier, & J. Kingdon), pp. 323–339. Cambridge: Cambridge University Press.

Cords, M. 1988b. Resolution of aggressive conflicts by immature male long-tailed macaques. Anim. Behav. 36, 1124–1135.

Cords, M. 1990. How immature long-tailed macaques cope with aggressive conflict. Paper presented at the 23rd Congress of the International Primatological Society, Kyoto, Japan.

Cords, M. 1992. Post-conflict reunions and reconciliation in long-tailed macaques. Anim. Behav. 44, 57–61.

Cords, M. 1993. On operationally defining reconciliation. Am. J. Primatol. 29, 255–267.

Crockett, C. M. 1984. Emigration by female red howler monkeys and the case for female competition. In: Female Primates: Studies by Women Primatologists (Ed. by M. F. Small), pp. 159–173. New York: Liss.

Crockett, C. M. 1985. Population studies of red howler monkeys (Alouatta seniculus). Natl. Geogr. Res. 1, 264–273.

Crockett, C. M. 1987. Diet, dimorphism and demography: Perspectives from howlers to hominids. In: The Evolution of Human Behavior: Primate Models (Ed. by W. G. Kinzey), pp. 115–135. Albany: State University of New York Press.

Crockett, C. M. 1991. Population growth and new troop formation in Venezuelan red howler monkeys. Am. J. Primatol. 24, 95.

Crockett, C. M., & Eisenberg, J. F. 1987. Howlers: Variations in group size and demography. In: Primate Societies (Ed. by B. B. Smuts, D. L. Cheney, R. M. Seyfarth, R. W. Wrangham, & T. T. Struhsaker), pp. 54–68. Chicago: University of Chicago Press.

Crockett, C. M., & Pope, T. 1988. Inferring patterns of aggression from red howler monkey injuries. Am. J. Primatol. 15, 289–308.

Crockett, C. M., & Rudran, R. 1987a. Red howler monkey birth data. I. Seasonal variation. Am. J. Primatol. 13, 347–368.

Crockett, C. M., & Rudran, R. 1987b. Red howler monkey birth data. II. Interannual, habitat, and sex comparisons. Am. J. Primatol. 13, 369–384.

Crockett, C. M., & Sekulic, R. 1982. Gestation length in red howler monkeys. Am. J. Primatol. 3, 291–294.

Crockett, C. M., & Sekulic, R. 1984. Infanticide in red howler monkeys (Alouatta seniculus). In: Infanticide: Comparative and Evolutionary Perspectives (Ed. by G. Hausfater & S. B. Hrdy), pp. 173–191. New York: Aldine.

Crompton, R. H. 1983. Age differences in locomotion of two subtropical Galaginae. Primates 24, 241–259.

Crompton, R. H., & Andau, P. M. 1987. Ranging, activity rhythms, and sociality in free-ranging Tarsius bancanus: A preliminary report. Int. J. Primatol. 8, 43–71.

Cronk, L. 1989. Low socioeconomic status and female-biased parental investment: The Mukogodo example. Am. Anthropol. 90, 414–29.

Crook, J. H. 1964. The evolution of social organisation and visual communication in the weaver birds (Ploceinae). Behav. Suppl. 10, 1–178.

Crook, J. H. 1965. The adaptive significance of avian social organization. Symp. Zool. Soc. London 14, 181–218.

Crook, J. H. 1970. Social organization and the environment: Aspects of contemporary social ethology. Anim. Behav. 18, 197–209.

Crook, J. H. 1972. Sexual selection, dimorphism, and social organization in the primates. In: Sexual Selection and the Descent of Man, 1871–1971 (Ed. by B. G. Campbell), pp. 231–281. Chicago: Aldine.

Crook, J. H., & Gartlan, J. S. 1966. Evolution of primate societies. Nature (London) 210, 1200–1203.

Crowl, T. A., & Covich, A. P. 1990. Predator-induced life-history shifts in a freshwater snail. Science 324, 58–60.

Cutler, G. B., Glenn, M., Bush, M., Hodgen, G. D., Graham, C. E., & Loriaux, D. L. 1978. Adrenarche: A survey of rodents, domestic animals, and primates. Endocrinology 103, 2112–2118.

Daly, M., & Wilson, M. 1978. Sex, Evolution, and Behavior. North Scituate, Mass.: Duxbury Press.

Daly, M., & Wilson, M. 1983. Sex, Evolution, and Behavior, 2nd ed. Boston: PWS.

Daly, M., & Wilson, M. 1984. A sociobiological analysis of human infanticide. In: Infanticide: Comparative and Evolutionary Perspectives (Ed. by G. Hausfater & S. B. Hrdy), pp. 487–502. New York: Aldine.

Dang, D. 1983. Female puberty in monkey Macaca fascicularis raised in laboratory: Menarche–copulation–gestation–fertility. Cahiers Anthropol. Biomet. Hum. 1, 33–45.

Daniel, D. 1978. A conversation with Suzanne Farrell. Ballet Rev. 7(1), 1–15.

Dare, R. J. 1975. The social behavior and ecology of spider monkeys, Ateles geoffroyi, on Barro Colorado Island. Ph.D. dissertation, University of Oregon.

Darwin, C. 1859. The Origin of Species by Means of Natural Selection. London: Murray.

Darwin, C. 1871. The Descent of Man and Selection in Relation to Sex. London: Murray.

Darwin, C. 1872. The Expression of Emotion in Man and Animals. London: Appleton.

Dasser, V. 1987. Slides of group members as representations of the real animals (Macaca fascicularis). Ethology 76, 65–73.

Dasser, V. 1988. A social concept in Java monkeys. Anim. Behav. 36, 225–230.

Datta, S. B. 1983a. Relative power and the acquisition of rank. In: Primate Social Relationships: An Integrated Approach (Ed. by R. A. Hinde), pp. 93–102. Oxford: Blackwell.

Datta, S. B. 1983b. Relative power and the maintenance of rank. In: Primate Social Relationships. An Integrated Approach (Ed. by R. A. Hinde), pp. 103–112. Oxford: Blackwell.

Datta, S. B. 1983c. Patterns of agonistic interference. In: Primate Social Relationships: An Integrated Approach (Ed. by R. A. Hinde), pp. 289–297. Oxford: Blackwell.

Datta, S. B. 1988. The acquisition of rank among free-ranging rhesus monkey siblings. Anim. Behav. 36, 754–772.

Deag, J. M., & Crook, J. H. 1971. Social behaviour and "agonistic buffering" in the wild barbary macaque Macaca sylvana. Folia Primatol. 15, 183–200.

Dennett, G., & Connell, J. 1988. Acculturation and health in the highlands of Papua New Guinea. Cur. Anthropol. 29, 273–299.

Deputte, B. L. 1983. Ontogenetic development of dyadic social relationships: Assessing individual roles. Am. J. Primatol. 4, 309–318.

de Ruiter, J. R. 1986. The influence of group size on predator scanning and foraging behaviour of wedgecapped capuchin monkeys (Cebus olivaceus). Behaviour 98, 240–258.

DeVore, I. 1963. Mother–infant relations in free-ranging baboons. In: Maternal Behavior in Mammals (Ed. by H. L. Rheingold), pp. 305–355. New York: Wiley.

DeVore, I., & Washburn, S. L. 1963. Baboon ecology and human evolution. In: African Ecology and Human Evolution (Ed. by F. Campbell & F. Bourliere), pp. 335–367. New York: Aldine.

de Waal, F. B. M. 1977. The organization of agonistic relations within two captive groups of Java-monkeys (Macaca fascicularis). Z. Tierpsychol. 44, 225–282.

de Waal, F. B. M. 1982. Chimpanzee Politics. London: Jonathan Cape.

de Waal, F. B. M. 1984. Coping with social tension: Sex differences in the effect of food provision to small rhesus monkey groups. Anim. Behav. 32, 765–773.

de Waal, F. B. M. 1986a. The integration of dominance and social bonding in primates. Quart. Rev. Biol. 61, 459–479.

de Waal, F. B. M. 1986b. The brutal elimination of a rival among captive male chimpanzees. Ethol. Sociobiol. 7, 237–251.

de Waal, F. B. M. 1986c. Class structure in a rhesus monkey group: The interplay between dominance and tolerance. Anim. Behav. 34, 1033–1040.

de Waal, F. B. M. 1986d. Conflict resolution in monkeys and apes. In: Primates: The Road to Self-Sustaining Populations (Ed. by K. Benirschke), pp. 341–350. New York: Springer.

de Waal, F. B. M. 1986e. Imaginative bonobo games. Zoonooz 59, 6–10.

de Waal, F. B. M. 1988. The communicative repertoire of captive bonobos (Pan paniscus), compared to that of chimpanzees. Behaviour 106, 183–251.

de Waal, F. B. M. 1989a. Peacemaking Among Primates. Cambridge, Mass.: Harvard University Press.

de Waal, F. B. M. 1989b. Dominance "style" and primate social organization. In: Comparative Socioecology, the Behavioural Ecology of Humans and Other Mammals (Ed. by V. Standen & R. Foley), pp. 243–263. Oxford: Blackwell.

de Waal, F. B. M., & Luttrell, L. M. 1985. The formal hierarchy of rhesus monkeys: An investigation of the bared-teeth display. Am. J. Primatol. 9, 73–85.

de Waal, F. B. M., & Luttrell, L. M. 1986. The similarity principle underlying social bonding among female rhesus monkeys. Folia Primatol. 46, 215–234.

de Waal, F. B. M., & Luttrell, L. M. 1989. Toward a comparative socioecology of the genus *Macaca:* Different dominance styles in rhesus and stumptail monkeys. Am. J. Primatol. 19, 83–109.

de Waal, F. B. M., & Ren, R. 1988. Comparison of the reconciliation behavior of stumptail and rhesus macaques. Ethology 78, 129–142.

de Waal, F. B. M., & van Roosmalen, A. 1979. Reconciliation and consolation among chimpanzees. Behav. Ecol. Sociobiol. 5, 55–66.

de Waal, F. B. M., & Yoshihara, D. 1983. Reconciliation and redirected affection in rhesus monkeys. Behaviour 85, 224–241.

Dewey, K. G. 1983. Nutrition survey in Tabasco, Mexico: Nutritional status of preschool children. Am. J. Clin. Nutr. 37, 1010–1019.

Dewsbury, D. A. 1984. Comparative Psychology in the Twentieth Century. Stroudsburg, Pa.: Hutchinson Ross.

Dice, L. R. 1945. Measures of the amount of ecologic association between species. Ecology 26, 297–302.

Dickemann, M. 1979. Female infanticide and reproductive strategies of stratified human societies. In: Evolutionary Biology and Human Social Behavior (Ed. by N. Chagnon & W. Irons), pp. 321–367. North Scituate, Mass.: Duxbury Press.

Dickerson, G. E. 1954. Hereditary mechanisms in animal growth. In: Dynamics of Growth Processes (Ed. by B. J. Boell), pp. 242–276. Princeton, N.J.: Princeton University Press.

Dienske, H. 1986. A comparative approach to the question of why human infants develop so slowly. In: Primate Ontogeny, Cognition and Social Behaviour (Ed. by J. G. Else & P. C. Lee), pp. 147–154. Cambridge: Cambridge University Press.

Dittus, W. P. J. 1977. The social regulation of population density and age–sex distribution in the toque monkey. Behaviour 63, 281–322.

Dittus, W. P. J. 1979. The evolution of behaviors regulating density and age-specific sex ratios

in a primate population. Behaviour 69, 265–302.

Dittus, W. P. J. 1980. The social regulation of primate populations: A synthesis. In: The Macaques (Ed. by D. G. Lindburg), pp. 263–286. New York: Van Nostrand.

Dobzhansky, T. 1937. Genetics and the Origin of Species. New York: Columbia University Press.

Dolhinow, P., & Krusko, N. 1984. Langur monkey females and infants: The female's point of view. In: Female Primates: Studies by Women Primatologists (Ed. by M. F. Small), pp. 37–57. New York: Liss.

Doran, D. M. 1989. Chimpanzee and pygmy chimpanzee positional behavior: The influence of environment, body size, morphology, and ontogeny on locomotion and posture. Ph.D. dissertation, State University of New York at Stony Brook.

Doyle, G. A. 1974. Behavior of prosimians. In: Behavior of Nonhuman Primates (Ed. by A. M. Schrier & F. Stollnitz), pp. 155–353. New York: Academic Press.

Doyle, G. A. 1979. Development of behavior in prosimians with special reference to the lesser bushbaby, *Galago senegalensis moholi.* In: The Study of Prosimian Behavior (Ed. by G. A. Doyle & R. D. Martin), pp. 157–206. New York: Academic Press.

Doyle, G. A., Andersson, A., & Bearder, S. K. 1969. Maternal behaviour in the lesser bushbaby (*Galago senegalensis moholi*) under semi-natural conditions. Folia Primatol. 11, 215–238.

Doyle, G. A., Pelletier, A., & Bekker, T. 1967. Courtship, mating and parturition in the lesser bushbaby (*Galago senegalensis moholi*) under semi-natural conditions. Folia Primatol. 7, 169–197.

Draper, P. 1975. Cultural pressure on sex differences. Am. Ethnol. 4, 602–615.

Draper, P. 1976. Social and economic constraints on child life among the !Kung. In: Kalahari Hunter Gatherers (Ed. by R. B. Lee & I. DeVore), pp. 199–217. Cambridge, Mass.: Harvard University Press.

Draper, P. 1989. African marriage systems: Perspectives from evolutionary ecology. Ethol. Sociobiol. 10, 145–170.

Draper, P., & Cashdan, E. 1988. Technological change and child behavior among the !Kung. Ethnology 27(4), 339–365.

Draper, P., & Harpending, H. C. 1982. Father absence and reproductive strategy: An evolutionary perspective. J. Anthropol. Res. 38, 255–273.

Drickamer, L. C. 1974. A ten-year summary of reproductive data for free-ranging *Macaca mulatta*. Folia Primatol. 21, 61–80.

Drickamer, L. C. 1982. Acceleration and delay of first vaginal oestrus in female mice by urinary chemosignals: Dose levels and mixing urine treatment sources. Anim. Behav. 30, 456–460.

Drickamer, L. C., & Hoover, J. E. 1979. Effects of urine from pregnant and lactating female house mice on sexual maturation of juvenile females. Dev. Psychobiol. 12, 545–551.

Ducharme, J. R. 1989. Normal puberty: Clinical manifestations and their endocrine control. In: Pediatric Endocrinology (Ed. by R. Collu, J. R. Ducharme, & H. J. Guyda), pp. 307–330. New York: Raven Press.

Dunbar, R. I. M. 1980. Demographic and life history variables of a population of gelada baboons (*Theropithecus gelada*). J. Anim. Ecol. 49, 485–506.

Dunbar, R. I. M. 1984. Infant-use by male gelada in agonistic contexts: Agonistic buffering, progeny protection or soliciting support? Primates 25, 28–35.

Dunbar, R. I. M. 1987. Demography and reproduction. In: Primate Societies (Ed. by B. B. Smuts, D. L. Cheney, R. M. Seyfarth, R. W. Wrangham, & T. T. Struhsaker), pp. 240–249. Chicago: University of Chicago Press.

Dunbar, R. I. M. 1988. Primate Social Systems. Ithaca, N.Y.: Cornell University Press.

Dunbar, R. I. M., & Dunbar, P. 1988. Maternal time budgets of gelada baboons. Anim. Behav. 36, 970–980.

Dunn, E. K. 1972. Effect of age on the fishing ability of sandwich terns *Sterna sandvicensis*. Ibis 114, 360–366.

Dyson, T. 1977. The demography of the Hadza in historical perspective. African Historical Demography, Centre for African Studies, University of Edinburgh.

Eaton, G. G. 1976. The social order of Japanese macaques. Sci. Am. 235(4), 96–106.

Eaton, G. G., Johnson, D. F., Glick, B. B., & Worlein, J. M. 1986. Japanese macaques (*Macaca fuscata*) social development: Sex differences in juvenile behavior. Primates 27, 141–150.

Edwards, C. P., & Whiting, B. B. 1980. Differential socialization of girls and boys in light of cross-cultural research. In: Anthropological Perspectives on Child Development (Ed. by C. Super & S. Harkness), pp. 45–58. San Francisco: Jossey-Bass.

Edwards, S. D. 1982. Social potential expressed in captive, group-living orang utans. In: The Orang Utan, Its Biology and Conservation (Ed. by L. E. M. deBoer), pp. 249–255. The Hague: Junk.

Ehardt-Steward C., & Bramblett, C. A. 1980. The structure of social space among a captive group of vervet monkeys. Folia Primatol. 34, 214–238.

Ehrlich, A. 1974. Infant development in two prosimian species: Greater galago and slow loris. Dev. Psychobiol. 7, 439–454.

Ehrlich, A. 1977. Social and individual behaviors in captive greater galagos. Behaviour 63, 195–214.

Ehrlich, A., & Macbride, L. 1989. Mother–infant interactions in captive slow lorises (*Nycticebus coucang*). Am. J. Primatol. 19, 217–228.

Ehrlich, A., & Macbride, L. 1990. Mother–infant interactions in captive thick-tailed galagos (*Galago garnetti*). J. Mammal. 72, 198–204.

Einon, D., & Potegal, M. 1991. Enhanced defense in adult rats deprived of playfighting experience as juveniles. Aggressive Behav. 17, 27–46.

Eisenberg, J. F. 1976. Communication Mechanisms and Social Integration in the Black Spider Monkey, *Ateles fusciceps robustus* and Related Species (Smithsonian Contributions to Zoology, 213). Washington, D.C.: Smithsonian Institution Press.

Eisenberg, J. F. 1979. Habitat, economy, and society: Some correlations and hypotheses for the neotropical primates. In: Primate Ecology and Human Origins (Ed. by I. S. Bernstein & E. O. Smith), pp. 215–262. New York: Garland STPM Press.

Eisenberg, J. F., & Kuehn, R. E. 1966. The Behavior of *Ateles geoffroyi* and Related Species (Smithsonian Miscellaneous Collections, 151). Washington, D.C.: Smithsonian Institution Press.

Eisenberg, J. F., Muckenhirn, N. A., & Rudran, R. 1972. The relation between ecology and social structure in primates. Science 176, 863–874.

Eklund, J., & Bradford, G. E. 1977. Longevity and lifetime body weight in mice selected for rapid growth. Nature (London) 265, 48–49.

Elardo, R., Solomons, H. C., & Snider, B. C. 1987. An analysis of accidents at a day care center. Am. J. Orthopsychiat. 57, 60–65.

Eley, R. M., Strum, S. C., Muchemi, G., & Reid, G. D. F. 1989. Nutrition, body condition, activity patterns, and parasitism of free-ranging troops of olive baboons (*Papio anubis*) in Kenya. Am. J. Primatol. 18, 209–219.

Ellison, P. T. 1981. Morbidity, mortality, and menarche. Hum. Biol. 53, 735–743.

Ellison, P. T. 1982. Skeletal growth, fatness, and menarcheal age: A comparison of two hypotheses. Ann. Hum. Biol. 54, 269–281.

Ember, C. 1973. Effects of feminine task assignment

on the social behavior of boys. Ethos 1, 424–439.

Emlen, S. T. 1984. Cooperative breeding in birds and mammals. In: Behavioural Ecology (Ed. by J. R. Krebs & N. B. Davies), pp. 305–339. Sunderland, Mass.: Sinauer.

Emlen, S. T., & Oring, L. W. 1977. Ecology, sexual selection, and the evolution of mating systems. Science 197, 215–223.

Estrada, A. 1984. Male–infant interactions among free-ranging stumptail macaques. In: Primate Paternalism (Ed. by D. M. Taub), pp. 56–87. New York: Van Nostrand Reinhold.

Etter, R. J. 1989. Life history variation in the intertidal snail Nucella lapillus across a wave-exposure gradient. Ecology 70, 1857–1876.

Evans, C. S., & Goy, R. W. 1968. Social behaviour and reproductive cycles in captive ringtailed lemurs (Lemur catta). J. Zool. London 156, 181–197.

Eveleth, P. B., & Tanner, J. M. 1976. Worldwide Variation in Human Growth. Cambridge: Cambridge University Press.

Fagen, R. 1980. Ontogeny of animal play behavior: Bimodal age schedules. Anim. Behav. 28, 1290.

Fagen, R. 1981. Animal Play Behavior. New York: Oxford University Press.

Fagen, R. 1984. Play and behavioral flexibility. In: Play in Animals and Humans (Ed. by P. K. Smith), pp. 159–173. Oxford: Blackwell.

Fagen, R. 1986. Play, games, and innovation: Sociobiological findings and unanswered questions. In: Sociobiology and Psychology: Ideas, Issues and Findings (Ed. by C. B. Crawford, M. F. Smith, & D. L. Krebs), pp. 253–268. Hillsdale, N.J.: Erlbaum.

Fagen, R. 1990. Playing with danger and dancing with strangers. Anthrozoös, 4, 4–6.

Fagen, R. 1993. Sociobiological considerations on exploration and play. In: Curiosity and Exploration (Ed. by H. Keller & K. Schneider). New York: Springer. In press.

Fairbanks, L. 1975. Communication of food quality in captive Macaca nemestrina and free-ranging Ateles geoffroyi. Primates 16, 181–190.

Fairbanks, L. A. 1980. Relationships among females in captive vervet monkeys: Testing a model of rank-related attractiveness. Anim. Behav. 28, 853–859.

Fairbanks, L. A. 1988a. Mother–infant behavior in vervet monkeys: Response to failure of last pregnancy. Behav. Ecol. Sociobiol. 23, 157–165.

Fairbanks, L. A. 1988b. Vervet monkey grandmothers: Interactions with infant grandoffspring. Int. J. Primatol. 9, 425–441.

Fairbanks, L. A. 1989. Early experience and cross-generational continuity of mother–infant contact in vervet monkeys. Dev. Psychobiol. 22, 669–681.

Fairbanks, L. A. 1990. Reciprocal benefits of allomothering for female vervet monkeys. Anim. Behav. 40, 553–562.

Fairbanks, L. A., & McGuire, M. T. 1984. Determinants of fecundity and reproductive success in captive vervets. Am. J. Primatol. 7, 27–38.

Fairbanks, L. A., & McGuire, M. T. 1985. Relationships of vervet mothers with sons and daughters from one through three years of age. Anim. Behav. 7, 27–38.

Fairbanks, L. A., & McGuire, M. T. 1986. Age, reproductive value and dominance-related behaviour in vervet monkey females: Cross-generational influences on social relationships and reproduction. Anim. Behav. 34, 1710–1721.

Fairbanks, L. A., & McGuire, M. T. 1988. Long-term effects of early mothering behavior on responsiveness to the environment in vervet monkeys. Dev. Psychobiol. 21, 711–724.

Farrell, S., & Bentley, T. 1990. Holding On to the Air: An Autobiography. New York: Summit Books.

Federer, S. S., & Dawe, H. C. 1964. Near-accidents and minor mishaps in the nursery school. J. Nursery Ed. 19, 188–193.

Fedigan, L. 1972. Social and solitary play in a colony of vervet monkeys (Cercopithecus aethiops). Primates 13, 347–364.

Fedigan, L. 1982. Dominance and reproductive success in primates. Yearbook Phys. Anthropol. 26, 91–129.

Festa-Bianchet, M. 1991. The social system of bighorn sheep: Grouping patterns, kinship and female dominance rank. Anim. Behav. 42, 71–82.

Findlay, C. S., Hansell, R. I. C., & Lumsden, C. J. 1989a. Behavioral evolution and biocultural games: Oblique and horizontal cultural transmission. J. Theor. Biol. 137, 45–269.

Findlay, C. S., Lumsden, C. J., & Hansell, R. I. C. 1989b. Behavioral evolution and biocultural games: Vertical cultural transmission. Proc. Natl. Acad. Sci. U.S.A. 86, 568–572.

Fischer, K. W., & Lazerson, A. 1984. Human Development. New York: Freeman.

Fisher, R. A. 1930. The Genetical Theory of Natural Selection. Oxford: Clarendon Press.

Fitz-Patrick, D. G., & Kimbuna, J. 1983. Bundi: The Culture of a Papua New Guinea People. Nerang, Qld.: Ryebuck.

Fleagle, J. G. 1985. Size and adaptation in primates. In: Size and Scaling in Primate Biology (Ed.

by W. L. Jungers), pp. 1–19. New York: Plenum.

Fleming, T. H., & Rauscher, R. J. 1978. On the evolution of litter size in *Peromyscus leucopus*. Evolution 32, 45–55.

Foerg, R. 1982. Reproduction in *Cheirogaleus medius*. Folia Primatol. 39, 49–62.

Fogden, M. P. L. 1974. A preliminary field study of the western tarsier, *Tarsius bancanus* Horsefield. In: Prosimian Biology (Ed. by R. D. Martin, G. A. Doyle, & A. C. Walker), pp. 151–165. London: Duckworth.

Fogel, A. 1984. Infancy: Infant, Family, and Society. St. Paul, Minn.: West.

Fogel, A., & Melson, G. F. 1986. Origins of Nurturance: Developmental, Biological, and Cultural Perspectives on Caregiving. Hillsdale, N.J.: Erlbaum.

Foighil, D. O. 1985. Form, function, and origin of temporary dwarf males in *Pseudopythina rugifera* (Carpenter, 1864) (Bivalvia: Galeommatacea). Veliger 27, 245–252.

Foley, R. 1982. A reconsideration of the role of predation on large mammals in tropical hunter-gatherer adaptation. Man (NS) 17, 383–402.

Fonseca, G. A. B. 1985. The vanishing Brazilian Atlantic forest. Biol. Conserv. 34, 17–34.

Forest, M. G., Sizonenko, P. C., Cathiard, A. M., & Bertrand, J. 1974. Hypophysogonadal function in human during the first year of life: Evidence for testicular activity in early infancy. J. Clin. Invest. 53, 819–28.

Fossey, D. 1979. Development of the mountain gorilla (*Gorilla gorilla beringei*): The first thirty-six months. In: The Great Apes (Ed. by D. A. Hamburg & E. R. McCown), pp. 139–186. Menlo Park, Calif.: Benjamin-Cummings.

Fossey, D. 1983. Gorillas in the Mist. Boston: Houghton Mifflin.

Fossey, D., & Harcourt, A. H. 1977. Feeding ecology of free-ranging mountain gorillas (*Gorilla gorilla beringei*). In: Primate Ecology: Studies of Feeding and Ranging Behaviour in Lemurs, Monkeys, and Apes (Ed. by T. H. Clutton-Brock), pp. 415–447. London: Academic Press.

Foster, A., Menken, J., Chowdhury, A., & Trussell, J. 1986. Female reproductive development: A hazards model analysis. Social Biol. 33, 183–198.

Fouts, D. H. 1989. Signing interactions between mother and infant chimpanzees. In: Understanding Chimpanzees (Ed. by P. G. Heltne & L. A. Marquardt), pp. 242–251. Cambridge, Mass.: Harvard University Press.

Fragaszy, D. M. 1986. Time budgets and foraging behavior in wedge-capped capuchins (*Cebus olivaceus*): Age and sex differences. In: Current Perspectives in Primate Social Dynamics (Ed. by D. Taub & F. King), pp. 159–174. New York: Van Nostrand Reinhold.

Fragaszy. D. M. 1990. Sex and age differences in the organization of behavior in wedge-capped capuchins. *Cebus olivaceus*. Behav. Ecol. 1, 81–94.

Fragaszy, D. M., Schwartz, S., & Shimosaka, D. 1982. Longitudinal observations of care and development of infant titi monkeys (*Callicebus moloch*). Am. J. Primatol. 2, 191–200.

Franklin, W. L. 1983. Contrasting socioecologies of South America's wild camelids: The vicuña and the guanaco. In: Advances in the Study of Mammalian Behavior (American Society of Mammalogists, Special Publication, 7). (Ed. by J. F. Eisenberg & D. G. Kleiman), pp. 573–629. Stillwater, Okla.: American Society of Mammalogists.

French, J. A., & Inglett, B. J. 1991. Responses to novel social stimuli in callitrichid monkeys: A comparative perspective. In: Primate Responses to Environmental Change (Ed. by H. O. Box), pp. 275–294. London: Chapman & Hall.

Frisch, R. E. 1984. Body fat, puberty, and fertility. Biol. Rev. 59, 161–88.

Fuller, E. M. 1948. Injury-prone children. Am. J. Orthopsychiat. 18, 702–723.

Furuichi, T. 1989. Social interactions and the life history of *Pan paniscus* in Wamba, Zaire. Intl. J. Primatol. 10, 173–189.

Gadgil, M., & Bossert, W. H. 1970. Life historical consequences of natural selection. Am. Nat. 104, 1–24.

Galdikas, B. M. F. 1979. Orangutan adaptation at Tanjung Puting Reserve: Mating behavior and ecology. In: The Great Apes (Ed. by D. A. Hamburg & E. R. McCown), pp. 195–234. Menlo Park, Calif.: Benjamin-Cummings.

Galdikas, B. M. F. 1981. Orangutan reproduction in the wild. In: Reproductive Biology of the Great Apes (Ed. by C. E. Graham), pp. 281–300. New York: Academic Press.

Galdikas, B. M. F. 1982. Orangutan tool-use at Tanjung Puting Reserve, Central Indonesian Borneo (Kalimantan Tengah). J. Hum. Evol. 10, 19–33.

Galdikas, B. M. F. 1984. Adult female sociality among wild orangutans at Tanjung Puting Reserve. In: Female Primates: Studies by Women Primatologists (Ed. by M. F. Small), pp. 217–235. New York: Liss.

Galdikas, B. M. F. 1985a. Subadult male orangutan sociality and reproductive behavior at Tanjung Puting. Am. J. Primatol. 8, 87–99.

Galdikas, B. M. F. 1985b. Orangutan sociality at Tanjung Puting. Am. J. Primatol. 9, 101–119.

Galdikas, B. M. F. 1985c. Adult male sociality and reproductive tactics among orangutans at Tanjung Puting. Folia Primatol. 45, 9–24.

Galdikas, B. M. F. 1988. Orangutan diet, range, and activity at Tanjung Puting, Central Borneo. Int. J. Primatol. 9, 1–35.

Galdikas, B. M. F. 1990. My life with orangutans. Int. Wildlife 20(2), 34–41.

Galdikas, B. M. F., & Teleki, G. 1981. Variations in subsistence activities of female and male pongids: New perspectives on the origins of hominid labor division. Curr. Anthropol. 22, 221–246.

Galdikas, B. M. F., & Wood, J. W. 1990. Birth spacing patterns in humans and apes. Am. J. Phys. Anthropol. 83, 185–192.

Galef, B. G. 1976. Social transmission of acquired behavior: A discussion of tradition and social learning in vertebrates. In: Advances in the Study of Behavior (Ed. by J. S. Rosenblatt, R. A. Hinde, E. Shaw, & C. Beer), pp. 77–100. New York: Academic Press.

Galef, B. G. 1981. The ecology of weaning: Parasitism and the achievement of independence by altricial mammals. In: Parental Care in Mammals (Ed. by D. J. Gubernik & P. H. Klopfer), pp. 211–241. New York: Plenum.

Galef, B. G. 1988. Imitation in animals: History, definition and interpretation of data from the psychological laboratory. In: Social Learning (Ed. by T. R. Zentall & B. G. Galef), pp. 3–28. Hillsdale, N.J.: Erlbaum.

Ganzhorn, J. U. 1988. Food partitioning among Malagasy primates. Oecologia 75, 436–450.

Ganzhorn, J. U. 1989. Niche separation of seven lemur species in the eastern rainforest of Madagascar. Oecologia 79, 279–286.

Gardner, R. A., & Gardner, B. T. 1989. A crossfostering laboratory. In: Teaching Sign Language to Chimpanzees (Ed. by R. A. Gardner, B. T. Gardner, & T. E. Van Cantfort), pp. 1–28. Albany: State University of New York Press.

Garfield, E. 1983. Child safety. Current Contents AB&ES 14(48), 5–12.

Garrettson, L. K., & Gallagher, S. S. 1985. Falls in children and youth. Pediat. Clin. N. Am. 32, 153–162.

Gauthier-Hion, A. 1971. Répertoire comportemental du Talapoin (Miopithecus talapoin). Biol. Gabon. 7, 295–391.

Gavan, J. A. 1982. Adolescent growth in nonhuman primates: An introduction. Hum. Biol. 54, 1–5.

Geist, V. 1971. Mountain Sheep. Chicago: University of Chicago Press.

Ghiglieri, M. 1987. Sociobiology of the great apes and the hominid ancestor. J. Hum. Evol. 16, 319–357.

Ghiselin, M. T. 1974. The Economy of Nature and the Evolution of Sex. Berkeley: University of California Press.

Ghiselin, M. T. 1982. On the evolution of play by means of artificial selection. Behav. Brain Sci. 5, 165.

Gibbons, J. W., & Semlitsch, R. D. 1982. Survivorship and longevity of a long-lived vertebrate species: How long do turtles live? J. Anim. Ecol. 51, 523–527.

Gibson, R. S., Heywood, A., Yaman, C., Sohlström, A., Thompson, L. U., & Heywood, P. 1991. Growth in children from the Wosera subdistrict, Papua New Guinea, in relation to energy and protein intakes and zinc status. Am. J. Clin. Nutr. 53, 782–789.

Gilmore, H. 1981. From Radcliffe-Brown to sociobiology: Some aspects of the rise of primatology within physical anthropology. Am. J. Phys. Anthropol. 56, 387–392.

Gilpin, M. E. 1987. Spatial structure and population vulnerability. In: Viable Populations for Conservation (Ed. by M. E. Soulé), pp. 125–139. Cambridge: Cambridge University Press.

Gittins, S. P. 1982. Feeding and ranging in the agile gibbon. Folia Primatol. 38, 39–71.

Gittleman, J. L. 1986. Carnivore life history patterns: Allometric, phylogenetic, and ecological associations. Am. Nat. 127, 744–771.

Glander, K. E. 1980. Reproduction and population growth in free-ranging mantled howling monkeys. Am. J. Phys. Anthropol. 53, 25–36.

Glander, K. E. 1981. Feeding patterns in mantled howling monkeys. In: Foraging Behaviour: Ecological, Ethological, and Psychological Approaches (Ed. by A. Kamil & T. D. Sargent), pp. 231–259. New York: Garland Press.

Glander, K. E., & Rabin, D. P. 1983. Food choice from endemic North Carolina tree species by captive prosimians (Lemur fulvus). Am. J. Primatol. 5, 221–229.

Glick, B. B., Eaton, G. G., Johnson, D. F., & Worlein, J. M. 1986. Development of partner preferences in Japanese macaques (Macaca fuscata): Effects of gender and kinship during the second year of life. Int. J. Primatol. 7, 467–479.

Goldfoot, D. A., & Neff, D. A. 1985. On measuring behavioral sex differences in social contexts. Handbook Behav. Neurobiol. 7, 767–783.

Goldizen, A. W. 1987. Tamarins and marmosets: Communal care of offspring. In: Primate Societies (Ed. by B. B. Smuts, D. L. Cheney, R. M. Seyfarth, R. W. Wrangham, & T. T. Struhsaker), pp. 34–43. Chicago: University of Chicago Press.

Goldstein, H. 1971. Factors influencing the height of

seven-year-old children: Results from the National Child Development Study. Hum. Biol. 43, 553–68.

Gomendio, M. 1988. The development of different types of play in gazelles: Implication for the nature and functions of play. Anim. Behav. 36, 825–836.

Goodall, J. 1968. The behaviour of free-living chimpanzees in the Gombe Stream Reserve. Anim. Behav. Monogr. 1, 165–311.

Goodall, J. 1971. In the Shadow of Man. London: Collins.

Goodall, J. 1973. Cultural elements in a chimpanzee community. In: Precultural Primate Behavior (Ed. by E. W. Menzel), pp. 144–184. Basel: Karger.

Goodall, J. 1983. Population dynamics during a 15-year period in one community of free-living chimpanzees in Gombe National Park, Tanzania. Z. Tierpsychol. 61, 1–60.

Goodall, J. 1986. The Chimpanzees of Gombe: Patterns of Behavior. Cambridge, Mass.: Harvard University Press.

Goodall, J. 1989a. Gombe: Highlights of current research. In: Understanding Chimpanzees (Ed. by P. G. Heltne & L. A. Marquardt), pp. 2–21. Cambridge, Mass.: Harvard University Press.

Goodall, J. 1989b. The Chimpanzee Family Book. Saxonville, Mass.: Picture Book Studio.

Goodall, J. 1990. Through a Window: My Thirty Years with the Chimpanzees of Gombe. Boston: Houghton Mifflin.

Goodall, J., Bandora, A., Bergmann, E., Busse, C., Matama, H., Mpongo, E., Pierce, A., & Riss, D. 1979. Intercommunity interactions in the chimpanzee population of the Gombe National Park. In: The Great Apes (Ed. by D. Hamburg & E. R. McCown), pp. 13–54. Menlo Park, Calif.: Benjamin-Cummings.

Goodman, D. 1974. Natural selection and a cost ceiling on reproductive effort. Am. Nat. 108, 247–268.

Gould, L. 1990. The social development of free-ranging infant Lemur catta at Berenty Reserve, Madagascar. Intl. J. Primatol. 11, 297–318.

Gould, S. J. 1977. Ontogeny and Phylogeny. Cambridge, Mass.: Harvard University Press.

Gouzoules, H. 1975. Maternal rank and early social interactions of infant stumptail macaques, Macaca arctoides. Primates 16, 405–418.

Gouzoules, H. 1984. Social relations of males and infants in a troop of Japanese macaques: A consideration of causal mechanisms. In: Primate Paternalism (Ed. by D. M. Taub), pp. 127–145. New York: Van Nostrand Reinhold.

Gouzoules, H., Gouzoules, S., & Fedigan, L. 1982. Behavioural dominance and reproductive success in female Japanese monkeys (Macaca fuscata). Anim. Behav. 30, 1138–1151.

Goy, R. W., Bercovitch, F. B., & McBrair, M. C. 1988. Behavioral masculinization is independent of genital masculinization in prenatally androgenized female rhesus macaques. Horm. Behav. 22, 552–571.

Goy, R. W., & McEwen, B. S. 1980. Sexual differentiation of the brain. Cambridge, Mass.: MIT Press.

Goy, R. W., & Resko, J. A. 1972. Gonadal hormones and behavior of normal and pseudohermaphroditic nonhuman female primates. Rec. Prog. Horm. Res. 28, 707–733.

Gratz, R. R. 1979. Accidental injury in childhood: A literature review on pediatric trauma. J. Trauma 19, 551–555.

Green, R. J. 1989. Play behaviour of chimpanzees at Taronga Zoo. Australian Primatol. 4, 24.

Greensher, J., & Mofenson, H. C. 1985. Injuries at play. Pediat. Clin. N. Am. 32, 127–140.

Greenwood, P. J. 1980. Mating systems, philopatry, and dispersal in birds and mammals. Anim. Behav. 28, 1140–1162.

Gross, M. R. 1984. Sunfish, salmon, and the evolution of alternative reproductive strategies. In: Fish Reproduction: Strategies and Tactics (Ed. by G. W. Potts & R. J. Wootton), pp. 55–75. New York: Academic Press.

Guilford, T. 1990. The secrets of aposematism: Unlearned responses to specific colors and patterns. Trends Ecol. Evol. 5, 323.

Gustaffson, L., & Part, T. 1990. Acceleration of senescence in the collared flycatcher (Ficedula albicollis) by reproductive costs. Nature (London) 347, 279–281.

Gustaffson, L., & Sutherland, W. J. 1988. The costs of reproduction in the collared flycatcher Ficedula albicollis. Nature (London) 335, 813–815.

Guyer, B., & Gallagher, S. S. 1985. An approach to the epidemiology of childhood injuries. Pediat. Clin. N. Am. 32, 5–16.

Haldane, J. B. S. 1932. The Causes of Evolution. London: Longmans, Green.

Hall, K. R. L. 1965. Experiment and quantification in the study of baboon behavior in its natural habitat. In: The Baboon in Medical Research (Ed. by H. Vagtborg), pp. 29–42. San Antonio: University of Texas Press.

Hall, K. R. L., & DeVore, I. 1965. Baboon social behavior. In: Primate Behavior (Ed. by I. DeVore), pp. 53–110. New York: Holt, Rinehart & Winston.

Halperin, S. D. 1979. Temporary association patterns in free ranging chimpanzees: an assessment of individual grouping preferences. In: The Great Apes (Ed. by D. A. Hamburg & E. R.

McCown), pp. 491–499. Menlo Park, Calif.: Benjamin-Cummings.

Hamilton, W. D. 1964. The genetical evolution of social behavior. I and II. J. Theor. Biol. 7, 1–52.

Hamilton, W. J., III. 1984. Significance of paternal investment by primates to the evolution of male–female associations. In: Primate Paternalism (Ed. by D. M. Taub), pp. 309–335. New York: Van Nostrand Reinhold.

Hamilton, W. J. 1985. Demographic consequences of a food and water shortage to desert chacma baboons, *Papio ursinus*. Int. J. Primatol. 6, 451–462.

Hamilton, W. J., III, & Bulger, J. B. 1990. Natal male baboon rank rises and successful challenges to resident alpha males. Behav. Ecol. Sociobiol. 26, 357–362.

Handen, C. E., & Rodman, P. S. 1980. Social development of bonnet macaques from six months to three years of age: A longitudinal study. Primates 21, 350–356.

Haraway, D. 1989. Primate Visions. New York: Routledge.

Harcourt, A. H. 1978. Strategies of emigration and transfer by gorillas, with particular reference to gorillas. Z. Tierpsychol. 48, 401–420.

Harcourt, A. H. 1979a. Social relationships among adult female mountain gorilla. Anim. Behav. 27, 251–264.

Harcourt, A. H. 1979b. Social relationships among adult male and female mountain gorillas in the wild. Anim. Behav. 27, 325–342.

Harcourt, A. H. 1981. Intermale competition and the reproductive behavior of the great apes. In: Reproduction Biology of the Great Apes (Ed. by C. E. Graham), pp. 301–318. New York: Academic Press.

Harcourt, A. H. 1987. Dominance and fertility among female primates. J. Zool. London 213, 471–487.

Harcourt, A. H. 1989. Social influences on competitive ability: Alliances and their consequences. In: Comparative Socioecology (Ed. by V. Standen & R. A. Foley), pp. 223–242. Oxford: Blackwell.

Harcourt, A. H., & de Waal, F. B. M. (Eds.). 1992. Coalitions and Alliances in Humans and other Animals. New York: Oxford University Press.

Harcourt, A. H., & Stewart, K. J. 1981. Gorilla male relationships: Can differences during immaturity lead to contrasting reproductive tactics in adulthood? Anim. Behav. 29, 206–210.

Harcourt, A. H., & Stewart, K. J. 1987. The influence of help in contests on dominance rank in primates: Hints from gorillas. Anim. Behav. 35, 182–190.

Harcourt, A. H., & Stewart, K. J. 1989. Functions of alliances in contests within wild gorilla groups. Behaviour 109, 176–190.

Harcourt, A. H., Stewart, K. J., & Fossey, D. 1981. Gorilla reproduction in the wild. In: Reproductive Biology of the Great Apes (Ed. by C. E. Graham), pp. 265–279. New York: Academic Press.

Harcourt, C. S. 1981. An examination of the function of urine washing in *Galago senegalensis*. Z. Tierpsychol. 55, 119–128.

Harcourt, C. S. 1986. Seasonal variation in the diet of South African galagos. Int. J. Primatol. 7, 491–506.

Harcourt, C. S., & Bearder, S. K. 1989. A comparison of *Galago moholi* in South Africa with *Galago zanzibaricus* in Kenya. Int. J. Primatol. 10, 35–46.

Harcourt, C. S., & Nash, L. T. 1986. Social organization of galagos in Kenyan coastal forests, I. *Galago zanzibaricus*. Am. J. Primatol. 10, 339–355.

Harkness, S. 1975. Child language socialization in a Kipsigis community of Kenya. Ph.D. dissertation, Harvard University.

Harlow, H. F. 1959. Love in infant monkeys. Sci. Am. 200, 68–74.

Harpending, H. C., & Draper, P. 1990. Comment on Solway & Lee. Curr. Anthropol. 31, 127–129.

Harpending, H. C., & Draper, P. 1992. Estimating parity of parents: Application to the history of infertility among the !Kung of southern Africa. Hum. Biol. 62, 195–203.

Harrison, M. J. S. 1983. Age and sex differences in the diet and feeding strategies of the green monkey, *Cercopithecus sabaeus*. Anim. Behav. 31, 969–977.

Hartnoll, R. G. 1982. Growth. In: The Biology of Crustacea, Vol. 2 (Ed. by L. G. Abele), pp. 111–196. New York: Academic Press.

Hartung, J. 1976. On natural selection and the inheritance of wealth. Curr. Anthropol. 17, 607–622.

Harvey, P. H. 1990. Life-history variation: Size and mortality patterns. In: Primate Life History and Evolution (Ed. by C. J. DeRousseau), pp. 81–88. New York: Wiley-Liss.

Harvey, P. H., & Clutton-Brock, T. H. 1985. Life history variation in primates. Evolution 39, 559–581.

Harvey, P. H., Martin, R. D., & Clutton-Brock, T. H. 1987. Life histories in comparative perspective. In: Primate Societies (Ed. by B. B. Smuts, D. L. Cheney, R. M. Seyfarth, R. W. Wrangham, & T. T. Struhsaker), pp. 181–196. Chicago: University of Chicago Press.

Harvey, P. H., Read, A. F., & Promislow, D. E. L.

1989. Life history variation in placental mammals: Unifying the data with theory. Oxford Surv. Evol. Biol. 6, 13–31.

Harvey, P. H., & Zammuto, R. M. 1985. Patterns of mortality and age at first reproduction in natural populations of mammals. Nature (London) 315, 318–329.

Hasegawa, T. 1990. Sex differences in ranging patterns. In: The Chimpanzees of Mahale (Ed. by T. Nishida), pp. 99–114. Tokyo: University of Tokyo Press.

Hasegawa, T., & Hiraiwa, M. 1980. Social interactions of orphans observed in a free-ranging troop of Japanese monkeys. Folia Primatol. 33, 129–158.

Hasegawa, T., & Hiraiwa-Hasegawa, M. 1983. Opportunistic and restrictive matings among wild chimpanzees in the Mahale Mountains, Tanzania. J. Ethol. 1, 75–85.

Hasegawa, T., & Hiraiwa-Hasegawa, M. 1990. Sperm competition and mating behavior. In: The Chimpanzees of Mahale (Ed. by T. Nishida), pp. 115–132. Tokyo: University of Tokyo Press.

Haskin, D. R. 1989. Management of imaginary play in a chimpanzee. Primatologist 4, 2–4.

Hauser, M. D. 1986. Parent–offspring conflict: Care elicitation behaviour and the "cry-wolf" syndrome. In: Primate Ontogeny, Cognition, and Social Behavior (Ed. by J. G. Else & P. G. Lee), pp. 193–203. Cambridge: Cambridge University Press.

Hauser, M. D. 1987. The behavioral ecology of free-ranging vervet monkeys: Proximate and ultimate levels of explanation. Ph.D. dissertation, University of California, Los Angeles.

Hauser, M. D. 1988. Invention and social transmission: New data from wild vervet monkeys. In: Machiavellian Intelligence (Ed. by R. W. Byrne & A. Whiten), pp. 327–343. Oxford: Clarendon Press.

Hausfater, G. 1972. Intergroup behavior of free-ranging rhesus monkeys (Macaca mulatta). Folia Primatol. 18, 78–107.

Hausfater, G. 1975. Dominance and reproduction in baboons (Papio cynocephalus): A quantitative analysis. Contrib. Primatol. 7, 1–150.

Hausfater, G. 1976. Predatory behavior of yellow baboons. Behaviour 56, 44–68.

Hausfater, G., Altmann, J., & Altmann, S. A. 1982. Long-term consistency of dominance relations among female baboons (Papio cynocephalus). Science 217, 752–755.

Hawkes, K., & O'Connell, J. F. 1982. Affluent hunters? Some comments in light of the Alyawara case. Am. Anthropol. 83, 622–626.

Hawkes, K., O'Connell, J. F., & Blurton Jones, N. G. 1989. Hardworking Hadza grand-

mothers. In: Comparative Socioecology (Ed. by V. Standen & R. Foley), pp. 341–366. Oxford: Blackwell.

Hawkes, K., O'Connell, J. F., & Blurton Jones, N. G. 1991. Hunting income patterns among the Hadza: Big game, common goods, foraging goals, and the evolution of the human diet. Phil. Trans. Roy. Soc. 334, 243–251.

Hayaki, H. 1983. The social interactions of juvenile Japanese monkeys on Koshima islet. Primates 24, 139–153.

Hayaki, H. 1985a. Social play of juvenile and adolescent chimpanzees in the Mahale Mountains National Park, Tanzania. Primates 26, 342–360.

Hayaki, H. 1985b. Copulation of adolescent male chimpanzees with special reference to the influence of adult males in the Mahale Mountains National Park, Tanzania. Folia Primatol. 44, 148–160.

Hayaki, H. 1988. Association patterns of young chimpanzees in the Mahale Mountain National Park, Tanzania. Primates 29, 147–161.

Henrich, B. 1989. Ravens in Winter. New York: Summit.

Hendy, H. 1986. Social interactions of free-ranging baboon infants. In: Primate Ontogeny, Cognition and Social Behavior (Ed. by J. G. Else & P. C. Lee), pp. 267–280. New York: Cambridge University Press.

Hertzog, J. D. 1974. The socialization of juveniles in primate and foraging societies: Implications for contemporary education. Anthropol. Educ. Quart. 5, 12–17.

Hewlett, B. S. 1988. Demography and child caretaking practices of tropical forest hunter-gatherers. Paper presented at the International Conference on Hunter-Gatherer Peoples, Darwin, Australia.

Hiernaux, J., & Hartono, C. B. 1980. Physical measurements of the adult Hadza of Tanzania. Ann. Hum. Biol. 7, 339–346.

Hill, D. A. 1986. Social relationships between adult male and immature rhesus macaques. Primates 27, 425–440.

Hill, K., & Kaplan, H. 1988. Tradeoffs in male and female reproductive strategies among the Ache. Parts 1 and 2. In: Human Reproductive Behavior: A Darwinian Perspective (Ed. by L. Betzig, M. Borgerhoff Mulder, & P. Turke), pp. 277–289, 291–305. Cambridge: Cambridge University Press.

Hill, W. C. O. 1936. Supplementary observations on the purple-faced leaf monkey (genus Kasi). Ceylon J. Sci. B. 20, 115–133.

Hinde, R. A. 1969. Analyzing the roles of the partners in a behavioral interaction—mother–infant relations in rhesus macaques. Ann. N.Y. Acad. Sci. 159, 651–667.

Hinde, R. A. 1974. Biological Bases of Human Social Behaviour. New York: McGraw-Hill.

Hinde, R. A. 1982. Ethology. New York: Oxford University Press.

Hinde, R. A. 1983. Primate Social Relationships. Sunderland, Mass.: Sinauer.

Hinde, R. A. 1985. Ethology in relation to other disciplines. In: Leaders in the Study of Animal Behavior (Ed. by D. A. Dewsbury), pp. 193–203. Lewisburg, Pa.: Bucknell University Press.

Hinde, R. A. 1987. Can non-human primates help us understand human behavior? In: Primate Societies (Ed. by B. B. Smuts, D. L. Cheney, R. M. Seyfarth, R. W. Wrangham, & T. T. Struhsaker), pp. 413–420. Chicago: University of Chicago Press.

Hinde, R. A., & Atkinson, S. 1970. Assessing the roles of social partners in maintaining mutual proximity, as exemplified by mother–infant relations in rhesus monkeys. Anim. Behav. 18, 169–176.

Hinde, R. A., & Rowell, T. E. 1962. Communication by postures and facial expressions in the rhesus monkey (Macaca mulatta). Proc. Zool. Soc. London 138, 1–21.

Hinde, R. A., & Spencer-Booth, Y. 1967. The behaviour of socially living rhesus monkeys in their first two and a half years. Anim. Behav. 15, 169–196.

Hiraiwa-Hasegawa, M. 1989. Sex differences in the behavioral development of chimpanzees at Mahale. In: Understanding Chimpanzees (Ed. by P. G. Heltne & L. A. Marquardt), pp. 104–115. Cambridge, Mass.: Harvard University Press.

Hladik, C. M., & Charles-Dominique, P. 1974. The behaviour and ecology of the sportive lemur (Lepilemur mustelinus) in relation to its dietary peculiarities. In: Prosimian Biology (Ed. by R. D. Martin, G. A. Doyle, & A. C. Walker), pp. 23–37. London: Duckworth.

Hladik, C. M., Charles-Dominique, P., & Petter, J. J. 1980. Feeding strategies of five nocturnal prosimians in the dry forest of the west coast of Madagascar. In: Nocturnal Malagasy Primates (Ed. by P. Charles-Dominique et al.), pp. 41–74. New York: Academic Press.

Hobson, W., Winter, J. S. D., Reyes, F. I., Fuller, G. B., & Faiman, C. 1980. Nonhuman primates as models for studies on puberty. In: Animal Models in Human Reproduction (Ed. by M. Serio & L. Martini), pp. 409–421. New York: Raven Press.

Hobson, W. C., Fuller, G. B., Winter, J. S. D., Faiman, C., & Reyes, F. I. 1981. Reproductive and endocrine development in the great apes. In: Reproductive Biology of the Great Apes: Comparative and Biomedical Perspectives (Ed. by C. E. Graham), pp. 83–103. New York: Academic Press.

Hofer, M. 1987. Early social relationships: A psychobiologist's view. Child Dev. 58, 633–647.

Hoffman, M. L. 1977. Sex differences in empathy and related behaviors. Psychol. Bull. 84(4), 712–722.

Holliday, M. A. 1986. Body composition and energy needs during growth. In: Human Growth, 2nd ed., Vol. 2 (Ed. by F. Falkner & J. M. Tanner), pp. 101–145. New York: Plenum.

Holman, S. D., & Goy, R. W. 1988. Sexually dimorphic transitions revealed in the relationships of yearling rhesus monkeys following the birth of siblings. Int. J. Primatol. 9, 113–133.

Holt, E. B. 1931. Animal Drive and the Learning Process. London: Williams & Norgate.

Honjo, S., Cho, F., & Terao, K. 1984. Establishing the cynomolgus monkey as a lab animal. Adv. Vet. Sci. Comp. Med. 28, 51–80.

Horn, H. S. 1978. Optimal tactics of reproduction and life history. In: Behavioural Ecology: An Evolutionary Approach (Ed. by J. R. Krebs & N. B. Davies), pp. 411–429. Sunderland, Mass.: Sinauer.

Horn, H. S., & Rubenstein, D. I. 1984. Behavioural adaptations and life history. In: Behavioural Ecology, 2nd ed. (Ed. by J. R. Krebs & N. B. Davies), pp. 279–290. Oxford: Blackwell..

Horr, D. A. 1977. Orang-utan maturation: Growing up in a female world. In: Primate Bio-Social Development: Biological, Social, and Ecological Determinants (Ed. by S. Chevalier-Skolnikoff & F. E. Poirier), pp. 289–321. New York: Garland Press.

Horrocks, J. A. 1986. Life-history characteristics of a wild population of vervets (Cercopithecus aethiops sabaeus) in Barbados, West Indies. Int. J. Primatol. 7, 31–47.

Horrocks, J. A., & Hunte, W. 1983a. Maternal rank and offspring rank in vervet monkeys: An appraisal of the mechanisms of rank acquisition. Anim. Behav. 31, 772–782.

Horrocks, J. A., & Hunte, W. 1983b. Rank relations in vervet sisters: A critique of the role of reproductive value. Am. Nat. 122, 417–421.

Horrocks, J. A., & Hunte, W. 1986. Sentinel behaviour in vervet monkeys: Who sees whom first? Anim. Behav. 34, 1566–1568.

Howell, N. 1979. Demography of the Dobe !Kung. New York: Academic Press.

Hrdy, S. B. 1976. Care and exploitation of nonhuman primate infants by conspecifics other than the mother. In: Advances in the Study of Behavior, Vol. 6 (Ed. by J. S. Rosenblatt, R. A. Hinde, E. Shaw, & C. Beer), pp. 101–158. New York: Academic Press.

Hrdy, S. B. 1977. The Langurs of Abu. Cambridge, Mass.: Harvard University Press.

Hrdy, S. B. 1989. Sex-based parental investment among primates and other mammals: A critical evaluation of the Trivers-Willard hypothesis. In: Offspring Abuse and Neglect in Biosocial Perspective (Ed. by R. Gelles & J. Lancaster), pp. 162–203. Hawthorne, N.Y.: Aldine.

Hrdy, S. B. 1990. Sex bias in nature and in history: A late 1980s reexamination of the "biological origins" argument. Yearbook Phys. Anthropol. 33, 25–37.

Huffman, M. A. 1984. Stone-play of Macaca fuscata in Arashiyama B Troop: Transmission of a non-adaptive behavior. J. Human Evol. 13, 725–735.

Humphrey, N. K. 1976. The social function of intellect. In: Growing Points in Ethology (Ed. by P. P. G. Bateson & R. A. Hinde), pp. 303–317. Cambridge: Cambridge University Press.

Hunte, W., & Horrocks, J. A. 1987. Kin and non-kin interventions in the aggressive disputes of vervet monkeys. Behav. Ecol. Sociobiol. 20, 257–263.

Hunziker, U. A., & Barr, R. G. 1986. Increased carrying reduces infant crying: A randomized controlled trial. Pediatrics 77, 641–648.

Hurtado, A. M., & Hill, K. R. 1990. Seasonality in a foraging society: Variation in diet, work effort, fertility, and the sexual division of labor among the Hiwi of Venezuela. J. Anthropol. Res. 46, 293–346.

Hurtado, A. M., & Hill, K. R. 1992. Paternal effect on offspring survivorship among Ache and Hiwi hunter-gatherers: Implications for modeling pair-bond stability. In: Father–Child Relations: Cultural and Biosocial Contexts (Ed. by B. Hewlett), pp. 31–76. Hawthorne, N.Y.: Aldine.

Imakawa, S. 1990. Playmate relationships of immature free-ranging Japanese monkeys at Katsuyama. Primates 31, 509–521.

Immelman, K., Barlow, G. W., Petrinovich, L., & Main, M. (Eds.). 1981. Behavioral Development. New York: Cambridge University Press.

Isbell, L. A. 1990. Sudden short-term increase in mortality of vervet monkeys (Cercopithecus aethiops) due to leopard predation in Amboseli National Park, Kenya. Am. J. Primatol. 21, 41–52.

Isbell, L. A., Cheney, D. L., & Seyfarth, R. M. 1990.

Costs and benefits of home range shifts among vervet monkeys (Cercopithecus aethiops) in Amboseli National Park, Kenya. Behav. Ecol. Sociobiol. 27, 351–358.

Itani, J. 1958. On the acquisition and propagation of a new food habit in the natural group of Japanese monkeys at Takasakiyama. J. Primatol. 1, 84–98.

Itani, J. 1959. Paternal care in the wild Japanese monkey Macaca fuscata fuscata. Primates 2, 61–93.

Iwano, T., & Iwakawa, C. 1988. Feeding behaviour of the aye-aye (Daubentonia madagascariensis) on nuts of ramy (Canarium madagascariensis). Folia Primatol. 50, 136–142.

Izard, M. K. 1987. Lactation length in three species of Galago. Am. J. Primatol. 13, 73–76.

Izard, M. K., & Nash, L. T. 1988. Contrasting reproductive parameters in Galago senegalensis braccatus and G. s. moholi. Int. J. Primatol. 9, 519–527.

Izard, M. K., & Rasmussen, D. T. 1985. Reproduction in the slender loris (Loris tardigradus malabaricus). Am. J. Primatol. 8, 153–165.

Izard, M. K., & Simons, E. L. 1986. Isolation of females prior to parturition reduces neonatal mortality in Galago. Am. J. Primatol. 10, 249–255.

Izard, M. K., Weisenseel, K. A., & Ange, R. L. 1988. Reproduction in the slow loris (Nycticebus coucang). Am. J. Primatol. 16, 331–339.

Jacklin, C. N. 1989. Female and male: Issues of gender. Am. Psychol. 44, 127–133.

Jain, A. K. 1969. Fecundability and its relationship to age in a sample of Taiwanese women. Pop. Studies 32, 169–185.

Janson, C. H. 1984. Female choice and mating system of the brown capuchin monkey Cebus apella (Primates: Cebidae). Z. Tierpsychol. 65, 177–200.

Janson, C. H. 1985a. Aggressive competition and individual food consumption in wild brown capuchin monkeys (Cebus apella). Behav. Ecol. Sociobiol. 18, 125–138.

Janson, C. H. 1985b. Social and ecological consequences of food competition in brown capuchin monkeys. Ph.D. dissertation, University of Washington.

Janson, C. H. 1986. The mating system as a determinant of social evolution in capuchin monkeys (Cebus). In: Primate Ecology and Conservation (Ed. by J. Else & P. Lee), pp. 169–179. Cambridge: Cambridge University Press.

Janson, C. H. 1988. Food competition in brown capuchin monkeys (Cebus apella): quantitative effects of group size and tree productivity. Behaviour 105, 53–76.

Janson, C. H. 1990a. Social correlates of individual spatial choice in foraging groups of brown capuchin monkeys, *Cebus apella.* Anim. Behav. 40, 910–921.

Janson, C. H. 1990b. Ecological consequences of individual spatial choice in foraging groups of brown capuchin monkeys, *Cebus apella.* Anim. Behav. 40, 922–934.

Janson, C. H., & Boinski, S. 1992. Morphological versus behavioral adaptations for foraging in generalist primates: The case of the cebines. Am. J. Phys. Anthropol, 88, 483–498.

Janson, C. H., & Terborgh, J. W. 1979. Age, sex, and individual specialization in foraging behavior of the brown capuchin (*Cebus apella*). Am. J. Phys. Anthropol. 50, 452.

Janus, M. 1989. Reciprocity in play, grooming, and proximity in sibling and nonsibling young rhesus monkeys. Int. J. Primatol. 10, 243–261.

Jay, P. C. 1963a. Mother–infant relations in langurs. In: Maternal Behavior in Mammals (Ed. by H. L. Rheingold), pp. 282–304. New York: Wiley.

Jay, P. 1963b. The social behavior of the langur monkey. Ph.D. dissertation, University of Chicago.

Jerison, H. J. 1973. Evolution of the Brain and Intelligence. New York: Academic Press.

Jewell, P. A., & Oates, J. F. 1969. Ecological observations on the lorisoid primates of African lowland forest. Zool. Africana 4, 231–248.

Johnson, C. J., Koerner, C. Estrin, M., & Duoos, D. 1980. Alloparental care and kinship in captive social groups of vervet monkeys (*Cercopithecus aethiops sabaeus*). Primates 21, 406–415.

Johnson, J. A. 1987. Dominance rank in juvenile olive baboons, *Papio anubis:* The influence of gender, size, maternal rank and orphaning. Anim. Behav. 35, 1694–1708.

Johnson, J. A. 1989. Supplanting by olive baboons: Dominance rank difference and resource value. Behav. Ecol. Sociobiol. 24, 277–283.

Johnston, T. D. 1982. Selective costs and benefits in the evolution of learning. Adv. Study Behav. 12, 65–106.

Johnston, T. D. 1987. The persistence of dichotomies in the study of behavioral development. Dev. Rev. 7, 149–182.

Johnston, T. D. 1988. Developmental explanation and the ontogeny of birdsong: Nature/nurture redux. Behav. Brain Sci. 11, 617–663.

Jolly, A. 1966. Lemur Behavior: A Madagascar Field Study. Chicago: University of Chicago Press.

Jolly, A. 1967. Breeding synchrony in wild Lemur

catta. In: Social Communication Among Primates (Ed. by S. A. Altmann), pp. 3–13. Chicago: University of Chicago Press.

Jolly, A. 1984. The puzzle of female feeding priority. In: Female Primates: Studies by Women Primatologists (Ed. by M. F. Small), pp. 197–215. New York: Liss.

Jolly, A. 1985. The Evolution of Primate Behavior. New York: Macmillan.

Jolly, A. 1991. Conscious chimpanzees? A review of recent literature. In: Cognitive Ethology: The Minds of Other Animals (Ed. by C. A. Ristau), pp. 231–252. Hillsdale, N.J.: Erlbaum.

Jolly, A., Gustafson, H., Oliver, W. L. R., & O'Connor, S. M. 1982. Population and troop ranges of *Lemur catta* and *Lemur fulvus* at Berenty, Madagascar: 1980 census. Folia Primatol. 40, 145–160.

Jones, K. C. 1983. Inter-troop transfer of *Lemur catta* males at Berenty, Madagascar. Folia Primatol. 40, 145–160.

Jonsson, B., & Hindar, K. 1982. Reproductive strategy of dwarf and normal arctic charr (*Salvelinus alpinus*) from Vangsvatnet Lake, Western Norway. Can. J. Fish. Aquat. Sci. 39, 1404–1413.

Judge, P. G. 1983. Reconciliation based on kinship in a captive group of pigtail macaques. Am. J. Primatol. 4, 346.

Judge, P. G. 1991. Dyadic and triadic reconciliation in pigtail macaques (*Macaca nemestrina*). Am. J. Primatol. 23, 225–237.

Jungers, W. L., & Susman, R. L. 1984. Body size and skeletal allometry in African apes. In: The Pygmy Chimpanzee (Ed. by R. L. Susman), pp. 131–177. New York: Plenum.

Kano, T. 1980. Social behavior of wild pygmy chimpanzees (*Pan paniscus*) of Wamba: A preliminary report. J. Hum. Evol. 9, 243–260.

Kano, T. 1989. The sexual behavior of pygmy chimpanzees. In: Understanding Chimpanzees (Ed. by P. G. Heltne & L. A. Marquardt), pp. 176–183. Cambridge, Mass.: Harvard University press.

Kano, T. 1990. The bonobos' peaceable kingdom. Nat. Hist. 11, 62–71.

Kaplan, H. 1993. Two theories of fertility: Empirical tests and directions for further theory development. In preparation.

Kaplan, H., & Dove, H. 1987. Infant development among the Ache of eastern Paraguay. Dev. Psychol. 23, 190–198.

Kaplan, J. 1977. Fight interference in free-ranging rhesus monkeys. Am. J. Phys. Anthropol. 47, 279–288.

Kaplan, J. R. 1987. Dominance and affiliation in the Cercopithecini and Papionini: A comparative examination. In: Comparative Behavior of African Monkeys (Ed. by E. L. Zucker), pp. 127–150. New York: Liss.

Kappeler, P. M. 1990a. Social status and scent-marking behaviour in Lemur catta. Anim. Behav. 40, 774–776.

Kappeler, P. M. 1990b. The evolution of sexual size dimorphism in prosimian primates. Am. J. Primatol. 21, 201–214.

Karssemeijer, G. J., Vos, D. R., & van Hoof, J. A. R. A. M. 1990. The effect of some non-social factors on mother–infant contact-time in long-tailed macaques (Macaca fascicularis). Behaviour 113, 273–291.

Katz, M. M., & Konner, M. J. 1981. The role of the father: An anthropological perspective. In: The Role of the Father in Child Development, 2nd ed. (Ed. by M. E. Lamb), pp. 155–186. New York: Wiley.

Katz, S. H., Hediger, M. L., Zemel, B. S., & Parks, J. S. 1985. Adrenal androgens, body fat and advanced skeletal age in puberty: New evidence for the relations of adrenarche and gonadarche in males. Hum. Biol. 57, 401–413.

Kaufmann, J. H. 1967. Social relations of adult males in a free-ranging band of rhesus monkeys. In: Social Communication Among Primates (Ed. by S. A. Altmann), pp. 73–98. Chicago: University of Chicago Press.

Kawai, M. 1958. On the system of social ranks in a natural troop of Japanese monkeys (1): Basic rank and dependent rank. Primates 1, 111–148. In: Japanese Monkeys, a Collection of Translations (Ed. by K. Imanishi & S. A. Altmann), pp. 66–86, Atlanta: Emory University Press, 1965.

Kawai, M. 1965. Newly acquired pre-cultural behavior of the natural troop of Japanese monkeys on Koshima Islet. Primates 6, 1–30.

Kawamura, S. 1958. Matriarchal social ranks in the Minoo-G troop: A study of the rank system of Japanese monkeys. Primates 1, 148–156. In: Japanese Monkeys, a Collection of Translations (Ed. by K. Imanishi & S. Altmann), pp. 105–112. Atlanta: Emory University Press, 1965.

Kawamura, S. 1959. The process of sub-culture propagation among Japanese macaques. Primates 2, 43–60.

Kawamura, S. 1965. Matriarchal social ranks in the Minoo-B troop: A study of the rank system of Japanese monkeys. In: Japanese Monkeys, a Collection of Translations (Ed. by K. Imanishi & S. A. Altmann), pp. 105–112. Atlanta: Emory University Press.

Kawanaka, K. 1984. Association, ranging, and the social unit in chimpanzees of the Mahale Mountains. Intl. J. Primatol. 5, 411–434.

Kay, R. F., & Scheine, W. S. 1979. On the relationship between chitin particle size and digestibility in the primate Galago senegalensis. Am. J. Phys. Anthropol. 50, 301–308.

Kaye, K. 1982. The Mental and Social Life of Babies. Chicago: University of Chicago Press.

Keddy Hector, A. C., Seyfarth, R. M., & Raleigh, M. J. 1989. Male parental care, female choice and the effect of an audience in vervet monkeys. Anim. Behav. 38, 262–271.

Kerr, G. R., Lozy, M. el, & Scheffler, G. 1975. Malnutrition studies in Macaca mulatta. IV. Energy and protein consumption during growth failure and "catch-up" growth. Am J. Clin. Nutr. 28, 1364–1376.

Keverne, E. B. 1976. Sexual receptivity and attractiveness in the female rhesus monkey. In: Advances in the Study of Behavior (Ed. by J. S. Rosenblatt, R. A. Hinde, E. Shaw, & C. Beer), pp. 155–200. New York: Academic Press.

Kiltie, R. 1988. Gestation as a constraint on the evolution of seasonal breeding in mammals. In: Evolution of Life Histories of Mammals (Ed. by M. S. Boyce), pp. 257–289. New Haven, Conn.: Yale University Press.

Kingdon, J. 1974. East African Mammals, Vol. 1. Chicago: University of Chicago Press.

Kingsley, S. 1982. Causes of non-breeding and the development of the secondary sexual characteristics in the male orangutan: A hormonal study. In: The Orang Utan (Ed. by L. E. M. de Boer), pp. 215–230. The Hague: Junk.

Kingsley, S. R. 1988. Physiological development of male orang-utans and gorillas. In: Orang-utan Biology (Ed. by J. H. Schwartz), pp. 123–132. New York: Oxford University Press.

Kirkevold, B. C., & Crockett, C. M. 1987. Behavioral development and proximity patterns in captive DeBrazza's monkeys. In: Comparative Behavior of African Monkeys (Ed. by E. L. Zucker), pp. 39–65. New York: Liss.

Kirkland, G., & Lawrence, G. 1987. Dancing on My Grave. New York: Doubleday.

Kirkland, G., & Lawrence, G. 1990. The Shape of Love. New York: Doubleday.

Kirkwood, T. B. L., & Rose, M. R. 1991. Evolution of senescence: Late survival sacrificed for reproduction. In: The Evolution of Reproductive Strategies (Ed. by P. H. Harvey, L. Partridge, & T. R. E. Southwood), pp. 15–24. Cambridge: Cambridge University Press.

Kleiber, M. 1961. The Fire of Life. New York: Wiley.

Klein, L. L., & Klein, D. B. 1977. Feeding behaviour of the Colombian spider monkey. In: Primate Ecology: Studies of Feeding and Ranging Behaviour in Lemurs, Monkeys and Apes (Ed.

by T. H. Clutton-Brock), pp. 153–182. London: Academic Press.

Klopfer, P. H. 1972. Patterns of maternal care in lemurs. II. Effects of group size and early separation. Z. Tierpsychol. 30, 277–296.

Klopfer, P. H. 1981. The naked ape reclothed. Am. J. Primatol. 1, 301–305.

Klopfer, P. H., & Boskoff, K. J. 1979. Maternal behavior in prosimians. In: The Study of Prosimian Behavior (Ed. by G. A. Doyle & R. D. Martin), pp. 123–156. New York: Academic Press.

Koford, C. 1966. Rhesus monkeys of Santiago Island, Puerto Rico. 16 mm cine film. USNAC, Washington, D.C.

Kohl-Larsen, L. 1958. Wildbeuter in Ost-Afrika: Die Tindiga, ein Jager- und Sammlervolk. Berlin: Dietrich Reimer.

Kohler, W. 1925. The Mentality of Apes. London: Routledge & Kegan Paul.

Kondo-Ikemura, K. 1988. The daily activity rhythm of a Japanese monkey troop at Arashiyama West, Texas, in the summer season. In: Research Reports of the Arashiyama West and East Groups of Japanese Monkeys, pp. 19–28. Laboratory of Ethological Studies, Osaka University, Faculty of Human Sciences, Suita, Japan.

Konner, M. J. 1972. Aspects of the developmental ethology of a foraging people. In: Ethological Studies of Child Behaviour (Ed. by N. G. Blurton Jones), pp. 285–303. London: Cambridge University Press.

Konner, M. J. 1975. Relations among infants and juveniles in comparative perspective. In: Friendship and Peer Relations (Ed. by M. Lewis & L. A. Rosenblum), pp. 99–124. New York: Wiley.

Konner, M. J. 1976. Maternal care, infant behavior and development among the !Kung. In: Kalihari Hunter-Gatherers (Ed. by R. B. Lee & I. DeVore), pp. 218–245. Cambridge, Mass.: Harvard University Press.

Konner, M. J. 1977a. Evolution of human behavior development. In: Culture and Infancy: Variations in the Human Experience (Ed. by P. H. Leiderman, S. R. Tulkin, & A. Rosenfeld), pp. 69–109. New York: Academic Press.

Konner, M. J. 1977b. Infancy among the Kalahari desert San. In: Culture and Infancy: Variation in the Human Experience (Ed. by P. H. Leiderman, S. R. Tulkin, & A. Rosenfeld), 287–328. New York: Academic Press.

Konner, M. J., & Shostak, M. 1986. Adolescent pregnancy and childbearing: An anthropological perspective. In: School-Age Pregnancy and Parenthood (Ed. by J. B. Lancaster & B. A. Hamburg), pp. 325–345. New York: Aldine.

Konner, M. J., & Worthman, C. 1980. Nursing frequency, gonadal function, and birth spacing among !Kung hunter gatherers. Science 207, 788–791.

Koyama, N. 1967. On dominance rank and kinship of a wild Japanese monkey troop in Arashiyama. Primates 8, 189–216.

Koyama, N. 1985. Playmate relationships among individuals of the Japanese monkey troop in Arashiyama. Primates 26, 390–406.

Koyama, N. 1990. Troop division and inter-troop relationships of ringtailed lemurs (*Lemur catta*) at Berenty, Madagascar. Paper presented at the 1990 Congress of the International Primatology Society, Nagoya, Japan.

Kozlowski, J., & Uchmanski, J. 1987. Optimal individual growth and reproduction in perennial species with indeterminate growth. Evol. Ecol. 1, 214–230.

Kraemer, H. C., Horvat, J. R., Doering, C., & McGinnis, P. R. 1982. Male chimpanzee development focusing on adolescence: Integration of behavioral with physiological changes. Primates 23, 393–405.

Kruuk, H. 1972. The Spotted Hyena. Chicago: University of Chicago Press.

Kuklin, S. 1989. Going to My Ballet Class. New York: Bradbury Press.

Kummer, H. 1968. Social Organization of Hamadryas Baboons: A Field Study. (Bibliotheca Primatologica, No. 6). Basel: Karger.

Kummer, H. 1971. Primate Societies: Group Techniques of Ecological Adaptation. Chicago: Aldine.

Kummer, H. 1978. On the value of social relationships to nonhuman primates; A heuristic scheme. Soc. Sci. Info. 17, 687–705.

Kummer, H., Abegglen, J. J., Bachmann, C., Falett, J., & Sigg, H. 1978. Grooming relationship and object competition among hamadryas baboons. In: Recent Advances in Primatology, Vol. 1. Behaviour (Ed. by D. Chivers & J. Herbert), pp. 31–38. London: Academic Press.

Kummer, H., Dasser, V., & Hoyningen-Huene, P. 1990. Exploring primate cognition: Some critical remarks. Behaviour 112, 84–98.

Kummer, J., & Goodall, J. 1985. Conditions of innovative behavior in primates. Phil. Trans. R. Soc. Ser. B. 308, 203–214.

Kurland, J. A. 1977. Kin Selection in the Japanese Monkey (Contributions to Primatology, 12). Basel: Karger.

Kuroda, S. 1980. Social behavior of the pygmy chimpanzees. Primates 21, 181–197.

Kuroda, S. 1989. Developmental retardation and behavioral characteristics of pygmy chimpanzees. In: Understanding Chimpanzees

(Ed. by P. G. Heltne & L. A. Marquardt), pp. 184–193. Cambridge, Mass.: Harvard University Press.

Küster, J., & Paul, A. 1988. Rank relations of juvenile and subadult natal males of Barbary macaques (*Macaca sylvanus*) at Affenberg, Salem. Folia Primatol. 51, 33–44.

Lacey, E. P. 1986. Onset of reproduction in plants: Size- versus age-dependency. Trends Ecol. Evol. 1, 72–75.

Lack, D. 1954. The Natural Regulation of Animal Numbers. Oxford: Clarendon Press.

Lack, D. 1968. Ecological Adaptations for Breeding in Birds. Oxford: Clarendon Press.

Lamb, M. E. 1984. Observational studies of father–child relationships in humans. In: Primate Paternalism (Ed. by D. M. Taub), pp. 407–430. New York: Van Nostrand Reinhold.

Lancaster, J. 1971. Play-mothering: The relations between juvenile females and young infants among free-ranging vervet monkeys (*Cercopithecus aethiops*). Folia Primatol. 15, 161–182.

Lancaster, J. B., & Lancaster, C. S. 1983. Parental investment: The hominid adaptation. In: How Humans Adapt (Ed. by D. Ortner), pp. 35–56. Washington, D.C.: Smithsonian Institution Press.

Landauer, T. K., & Whiting, J. W. M. 1981. Correlates and consequences of stress in infancy. In: Handbook of Cross-Cultural Research in Human Development (Ed. by R. H. Munroe, R. L. Munroe, & B. B. Whiting), pp. 355–375, New York: Garland Press.

Lande, R. 1982. A quantitative genetic theory of life history evolution. Ecology 63, 607–615.

Lawlor, L. R. 1976. Molting, growth and reproductive strategies in the terrestrial isopod *Armadillidium vulgare*. Ecology 57, 1179–1194.

Laws, J. W., & Von der Haar Laws, J. 1984. Social interaction among adult langurs (*Presbytis entellus*) at Rajaji Wildlife Sanctuary. Int. J. Primatol. 5, 31–50.

Lazarus, J., & Inglis, I. R. 1986. Shared and unshared parental investment, parent–offspring conflict and brood size. Anim. Behav. 34, 1791–1804.

LeBoeuf, B. J., & Reiter, J. 1988. Lifetime reproductive success in northern elephant seals. In: Reproductive Success (Ed. by T. H. Clutton-Brock), pp. 344–362. Chicago: University of Chicago Press.

Lee, P. A., & Migeon, C. J. 1975. Puberty in boys: Correlation of plasma levels of gonadotropins (LH, FSH), androgens (testosterone, androstenedione, dehydroepiandrosterone and its sulfate), estrogens (estrone and estradiol) and progestins (progesterone and 17-hydroxyprogesterone). J. Clin. Endocrinol. Metab. 41, 556–62.

Lee, P. A., Xanakis, T., Winer, J., & Matsenbaugh, S. 1976. Puberty in girls: Correlation of serum levels of gonadotropins, prolactin, androgens, estrogens, and progestins with physical changes. J. Clin. Endocrinol. Metab. 43, 775–784.

Lee, P. C. 1983a. Play as a means for developing relationships. In: Primate Social Relationships (Ed. by R. A. Hinde), pp. 82–89. Sunderland, Mass.: Sinauer.

Lee, P. C. 1983b. Context-specific unpredictability in dominance interactions. In: Primate Social Relationships: An Integrated Approach (Ed. by R. A. Hinde), pp. 35–44. Oxford: Blackwell.

Lee, P. C. 1984. Ecological constraints on the social development of vervet monkeys. Behaviour 91, 245–262.

Lee, P. C. 1986. Early social development among African elephant calves. Natl. Geog. Res., 2, 388–401.

Lee, P. C. 1989. Family structure, communal care and female reproductive effort. In: Comparative Socioecology (Ed. by V. Standen & R. Foley), pp. 323–340. Oxford: Blackwell.

Lee, P. C., & Oliver, J. I. 1979. Competition, dominance and the acquisition of rank in juvenile yellow baboons (*Papio cynocephalus*). Anim. Behav. 27, 576–585.

Lee, R. B. 1969. !Kung bushman subsistence: An input–output analysis. In: Environment and Cultural Behavior (Ed. by A. P. Vayda), pp. 47–79. Garden City, N.Y.: Natural History Press.

Lee, R. B. 1972. Population growth and the beginnings of sedentary life among the !Kung bushmen. In: Population Growth: Anthropological Implications (Ed. by B. Spooner), pp. 47–79. Cambridge, Mass.: MIT Press.

Lee, R. B. 1979. The !Kung San. London: Cambridge University Press.

Lee, R. B., & DeVore, I. (Eds.). 1976. Kalahari Hunter-Gatherers. Cambridge, Mass.: Harvard University Press.

Lehrman, D. S. 1970. Semantic and conceptual issues in the nature-nurture problem. In: Development and Evolution of Behavior (Ed. by L. R. Aronson, E. Tobach, D. S. Lehrman, & J. S. Rosenblatt), pp. 17–52. San Francisco: Freeman.

Leigh, E. G., Jr., Rand, A. S., & Windsor, D. M. (Eds.). 1982. The Ecology of a Tropical Forest: Seasonal Rhythms and Long-Term Changes. Washington, D.C.: Smithsonian Institution Press.

Lemarchand-Béraud, T., Zufferey, M.-M., Reymond, M., & Rey, I. 1982. Maturation of the hypothalamo-pituitary-ovarian axis in adolescent girls. J. Clin. Endocrinol. Metab. 54, 241–246.

Lemos de Sa, R. M. 1988. Situacão de uma populacão de Monocarvoeiro (*Brachyteles arachnoides*), em fragmento de mata Atlantica (M.G.), e implicacoes para sua conservacão. M.A. thesis, Universidade de Brasilia.

Leonard, W. R., & Thomas, R. B. 1989. Biosocial responses to seasonal food stress in highland Peru. Hum. Biol. 61, 65–85.

Leresche, L. A. 1976. Dyadic play in Hamadryas baboons. Behaviour 57, 190–205.

Leridon, H. 1977. Human Fertility: The Basic Components. Chicago: University of Chicago Press.

Lethmate, J. 1976. Versuche zur Doppelstockhandlung mit einem jungen Orang-Utan. Zool. Anz. 197, 264–271.

Lethmate, J. 1977. Werkzeugenstellen eines jungen Orang-Utans. Behaviour 62, 174–189.

Levine, D. N. 1965. Wax and Gold: Tradition and Innovation in Ethiopian Culture. Chicago: University of Chicago Press.

LeVine, R. A. 1977. Child rearing as cultural adaptation. In: Culture and Infancy: Variations in the Human Experience (Ed. by P. H. Leiderman, S. R. Tulkin, & A. Rosenfeld), pp. 15–27. New York: Academic Press.

Levine, S., Haltmeyer, G., Karas, G., & Denenberg, V. H. 1967. Physiological and behavioral effects of infantile stimulation. Physiol. Behav. 2, 55–59.

Levy, J. 1979. Play behavior and its decline during development in rhesus monkeys (*Macaca mulatta*). Ph.D. dissertation, University of Chicago.

Lewontin, R. C. 1965. Selection for colonizing ability. In: The Genetics of Colonizing Species (Ed. by H. G. Baker & G. L. Stebbins), pp. 77–94. New York: Academic Press.

Lickliter, R. E. 1984. Mother–infant spatial relationships in domestic goats. Appl. Anim. Behav. Sci. 13, 93–100.

Lindstedt, S. L. 1985. Birds. In: Non-mammalian Models for Research in Ageing (Ed. by F. A. Lints), pp. 1–21. Basel: Karger.

Lindstedt, S. L., & Calder, W. A. 1981. Body size, physiological time, and longevity of homeothermic animals. Quart. Rev. Biol. 56, 1–16.

Lindstedt, S. L., & Swain, S. D. 1988. Body size as a constraint of design and function. In: Evolution of Life Histories of Mammals (Ed. by M. S. Boyce), pp. 93–105. New Haven, Conn.: Yale University Press.

Lively, C. M. 1986a. Predator-induced shell dimorphism in the acorn barnacle. Evolution 40, 232–242.

Lively, C. M. 1986b. Competition, comparative life histories, and maintenance of shell dimorphism in a barnacle. Ecology 67, 858–864.

Lloyd, L. E., McDonald, B. E., & Crampton, E. W. 1978. Fundamentals of Nutrition. San Francisco: Freeman.

Lloyd, M., & Dybas, H. S. 1966a. The periodical cicada problem. I. Population ecology. Evolution 20, 133–149.

Lloyd, M., & Dybas, H. S. 1966b. The periodical cicada problem. II. Evolution. Evolution 20, 466–505.

Lord, R. D., Jr. 1960. Litter size and latitude in North American mammals. Am. Midl. Natl. 64, 488–499.

Lorenz, K. 1950. The comparative method in studying innate behaviour patterns. Sym. Soc. Exp. Biol. 4, 221–268.

Lotka, A. J. 1925. Elements of Physical Biology. Baltimore: Williams & Wilkins.

Loy, J., Loy, K., Keifer, G., & Conaway, C. 1984. The behavior of gonadectomized rhesus monkeys. (Contributions to Primatology, 20). Basel: Karger.

Loy, J. D., & Peters, C. B. 1991. Mortifying reflections: Primatology and the human disciplines. In: Understanding Behavior: What Primate Studies Tell Us About Human Behavior (Ed. by J. D. Loy & C. B. Peters), pp. 3–16. New York: Oxford University Press.

Lozoff, B. 1983. Birth and "bonding" in nonindustrial societies. Dev. Med. Child Neurol. 25, 595–600.

Luckinbill, L. S., Arking, R., Clare, M. J., Cirocco, W. C., & Buck, S. A. 1984. Selection for delayed senescence in *Drosophila melanogaster*. Evolution 38, 996–1003.

Luckinbill, L. S., & Clare, M. J. 1986. A density threshold for the expression of longevity in *Drosophila melanogaster*. Heredity 56, 329–335.

Lyon, B. E., & Montgomerie, R. D. 1986. Delayed plumage maturation in passerine birds: Reliable signalling by subordinate males? Evolution 40, 605–615.

MacArthur, R. H., & Wilson, E. O. 1967. The Theory of Island Biogeography. Princeton, N.J.: Princeton University Press.

Maccoby, E. E. 1988. Gender as a social category. Dev. Psychol. 24, 755–765.

Maccoby, E. E. 1990. The role of gender identity and gender constancy in sex-differentiated development. In: The Legacy of Lawrence Kohlberg (Ed. by D. Schrader), pp. 5–20. San Francisco: Jossey-Bass.

Maccoby, E. E., & Jacklin, C. N. 1974. The Psychology of Sex Differences. Stanford, Calif.: Stanford University Press.

MacDonald, K. (Ed.). 1988. Sociobiological Perspectives on Human Development. New York: Springer.

Mace, G. M. 1979. The evolutionary ecology of small mammals. D. Phil. dissertation, University of Sussex.

Macedonia, J. M. 1991. Vocal communication and anti-predator behavior in the ring-tailed lemur (*Lemur catta*) with a comparison to the ruffed lemur (*Varecia variagatta*). Ph.D. dissertation, Duke University.

Macedonia, J. M., & Polak, J. F. 1989. Visual assessment of avian threat in semi-captive ringtailed lemurs (*Lemur catta*). Behaviour 111, 291–304.

Mack, D. 1979. Growth and development of infant red howling monkeys (*Alouatta seniculus*) in a free ranging population. In: Vertebrate Ecology in the Northern Neotropics (Ed. by J. F. Eisenberg), pp. 127–136. Washington, D.C.: Smithsonian Institution Press.

MacKinnon, J. 1971. The orang-utan in Sabah today. Oryx 11, 141–191.

MacKinnon, J. 1974. The behaviour and ecology of wild orangutans, *Pongo pygmaeus*. Anim. Behav. 22, 3–74.

MacKinnon, J. 1978. The Ape Within Us. New York: Holt, Rinehart & Winston.

MacKinnon, J. 1979. Reproductive behavior in wild orangutan populations. In: The Great Apes (Ed. by D. Hamburg & E. R. McCown), pp. 257–274. Menlo Park, Calif.: Benjamin-Cummings.

MacKinnon, J., & MacKinnon, K. 1980. The behavior of wild spectral tarsiers. Int. J. Primatol. 1, 361–379.

Malcolm, L. A. 1970. Growth and development in New Guinea. A study of Bundi people of the Madang district. Inst. Hum. Biol., Monogr. Series, no. 1. Madang.

Malenky, R. R. 1990. Ecological factors affecting food choice and social organization in *Pan paniscus*. Ph.D. dissertation, State University of New York at Stony Brook.

Maple, T. L. 1980. Orangutan Behavior. New York: Van Nostrand Reinhold.

Maple, T. L. 1982. Orang utan behavior and its management in captivity. In: The Orang Utan: Its Biology and Conservation (Ed. by L. E. M. de Boer), pp. 257–268. The Hague: Junk.

Markham, R., & Groves, C. P. 1990. Brief communication: Weights of wild orangutans. Am. J. Phys. Anthropol. 81, 1–3.

Marler, P., & Peters, S. 1982. Subsong and plastic song: Their role in the vocal learning process.

In: Acoustic Communication in Birds, Vol. 2 (Ed. by D. E. Kroodsma & E. H. Miller), pp. 25–50. New York: Academic Press.

Marsh, C. W. 1979. Comparative aspects of social organisation in the Tana River red colobus, *Colobus badius rufomitratus*. Z. Tierpsychol. 51, 337–362.

Marshall, W. A., & Tanner, J. M. 1969. Variation in pattern of pubertal changes in girls. Arch. Dis. Child. 44, 291–303.

Marshall, W. A., & Tanner, J. M. 1970. Variations in the pattern of pubertal changes in boys. Arch. Dis. Child. 45, 13–23.

Martin, P. 1984. The (four) whys and wherefores of play in cats: A review of functional, evolutionary, developmental and causal issues. In: Play: In Animals and Humans (Ed. by P. K. Smith), pp. 71–94. Oxford: Blackwell.

Martin, P., & Caro, T. M. 1985. On the functions of play and its role in behavioural development. Adv. Study Behav. 15, 59–103.

Martin, R. D. 1972. A preliminary field-study of the lesser mouse lemur (*Microcebus murinus* J. F. Miller 1777). In: Behavior and Evolution of Nocturnal Prosimians: Advances in Ethology 9 (Ed. by P. Charles-Dominique & R. D. Martin), pp. 43–89. Berlin: Paul Pareg.

Martin, R. D. 1973. A review of the behaviour and ecology of the lesser mouse lemur (*Microcebus murinus* J. F. Miller 1777). In: Comparative Ecology and Behaviour of Primates (Ed. by R. P. Michael & J. H. Crook), pp. 1–68. New York: Academic Press.

Martin, R. D. 1981. Relative brain size and basal metabolic rate in terrestrial vertebrates. Nature (London) 293, 56–60.

Martin, R. D., & Harvey, P. H. 1985. Brain size allometry: Ontogeny and phylogeny. In: Size and Scaling in Primate Biology (Ed. by W. L. Jungers), pp. 147–173. New York: Plenum.

Martin, R. D., & MacLarnon, A. M. 1990. Reproductive patterns in primates and other mammals: The dichotomy between altricial and precocial offspring. In: Primate Life History Evolution (Ed. by C. J. deRousseau), pp. 47–79. New York: Wiley-Liss.

Martorell, R., & Habicht, J. P. 1986. Growth in early childhood in developing countries. In: Human Growth, 2nd ed., Vol. 3 (Ed. by F. Falkner & J. M. Tanner), pp. 241–262. New York: Plenum.

Mascie-Taylor, C. G. N. 1991. Biosocial influences on stature: A review. J. Biosoc. Sci. 23, 113–128.

Maslow, A. H. 1940. Dominance quality and social behavior in infra-human primates. J. Soc. Psychol. 11, 131–324.

Mason, W. A. 1979a. Ontogeny of social behavior.

In: Handbook of Behavioral Neurobiology, Vol. 3. Social Behavior and Communication (Ed. by P. Marler & J. G. Vandenbergh), pp. 1–28. New York: Plenum.

Mason, W. A. 1979b. Wanting and knowing: A biological perspective on maternal deprivation. In: Origins of the Infant's Social Responsiveness (Ed. by E. B. Thoman), pp. 225–249. Hillsdale, N.J.: Erlbaum.

Massey, A. 1977. Agonistic aids and kinship in a group of pigtail macaques. Behav. Ecol. Sociobiol. 2, 31–40.

Mathur, R., & Manohar, R. 1991. Departure of juvenile male *Presbytis entellus* from the natal troop. Int. J. Primatol. 12, 39–43.

May, R. M., & Rubenstein, D. I. 1985. Reproductive Strategies. In: Reproductive Fitness (Ed. by R. V. Short & M. V. Austin), pp. 1–23. Cambridge: Cambridge University Press.

Maynard Smith, J. 1964. Group selection and kin selection. Nature (London) 201, 1145–1147.

Maynard Smith, J. 1986. Problems of Biology. Oxford: Oxford University Press.

Maynard Smith, J., Burian, R., Kauffman, S., Alberch, P., Campbell, J., Goodwin, B., Lande, R., Raup, D., & Wolpert, L. 1985. Developmental constraints and evolution. Quart. Rev. Biol. 60, 265–287.

Mayr, E. 1963. Animal Species and Evolution. Cambridge, Mass.: Harvard University Press.

McGinley, M. A., & Charnov, E. L. 1988. Multiple resources and the optimal balance between size and number of offspring. Evol. Ecol. 2, 77–84.

McGinnis, P. R. 1979. Sexual behavior in free-living chimpanzees: Consort relationships. In: The Great Apes (Ed. by D. A. Hamburg & E. R. McCown), pp. 429–440. Menlo Park, Calif.: Benjamin-Cummings.

McKee, L. (Ed.). 1984. Child survival and sex differences in the treatment of children. Med. Anthropol. 8, 81–144.

McKenna, J. J. 1979. The evolution of allomothering behavior among colobine monkeys: Function and opportunism in evolution. Am. Anthropol. 81, 818–840.

McKenna, J. J. 1982. The evolution of primate societies, reproduction, and parenting. In: Primate Behavior (Ed. by J. L. Fobes & J. E. King), pp. 87–133. New York: Academic Press.

McKenna, J. J. 1987. Parental supplements and surrogates among primates: Cross-species and cross-cultural comparisons. In: Parenting Across the Life Span: Biosocial Dimensions (Ed. by J. B. Lancaster, J. Altmann, A. S. Rossi, & L. R. Sherrod), pp. 143–184. New York: Aldine.

McNab, B. K., & Eisenberg, J. F. 1989. Brain size and its relation to the rate of metabolism in mammals. Am. Nat. 133, 157–167.

Meaney, M. J. 1988. The sexual differentiation of social play. Trends NeuroSci. 11, 54–57.

Meaney, M. J., Stewart, J., & Beatty, W. W. 1985. Sex differences in social play—the socialization of sex roles. Adv. Study Behav. 15, 1–58.

Medawar, P. B. 1946. Old age and natural death. Mod. Quart. 1, 30–56.

Medawar, P. B. 1952. An Unsolved Problem of Biology. London: Lewis.

Meikle, D. B., & Vessey, S. H. 1981. Nepotism among rhesus monkey brothers. Nature (London) 294, 160–161.

Menard, N. 1985. Le Régime alimentaire de *Macaca sylvanus* dans differents habitats d'Algerie. I. Régime en chenaie decidue. Rev. Ecol. (Terre Vie) 40, 451–466.

Menard, N., & Vallet, D. 1986. Le régime alimentaire de *Macaca sylvanus* dans differents habitats d'Algerie. II. Régime en foret sempervirente et sur les sommets rocheux. Rev. Ecol. (Terre Vie) 41, 173–192, 451–466.

Mendes, S. L. 1989. Estudo ecológico de *Alouatta fusca* (Primates: Cebidae) na Estação Biológica de Caratinga, MG. Rev. Nordestina Biol. 6, 71–104.

Meredith, H. V. 1976. Findings from Asia, Australia, Europe, and North America on secular change in mean height of children, youths, and young adults. Am. J. Phys. Anthropol. 44, 315–326.

Mertl-Milhollen, A. 1988. Olfactory demarcation of territorial but not home range boundaries by *Lemur catta*. Folia Primatol. 50, 175–187.

Mertz, D. B. 1971. Life history phenomena in increasing and decreasing populations. In: Statistic Ecology, Vol. 2 (Ed. by G. P. Patil, E. G. Pielou, & W. E. Waters), pp. 361–399. University Park: Pennsylvania State University Press.

Millar, J. S. 1977. Adaptive features of mammalian reproduction. Evolution 31, 370–386.

Millar, J. S. 1981. Pre-partum reproductive characteristics of eutherian mammals. Evolution 35, 1149–1163.

Millar, J. S., & Zammuto, R. M. 1983. Life histories of mammals: An analysis of life tables. Ecology 64, 631–635.

Miller, M. K., & Stokes, C. S. 1985. Teenage fertility, socioeconomic status and infant mortality. J. Biosoc. Sci. 17, 147–155.

Milton, K. 1979. Factors influencing leaf choice by howler monkeys: A test of some hypotheses of food selection by generalist herbivores. Am. Nat. 114, 362–378.

Milton, K. 1980. The Foraging Strategy of Howler

Monkeys. New York: Columbia University Press.

Milton, K. 1981a. Estimates of reproductive parameters for free-ranging *Ateles Geoffroyi*. Primates 22, 574–579.

Milton, K. 1981b. Diversity of plant foods in tropical forests as a stimulus to mental development in primates. Am. Anthropol. 83, 534–548.

Milton, K. 1984a. The role of food-processing factors in primate food choice. In: Adaptations for Foraging in Nonhuman Primates (Ed. by P. S. Rodman & J.G.H. Cant), pp. 249–279. New York: Columbia University Press.

Milton, K. 1984b. Habitat, diet, and activity patterns of free-ranging woolly spider monkeys (*Brachyteles arachnoides* E Geoffroy 1806). Int. J. Primatol. 5, 491–514.

Milton, K. 1985. Mating patterns of woolly spider monkeys, *Brachyteles arachnoides:* Implications for female choice. Behav. Ecol. Sociobiol. 17, 53–59.

Milton, K. 1991. Leaf change and fruit production in six neotropical Moraceae species. J. Ecol. 79, 1–26.

Milton, K., Casey, T. M., & Casey, K. K. 1979. The basal metabolism of mantled howler monkeys (*Alouatta palliata*). J. Mammal. 60, 373–376.

Mineka, S., & Cook, M. 1988. Social learning and the acquisition of fear in monkeys. In: Social Learning: Psychological and Biological Perspectives (Ed. by T. R. Zentall & B. G. Galef), pp. 51–73. Hillsdale, N.J.: Erlbaum.

Mineka, S., Davidson, M., Cook, M., & Keir, R. 1984. Observational conditioning of snake fear in rhesus monkeys. J. Abnormal Psychol. 93, 355–372.

Missakian, E. S. 1972. Genealogical and cross-genealogical dominance relations in a group of free-ranging rhesus monkeys (*Macaca mulatta*) on Cayo Santiago. Primates 13, 169–180.

Mitani, J. C. 1985. Mating behavior of male orangutans in the Kutai Game Reserve, Indonesia. Anim. Behav. 33, 392–402.

Mitani, J. C. 1989. Orangutan activity budgets: Monthly variations and the effects of body size, parturition, and sociality. Am. J. Primatol. 18, 87–100.

Mitchell, G. 1979. Behavioral Sex Differences in Nonhuman Primates. New York: Van Nostrand Reinhold.

Mitchell, R. W. 1990. A theory of play. In: Interpretation and Explanation in the Study of Animal Behavior, Vol. 1 (Ed. by M. Bekoff & D. Jamieson), pp. 197–227. Boulder, Colo.: Westview Press.

Mitchell, R. W., & Thompson, N. S. 1990. The effects of familiarity on dog–human play. Anthrozoos 4, 24–43.

Mittermeier, R. A., Valle, C. M. C., Alves, M. C., Santos, I. B., Pinto, C. A. M., Strier, K. B., Young, A. L., Veado, E. M., Constable, I. D., Paccagnella, S. G., & Lemos de Sa, R. M. 1987. Current distribution of the muriqui in Atlantic forest region of eastern Brazil. Primate Conserv. 8, 143–149.

Mofenson, H. C., & Greensher, J. 1985. Management of the choking child. Pediatric Clinics N. Am. 32, 183–192.

Mohnot, S. M. 1971. Some aspects of social changes and infant-killing in the Hanuman langur, *Presbytis entellus entellus* (Primates: Cercopithecidae), in western India, Mammalia 35, 175–198.

Mohnot, S. M. 1974. Ecology and behaviour of the common Indian langur, *Presbytis entellus*. Ph.D. dissertation, University of Jodhpur.

Mohnot, S. M. 1977. Interactions and social changes in troops of the langur, *Presbytis entellus*, in India. In: The Natural Resources of Rajasthan (Ed. by M. L. Roonwal), pp. 505–514. Jodhpur: University of Jodhpur Press.

Mohnot, S. M. 1978. Peripheralization of weaned male juveniles in *Presbytis entellus*. In: Recent Advances in Primatology (Ed. by D. J. Chivers & J. Herbert), pp. 87–91. New York: Academic Press.

Mohnot, S. M., Dave, V. K., Agoramoorthy, G., & Rajpurohit, L. S. 1984. Ecobehavioral studies of Hanuman langur, *Presbytis entellus*. In: Proceedings of the 2nd Annu. Workshop on MAB Projects, pp. 96–98. Government of India, New Delhi.

Moore, J. 1982. Coalitions in langur all-male bands. Int. J. Primatol. 3, 314.

Moore, J. 1984a. Age and grooming in langur male bands (*Presbytis entellus*). In: Current Primate Researches (Ed. by M. L. Roonwal, S. M. Mohnot, & N. S. Rathore), pp. 381–388. Jodhpur: University of Jodhpur Press.

Moore, J. 1984b. Female transfer in primates. Int. J. Primatol. 5, 537–589.

Moore, J. 1985. Demography and Sociality in Primates. Ph.D. dissertation, Harvard University.

Moore, M., & Ali, R. 1984. Are dispersal and inbreeding avoidance related? Anim. Behav. 32, 94–112.

Morgan, B. J. T., Simpson, M. J. A., Hanby, J. P., & Hall-Craggs, J. 1976. Visualising interactions and sequential data in animal behaviour: Theory and application of cluster analysis. Behaviour 56, 1–43.

Mori, A. 1979. Analysis of population changes by measurement of body weight in the Koshima troop of Japanese monkeys. Primates 20, 371–397.

Mori, U. 1974. The inter-individual relationships observed in social play of the young Japanese monkeys of the natural troop in Koshima Islet. J. Anthropol. Soc. Nippon 82, 303–318.

Mori, U. 1979. Individual relationships within a unit. In: Ecological and Sociological Studies of Gelada Baboons (Ed. by M. Kawai), pp. 93–124. Basel: Karger.

Moses, L. E., Gale, L. C., & Altmann, J. 1992. Methods for analysis of unbalanced, longitudinal, growth data. Am. J. Primatol. 28, 49–59.

Mountjoy, P. T. 1980. An historical approach to comparative psychology. In: Comparative Psychology: An Evolutionary Analysis of Behavior (Ed. by M. R. Denny), pp. 128–152. New York: Wiley.

Muckenhirn, N. A. 1972. Leaf-eaters and their predators in Ceylon: Ecological roles of gray langurs, *Presbytis entellus,* and leopards. Thesis, University of Maryland.

Muller, H. 1970. Beitrage zur Biologie des Hermelins, *Mustela erminea* Linne 1758. Saugetierkunkliche Mitteilungen 18, 293–380.

Munroe, R. H., Munroe, R. L., & Shimmin, H. S. 1984. Children's work in four cultures: Determinants and consequences. Am. Anthopol. 86, 369–379.

Mussen, P., & Eisenberg-Berg, N. 1977. Roots of Caring, Sharing, and Helping: The Development of Prosocial Behavior in Children. San Francisco: Freeman.

Myers, P., & Masters, L. L. 1983. Reproduction by *Peromyscus maniculatus:* Size and compromise. J. Mammal. 64, 1–18.

Nadler, R. D. 1988. Sexual and reproductive behavior. In: Orang-utan Biology (Ed. by J. H. Schwartz), pp. 105–116. New York: Oxford University Press.

Nakamichi, M. 1989. Sex differences in social development during the first 4 years in a free-ranging group of Japanese monkeys, *Macaca fuscata*. Anim. Behav. 38, 737–748.

Napier, J. R., & Napier, P. H. 1967. A Handbook of Living Primates. London: Academic Press.

Nash, L. T. 1983. Reproductive patterns in galagos (*Galago zanzibaricus* and *Galago garnettii*) in relation to climatic variability. Am. J. Primatol. 5, 181–196.

Nash, L. T. 1990. Development of food sharing in captive infant *Galago senegalensis braccatus*. Paper presented at the 23rd Congress of the International Primatological Society, Nagoya, Japan.

Nash, L. T., Bearder, S. K., & Olson, T. R. 1989. Synopsis of galago species characteristics. Int. J. Primatol. 10, 57–80.

Nash, L. T., & Flinn, L. 1978. Group formation in captive lesser galagos (*Galago senegalensis*). Primates 19, 493–503.

Nash, L. T., & Harcourt, C. S. 1986. Social organization in galagos in Kenyan coastal forests. II. *Galago garnettii*. Am. J. Primatol. 10, 357–369.

Nash, L. T., & Whitten, P. L. 1989. Preliminary observations in the role of acacia gum chemistry in acacia utilization by *Galago senegalensis* in Kenya. Am. J. Primatol. 17, 27–39.

Netto, W. J., & van Hooff, J. 1986. Conflict interference and the development of dominance relationships in immature *Macaca fascicularis*. In: Primate Ontogeny, Cognition and Social Behavior (Ed. by J. G. Else & P. C. Lee), pp. 291–300. New York: Cambridge University Press.

Neuringer, M., Kosobud, A., & Cochrane, G. 1981. Visual acuity of *Lemur catta*, a diurnal prosimian. Investig. Ophthamol. Visual Sci. 20(Suppl.3), 49.

Newton, P. N. 1988. The variable social organization of Hanuman langurs (*Presbytis entellus*). Infanticide, and the monopolization of females. Int. J. Primatol. 9, 59–77.

Nicolson, N. A. 1982. Weaning and the development of independence in olive baboons. Ph.D. dissertation, Harvard University.

Nicolson, N. A. 1987. Infants, mothers, and other females. In: Primate Societies (Ed. by B. B. Smuts, D. L. Cheney, R. M. Seyfarth, R. W. Wrangham, & T. T. Struhsaker), pp. 330–342. Chicago: University of Chicago Press.

Nicolson, N., & Demment, M. W. 1982. The transition from suckling to independent feeding in wild baboon infants. Int. J. Primatol. 3, 318.

Niemitz, C. 1974. A contribution to the postnatal behavioral development of *Tarsius bancanus*, Horsfield, 1821, studied in two cases. Folia Primatol. 21, 250–276.

Niemitz, C. 1984a. Taxonomy and distribution of the genus *Tarsius* Storr, 1780. In: Biology of Tarsiers (Ed. by C. Niemitz), pp. 1–16. New York: Gustav Fischer Verlag.

Niemitz, C. 1984b. An investigation and review of the territorial behaviour and social organization of the spectral tarsiers. In: Biology of Tarsiers (Ed. by C. Niemitz), pp. 117–127. New York: Gustav Fischer Verlag.

Niemitz, C. 1984c. Vocal communication of two tarsier species (*Tarsius bancanus* and *Tarsius spectrum*). In: Biology of Tarsiers (Ed. by C. Niemitz), pp. 129–141. New York: Gustav Fischer Verlag.

Niemitz, C. 1984d. Synecological relationships and feeding behaviour of the genus *Tarsius*. In: Biology of Tarsiers (Ed. by C. Niemitz), pp. 59–75. New York: Gustav Fischer Verlag.

Niemitz, C. 1991. *Tarsius dianae:* A new primate species from Central Sulawesi (Indonesia). Folia Primatol. 56, 105–116.

Nishida, T. 1979. The social structure of chimpanzees in the Mahale Mountains. In: The Great Apes (Ed. by D. Hamburg & E. R. McKown), pp. 73–122. Menlo Park, Calif.: Benjamin-Cummings.

Nishida, T. 1983. Alloparental behavior in wild chimpanzees of the Mahale Mountains, Tanzania. Folia Primatol. 41, 1–33.

Nishida, T. 1988. Development of social grooming between mother and offspring in wild chimpanzees. Folia Primatol. 50, 109–123.

Nishida, T. 1989. Social interactions between immigrant and resident female chimpanzees. In: Understanding Chimpanzees (Ed. by P. G. Heltne & L. A. Marquard), pp. 68–89. Cambridge, Mass.: Harvard University Press.

Nishida, T., & Hiraiwa-Hasegawa, M. 1987. Chimpanzees and bonobos: Cooperative relationships among males. In: Primate Societies (Ed. by B. B. Smuts, D. L. Cheney, R. M. Seyfarth, R. W. Wrangham, & T. T. Struhsaker), pp. 165–177. Chicago: University of Chicago Press.

Nishida, T., Takasaki, H., & Takahata, Y. 1990. Demography and reproductive profiles. In: The Chimpanzees of Mahale (Ed. by T. Nishida), pp. 63–98. Tokyo: University of Tokyo Press.

Nishimura, A. 1990. A sociological and behavioral study of woolly monkeys, *Lagothrix lagotricha,* in the Upper Amazon. Sci. Eng. Rev. Doshisha Univ. 31, 87–121.

Nissen, H. W. 1931. A field study of the chimpanzee: Observations of chimpanzee behavior and environment in western French Guinea. Comp. Psychol. Monog. 8, 1–122.

Noë, R. 1989. Coalition formation among male baboons. Ph.D. dissertation, Rijksuniversiteit te Utrecht.

Noë, R. 1990. A Veto game played by baboons: A challenge to the use of the Prisoner's Dilemma as a paradigm for reciprocity and cooperation. Anim. Behav. 39, 78–90.

Noë, R., de Waal, F. B. M., & van Hooff, J. A. R. A. M. 1980. Types of dominance in a chimpanzee colony. Folia Primatol. 34, 90–110.

Nottelmann, E. D., Susman, E. J., Dorn, L. D., Inoff-Germain, G., Loriaux, D. L., Cutler, G. B., & Chrousos, G. P. 1987. Developmental processes in early adolescence: Relations among chronologic age, pubertal stage, height, weight, and serum levels of gonadotrophins, sex steroids, and adrenal androgens. J. Adol. Health Care 8, 246–260.

Nur, N. 1988a. The cost of reproduction in birds: An examination of the evidence. ARDEA 76, 155–168.

Nur, N. 1988b. The consequences of brood size for breeding blue tits. III. Measuring costs of reproduction: Survival, future fecundity and differential dispersal. Evolution 42, 351–362.

Oates, J. F. 1977. The social life of a black-and-white colobus monkey (*Colobus guereza*). Z. Tierpsychol. 45, 1–60.

Oates, J. F. 1984. The niche of the potto, *Perodicticus potto.* Int. J. Primatol. 5, 51–61.

O'Brien, T. G. 1990. Determinants and consequences of social structure in a neotropical primate, Cebus olivaceus. Ph.D. dissertation, University of Florida.

O'Brien, T. G. 1991. Female–male social interactions in wedge-capped capuchin monkeys: Benefits and costs of group living. Anim. Behav. 41, 555–567.

Obst, E. 1912. Von Mkalama ins Land der Wakindiga. Mitteilung. Geograph. Gesell. Hamburg 26, 3–45.

O'Connell, R. J., Singer, A. G., Stern, F. L., Jesmajian, S., & Agosta, W. C. 1981. Cyclic variation in concentration of sex attractant pheromone in hamster vaginal discharge. Behav. Neural. Biol. 31, 457–464.

Oftedal, O. T., Boness, D. J., & Tedman, R. A. 1987. The behavior, physiology and anatomy of lactation in the Pinnipedia. Curr. Mammal. 1, 175–245.

Ohman, A. Dimberg, U., & Ost, L. G. 1985. Biological constraints on the learned fear response. In: Theoretical Issues in Behavior Therapy (Ed. by S. Reiss & R. Bootzin), pp. 123–175. New York: Academic Press.

O'Keefe, R. T., Lifshitz, K., & Linn, G. 1983. Relationships among dominance, interanimal spatial proximity and affiliative social behavior in stumptail macaques (*Macaca arctoides*). Appl. Anim. Ethol. 9, 331–339.

Ollason, J. C., & Dunnet, G. M. 1988. Variation in breeding success in fulmars. In: Reproductive Success (Ed. by T. H. Clutton-Brock), pp. 263–278. Chicago: University of Chicago Press.

Olson, T. 1979. Studies on aspects of the morphology and systematics of the genus *Otolemur* Coquerel, 1859 (Primates: Galagidae). Ph.D. dissertation, University of London.

Oppenheimer, J. R. 1977. *Presbytis entellus,* the Hanuman langur. In: Primate Conservation (Ed. by H. S. H. Rainier & G. H. Bourne), pp. 469–512. New York: Academic Press.

Oster, G., & Alberch, P. 1982. Evolution and bifurcation of developmental programs. Evolution 36, 444–459.

Owens, N. W. 1975. Social play behaviour in free-living baboons, *Papio anubis.* Anim. Behav. 23, 387–408.

Oxnard, C. E. 1984. The Order of Man. New Haven, Conn.: Yale University Press.

Packer, C. 1977. Reciprocal altruism in *Papio anubis*. Nature (London) 265, 441–443.

Packer, C. 1979a. Inter-troop transfer and inbreeding avoidance in *Papio anubis*. Anim. Behav. 27, 1–36.

Packer, C. 1979b. Male dominance and reproductive activity in *Papio anubis*. Anim. Behav. 27, 37–45.

Packer, C. 1980. Male care and exploitation of infants in *Papio anubis*. Anim. Behav. 28, 512–520.

Packer, C., & Pusey, A. E. 1979. Female aggression and male membership in troops of Japanese macaques and olive baboons. Folia Primatol. 31, 212–218.

Packer, C., & Pusey, A. E. 1982. Cooperation and competition within coalitions of male lions: Kin selection or game theory? Nature (London) 296, 740–742.

Pagel, M. D., & Harvey, P. H. 1988a. Recent developments in the analysis of comparative data. Quart. Rev. Biol. 63, 413–440.

Pagel, M. D., & Harvey, P. H. 1988b. How mammals produce large-brained offspring. Evolution 42, 948–957.

Pages, E. 1980. Ethoecology of *Microcebus coquereli* during the dry season. In: Nocturnal Malagasy Primates: Ecology, Physiology, and Behavior (Ed. by P. Charles-Dominique et al.), pp. 97–116. New York: Academic Press.

Pages, E. 1982. Jeu et socilisation: Aspect descriptif et théorique de l'ontogenèse chez un Microcèbe prosimien Malgache. J. Psychol. 3, 241–262.

Pages, E. 1983. Identification, caractérisation et rôle du jeu social chez un prosemien nocturne, *Microcebus coquereli*. Biol. Behav. 8, 319–343.

Pages-Feuillade, E. 1988. Spatial distribution and interindividual relationships in a nocturnal Malagasy lemur *Microcebus murinus*. Folia Primatol. 50, 319–343.

Paglia, C. 1991. Ninnies, pedants, tyrants, and other academics. New York Times Book Review, May 5, 1, 29, 33.

Parker, S. T. 1990. Origins of comparative developmental evolutionary studies of primate mental abilities. In: "Language" and Intelligence in Monkeys and Apes (Ed. by S. T. Parker & K. R. Gibson), pp. 3–64. New York: Cambridge University Press.

Partridge, L. 1988. Lifetime reproductive success in *Drosophilia*. In: Reproductive Success (Ed. by T. H. Clutton-Brock), pp. 11–23. Chicago: University of Chicago Press.

Partridge, L., & Harvey, P. H. 1988. The ecological context of life history evolution. Science 241, 1449–1455.

Paul, A., & Küster, J. 1985. Inter-group transfer and incest avoidance in semi-free-ranging Barbary macaques (*Macaca sylvana*). Folia Primatol. 42, 2–16.

Paul, A., & Küster, J. 1987. Dominance, kinship and reproductive value in female Barbary Macaques (*Macaca sylvanus*) at Affenberg Salem. Behav. Ecol. Sociobiol. 21, 323–331.

Pauly, D. 1980. On the interrelationships between natural mortality, growth parameters and mean environmental temperature in 175 fish stocks. J. Cons. Int. Explor. Mer. 39, 175–192.

Pellegrini, A. D. 1989. What is a category? The case of rough-and-tumble play. Ethol. Sociobiol. 10, 331–342.

Pellis, S. 1981. A description of social play by the Australian magpie *Gymnorhina tibicien* based on Eshkol-Wachmann notation. Bird Behav. 3, 61–79.

Pellis, S. 1988. Agonistic versus amicable targets of attack and defense: Consequences for the origin, function, and descriptive classification of play-fighting. Aggressive Behav. 14, 85–104.

Pennington, R., & Harpending, H. 1988. Fitness and fertility among Kalahari !Kung. Am. J. Phys. Anthropol. 77, 303–319.

Pereira, M. E. 1984. Age changes and sex differences in the social behavior of juvenile yellow baboons (*Papio cynocephalus*). Ph.D. dissertation, University of Chicago.

Pereira, M. E. 1988a. Effects of age and sex on intragroup spacing behavior in juvenile savanna baboons, *Papio cynocephalus cynocephalus*. Anim. Behav. 36, 184–204.

Pereira, M. E. 1988b. Agonistic interactions of juvenile savanna baboons. I. Fundamental features. Ethology 79, 195–217.

Pereira, M. E. 1989. Agonistic interactions of juvenile savanna baboons. II. Agonistic support and rank acquisition. Ethology 80, 152–171.

Pereira, M. E. 1991. Asynchrony within estrous synchrony among ringtailed lemurs (Primates: Lemuridae). Physiol. Behav. 49, 47–52.

Pereira, M. E. 1992. The development of dominance relations before puberty in cercopithecine societies. In: Aggression and Peacefulness in Humans and Other Primates (Ed. by J. Silverberg & P. Gray), pp. 117–149. New York: Oxford University Press.

Pereira, M. E. 1993. Seasonal adjustment of growth rate and adult body weight in ringtailed lemurs. In: Lemur Social Systems and Their Ecological Bases (Ed. by P. M. Kappeler & J. U. Ganzhorn). New York: Plenum. In press.

Pereira, M. E., & Altmann, J. 1985. Development of social behavior in free-living nonhuman primates. In: Nonhuman Primate Models for

of primates. In: Primate Ethology (Ed. by D. Morris), pp. 283–305. London: Weidenfeld & Nicolson.

Rowell, T. E. 1988. The social system of guenons, compared with baboons, macaques, and mangabeys. In: A Primate Radiation: Evolutionary Biology of the African Guenons (Ed. by A. Gautier-Hion, F. Bourliere, J-P. Gautier, and J. Kingdon), pp. 439–451. Cambridge: Cambridge University Press.

Rowell, T. E., & Chism, J. 1986. The ontogeny of sex differences in the behavior of patas monkeys. Int. J. Primatol. 7, 83–107.

Rowell, T. E., & Hinde, R. A. 1962. Vocal communication by the rhesus monkey (*Macaca mulatta*). Proc. Zool. Soc. London 138, 279–294.

Rowell, T. E., Hinde, R. A., & Spencer-Booth, Y. 1964. "Aunt"–infant interaction in captive rhesus monkeys. Anim. Behav. 12, 219–226.

Rubenstein, D. I. 1982. Reproductive value and behavioral strategies: Coming of age in monkeys and horses. In: Perspectives in Ethology, vol. 5 (Ed. by P. P. G. Bateson & P. H. Klopfer), pp. 469–487. New York: Plenus.

Rubenstein, D. I. 1986. Ecology and sociality in horses and zebras. In: Ecological Aspects of Social Evolution (Ed. by D. I. Rubenstein & R. W. Wrangham), pp. 282–302. Princeton N.J.: Princeton University Press.

Rubenstein, D. I. 1993. The ecology of female social behavior in horses, zebras and asses. In: Animal Societies: Individuals, Interactions and Organization (Ed. by P. Jarman & A. Rossiter). Tokyo: Blackwell. In press.

Rudran, R. 1978. Socioecology of the blue monkeys (*Cercopithecus mitis stuhlmanni*) of the Kibale Forest, Uganda. Smithsonian Contr. Zool. 249, 1–88.

Rudran, R. 1979. The demography and social mobility of a red howler (*Alouatta seniculus*) population in Venezuela. In: Vertebrate Ecology in the Northern Neotropics (Ed. by J. F. Eisenberg), pp. 107–126. Washington, D.C.: Smithsonian Institution Press.

Rumiz, D. I. 1990. *Alouatta caraya:* Population density and demography in northern Argentina. Am. J. Primatol. 21, 279–294.

Russell, R. J. 1977. The behavior, ecology and environmental physiology of a nocturnal primate, *Lepilemur mustelinus,* (Strepsirhini, Lemuriformes, Lemuridae). Ph.D. dissertation, Duke University.

Rutberg, A. T. 1987. Adaptive hypotheses of birth synchrony in ruminants: An interspecific test. Am. Nat. 130, 692–710.

Saavedra, C. J. 1984. Spatial and social relationships of males in two groups of red howler monkeys (*Alouatta seniculus*). M.S. thesis, University of Florida.

Sacher, G. A. 1959. Relationship of lifespan to brain weight and body weight in mammals. In: The Lifespan of Animals (CIBA Foundation Colloquia on Ageing, 5). (Ed. by G. E. W. Wolstenholme & M. O'Connor), pp. 115–133. London: Churchill.

Sacher, G. A. 1975. Maturation and longevity in relation to cranial capacity in hominid evolution. In: Primate Functional Morphology and Evolution (Ed. by R. H. Tuttle), pp. 417–441. The Hague: Mouton.

Sacher, G. A. 1982. The role of brain maturation in the evolution of the primates. In: Primate Brain Evolution (Ed. by E. Armstrong & D. Falk), pp. 97–112. New York: Plenum.

Sacher, G. A., & Staffeldt, E. F. 1974. Relationship of gestation time to brain weight for placental mammals: Implications for the theory of vertebrate growth. Am. Nat. 108, 593–615.

Sacks, J. J., Smith, J. D., Kaplan, K. M., Lambert, D. A., Sattin, R. W., & Sikes, R. K. 1989. The epidemiology of injuries in Atlanta daycare centers. J. Am. Med. Assoc. 262, 1641–1645.

Sade, D. S. 1967. Determinants of dominance in a group of free-ranging rhesus monkeys. In: Social Communication Among Primates (Ed. by S. Altmann), pp. 99–114. Chicago: University of Chicago Press.

Sade, D. S. 1972a. A longitudinal study of social behavior of rhesus monkeys. In: The Functional and Evolutionary Biology of Primates (Ed. by R. Tuttle), pp. 378–398. Chicago: Aldine-Atherton.

Sade, D. S. 1972b. Sociometrics of *Macaca mulatta.* 1. Linkages and cliques in grooming matrices. Folia Primatol. 18, 196–223.

Sadlier, R. M. F. S. 1969. The Ecology of Reproduction in Wild and Domestic Mammals. London: Methuen.

Saether, B.-E. 1988. Pattern of covariation between life-history traits of European birds. Nature (London) 331, 616–617.

Saether, B.-E. 1990. Age-specific variation in reproductive performance of birds. In: Current Ornithology (Ed. by D. M. Power), pp. 251–283. New York: Plenum.

Saether, B.-E., & Haagenrud, H. 1983. Life history of the moose (*Alces alces*): Fecundity rates in relation to age and carcass weight. J. Mammal. 64, 226–232.

Salafsky, N. 1988. The foraging patterns and socioecology of the kelasi (*Presbytis rubicunda*). Senior honors thesis, Harvard University.

Samuels, A., Silk, J. B., & Rodman, P. S. 1984. Changes in the dominance rank and reproductive behaviour of male bonnet macaques (*Macaca radiata*). Anim. Behav. 32, 994–1003.

Samuels, A., Silk, J. B., & Altmann, J. 1987. Continuity and change in dominance relations among female baboons. Anim. Behav. 35, 785–793.

Sandler, D. P., Wilcox, A. J., & Horney, L. F. 1984. Age at menarche and subsequent reproductive events. Am. J. Epidemiol. 199, 765–774.

Sastry, A. N. 1983. Ecological aspects of reproduction. In: The Biology of Crustacea, Vol. 8 (Ed. by F. J. Vernberg & W. B. Vernberg), pp. 179–270. New York: Academic Press.

Sauer, E. G. F., & Sauer, E. M. 1963. The South West African bush-baby of the *Galago senegalensis* group. J. S. W. Af. Sci. Soc. 16, 5–36.

Saunders, C. D., & Hausfater, G. 1978. Sexual selection in baboons (*Papio cynocephalus*): A computer simulation of differential reproduction with respect to dominance rank in males. In: Recent Advances in Primatology, Vol. 1 (Ed. by D. J. Chivers & J. Herbert), pp. 565–571. London: Academic Press.

Sauther, M. S. 1991. Reproductive behavior of free-ranging *Lemur catta* at Beza Mahafaly Special Reserve, Madagascar. Am. J. Phys. Anthropol. 84, 463–477.

Scammon, R. E. 1930. The measurement of the body in childhood. In: The Measurement of Man (Ed. by J. A. Harris, C. M. Jackson, D. G. Paterson, & R. E. Scammon), pp. 171–215. Minneapolis: University of Minnesota Press.

Schaffer, W. M. 1974. Selection for optimal life histories: The effects of age structure. Ecology 55, 291–303.

Schaffer, W. M. 1979. Equivalence of maximizing reproductive value and fitness in the case of reproductive strategies. Proc. Natl. Acad. Sci. U.S.A. 76, 3567–3569.

Schaffer, W. M., & Elson, P. F. 1975. The adaptive significance of variations in life history among local populations of Atlantic salmon in N. America. Ecology 56, 577–590.

Schaller, G. B. 1972. The Serengeti Lion. Chicago: University of Chicago Press.

Schaller, G. S. 1963. The Mountain Gorilla. Chicago: University of Chicago Press.

Schanberg, S. M., & Field, T. M. 1987. Sensory deprivation stress and supplemental stimulation in the rat pup and preterm human neonate. Child Dev. 58, 1431–47.

Schmidt-Nielsen, K. 1984. Scaling: Why Is Animal Size So Important? New York: Cambridge University Press.

Schneirla, T. C. 1957. The concept of development in comparative psychology. In: The Concept of Development: An Issue in the Study of Human Behavior (Ed. by D. B. Harris), pp. 78–108. Minneapolis: University of Minnesota Press.

Schrire, C. 1980. An enquiry into the evolutionary status and apparent identity of San hunter-gatherers. Hum. Ecol. 8, 9–32.

Schulman, S. R., & Chapais, B. 1980. Reproductive value and rank relations among macaque sisters. Am. Nat. 115, 580–593.

Schulz, A. H. 1936. Characters common to higher primates and characters specific for man. Quart. Rev. Biol. 11, 259–283.

Schulz, A. H. 1956. Postembryonic age changes. Primatologia 1, 887–964.

Schultz, A. H. 1969. The Life of Primates. New York: Universe Books.

Schürmann, C. L. 1982. Mating behaviour of wild orang-utans. In: The Orang utan (Ed. by L. E. M. de Boer), pp. 269–284. The Hague: Junk.

Schürmann, C. L., & van Hooff, J. A. R. A. M. 1986. Reproductive strategies of the orang-utan: New data and a reconsideration of existing sociosexual models. Int. J. Primatol. 7, 265–287.

Scollay, P. A., & DeBold, P. 1980. Allomothering in a captive colony of hanuman langurs (*Presbytis entellus*). Ethol. Sociobiol. 1, 291–299.

Seger, J. 1977. A numerical method for estimating coefficients of relationship in a langur troop. In: The Langurs of Abu (S. B. Hrdy), appendix 3, pp. 317–326. Cambridge, Mass.: Harvard University Press.

Sekulic, R. 1982a. Behavior and ranging patterns of a solitary female red howler (*Alouatta seniculus*). Folia Primatol. 38, 217–232.

Sekulic, R. 1982b. Daily and seasonal patterns of roaring and spacing in four red howler *Alouatta seniculus* troops. Folia Primatol. 39, 22–48.

Sekulic, R. 1982c. The function of howling in red howler monkeys (*Alouatta seniculus*). Behaviour 81, 38–54.

Sekulic, R. 1983. Male relationships and infant deaths in red howler monkeys (*Alouatta seniculus*). Z. Tierpsychol. 61, 185–202.

Semler, D. E. 1971. Some aspects of adaptation in a polymorphism for breeding colours in the three-spined stickleback (*Gasterosteus aculeatus*). J. Zool. London 165, 291–302.

Seyfarth, R. M. 1977. A model of social grooming among adult female monkeys. J. Theor. Biol. 65, 671–698.

Seyfarth, R. M. 1978. Social relationships among adult male and female baboons. 2. Behaviour throughout the female reproductive cycle. Behaviour 64, 227–247.

Seyfarth, R. M. 1980. The distribution of grooming and related behaviours among adult female vervet monkeys. Anim. Behav. 28, 798–813.

Seyfarth, R. M., & Cheney, D. L. 1980. The ontogeny of vervet monkey alarm-calling behaviour: A preliminary report. Z. Tierpsychol. 54, 37–56.

Seyfarth, R. M., & Cheney, D. L. 1986. Vocal development in vervet monkeys. Anim. Behav. 34, 1450–1468.

Seymour, S. 1988. Expressions of responsibility among Indian children: Some precursors of adult status and sex roles. Ethos 16, 355–370.

Shapiro, D. Y. 1984. Sex reversal and sociodemographic processes in coral reef fishes. In: Fish Reproduction: Strategies and Tactics (Ed. by G. W. Potts & R. J. Wooton), pp. 103–118. New York: Academic Press.

Shea, B. T. 1983. Allometry and heterochrony in the African apes. Am. J. Phys. Anthropol. 62, 275–289.

Shea, B. T. 1985. The ontogeny of sexual dimorphism in the African apes. Am. J. Primatol. 8, 183–188.

Shea, B. T. 1986. Ontogenetic approaches to sexual dimorphism in anthropoids. Hum. Evol. 1, 97–110.

Shively, C. 1985. The evolution of dominance hierarchies in a nonhuman primate society. In: Power, Dominance and Nonverbal Behaviour (Ed. by S. Ellyson & I. Dividio), pp. 67–87. Berlin: Springer-Verlag.

Short, R. V. 1976. The evolution of human reproduction. Proc. R. Soc. London Ser. B. 195, 3–24.

Shostak, M. 1981. Nisa. The Life and Words of a !Kung Woman. Cambridge, Mass.: Harvard University Press.

Shuster, S. M. 1989. Male alternative reproductive behaviors in a marine isopod crustacean (Paracerceis sculpta): Use of genetic markers to measure differences in fertilization success among α, β, and γ-males. Evolution 34, 1683–1698.

Shuster, S. M., & Wade, M. J. 1991. Equal mating success among male reproductive strategies in a marine isopod. Nature (London) 350, 606–610.

Sibly, R. M., & Calow, P. 1986. Physiological Ecology of Animals. Boston: Blackwell.

Sieff, D. 1990. Explaining biased sex ratios in human populations. Curr. Anthropol. 31, 25–48.

Siegel, S., & Castellan, N. J., Jr. 1988. Nonparametric Statistics for the Behavioral Sciences. New York: McGraw-Hill.

Sieley, S. J. 1975. Environmental influences on the cognitive development of rural children: A study of a Kipsigis community in Western Kenya. Ed.D. dissertation, Harvard University.

Sih, A., & Milton, K. A. 1982. Optimal diet theory: Should the !Kung eat mongongos? Am. Anthropol. 87, 395–401.

Silk, J. B. 1978. Patterns of food sharing among mother and infant chimpanzees at Gombe National Park, Tanzania. Folia Primatol. 29, 129–141.

Silk, J. B. 1980. Kidnapping and female competition among captive bonnet macaques. Primates 21, 100–110.

Silk, J. B. 1982. Altruism among female Macaca radiata: Explanations and analysis of patterns of grooming and coalition formation. Behaviour 79, 162–188.

Silk, J. B. 1987. Social behavior in evolutionary perspective. In: Primate Societies (Ed. by B. B. Smuts, D. L. Cheney, R. M. Seyfarth, R. W. Wrangham, & T. T. Struhsaker), pp. 318–329. Chicago: University of Chicago Press.

Silk, J. B., Samuels, A., & Rodman, P. S. 1981. The influence of kinship, rank, and sex on affiliation and aggression between adult female and immature bonnet macaques (Macaca radiata) Behaviour 78, 111–137.

Silverberg, J., & Gray, P. (Eds.). 1992. Aggression and Peacefulness in Humans and Other Primates. New York: Oxford University Press.

Simpson, M. J. A. 1978. Tactile experience and social behavior: Aspects of development with special reference to primates. In: Biological Determinants of Social Behaviour (Ed. by J. B. Hutchison), pp. 785–807. London: Wiley.

Singer, A. G., Macrides, F., & Agosta, W. C. 1980. Chemical studies of hamster pheromones. In: Chemical Signals: Vertebrates and Aquatic Invertebrates (Ed. by D. Muller-Schwarze & R. M. Silverstein), pp. 365–375. New York: Plenum.

Sizonenko, P. C., & Paunier, L. 1975. Hormonal changes in puberty. III. Correlation of plasma dehydroepiandrosterone, testosterone, FSH, and LH with stages of puberty and bond age in normal boys and girls and in patients with Addison's disease of hypogonadism or with premature or late adrenarche. J. Clin. Endocrinol. Metabl. 51, 894–904.

Skutch, A. F. 1976. Parent Birds and Their Young. Austin: University of Texas Press.

Smail, P. J., Faiman, C., Hobson, W. C., Fuller, G. B., & Winter, J. D. 1982. Further studies on adrenarche in nonhuman primates. Endocrinology 111, 844–848.

Small, M. F. 1989. Female choice in nonhuman primates. Yearbook Phys. Anthropol. 32, 103–127.

Small, M. F., & Smith, D. G. 1981. Interactions with infants by full siblings, paternal half-siblings, and nonrelatives in a captive group of rhesus macaques (*Macaca mulatta*). Am. J. Primatol. 1, 91–94.

Smith, C. C., & Fretwell, S. D. 1974. The optimal balance between size and number of offspring. Am. Nat. 108, 499–506.

Smith, C. L. 1975. The evolution of hermaphroditism in fishes. In: Intersexuality in the Animal Kingdom (Ed. by R. Reinboth), pp. 295–310. Berlin: Springer-Verlag.

Smith, E. O. 1978a. Social Play in Primates. New York: Academic Press.

Smith, E. O. 1978b. A historical view on the study of play: Statement of the problem. In: Social Play in Primates (Ed. by E. O. Smith), pp. 1–32. New York: Academic Press.

Smith, P. K. (Ed.). 1984. Play in Animals and Humans. Oxford: Blackwell.

Smith-Gill, S. J. 1983. Developmental plasticity: Developmental conversions versus phenotypic modulation. Am. Zool. 23, 47–55.

Smuts, B. B. 1985. Sex and Friendship in Baboons. Hawthorne, N.Y.: Aldine.

Smuts, B. 1987. Gender, aggression, and influence. In: Primate Societies (Ed. by B. B. Smuts, D. L. Cheney, R. M. Seyfarth, R. W. Wrangham, & T. T. Struhsaker), pp. 400–412. Chicago: University of Chicago Press.

Smuts, B. B., Cheney, D. L., Seyfarth, R. M., Wrangham, R. W., & Struhsaker, T. T. (Eds.). 1987. Primate Societies. Chicago: University of Chicago Press.

Snowden, C. T., & Suomi, S. J. 1982. Paternal behavior in primates. In: Child Nurturance, Vol. 3. Studies of Development in Nonhuman Primates (Ed. by H. E. Fitzgerald, J. R. Mullins, & P. Gage), pp. 63–108. New York: Plenum.

Sokal, R. R., & Rohlf, F. J. 1981. Biometry. San Francisco: Freeman.

Solomons, H. C. 1982. Is day care safe for children? Accident records reviewed. Child Hlth. Care 10, 90–93.

Solway, J. S., & Lee, R. B. 1990. Foragers, genuine or spurious? Curr. Anthropol. 31, 109–146.

Sommer, V. 1985. Weibliche und männliche Reproduktionsstrategien der Hanuman-Languren (*Presbytis entellus*) von Jodhpur, Rajasthan/Indien. Ph.D. dissertation, Georg-August-Universität, Göttingen.

Sommer, V. 1987. Infanticide among free-ranging langurs (*Presbytis entellus*) at Jodhpur (Rajasthan/India): Recent observations and a reconsideration of hypotheses. Primates 28, 163–197.

Sommer, V. 1988. Male competition and coalition in langurs (*Presbytis entellus*) at Jodhpur, Rajasthan, India. Hum. Evol. 3, 261–278.

Sommer, V. 1993. Infanticide among the langurs of Jodhpur—testing the sexual selection hypothesis with a long-term record. In: Infanticide an Parentall Care (Ed. S. Parmigiani, & F. vom Saal). London: Harwood. In press.

Sommer, V., & Rajpurohit, L. A. 1989. Male reproductive success in harem troops of Hanuman langurs (*Presbytis entellus*). Int. J. Primatol. 10, 293–317.

Sommer, V., Srivastava, A., & Borries, C. 1992. Cycles, sexuality, and conception in free-ranging langurs (*Presbytis entellus*). Am. J. Primatol. 28, 9–27.

Spencer, A. W., & Steinhof, H. W. 1968. An explanation of geographical variation in litter size. J. Mammal. 49, 281–286.

Spiro, M. 1980. Gender and Culture: Kibbutz Women Revisited. New York: Schocken.

Srikosamatara, S. 1987. Group size in the wedge-capped capuchin monkey (*Cebus olivaceus*): Vulnerability to predators, intragroup and intergroup feeding competition. Ph.D. dissertation, University of Florida.

Srivastava, A. 1989. Feeding ecology and behaviour of Hanuman langur, *Presbytis entellus*. Ph.D. dissertation, University of Jodhpur.

Srivastava, A., Mohnot, S. M., & Rajpurohit, L. S. 1986. Existence of multi-male bisexual troops of Hanuman langur (*Presbytis entellus*) in a predominantly one-male troop habitat. In: Abstracts, International Symposium on Primates—The New Revolution, pp. 41–42. 26–31 December. New Delhi, India.

Stamm, C. 1990. Koalitionsmuster und Verwandschaft in einer Kolonie Javaner-Makaken (*Macaca fascicularis*). Diplomarbeit, University of Zurich.

Stanhope, R. 1989. The endocrine control of puberty. In: The Physiology of Human Growth (Ed. by J. M. Tanner & M. A. Preece), pp. 191–199. Cambridge: Cambridge University Press.

Stanhope, R., & Brooke, C. G. D. 1988. An evaluation of hormonal changes at puberty in man. J. Endocrinal. 116, 301–305.

Starin, E. D. 1978. A preliminary investigation of home range use in the Gir Forest langur. Primates 19, 551–568.

Starin, E. D. 1990. Object manipulation by wild red colobus monkeys living in the Abuko Nature Reserve, The Gambia. Primates 31, 385–391.

Stearns, S. C. 1976. Life-history tactics: A review of the ideas. Quart. Rev. Biol. 51, 3–47.

Stearns, S. C. 1977. The evolution of life-history traits: A critique of the theory and a review of the data. Annu. Rev. Ecol. Syst. 8, 145–171.

Stearns, S. C. 1984. Adaptations of colonizers. Science 223, 693–694.

Stearns, S. C., & Crandall, R. E. 1981. Quantitative predictions of delayed maturity. Evolution 35, 455–463.

Stearns, S. C., & Crandall, R. E. 1984. Plasticity for age and size at maturity: A life-history response to unavoidable stress. In: Fish Reproduction: Strategies and Tactics (Ed. by G. W. Potts & R. J. Wooton), pp. 13–33. New York: Academic Press.

Stearns, S. C., & Koella, J. C. 1986. The evolution of phenotypic plasticity in life-history traits: Predictions of reaction norms for age and size at maturity. Evolution 40, 893–913.

Stein, D. M. 1984a. The Sociobiology of Infant and Adult Male Baboons. Norwood, N.J.: Albex.

Stein, D. M. 1984b. Ontogeny of infant–adult male relationships during the first year of life for yellow baboons (Papio cynocephalus). In: Primate Paternalism (Ed. by D. M. Taub), pp. 213–243. New York: Van Nostrand Reinhold.

Steinberg, L. 1988. Reciprocal relation between parent–child distance and pubertal maturation. Dev. Psychol. 24, 122–128.

Stern, D. N., Barnett, R. K., & Spieker, S. 1983. Early transmission of affect: Some research issues. In: Frontiers of Infant Psychiatry, Vol. 1 (Ed. by J. D. Call, E. Galenson, & R. L. Tyson), pp. 74–85. New York: Basic Books.

Stevenson, M. F., & Poole, T. B. 1982. Playful interactions in family groups of the common marmoset (Callithrix jacchus jacchus). Anim. Behav. 30, 886–900.

Stewart, K. J., & Harcourt, A. H. 1987. Gorillas: Variation in female relationships. In: Primate Societies (Ed. by B. B. Smuts, D. L. Cheney, R. M. Seyfarth, R. W. Wrangham, & T. T. Struhsaker), pp. 155–164. Chicago: University of Chicago Press.

Stickel, L. F. 1978. Changes in a box turtle population during three decades. Copeia 1978, 221–225.

Stini, W. A. 1975. Adaptive strategies of human populations under nutritional stress. In: Biosocial Interrelations in Population Adaptation (Ed. by E. S. Watts, F. E. Johnston, & G. W. Lasker), pp. 19–41. The Hague: Mouton.

Stini, W. A. 1982. Sexual dimorphism and nutrient reserves. In: Sexual Dimorphism in Homo sapiens (Ed. by R. L. Hall), pp. 391–419. New York: Praeger.

Stinson, S. 1985. Sex differences in environmental sensitivity during growth and development. Yearbook Phy. Anthropol. 28, 123–147.

Stradling, D. J. 1978. The growth and maturation of the "tarantula" Avicularia avicularia L. Biol. J. Linn. Soc. 62, 291–303.

Strayer, F., & Noël, J. 1986. The prosocial and antisocial functions of preschool aggression: An ethological study of triadic conflict among young children. In: Altruism and Aggression (Ed. by C. Zahn-Waxler, E. Cummings, & R. Iannotti), pp. 107–131. Cambridge: Cambridge University Press.

Strier, K. B. 1986. The behavior and ecology of the woolly spider monkey, or muriqui (Brachyteles arachnoides E. geoffroy 1806). Ph.D. dissertation, Harvard University Press.

Strier, K. B. 1987a. Reproducão de Brachyteles arachnoides. In: A Primatologia no Brasil—no. 2 (Ed. by M. Thiago de Mello), pp. 163–175. Brasilia: Sociedade Brasiliera de Primatologica.

Strier, K. B. 1987b. Activity budgets of woolly spider monkeys, or muriquis (Brachyteles arachnoides). Am. J. Primatol. 13, 385–395.

Strier, K. B. 1987c. Demographic patterns in one group of muriquis. Prim. Cons. 8, 73–74.

Strier, K. B. 1989. Effects of patch size on feeding associations in muriquis (Brachyteles arachnoides). Folia Primatol. 52, 70–77.

Strier, K. B. 1990. New World primates, new frontiers: Insights from the woolly spider monkey, or muriqui (Brachyteles arachnoides). Int. J. Primatol. 11, 7–19.

Strier, K. B. 1991. Demography and conservation in an endangered primate, Brachyteles arachnoides. Conserv. Biol. 5, 214–218.

Strier, K. B. 1992a. Causes and consequences of nonaggression in woolly spider monkeys. In: Aggression and Peacefulness in Humans and Other Primates (Ed. by J. Silverberg & J. P. Gray), pp. 100–116. New York: Oxford University Press.

Strier, K. B. 1992b. Faces in the Forest: The Endangered Muriqui Monkeys of Brazil. New York: Oxford University Press.

Strier, K. B. 1993. Subtle cues of social relations in muriquis: Relevance for understanding social relationships in humans. In: New World Primates: Ecology, Evolution, and Behavior (Ed. by W. G. Kinzey). New York: Aldine. In press.

Strier, K. B., Mendes, F. D. C., Rimoli, J., & Rimoli, A. O. 1993. Demography and social structure in one group of muriquis (Brachyteles arachnoides). Int. J. Primatol. In press.

Struhsaker, T. T. 1967a. Ecology of vervet monkeys (Cercopithecus aethiops) in the Masai-Amboseli Game Reserve, Kenya. Ecology 48, 891–904.

Struhsaker, T. T. 1967b. Social structure among vervet monkeys (Cercopithecus aethiops). Behaviour 29, 83–121.

Struhsaker, T. T. 1971. Social behavior of mother and infant vervet monkeys (*Cercopithecus aethiops*). Anim. Behav. 19, 233–250.

Struhsaker, T. T. 1973. A recensus of vervet monkeys in the Masai-Amboseli Game Reserve, Kenya. Ecology 54, 930–932.

Struhsaker, T. T. 1975. The Red Colobus Monkey. Chicago: University of Chicago Press.

Struhsaker, T. T., & Leland, L. 1979. Socioecology of five sympatric monkey species in the Kibale Forest, Uganda. Adv. Study Behav. 9, 159–228.

Strum, S. C. 1983. Use of females by male olive baboons (*Papio anubis*). Am. J. Primatol. 5, 93–109.

Strum, S. C. 1984. Why males use infants. In: Primate Paternalism (Ed. by D. M. Taub), pp. 146–185. New York: Van Nostrand Reinhold.

Strum, S. C. 1985. Baboons may be smarter than people. Anim. Kingdom 88(2), 12–25.

Strum, S. C. 1987. Almost Human: A Journey into the World of Baboons. New York: Random House.

Sugardjito, J. 1983. Selecting nest-sites of Sumatran orang-utans, *Pongo pygmaeus abelii* in the Gunung Leuser National Park, Indonesia. Primates 24, 467–474.

Sugardjito, J., & van Hooff, J. A. R. A. M. 1986. Age–sex class differences in postural behavior of the Sumatran orangutan (*Pongo pygmaeus abelii*) in the Gunung Leuser National Park, Indonesia. Folia Primatol. 47, 14–25.

Sugardjito, J., te Boekhorst, I. J. A., & van Hooff, J. A. R. A. M. 1987. Ecological constraints on the grouping of wild orang-utans (*Pongo pygmaeus*) in the Gunung Leuser National Park, Sumatra, Indonesia. Int. J. Primatol. 8, 17–41.

Sugiyama, Y. 1965. On the social change of Hanuman langurs (*Presbytis entellus*) in their natural conditions. Primates 6, 381–417.

Sugiyama, Y. 1967. Social organisation of hanuman langurs. In: Social Communication Among Primates (Ed. by S. Altmann). Chicago: University of Chicago Press.

Sugiyama, Y. 1976a. Characteristics of the ecology of the Himalayan langurs. J. Hum. Evol. 5, 249–277.

Sugiyama, Y. 1976b. Life history of male Japanese macaques. Adv. Study Behav. 7, 255–284.

Suomi, S. J. 1982. Why does play matter? Behav. Brain Sci. 5, 169–170.

Suomi, S. J., & Leroy, H. A. 1982. In memoriam: Harry F. Harlow (1905–1981). Am. J. Primatol. 2, 319–342.

Surbey, M. 1990. Family composition, stress, and the timing of human menarche. In: Socioen-

docrinology of Primate Reproduction (Ed. by T. E. Ziegler & F. B. Bercovitch), pp. 11–32. New York: Wiley-Liss.

Sussman, R. W. 1991. Demography and social organization of free-ranging *Lemur catta* in the Beza Mahafaly Reserve, Madagascar. Am. J. Phys. Anthropol. 84, 43–58.

Sutherland, W. J., Grafen, W. J., & Harvey, P. H. 1986. Life history correlations and demography. Nature (London) 320, 88.

Symington, M. M. 1988. Demography, ranging patterns and activity budgets of black spider monkeys (*Ateles paniscus chamek*) in the Manu National Park, Peru. Am. J. Primatol. 15, 45–67.

Symington, M. M. 1990. Fission–fusion social organization in *Ateles* and *Pan*. Int. J. Primatol. 11, 47–62.

Symons, D. 1978. Play and Aggression: A Study of Rhesus Monkeys. New York: Columbia University Press.

Symons, D., & Bishop, J. M. 1977. Rhesus play. 16 mm cine film. University of California Educational Media Center, Berkeley.

Taber, S., & Thomas, P. 1982. Calf development and mother–calf spatial relationships in southern right whales. Anim. Behav. 30, 1072–1083.

Takahata, Y. 1990. Social relationships among adult males. In: The Chimpanzees of the Mahale Mountains (Ed. by T. Nishida), pp. 149–170. Tokyo: Tokyo University Press.

Takasaki, H. 1985. Female life history and mating patterns among the M group chimpanzees of the Mahale National Park, Tanzania. Primates 26, 121–129.

Tanner, J. M. 1978. Fetus into Man. Cambridge, Mass.: Harvard University Press.

Tanner, J. M., & Whitehouse, R. H. 1976. Clinical longitudinal standards for height, weight, height velocity and weight velocity and the stages of puberty. Arch. Dis. Child 51, 170–179.

Tanner, J. M., Wilson, M. E., & Rudman, C. G. 1990. Pubertal growth spurt in the female rhesus monkey: Relation to menarche and skeletal maturation. Am. J. Hum. Biol. 2, 101–106.

Tattersall, I. 1982. The Primates of Madagascar. New York: Columbia University Press.

Taub, D. M. (Ed.). 1984a. Primate Paternalism. New York: Van Nostrand Reinhold.

Taub, D. M. 1984b. Male caretaking behaviour among wild Barbary macaques (*Macaca sylvanus*). In: Primate Paternalism (Ed. by D. M. Taub), pp. 20–55. New York: Van Nostrand Reinhold.

Taylor, L. L. 1986. Kinship, dominance and social organization in a semi-free-ranging group of ringtailed lemurs (*Lemur catta*). Ph.D. dissertation, Washington University.

Taylor, L. L., & Sussman, R. W. 1985. A preliminary study of kinship and social organization in a semi-free-ranging group of *Lemur catta*. Int. J. Primatol. 6, 601–614.

te Boekhoerst, I. J. A., Schurmann, C. L., & Sugardjito, J. 1990. Residential status and seasonal movements of wild orang-utans in the Gunung Leuser Reserve (Sumatra, Indonesia). Anim. Behav. 39, 1098–1109.

Terborgh, J. W. 1983. Five New World Primates: A Study in Comparative Ecology. Princeton, N.J.: Princeton University Press.

Terborgh, J. W. 1986. Community aspects of frugivory in tropical forests. In: Frugivores and Seed Dispersal (Ed. by A. Estrada & T. Fleming), pp. 371–384. Dordrecht: Junk.

Terborgh, J., & Janson, C. H. 1986. The socioecology of primate groups. Annu. Rev. Ecol. Syst. 17, 111–135.

Terrace, H. S. 1979. Nim. New York: Knopf.

Terrell, T. R., & Mascie-Taylor, C. G. N. 1991. Biosocial correlates of stature in a 16-year-old British cohort. J. Biosoc. Sci. 23, 401–408.

Thelen, E. 1989. The (re)discovery of motor development: Learning new things from an old field. Dev. Psychol. 25, 946–949.

Thelen, E., & Fogel, A. 1989. Towards an action-based theory of infant development. In: Action in Social Context (Ed. by J. Lockman & N. Hazen), pp. 23–63. New York: Plenum.

Thierry, B. 1986. A comparative study of aggression and response to aggression in three species of macaque. In: Primate Ontogeny, Cognition and Social Behavior (Ed. by J. G. Else & P. C. Lee), pp. 307–313. Cambridge: Cambridge University Press.

Thomas, C. S., & Coulson, J. C. 1988. Reproductive success of kittiwake gulls, *Rissa tridactyla*. In: Reproductive Success (Ed. by T. H. Clutton-Brock), pp. 251–262. Chicago: University of Chicago Press.

Thorington, R. W., Jr., Rudran, R., & Mack, D. 1979. Sexual dimorphism of *Alouatta seniculus* and observations on capture techniques. In: Vertebrate Ecology in the Northern Neotropics (Ed. by J. F. Eisenberg), pp. 97–106. Washington, D.C.: Smithsonian Institution Press.

Tilley, S. G. 1977. Studies of life histories and reproduction in North American plethodontid salamanders. In: The Reproductive Biology of Amphibians (Ed. by D. H. Taylor & S. I. Guttman), pp. 1–41. New York: Plenum.

Tilley, S. G. 1980. Life histories and comparative demography of two salamander populations. Copeia 1980, 806–821.

Tilson, R. 1981. Family formation strategies of Kloss gibbons (*Hylobates klossi*). Folia Primatol. 35, 259–287.

Tinbergen, N. 1951. The Study of Instinct. Oxford: Oxford University Press.

Tobach, E., Greenberg, G., Radell, P., & McCarthy, T. 1989. Social behavior in a group of orang-utans (*Pongo pygmaeus abelii*) in a zoo setting. Appl. Anim. Behav. Sci. 23, 141–154.

Tooby, J., & DeVore, I. 1987. The reconstruction of hominid behavioral evolution through strategic modeling. In: The Evolution of Human Behavior: Primate Models (Ed. by W. G. Kinzey). Albany: State University of New York Press.

Trivers, R. L. 1972. Parental investment and sexual selection. In: Sexual Selection and the Descent of Man 1871–1971 (Ed. by B. Campbell), pp. 136–179. Chicago: Aldine.

Trivers, R. L. 1974. Parent–offspring conflict. Am. Zool. 14, 249–264.

Tronick, E. Z., Morelli, G. A., & Winn, S. 1987. Multiple caretaking of Efe (Pygmy) infants. Am. Anthropol. 89, 96–106.

Troth Ovrebo, R. G. 1979. Vegetational types on a ranch in the Central Llanos of Venezuela. In: Vertebrate Ecology of the Northern Neotropics (Ed. by J. F. Eisenberg), pp. 17–30. Washington, D.C.: Smithsonian Institution Press.

Tuan, Y-F. 1984. Dominance and Affection. New Haven, Conn.: Yale University Press.

Turke, P. W. 1987. Helpers at the nest: Childcare networks on Ifaluk. In: Human Reproductive Behavior (Ed. by L. L. Betzig, M. Borgerhoff-Mulder, & P. W. Turke), pp. 173–188. London: Cambridge University Press.

Turke, P. W. 1989. Evolution and the demand for children. Pop. Dev. Rev. 15, 61–90.

Turner, F. B., Medica, P. A., & Lyons, C. L. 1984. Reproduction and survival of the desert tortoise (*Scaptochelys agassizii*) in Ivanpah Valley, California. Copeia 1984, 811–820.

Tutin, C. E. G. 1979. Mating patterns and reproductive strategies in a community of wild chimpanzees (*Pan troglodytes schweinfurthii*). Behav. Ecol. Sociobiol. 6, 29–38.

Tutin, C. E. G., & McGinnis, P. R. 1981. Chimpanzee reproduction in the wild. In: Reproductive Biology of the Great Apes (Ed. by C. E. Graham), pp. 239–264. New York: Academic Press.

Udry, J. R., & Cliquet, R. L. 1982. A cross-cultural examination of the relationship between ages at menarche, marriage, and first birth. Demography 19, 53–63.

Uehara, S., & Nishida, T. 1987. Body weights of wild chimpanzees (*Pan troglodytes schwein-*

furthii) of the Mahale Mountains National Park, Tanzania. Am. J. Phys. Anthropol. 72, 315–321.

UNICEF. 1985. The State of the World's Children. New York: Oxford University Press.

Vandenbergh, J. G., & Coppola, D. M. 1986. The physiology and ecology of puberty modulation by primer pheromones. Adv. Study Behav. 16, 71–107.

Vandenbergh, J. G., Drickamer, L. C., & Colby, D. R. 1972. Social and dietary factors in the sexual maturation of female mice. J. Reprod. Fertil. 28, 515–523.

van de Rijt-Plooij, H. H. C., & Plooij, F. X. 1987. Growing independence, conflict and learning in mother–infant relations in free-ranging chimpanzees. Behaviour 101, 1–86.

van Hooff, J. A. R. A. M. 1962. Facial expressions in higher primates. Symp. Zool. Soc. London 8, 97–125.

van Hooff, J. A. R. A. M. 1970. A component analysis of the structure of the social behaviour of a semi-captive Chimpanzee group. Experientia 26, 549–550.

van Hooff, J. A. R. A. M. 1989. Laughter and humor, and the "duo-in-uno" of nature and culture. In: The Nature of Culture (Ed. by W. A. Koch), pp. 120–149. Bochum: BPX.

van Horn, R., & Eaton, G. 1979. Reproductive physiology and behavior in prosimians. In: The Study of Prosimian Behavior (Ed. by G. A. Doyle & R. D. Martin), pp. 79–122. New York: Academic Press.

van Noordwijk, M. A., & van Schaik, C. P. 1985. Male migration and rank acquisition in wild long-tailed macaques (*Macaca fascicularis*). Anim. Behav. 33, 849–861.

van Noordwijk, M. A., & van Schaik, C. P. 1987. Competition among female long-tailed macaques, *Macaca fascicularis*. Anim. Behav. 35, 577–589.

van Noordwijk, M. A., & van Schaik, C. P. 1988. Male careers in Sumatran long-tailed macaques (*Macaca fascicularis*). Behaviour 107, 24–43.

van Rhijn, J. G. 1973. Behavioral dimorphism in male ruffs, *Philomacus pugnax* (L.). Behaviour 47, 153–229.

van Roosmalen, M. G. M. 1985. Habitat preferences, diet, feeding strategy and social organization of the black spider monkey (*Ateles paniscus paniscus* Linnaeus 1758) in Surinam. Act. Amazon. 15 (3–4). Suppl.

van Schaik, C. P. 1983. Why are diurnal primates living in groups? Behaviour 87, 120–144.

van Schaik, C. P. 1985. The socio-ecology of Sumatran long-tailed macaques (*Macaca fascicularis*). Ph.D. dissertation, University of Utrecht.

van Schaik, C. P. 1989. The ecology of social relationships amongst female primates. In: Comparative Socioecology (Ed. by V. Standen & R. A. Foley), pp. 195–218. Oxford: Blackwell.

van Schaik, C. P., & de Visser, J.A.G.M. 1990. Fragile sons or harassed daughters? Sex differences in morality among juvenile primates. Folia Primatol. 55, 10–23.

van Schaik, C. P., Ijsseldijk, A. T. H. J., & de Ruiter, J. R. In preparation. Sex differences in growth and development of long-tailed macaques (*Macaca fascicularis*): The effect of food supply.

van Schaik, C. P., & Mirmanto, E. 1985. Spatial variation in the structure and litterfall of a Sumatran rain forest. Biotropica 17, 196–205.

van Schaik, C. P., & Mitrasetia, T. 1990. Changes in the behaviour of wild long-tailed macaques (*Macaca fascicularis*) after encounters with a model python. Folia Primatol. 55, 104–108.

van Schaik, C. P., & van Hooff, J. 1983. On the ultimate causes of primate social systems. Behaviour 85, 91–117.

van Schaik, C. P., & van Noordwijk, M. A. 1985a. Evolutionary effect of the absence of felids on the social organization of the macaques on the island of Simeulue (*Macaca fascicularis fusca*, Miller 1903). Folia Primatol. 44, 138–47.

van Schaik, C. P., & van Noordwijk, M. A. 1985b. Interannual variability in fruit abundance and reproductive seasonality of Sumatran long-tailed macaques. J. Zool. 206, 533–549.

van Schaik, C. P., & van Noordwijk, M. A. 1986. The hidden costs of sociality: Intra-group variation in feeding strategies in Sumatran long-tailed macaques (*Macaca fascicularis*). Behaviour 99, 296–315.

van Schaik, C. P., & van Noordwijk, M. A. 1988. Scramble and contest in feeding competition among female long-tailed macaques (*Macaca fascicularis*). Behaviour 105, 77–98.

van Schaik, C. P., & van Noordwijk, M. A. 1989. The special role of male Cebus monkeys in predation avoidance and its effect on group composition. Behav. Ecol. Sociobiol. 24, 265–276.

van Schaik, C. P., van Noordwijk, M. A., de Boer, R. J., & den Tonkelaar, I. 1983a. The effect of group size on time budgets and social behaviour in wild long-tailed macaques (*Macaca fascicularis*). Behav. Ecol. Sociobiol. 13, 173–181.

van Schaik, C. P., van Noordwijk, M. A., Warsono, B., & Sutriono, E. 1983b. Party size and early detection of predators in Sumatran forest primates. Primates 24, 211–221.

van Wagenen, G. 1947. Maturity induced by testosterone in the young male monkey. Fed. Proc. 6, 219.

van Wagenen, G. 1949. Accelerated growth with sexual precocity in female monkeys receiving testosterone propionate. Endocrinology 45, 554.

van Wagenen, G., & Catchpole, H. R. 1956. Physical growth of the rhesus monkey (*Macaca mulatta*). Am. J. Phys. Anthropol. 14, 245–273.

van Wieringen, J. C. 1986. Secular growth changes. In: Human Growth, 2nd ed., Vol. 3 (Ed. by F. Falkner & J. M. Tanner), pp. 307–331. New York: Plenum.

Vehrencamp, S. A. 1983. A model for the evolution of despotic versus egalitarian societies. Anim. Behav. 31, 667–682.

Vessey, S. H., & Meikle, D. B. 1984. Free-living rhesus monkeys: Adult male interactions with infants and juveniles. In: Primate Paternalism (Ed. by D. M. Taub), pp. 113–126. New York: Van Nostrand Reinhold.

Vick, L. G. 1977. The role of interindividual relationships in two troops of captive *Lemur fulvus*. Ph.D. dissertation, University of North Carolina.

Vick, L. G., & Pereira, M. E. 1989. Episodic targeting aggression and the histories of Lemur social groups. Behav. Ecol. Sociobiol. 25, 3–12.

Vihko, R., & Apter, D. 1980. The role of androgens in adolescent cycles. J. Ster. Biochem. 12, 369–373.

Visalberghi, E. 1990. Tool use in *Cebus*. Folia Primatol. 54, 146–154.

Visalberghi, E., & Fragaszy, D. M. 1990a. Do monkeys ape? In: Language and Intelligence in Monkeys and Apes (Ed. by S. T. Parker & K. R. Gibson), pp. 247–273. Cambridge: Cambridge University Press.

Visalberghi, E., & Fragaszy, D. M. 1990b. Food-washing behaviour in tufted capuchin monkeys, *Cebus apella,* and crabeating macaques, *Macaca fascicularis*. Anim. Behav. 40, 829–836.

Vogel, C. 1977. Ecology and sociology of *Presbytis entellus*. In: Use of Non-human Primates in Biomedical Research (Ed. by M.R.N. Prasad & T. C. Anand Kumar), pp. 24–45. New Delhi: Indian National Science Academy.

Vogel, C. 1979. Der Hanuman Langur (*Presbytis entellus*), ein Parade-Exempel fur die theoretischen Konzepte der Soziobiologie? Verhandl. Dtsch. Zool. Ges. 77, 73–89.

Vogel, C., & Loch, H. 1984. Reproductive parameters, adult-male replacements, and infanticide among free-ranging langurs (*Presbytis entellus*) at Jodhpur (Rajasthan), India. In: Infanticide: Comparative and Evolutionary Perspectives (Ed. G. Hausfater & S. B. Hrdy), pp. 237–255. New York: Adline.

Voigt, J.-P. 1984. Verhaltensbeobachtungen an einer Gruppe junger Orang-Utans (*Pongo pygmaeus abelii* Lesson, 1827). Zool. Anz. 213, 258–274.

Vollrath, F. 1987. Growth, foraging and reproductive success. In: Ecophysiology of Spiders (Ed. by W. Nentwig), pp. 357–370. New York: Springer.

Waddington, C. H. 1940. Organizers and Genes. Cambridge: Cambridge University Press.

Walters, J. R. 1980. Interventions and the development of dominance relationships in female baboons. Folia Primatol. 34, 61–89.

Walters, J. R. 1981. Inferring kinship from behaviour: Maternity determinants in yellow baboons. Anim. Behav. 29, 126–136.

Walters, J. R. 1987a. Transition to adulthood. In: Primate Societies (Ed. by B. B. Smuts, D. L. Cheney, R. M. Seyfarth, R. W. Wrangham, & T. T. Struhsaker), pp. 358–369. Chicago: University of Chicago Press.

Walters, J. R. 1987b. Kin recognition in non-human primates. In: Kin Recognition in Animals (Ed. by D. J. C. Fletcher & C. D. Michener), pp. 359–393. New York: Wiley.

Walters, J. R., & Seyfarth, R. M. 1987. Conflict and cooperation. In: Primate Societies. (Ed. by B. B. Smuts, D. L. Cheney, R. M. Seyfarth, R. W. Wrangham, & T. T. Struhsaker), pp. 306–317. Chicago: University of Chicago Press.

Warner, R. R. 1984. Deferred reproduction as a response to sexual selection in a coral reef fish: A test of the life historical consequences. Evolution 38, 148–162.

Warner, R. R., & Hoffman, S. G. 1980. Local population size as a determinant of mating system and sexual composition in two tropical reef fishes (*Thalassoma* sp.). Evolution 34, 508–518.

Warner, R. R., Robertson, D. R., & Leigh, E. G. 1975. Sex change and sexual selection. Science 190, 633–638.

Waser, P. 1977a. Feeding, ranging and group size in the mangabey, *Cercocebus albigena*. In: Primate Ecology (Ed. by T. H. Clutton-Brock), pp. 183–222. London: Academic Press.

Waser, P. 1977b. Individual recognition, intragroup cohesion and intergroup spacing: Evidence from sound playback to forest monkeys. Behaviour 60, 28–74.

Waser, P. M., & Jones, W. T. 1983. Natal philopatry among solitary mammals. Quart. Rev. Biol. 58, 355–390.

Washburn, S. L. 1951. The new physical anthropology. Trans. N.Y. Acad. Sci. Ser. II 13, 298–304.

Washburn, S. L., & Avis, V. 1958. The evolution of human behavior. In: Behavior and Evolution (Ed. by A. Roe & G. G. Simpson), pp. 421–

436. New Haven, Conn.: Yale University Press.

Washburn, S. L., & DeVore, I. 1961. The social life of baboons. Sci. Am. 204, 62–71.

Washburn, S. L., & Hamburg, D. A. 1965. The implications of primate research. In: Primate Behavior (Ed. by I. DeVore), pp. 607–622. New York: Holt, Rinehart & Winston.

Washburn, S. L., & Hamburg, D. A. 1968. Aggressive behavior in Old World monkeys and apes. In: Primate Patterns (Ed. by P. Dolhinow), pp. 276–296. New York: Holt, Rinehart & Winston.

Watts, D. P. 1985a. Observations on the ontogeny of feeding behavior in mountain gorillas (*Gorilla gorilla beringei*). Am. J. Primatol. 8, 1–10.

Watts, D. P. 1985b. Relations between group size and composition and feeding competition in mountain gorilla groups. Anim. Behav. 33, 73–85.

Watts, D. P. 1988. Environmental influences on mountain gorilla time budgets. Am. J. Primatol. 15, 195–211.

Watts, D. P. 1989. Infanticide in mountain gorillas: New cases and a reconsideration of the evidence. Ethology 81, 1–18.

Watts, D. P. 1990a. Ecology of gorillas and its relation to female transfer in mountain gorillas. Intl. J. Primatol. 11, 21–45.

Watts, D. P. 1990b. Reproduction, life history tactics, and sociosexual behavior of mountain gorillas and some implications for captive husbandry. Zoo Biol. 9, 185–200.

Watts, D. P. 1991a. Mountain gorilla reproduction and sexual behavior. Am. J. Primatol. 24, 211–226.

Watts, D. P. 1991b. Harrassment of immigrant female mountain gorillas by resident females. Ethology 89, 135–153.

Watts, D. P. 1992. Social relationships of immigrant and resident female mountain gorillas. I. Male–female relationships. Am. J. Primatol. 28, 159–181.

Watts, E. S. 1986. Evolution of the human growth curve. In: Human Growth, 2nd ed., Vol. 1 (Ed. by F. Falkner & J. M. Tanner), pp. 153–66. New York: Plenum.

Watts, E. S. 1990. Evolutionary trends in primate growth and development. In: Primate Life History and Evolution (Ed. by C. J. deRousseau), pp. 89–104. New York: Wiley-Liss.

Watts, E. S., & Gavan, J. A. 1982. Postnatal growth of nonhuman primates: The problem of the adolescent spurt. Hum. Biol. 54, 53–70.

Wcislo, W. T. 1989. Behavioral environments and evolutionary change. Annu. Rev. Ecol. Syst. 20, 137–169.

Weigel, R. M. 1980. Dyadic spatial relationships in pigtail and stumptail macaques: A multiple regression analysis. Int. J. Primatol. 1, 287–321.

Weirman, M. E., Beardsworth, D. E., Crawford, J. D., Crigler, J. F., Mansfield, M. J., Bode, H. H., Boepple, P. A., Kushner, D. C., & Crowley, W. F. 1986. Adrenarche and skeletal maturation during luteinizing hormone releasing hormone analogue suppression of gonadarche. J. Clin. Invest. 77, 121–6.

Weirman, M. E., & Crowley, W. F. 1986. Neuroendocrine control of the onset of puberty. In: Human Growth 2nd ed., Vol. 2 (Ed. by F. Falkner & J. M. Tanner), pp. 225–241. New York: Plenum.

Weisbard, C., & Goy, R. W. 1976. Effect of parturition and group composition on competitive order in stumptail macaques (*Macaca arctoides*). Folia Primatol. 25, 95–121.

Weisner, T. S. 1987. Socialization for parenthood in sibling caretaking societies. In: Parenting Across the Life Span: Biosocial Dimensions (Ed. by J. B. Lancaster, J. Altmann, A. S. Rossi, & L. R. Sherrod), pp. 237–270. New York: Aldine.

Welker, C., Prange, E., & Witt, C. 1982. Preliminary observations on the social behaviour of the greater galago, *Galago agisymbanus* Coquerel, 1959, in captivity. Anthropol. Anz. 40, 193–203.

Wenger, M. 1983. Gender role socialization in an East African community: Social interaction between 2- to 3-year-olds and older children in social ecological perspective. Ed.D. dissertation, Harvard University.

Wenger, M. 1989. Work, play, and social relationships among the children in a Giriama community. In: Children's Social Networks and Social Supports (Ed. by D. Belle), pp. 91–118. New York: Wiley.

Werner, E. 1989. High-risk children in young adulthood: A longitudinal study from birth to 32 years. Am. J. Orthopsychiat. 59, 72–81.

Werner, E. E. 1985. Stress and protective factors in children's lives. In: Longitudinal Studies in Child Psychology and Psychiatry (Ed. by A. R. Nicol), pp. 335–355. Chichester: Wiley.

Werner, E. E., & Smith, R. S. 1977. Kauai's Children Come of Age. Honolulu: University Press of Hawaii.

Werner, E. E., & Smith, R. S. 1982. Vulnerable but Invincible. New York: McGraw-Hill.

West-Eberhard, M. J. 1989. Phenotypic plasticity and the origins of diversity. Annu. Rev. Ecol. Syst. 20, 249–278.

Western, D. 1979. Size, life history, and ecology in mammals. Afr. J. Ecol. 17, 185–204.

Western, D., & Ssemakula, J. 1982. Life history patterns in birds and mammals and their evolutionary significance. Oecologia 54, 281–290.

White, F. J. 1988. Party composition and dynamics in *Pan paniscus*. Intl. J. Primatol. 9, 179–193.

White, F. J. 1989. Social organization of pygmy chimpanzees. In: Understanding Chimpanzees (Ed. by P. G. Heltne & L. A. Marquard), pp. 194–207. Cambridge, Mass.: Harvard University Press.

White, F. J. In press. The pygmy chimpanzee. In: Illustrated Monographs of Living Primates (Ed. by J. B. Kaiser, M. G. M. van Roosmalen, R. A. Mittermeier, & W. L. R. Oliver).

White, F. J., & Burgman, M. A. 1990. Social organization of the pygmy chimpanzee (*Pan paniscus*): Multivariate analysis of intracommunity associations. Am. J. Phys. Anthropol. 83, 193–202.

White, F. J., & Wrangham, R. W. 1988. Feeding competition and patch size in the chimpanzee species, *Pan paniscus* and *Pan troglodytes*. Behaviour 105, 148–164.

Whitehead, J. M. 1986. Development of feeding selectivity in mantled howling monkeys, *Alouatta palliata*. In: Primate Ontogeny, Cognition, and Social Behaviour (Ed. by J. G. Else & P. C. Lee), pp. 105–117. Cambridge: Cambridge University Press.

Whitehead, J. M. 1989. The effect of the location of a simulated intruder on responses to long-distance vocalizations of mantled howling monkeys, *Alouatta palliata palliata*. Behaviour 108, 73–103.

Whiting, B. B. 1980. Culture and social behavior: A model for the development of social behavior. Ethos 8(2), 95–116.

Whiting, B. B., & Edwards, C. P. 1973. A cross-cultural analysis of sex differences in the behavior of children aged three through eleven. J. Soc. Psychol. 91, 171–188.

Whiting, B. B., & Edwards, C. P. 1988. Children of Different Worlds: The Formation of Social Behavior. Cambridge, Mass.: Harvard University Press.

Whiting, B. B., & Whiting, J. W. M. 1975. Children of Six Cultures: A Psycho-Cultural Analysis. Cambridge, Mass.: Harvard University Press.

Whiting, J. W. M. 1981. Environmental constraints on infant care practices. In: Handbook of Cross-Cultural Human Development (Ed. by R. H. Munroe, R. L. Munroe, & B. B. Whiting), pp. 155–79. New York: Garland Press.

Whitten, A. J. 1980. The Kloss gibbon in Siberut rain forest. Ph.D. dissertation, Cambridge University.

Whitten, P. L. 1983. Diet and dominance among female vervet monkeys (*Cercopithecus aethiops*). Am. J. Primatol. 5, 139–159.

Whitten, P. L. 1987. Infants and adult males. In: Primate Societies (Ed. by B. B. Smuts, D. L. Cheney, R. M. Seyfarth, R. W. Wrangham, & T. T. Struhsaker), pp. 343–357. Chicago: University of Chicago Press.

Whitten, W. K. 1959. Occurrence of anoestrous in mice caged in groups. J. Endocrinol. 18, 102–107.

Wiesel, T. N. 1982. Postnatal development of the visual cortex and the influence of environment. Nature (London) 299, 583–91.

Wilbur, H. M. 1975. A growth model for the turtle *Chrysemys picta*. Copeia 1975, 337–343.

Wilbur, H. M., & Collins, J. P. 1973. Ecological aspects of amphibian metamorphosis. Science 182, 1305–1314.

Wilbur, H. M., & Morin, P. J. 1988. Life history evolution in turtles. In: Biology of the Reptilia, Vol. 16 (Ed. by C. Gans & R. B. Huey), pp. 387–439. New York: Liss.

Wiley, R. H. 1974a. Effects of delayed reproduction on survival, fecundity, and the rate of population increase. Am. Nat. 108, 705–709.

Wiley, R. H. 1974b. Evolution of social organization and life-history patterns among grouse. Quart. Rev. Biol. 49, 201–227.

Wiley, R. H. 1981. Social structure and individual ontogenies: Problems of description, mechanism, and evolution. In: Perspectives in Ethology (Ed. by P. P. G. Bateson & P. H. Klopfer), pp. 105–133. New York: Plenum.

Wiley, R. H., & Harnett, S. A. 1976. Effects of interactions with older males on behavior and reproductive development in first-year male red-winged blackbirds *Agelauis phoeniceus*. J. Exp. Zool. 196, 231–242.

Wiley, R. H., & Rabenold, K. N. 1984. The evolution of cooperative breeding by delayed reciprocity and queuing for favorable social positions. Evolution 38, 609–621.

Williams, G. C. 1957. Pleiotropy, natural selection, and the evolution of senescence. Evolution 11, 298–411.

Williams, G. C. 1966. Adaptation and Natural Selection. Princeton, N.J.: Princeton University Press.

Williamson, E., Tutin, C. E. G., Rogers, M. E., & Fernandez, M. 1990. Composition of the diet of lowland gorillas at Lope in Gabon. Am. J. Primatol. 21, 265–278.

Wilmsen, E. N. 1989. Land Filled with Flies: A Political Economy of the Kalahari. Chicago: University of Chicago Press.

Wilson, A. P., & Boelkins, R. C. 1970. Evidence for seasonal variation in aggressive behaviour by *Macaca mulatta*. Anim. Behav. 18, 719–724.

Wilson, E. O. 1975. Sociobiology: The New Synthesis. Cambridge, Mass.: Harvard University Press.

Wilson, M., & Daly, M. 1985. Competitiveness, risk taking, and violence: The young male syndrome. Ethol. Sociobiol. 6, 59–73.

Wilson, M. E., Gordon, T. P., Rudman, C. G., & Tanner, J. M. 1988. Effects of a natural versus artificial environment on the tempo of maturation in female rhesus monkeys. Endocrinology 123, 2553–2561.

Winkler, P. 1981. Zur Öko-Ethologie freilebender Hanuman-Languren (*Presbytis entellus entellus* Dufresne, 1797) in Jodhpur (Rajasthan), Indian. Ph.D. dissertation, Georg-August-Universität, Göttingen.

Winkler, P., Loch, H., & Vogel, C. 1984. Life history of Hanuman langurs (*Presbytis entellus*): Reproductive parameters, infant mortality, and troop development. Folia Primatol. 43, 1–23.

Winn, R. M. 1989. The aye-ayes, *Daubentonia madagascariensis*, at the Paris Zoological Garden: Maintenance and preliminary behavioural observations. Folia Primatol. 52, 109–123.

Wise, D. H. 1976. Variable rates of maturation of the spider, *Neriene radiata* (*Linyphia marginata*). Am. Midl. Nat. 96, 66–75.

Wise, D. H. 1984. Phenology and life history of the filmy dome spider (Araneae: Linyphiidae) in two local Maryland populations. Psyche 91, 267–288.

Wood, J. W., Johnson, P. L., & Campbell, K. L. 1985. Demographic and endocrinological aspects of low natural fertility in highland New Guinea. J. Biosoc. Sci. 17, 57–79.

Woodburn, J. C. 1968a. An introduction to Hadza ecology. In: Man the Hunter (Ed. by R. B. Lee & I. DeVore), pp. 49–55. Chicago: Aldine.

Woodburn, J. C. 1968b. Stability and flexibility in Hadza residential groupings. In: Man the Hunter (Ed. by R. B. Lee & I. DeVore), pp. 103–110. Chicago: Aldine.

Wooller, R. D., & Coulson, J. C. 1977. Factors affecting the age of first breeding of the kittiwake *Rissa tridactyla*. Ibis 119, 339–349.

Worthman, C. M. 1986. Later-maturing populations and control of the onset of puberty (abstract). Am. J. Phys. Anthropol. 69, 282.

Worthman, C. M. 1987a. Developmental dyssynchrony as normative experience: Kikuyu adolescents. In: School-age Pregnancy and Parenthood: Biosocial Dimensions (Ed. by J. B. Lancaster & B. A. Hamburg), pp. 95–112. New York: Aldine.

Worthman, C. M. 1987b. Interactions of physical maturation and cultural practice in ontogeny: Kikuyu adolescents. Cult. Anthropol. 2, 29–38.

Worthman, C. M. 1990. Socioendocrinology: Key to a fundamental synergy. In: Socioen-

docrinology of Primate Reproduction (Ed. by T. E. Ziegler & F. B. Bercovitch), pp. 187–212. New York: Wiley-Liss.

Worthman, C. M. 1993. Not quite themselves: The constitution of adolescents in the Pacific. In: Adolescence in Pacific Island Societies (Ed. by G. Herdt). Cambridge: Cambridge University Press. In press.

Worthman, C. M., & Whiting, J. W. M. 1987. Social change in sexual behavior and mate selection in a Kikuyu community. Ethos 15, 145–165.

Worthy, G. A. J., & Lavigne, D. M. 1983. Changes in energy stores during postnatal development of the harp seal, *Phoca groenlandica*. J. Mammal. 64, 89–96.

Wrangham, R. W. 1977. Feeding behaviour of chimpanzees in Gombe National Park, Tanzania. In: Primate Ecology (Ed. by T. H. Clutton-Brock), pp. 503–538. London: Academic Press.

Wrangham, R. W. 1979a. On the evolution of ape social systems. Soc. Sci. Info. 18, 335–368.

Wrangham, R. W. 1979b. Sex differences in chimpanzee dispersion. In: The Great Apes, (Ed. by D. A. Hamburg & E. R. McCown), pp. 481–489. Menlo Park, Calif.: Benjamin-Cummings.

Wrangham, R. W. 1980. An ecological model of female-bonded primate groups. Behaviour 75, 262–300.

Wrangham, R. W. 1981. Drinking competition in vervet monkeys. Anim. Behav. 29, 904–910.

Wrangham, R. W., & Rubenstein, D. I. 1986. Social evolution in birds and mammals. In: Ecological Aspects of Social Evolution (Ed. by D. I. Rubenstein & R. W. Wrangham), pp. 452–470. Princeton, N.J.: Princeton University Press.

Wrangham, R. W., & Smuts, B. B. 1980. Sex differences in the behavioral ecology of chimpanzees in the Gombe National Park, Tanzania. J. Reprod. Fertil. (Suppl. 28), 13–31.

Wright, P. C., Toyama, L. M., & Simons, E. L. 1986. Courtship and copulation in *Tarsius bancanus*. Folia Primatol. 46, 142–148.

Wright, S. 1931. Evolution in Mendelian populations. Genetics 16, 97–159.

Wrigley, J. In preparation. Servants and cultural transmission within English families.

Wurtman, R. L., & Waldhauser, F. 1986. Melatonin in humans. J. Neural Trans. (Suppl.) 21, 1–475.

Yamada, M. 1963. A study of blood relationships in the natural society of the Japanese macaque: An analysis of co-feeding, grooming, and playmate relationship in the Minoo-B troop. Primates 4, 44–65.

Yamagiwa, J. 1983. Diachronic changes in two eastern lowland gorilla groups (*Gorilla gorilla graueri*) in the Mt. Kahuzi region, Zaire. Primates 24, 174–183.

Ydenberg, R. C. 1989. Growth-mortality trade-offs and the evolution of juvenile life histories in the Alcidae. Ecology 70, 1494–1506.

Yellen, J. E., & Lee, R. B. 1976. The Dobe-/Du/da environment. In: Kalahari Hunter Gatherers (Ed. by R. B. Lee & I. DeVore), pp. 27–46. Cambridge, Mass.: Harvard University Press.

Yerkes, R. M. 1916. Provision for the study of monkeys and apes. Science 43, 231–234.

Yerkes, R. M. 1932. Yale laboratories of comparative psychobiology. Comp. Psychol. Monog. 8, 1–33.

Yerkes, R. M., & Yerkes, A. W. 1929. The Great Apes: A Study of Anthropoid Life. New Haven, Conn.: Yale University Press.

York, A. D., & Rowell, T. E. 1988. Reconciliation following aggression in patas monkeys, *Erythrocebus patas*. Anim. Behav. 36, 502–509.

Young, A. L., Richard, A. F., & Aiello, L. C. 1990. Female dominance and maternal investment in strepsirhine primates. Am. Nat. 135, 473–488.

Zahavi, A. 1977. The testing of a bond. Anim. Behav. 25, 246–247.

Zahorik, D. M., & Houpt, K. A. 1981. Species differences in feeding strategies, food hazards, and the ability to learn food aversions. In: Foraging Behaviour: Ecology, Ethology, and Psychological Approaches (Ed. by A. C. Kamil & T. D. Sargent), pp. 289–310. New York: Garland Press.

Zemel, B. 1989. Dietary change and adolescent growth among the Bundi (Gende-speaking) people of Papua New Guinea. Ph.D. dissertation, University of Pennsylvania.

Zemel, B., Worthman, C. M., & Jenkins, C. L. 1993. Differences in endocrine status associated with urban–rural patterns of growth and maturation in Bundi (Gende-speaking) adolescents of Papua New Guinea. In: Urban Health and Ecology in the Third World (Ed. by L. M. Schell, M. T. Smith, & A. Bilsborough). Cambridge: Cambridge University Press. In press.

Zeveloff, S. I., & Boyce, M. S. 1988. Body size patterns in North American faunas. In: Evolution of Life Histories of Mammals (Ed. by M. S. Boyce), pp. 123–146. New Haven, Conn.: Yale University Press.

Zimmerman, E. 1989a. Aspects of reproduction and behavioral and vocal development in senegal bushbabies (*Galago senegalensis*). Int. J. Primatol. 10, 1–16.

Zimmerman, E. 1989b. Reproduction, physical growth and behavioral development in slow loris (*Nycticebus coucang*, Lorisidae). Hum. Evol. 4, 171–179.

Zimmermann, R. R., Geist, C. R., & Ackles, P. K. 1975a. Changes in the social behavior of rhesus monkeys during rehabilitation from prolonged protein-calorie malnutrition. Behav. Biol. 14, 325–334.

Zimmermann, R. R., Strobel, D. A., Maguire, D., Steere, R. R., & Hom, H. L. 1976. The effects of protein deficiency on activity, learning, manipulative tasks, curiosity, and social behaviour of monkeys. In: Play—Its Role in Development and Evolution (Ed. by J. S. Bruner, A. Jolly, & K. Sylva), pp. 496–511. New York: Basic Books.

Zimmermann, R. R., Strobel, D. A., Steere, P., & Geist, C. R. 1975b. Behavior and malnutrition in the rhesus monkey. In: Primate Behavior, Vol. 4 (Ed. by L. A. Rosenblum), pp. 241–306. New York: Academic Press.

Zucker, E. L. 1987. Social status and the distribution of social behavior by adult female patas monkeys: A comparative perspective. In: Comparative Behavior of African Monkeys (Ed. by E. L. Zucker), pp. 151–173. New York: Liss.

Zuckerman, S. 1932. The Social Life of Monkeys and Apes. London: Routledge & Kegan Paul.

Index

INDEX